T0135283

**Lecture Notes on Data Engineering and Communications Technologies** **203**

The aim of the book series is to present cutting edge engineering approaches to data technologies and communications. It will publish latest advances on the engineering task of building and deploying distributed, scalable and reliable data infrastructures and communication systems.

The series will have a prominent applied focus on data technologies and communications with aim to promote the bridging from fundamental research on data science and networking to data engineering and communications that lead to industry products, business knowledge and standardisation.

Indexed by SCOPUS, INSPEC, EI Compendex.

All books published in the series are submitted for consideration in Web of Science.

Leonard Barolli
Editor

# Advanced Information Networking and Applications

## Proceedings of the 38th International Conference on Advanced Information Networking and Applications (AINA-2024), Volume 5

*Editor*
Leonard Barolli
Department of Information and Communication Engineering
Fukuoka Institute of Technology
Fukuoka, Japan

ISSN 2367-4512 ISSN 2367-4520 (electronic)
Lecture Notes on Data Engineering and Communications Technologies
ISBN 978-3-031-57930-1 ISBN 978-3-031-57931-8 (eBook)
https://doi.org/10.1007/978-3-031-57931-8



This Springer imprint is published by the registered company Springer Nature Switzerland AG
The registered company address is: Gewerbestrasse 11, 6330 Cham, Switzerland

Paper in this product is recyclable.

# Welcome Message from AINA-2024 Organizers

Welcome to the 38th International Conference on Advanced Information Networking and Applications (AINA-2024). On behalf of AINA-2024 Organizing Committee, we would like to express to all participants our cordial welcome and high respect.

AINA is an International Forum, where scientists and researchers from academia and industry working in various scientific and technical areas of networking and distributed computing systems can demonstrate new ideas and solutions in distributed computing systems. AINA is a very open society and is always welcoming international volunteers from any country and any area in the world.

AINA International Conference is a forum for sharing ideas and research work in the emerging areas of information networking and their applications. The area of advanced networking has grown very rapidly and the applications have experienced an explosive growth, especially in the area of pervasive and mobile applications, wireless sensor and ad-hoc networks, vehicular networks, multimedia computing, social networking, semantic collaborative systems, as well as IoT, big data, cloud computing, artificial intelligence, and machine learning. This advanced networking revolution is transforming the way people live, work, and interact with each other and is impacting the way business, education, entertainment, and health care are operating. The papers included in the proceedings cover theory, design and application of computer networks, distributed computing, and information systems.

Each year AINA receives a lot of paper submissions from all around the world. It has maintained high-quality accepted papers and is aspiring to be one of the main international conferences on the information networking in the world.

We are very proud and honored to have two distinguished keynote talks by Prof. Fatos Xhafa, Technical University of Catalonia, Spain, and Dr. Juggapong Natwichai, Chiang Mai University, Thailand, who will present their recent work and will give new insights and ideas to the conference participants.

An international conference of this size requires the support and help of many people. A lot of people have helped and worked hard to produce a successful AINA-2024 technical program and conference proceedings. First, we would like to thank all authors for submitting their papers. We are indebted to Program Track Co-chairs, Program Committee Members and Reviewers, who carried out the most difficult work of carefully evaluating the submitted papers.

We would like to thank AINA-2024 General Co-chairs, PC Co-chairs, Workshops Organizers for their great efforts to make AINA-2024 a very successful event. We have special thanks to the Finance Chair and Web Administrator Co-chairs.

We do hope that you will enjoy the conference proceedings and readings.

# AINA-2024 Organizing Committee

## Honorary Chair

| | |
|---|---|
| Makoto Takizawa | Hosei University, Japan |

## General Co-chairs

| | |
|---|---|
| Minoru Uehara | Toyo University, Japan |
| Euripides G. M. Petrakis | Technical University of Crete (TUC), Greece |
| Isaac Woungang | Toronto Metropolitan University, Canada |

## Program Committee Co-chairs

| | |
|---|---|
| Tomoya Enokido | Rissho University, Japan |
| Mario A. R. Dantas | Federal University of Juiz de Fora, Brazil |
| Leonardo Mostarda | University of Perugia, Italy |

## International Journals Special Issues Co-chairs

| | |
|---|---|
| Fatos Xhafa | Technical University of Catalonia, Spain |
| David Taniar | Monash University, Australia |
| Farookh Hussain | University of Technology Sydney, Australia |

## Award Co-chairs

| | |
|---|---|
| Arjan Durresi | Indiana University Purdue University in Indianapolis (IUPUI), USA |
| Fang-Yie Leu | Tunghai University, Taiwan |
| Marek Ogiela | AGH University of Science and Technology, Poland |
| Kin Fun Li | University of Victoria, Canada |

## Publicity Co-chairs

| | |
|---|---|
| Markus Aleksy | ABB Corporate Research Center, Germany |
| Flora Amato | University of Naples "Federico II", Italy |
| Lidia Ogiela | AGH University of Science and Technology, Poland |
| Hsing-Chung Chen | Asia University, Taiwan |

## International Liaison Co-chairs

| | |
|---|---|
| Wenny Rahayu | La Trobe University, Australia |
| Nadeem Javaid | COMSATS University Islamabad, Pakistan |
| Beniamino Di Martino | University of Campania "Luigi Vanvitelli", Italy |

## Local Arrangement Co-chairs

| | |
|---|---|
| Keita Matsuo | Fukuoka Institute of Technology, Japan |
| Tomoyuki Ishida | Fukuoka Institute of Technology, Japan |

## Finance Chair

| | |
|---|---|
| Makoto Ikeda | Fukuoka Institute of Technology, Japan |

## Web Co-chairs

| | |
|---|---|
| Phudit Ampririt | Fukuoka Institute of Technology, Japan |
| Ermioni Qafzezi | Fukuoka Institute of Technology, Japan |
| Shunya Higashi | Fukuoka Institute of Technology, Japan |

## Steering Committee Chair

| | |
|---|---|
| Leonard Barolli | Fukuoka Institute of Technology, Japan |

## Tracks Co-chairs and Program Committee Members

## 1. Network Architectures, Protocols and Algorithms

### Track Co-chairs

| | |
|---|---|
| Spyropoulos Thrasyvoulos | Technical University of Crete (TUC), Greece |
| Shigetomo Kimura | University of Tsukuba, Japan |
| Darshika Perera | University of Colorado at Colorado Springs, USA |

### TPC Members

| | |
|---|---|
| Thomas Dreibholz | Simula Metropolitan Center for Digital Engineering, Norway |
| Angelos Antonopoulos | Nearby Computing SL, Spain |
| Hatim Chergui | i2CAT Foundation, Spain |
| Bhed Bahadur Bista | Iwate Prefectural University, Japan |
| Chotipat Pornavalai | King Mongkut's Institute of Technology Ladkrabang, Thailand |
| Kenichi Matsui | NTT Network Innovation Center, Japan |
| Sho Tsugawa | University of Tsukuba, Japan |
| Satoshi Ohzahata | University of Electro-Communications, Japan |
| Haytham El Miligi | Thompson Rivers University, Canada |
| Watheq El-Kharashi | Ain Shams University, Egypt |
| Ehsan Atoofian | Lakehead University, Canada |
| Fayez Gebali | University of Victoria, Canada |
| Kin Fun Li | University of Victoria, Canada |
| Luis Blanco | CTTC, Spain |

## 2. Next Generation Mobile and Wireless Networks

### Track Co-chairs

| | |
|---|---|
| Purav Shah | School of Science and Technology, Middlesex University, UK |
| Enver Ever | Middle East Technical University, Northern Cyprus |
| Evjola Spaho | Polytechnic University of Tirana, Albania |

**TPC Members**

| | |
|---|---|
| Burak Kizilkaya | Glasgow University, UK |
| Muhammad Toaha | Middle East Technical University, Turkey |
| Ramona Trestian | Middlesex University, UK |
| Andrea Marotta | University of L'Aquila, Italy |
| Adnan Yazici | Nazarbayev University, Kazakhstan |
| Orhan Gemikonakli | Final International University, Cyprus |
| Hrishikesh Venkataraman | Indian Institute of Information Technology, Sri City, India |
| Zhengjia Xu | Cranfield University, UK |
| Mohsen Hejazi | University of Kashan, Iran |
| Sabyasachi Mukhopadhyay | IIT Kharagpur, India |
| Ali Khoshkholghi | Middlesex University, UK |
| Admir Barolli | Aleksander Moisiu University of Durres, Albania |
| Makoto Ikeda | Fukuoka Institute of Technology, Japan |
| Yi Liu | Oita National College of Technology, Japan |
| Testuya Oda | Okayama University of Science, Japan |
| Ermioni Qafzezi | Fukuoka Institute of Technology, Japan |

## 3. Multimedia Networking and Applications

**Track Co-chairs**

| | |
|---|---|
| Markus Aleksy | ABB Corporate Research Center, Germany |
| Francesco Orciuoli | University of Salerno, Italy |
| Tomoyuki Ishida | Fukuoka Institute of Technology, Japan |

**TPC Members**

| | |
|---|---|
| Hadil Abukwaik | ABB Corporate Research Center, Germany |
| Thomas Preuss | Brandenburg University of Applied Sciences, Germany |
| Peter M. Rost | Karlsruhe Institute of Technology (KIT), Germany |
| Lukasz Wisniewski | inIT, Germany |
| Angelo Gaeta | University of Salerno, Italy |
| Angela Peduto | University of Salerno, Italy |
| Antonella Pascuzzo | University of Salerno, Italy |
| Roberto Abbruzzese | University of Salerno, Italy |
| Tetsuro Ogi | Keio University, Japan |

| | |
|---|---|
| Yasuo Ebara | Osaka Electro-Communication University, Japan |
| Hideo Miyachi | Tokyo City University, Japan |
| Kaoru Sugita | Fukuoka Institute of Technology, Japan |

## 4. Pervasive and Ubiquitous Computing

### Track Co-chairs

| | |
|---|---|
| Vamsi Paruchuri | University of Central Arkansas, USA |
| Hsing-Chung Chen | Asia University, Taiwan |
| Shinji Sakamoto | Kanazawa Institute of Technology, Japan |

### TPC Members

| | |
|---|---|
| Sriram Chellappan | University of South Florida, USA |
| Yu Sun | University of Central Arkansas, USA |
| Qiang Duan | Penn State University, USA |
| Han-Chieh Wei | Dallas Baptist University, USA |
| Ahmad Alsharif | University of Alabama, USA |
| Vijayasarathi Balasubramanian | Microsoft, USA |
| Shyi-Shiun Kuo | Nan Kai University of Technology, Taiwan |
| Karisma Trinanda Putra | Universitas Muhammadiyah Yogyakarta, Indonesia |
| Cahya Damarjati | Universitas Muhammadiyah Yogyakarta, Indonesia |
| Agung Mulyo Widodo | Universitas Esa Unggul Jakarta, Indonesia |
| Bambang Irawan | Universitas Esa Unggul Jakarta, Indonesia |
| Eko Prasetyo | Universitas Muhammadiyah Yogyakarta, Indonesia |
| Sunardi S. T. | Universitas Muhammadiyah Yogyakarta, Indonesia |
| Andika Wisnujati | Universitas Muhammadiyah Yogyakarta, Indonesia |
| Makoto Ikeda | Fukuoka Institute of Technology, Japan |
| Tetsuya Oda | Okayama University of Science, Japan |
| Evjola Spaho | Polytechnic University of Tirana, Albania |
| Tetsuya Shigeyasu | Hiroshima Prefectural University, Japan |
| Keita Matsuo | Fukuoka Institute of Technology, Japan |
| Admir Barolli | Aleksander Moisiu University of Durres, Albania |

## 5. Web-Based Systems and Content Distribution

### Track Co-chairs

| | |
|---|---|
| Chrisa Tsinaraki | Technical University of Crete (TUC), Greece |
| Yusuke Gotoh | Okayama University, Japan |
| Santi Caballe | Open University of Catalonia, Spain |

### TPC Members

| | |
|---|---|
| Nikos Bikakis | Hellenic Mediterranean University, Greece |
| Ioannis Stavrakantonakis | Ververica GmbH, Germany |
| Sven Schade | European Commission, Joint Research Center, Italy |
| Christos Papatheodorou | National and Kapodistrian University of Athens, Greece |
| Sarantos Kapidakis | University of West Attica, Greece |
| Manato Fujimoto | Osaka Metropolitan University, Japan |
| Kiki Adhinugraha | La Trobe University, Australia |
| Tomoki Yoshihisa | Shiga University, Japan |
| Jordi Conesa | Open University of Catalonia, Spain |
| Thanasis Daradoumis | Open University of Catalonia, Spain |
| Nicola Capuano | University of Basilicata, Italy |
| Victor Ströele | Federal University of Juiz de Fora, Brazil |

## 6. Distributed Ledger Technologies and Distributed-Parallel Computing

### Track Co-chairs

| | |
|---|---|
| Alfredo Navarra | University of Perugia, Italy |
| Naohiro Hayashibara | Kyoto Sangyo University, Japan |

### TPC Members

| | |
|---|---|
| Serafino Cicerone | University of L'Aquila, Italy |
| Ralf Klasing | LaBRI Bordeaux, France |
| Giuseppe Prencipe | University of Pisa, Italy |
| Roberto Tonelli | University of Cagliari, Italy |
| Farhan Ullah | Northwestern Polytechnical University, China |

| | |
|---|---|
| Leonardo Mostarda | University of Perugia, Italy |
| Qiong Huang | South China Agricultural University, China |
| Tomoya Enokido | Rissho University, Japan |
| Minoru Uehara | Toyo University, Japan |
| Lucian Prodan | Polytechnic University of Timisoara, Romania |
| Md. Abdur Razzaque | University of Dhaka, Bangladesh |

## 7. Data Mining, Big Data Analytics and Social Networks

### Track Co-chairs

| | |
|---|---|
| Pavel Krömer | Technical University of Ostrava, Czech Republic |
| Alex Thomo | University of Victoria, Canada |
| Eric Pardede | La Trobe University, Australia |

### TPC Members

| | |
|---|---|
| Sebastián Basterrech | Technical University of Denmark, Denmark |
| Tibebe Beshah | University of Addis Ababa, Ethiopia |
| Nashwa El-Bendary | Arab Academy for Science, Egypt |
| Petr Musilek | University of Alberta, Canada |
| Varun Ojha | Newcastle University, UK |
| Alvaro Parres | ITESO, Mexico |
| Nizar Rokbani | ISSAT-University of Sousse, Tunisia |
| Farshid Hajati | Victoria University, Australia |
| Ji Zhang | University of Southern Queensland, Australia |
| Salimur Choudhury | Lakehead University, Canada |
| Carson Leung | University of Manitoba, Canada |
| Syed Mahbub | La Trobe University, Australia |
| Osama Mahdi | Melbourne Institute of Technology, Australia |
| Choiru Zain | La Trobe University, Australia |
| Rajalakshmi Rajasekaran | La Trobe University, Australia |
| Nawfal Ali | Monash University, Australia |

## 8. Internet of Things and Cyber-Physical Systems

### Track Co-chairs

| | |
|---|---|
| Tomoki Yoshihisa | Shiga University, Japan |
| Winston Seah | Victoria University of Wellington, New Zealand |
| Luciana Pereira Oliveira | Instituto Federal da Paraiba (IFPB), Brazil |

### TPC Members

| | |
|---|---|
| Akihiro Fujimoto | Wakayama University, Japan |
| Akimitsu Kanzaki | Shimane University, Japan |
| Kazuya Tsukamoto | Kyushu Institute of Technology, Japan |
| Lei Shu | Nanjing Agricultural University, China |
| Naoyuki Morimoto | Mie University, Japan |
| Teruhiro Mizumoto | Chiba Institute of Technology, Japan |
| Tomoya Kawakami | Fukui University, Japan |
| Adrian Pekar | Budapest University of Technology and Economics, Hungary |
| Alvin Valera | Victoria University of Wellington, New Zealand |
| Chidchanok Choksuchat | Prince of Songkla University, Thailand |
| Jyoti Sahni | Victoria University of Wellington, New Zealand |
| Murugaraj Odiathevar | Sungkyunkwan University, South Korea |
| Normalia Samian | Universiti Putra Malaysia, Malaysia |
| Qing Gu | University of Science and Technology Beijing, China |
| Tao Zheng | Beijing Jiaotong University, China |
| Wenbin Pei | Dalian University of Technology, China |
| William Liu | Unitec, New Zealand |
| Wuyungerile Li | Inner Mongolia University, China |
| Peng Huang | Sichuan Agricultural University, PR China |
| Ruan Delgado Gomes | Instituto Federal da Paraiba (IFPB), Brazil |
| Glauco Estacio Goncalves | Universidade Federal do Pará (UFPA), Brazil |
| Eduardo Luzeiro Feitosa | Universidade Federal do Amazonas (UFAM), Brazil |
| Paulo Ribeiro Lins Júnior | Instituto Federal da Paraiba (IFPB), Brazil |

## 9. Intelligent Computing and Machine Learning

### Track Co-chairs

| | |
|---|---|
| Takahiro Uchiya | Nagoya Institute of Technology, Japan |
| Flavius Frasincar | Erasmus University Rotterdam, The Netherlands |
| Miltos Alamaniotis | University of Texas at San Antonio, USA |

### TPC Members

| | |
|---|---|
| Kazuto Sasai | Ibaraki University, Japan |
| Shigeru Fujita | Chiba Institute of Technology, Japan |
| Yuki Kaeri | Mejiro University, Japan |
| Jolanta Mizera-Pietraszko | Military University of Land Forces, Poland |
| Ashwin Ittoo | University of Liège, Belgium |
| Marco Brambilla | Politecnico di Milano, Italy |
| Alfredo Cuzzocrea | University of Calabria, Italy |
| Le Minh Nguyen | JAIST, Japan |
| Akiko Aizawa | National Institute of Informatics, Japan |
| Natthawut Kertkeidkachorn | JAIST, Japan |
| Georgios Karagiannis | Durham University, UK |
| Leonidas Akritidis | International Hellenic University, Greece |
| Athanasios Fevgas | University of Thessaly, Greece |
| Yota Tsompanopoulou | University of Thessaly, Greece |
| Yuvaraj Munian | Texas A&M-San Antonio, USA |

## 10. Cloud and Services Computing

### Track Co-chairs

| | |
|---|---|
| Salvatore Venticinque | University of Campania "Luigi Vanvitelli", Italy |
| Shigenari Nakamura | Tokyo Denki University, Japan |
| Sajal Mukhopadhyay | National Institute of Technology, Durgapur, India |

### TPC Members

| | |
|---|---|
| Giancarlo Fortino | University of Calabria, Italy |
| Massimiliano Rak | University of Campania "Luigi Vanvitelli", Italy |
| Jason J. Jung | Chung-Ang University, Korea |

| | |
|---|---|
| Dimosthenis Kyriazis | University of Piraeus, Greece |
| Geir Horn | University of Oslo, Norway |
| Dario Branco | University of Campania "Luigi Vanvitelli", Italy |
| Dilawaer Duolikun | Cognizant Technology Solutions, Hungary |
| Naohiro Hayashibara | Kyoto Sangyo University, Japan |
| Tomoya Enokido | Rissho University, Japan |
| Sujoy Saha | NIT Durgapur, India |
| Animesh Dutta | NIT Durgapur, India |
| Pramod Mane | IIM Rohtak, India |
| Nanda Dulal Jana | NIT Durgapur, India |
| Banhi Sanyal | NIT Kurukshetra, India |

## 11. Security, Privacy and Trust Computing

### Track Co-chairs

| | |
|---|---|
| Ioannidis Sotirios | Technical University of Crete (TUC), Greece |
| Michail Alexiou | Georgia Institute of Technology, USA |
| Hiroaki Kikuchi | Meiji University, Japan |

### TPC Members

| | |
|---|---|
| George Vasiliadis | Hellenic Mediterranean University, Greece |
| Antreas Dionysiou | University of Cyprus, Cyprus |
| Apostolos Fouranaris | Athena Research Center, Greece |
| Panagiotis Ilia | Technical University of Crete, Greece |
| George Portokalidis | IMDEA, Spain |
| Nikolaos Gkorgkolis | University of Crete, Greece |
| Zeezoo Ryu | Georgia Institute of Technology, USA |
| Muhammad Faraz Karim | Georgia Institute of Technology, USA |
| Yunjie Deng | Georgia Institute of Technology, USA |
| Anna Raymaker | Georgia Institute of Technology, USA |
| Takamichi Saito | Meiji University, Japan |
| Kazumasa Omote | University of Tsukuba, Japan |
| Masakatsu Nishigaki | Shizuoka University, Japan |
| Mamoru Mimura | National Defense Academy of Japan, Japan |
| Chun-I Fan | National Sun Yat-sen University, Taiwan |
| Aida Ben Chehida Douss | National School of Engineers of Tunis, ENIT Tunis, Tunisia |
| Davinder Kaur | IUPUI, USA |

## 12. Software-Defined Networking and Network Virtualization

### Track Co-chairs

| | |
|---|---|
| Flavio de Oliveira Silva | Federal University of Uberlândia, Brazil |
| Ashutosh Bhatia | Birla Institute of Technology and Science, Pilani, India |

### TPC Members

| | |
|---|---|
| Rui Luís Andrade Aguiar | Universidade de Aveiro (UA), Portugal |
| Ivan Vidal | Universidad Carlos III de Madrid, Spain |
| Eduardo Coelho Cerqueira | Federal University of Pará (UFPA), Brazil |
| Christos Tranoris | University of Patras (UoP), Greece |
| Juliano Araújo Wickboldt | Federal University of Rio Grande do Sul (UFRGS), Brazil |
| Haribabu K. | BITS Pilani, India |
| Virendra Shekhavat | BITS Pilani, India |
| Makoto Ikeda | Fukuoka Institute of Technology, Japan |
| Farookh Hussain | University of Technology Sydney, Australia |
| Keita Matsuo | Fukuoka Institute of Technology, Japan |

## AINA-2024 Reviewers

Admir Barolli
Aida ben Chehida Douss
Akimitsu Kanzaki
Alba Amato
Alberto Postiglione
Alex Thomo
Alfredo Navarra
Amani Shatnawi
Anas AlSobeh
Andrea Marotta
Angela Peduto
Anne Kayem
Antreas Dionysiou
Arjan Durresi
Ashutosh Bhatia
Beniamino Di Martino
Bhed Bista
Burak Kizilkaya
Carson Leung
Chidchanok Choksuchat
Christos Tranoris
Chung-Ming Huang
Dario Branco
David Taniar
Elinda Mece
Enver Ever
Eric Pardede
Euripides Petrakis
Evjola Spaho
Fabrizio Messina
Feilong Tang
Flavio Silva
Francesco Orciuoli
George Portokalidis

Giancarlo Fortino
Giorgos Vasiliadis
Glauco Gonçalves
Hatim Chergui
Hiroaki Kikuchi
Hiroki Sakaji
Hiroshi Maeda
Hiroyuki Fujioka
Hyunhee Park
Isaac Woungang
Jana Nowaková
Jolanta Mizera-Pietraszko
Junichi Honda
Jyoti Sahni
Kazunori Uchida
Keita Matsuo
Kenichi Matsui
Kiki Adhinugraha
Kin Fun Li
Kiyotaka Fujisaki
Leonard Barolli
Leonardo Mostarda
Leonidas Akritidis
Lidia Ogiela
Lisandro Granville
Lucian Prodan
Luciana Oliveira
Mahmoud Elkhodr
Makoto Ikeda
Mamoru Mimura
Manato Fujimoto
Marco Antonio To
Marek Ogiela
Masaki Kohana
Minoru Uehara
Muhammad Karim
Muhammad Toaha Raza Khan
Murugaraj Odiathevar
Nadeem Javaid
Naohiro Hayashibara
Nobuo Funabiki
Nour El Madhoun
Omar Darwish
Panagiotis Ilia
Petr Musilek
Philip Moore
Purav Shah
R. Madhusudhan
Raffaele Guarasci
Ralf Klasing
Roberto Tonelli
Ronald Petrlic
Sabyasachi Mukhopadhyay
Sajal Mukhopadhyay
Salvatore D'Angelo
Salvatore D'Angelo
Salvatore D'Angelo
Salvatore Venticinque
Santi Caballé
Satoshi Ohzahata
Serafino Cicerone
Shigenari Nakamura
Shinji Sakamoto
Sho Tsugawa
Sriram Chellappan
Stephane Maag
Takayuki Kushida
Tetsuya Oda
Thomas Dreibholz
Tomoki Yoshihisa
Tomoya Enokido
Tomoya Kawakami
Tomoyuki Ishida
Vamsi Paruchuri
Victor Ströele
Vikram Singh
Wei Lu
Wenny Rahayu
Winston Seah
Yong Zheng
Yoshitaka Shibata
Yusuke Gotoh
Yuvaraj Munian
Zeezoo Ryu
Zhengjia Xu

# AINA-2024 Keynote Talks

# Agile Edge: Harnessing the Power of the Intelligent Edge by Agile Optimization

Fatos Xhafa

Technical University of Barcelona, Barcelona, Spain

**Abstract.** The digital cloud ecosystem comprises various degrees of computing granularity from large cloud servers and data centers to IoT devices, leading to the cloud-to-thing continuum computing paradigm. In this context, the intelligent edge aims at placing intelligence to the end devices, at the edges of the Internet. The premise is that collective intelligence from the IoT data deluge can be achieved and used at the edges of the Internet, offloading the computation burden from the cloud systems and leveraging real-time intelligence. This, however, comes with the challenges of processing and analyzing the IoT data streams in real time. In this talk, we will address how agile optimization can be useful for harnessing the power of the intelligent edge. Agile optimization is a powerful and promising solution, which differently from traditional optimization methods, is able to find optimized and scalable solutions under real-time requirements. We will bring real-life problems and case studies from Smart City Open Data Repositories to illustrate the approach. Finally, we will discuss the research challenges and emerging vision on the agile intelligent edge.

# Challenges in Entity Matching in AI Era

Juggapong Natwichai

Chiang Mai University, Chiang Mai, Thailand

**Abstract.** Entity matching (EM) is to identify and link entities originating from various sources that correspond to identical real-world entities, thereby constituting a foundational component within the realm of data integration. For example, in order to counter-fraud detection, the datasets from sellers, financial services providers, or even IT infrastructure service providers might be in need for data integration, and hence, the EM is highly important here. This matching process is also recognized for its pivotal role in data augmenting to improve the precision and dependability of subsequent tasks within the domain of data analytics. Traditionally, the EM procedure composes of two integral phases, namely blocking and matching. The blocking phase associates with the generation of candidate pairs and could affect the size and complexity of the data. Meanwhile, the matching phase will need to trade-off between the accuracy and the efficiency. In this talk, the challenges of both components are thoroughly explored, particularly with the aid of AI techniques. In addition, the preliminary experiment results to explore some important factors which affect the performance will be presented.

# Contents

# Application of Soft Computing Techniques for Clustering in Vehicular Ad Hoc Networks: A Survey

Evjola Spaho(✉) and Orjola Jaupi

Department of Electronics and Telecommunications, Faculty of Information Technology, Polytechnic University of Tirana, Tirana, Albania
{espaho,orjola.jaupi}@fti.edu.al

**Abstract.** Vehicular Ad Hoc Networks (VANETs) are dynamic mobile networks that use multi-hop communication. VANETs are a very important part of the Intelligent Transportation System (ITS) and can be used in a wide range of applications. In VANETs vehicles exchange information and use this information to make intelligent decisions to optimize travel time and traffic management to minimize congestion. Due to the increased number of vehicles, clustering is very important in VANETs. Clustering can be used to improve network management and scalability in VANETs. Creation and stability of clusters is a challenge in VANETs. In this paper we have presented a survey on different works where soft computing techniques are used for clustering in VANETs and discuss their advantages and limitations.

## 1 Introduction

Vehicular Ad Hoc Networks (VANETs) are composed of vehicles equipped with on board units and wireless transceivers and are part of the Intelligent Transportation System (ITS) [1–4]. Connected and Autonomous Vehicles (CAVs) [5, 6] can be part of VANETs and exchange data to safely operate. VANETs can be used in a wide range of applications [7, 8] as presented in Fig. 1. Vehicles in VANETs exchange information according to the application. To exchange information there are two types of communication: Vehicle to Vehicle (V2V) and Vehicle to Roadside Infrastructure (V2I).

Because the number of cars has increased a lot, and VANETs are large scale, to better manage, cars are divided into groups called clusters. Clustering divides the network into smaller parts and is an effective approach for node management. In Fig. 2 is presented clustering in VANETs and the main elements of a cluster.

The traditional clustering techniques usually decide if a vehicle will be in a cluster or not by using a threshold value for the distance. However, due to the very dynamic and the unpredictable nature of VANETs, new adaptive clustering techniques that do not only consider the distance, but also other parameters are emerging.

Soft computing techniques [9] consider uncertainty and imprecision and multiple criteria when they take decisions for cluster formations. Soft computing techniques are very important because they can be used for imprecise data deriving from dynamic


L. Barolli (Ed.): AINA 2024, LNDECT 203, pp. 1–11, 2024.
https://doi.org/10.1007/978-3-031-57931-8_1

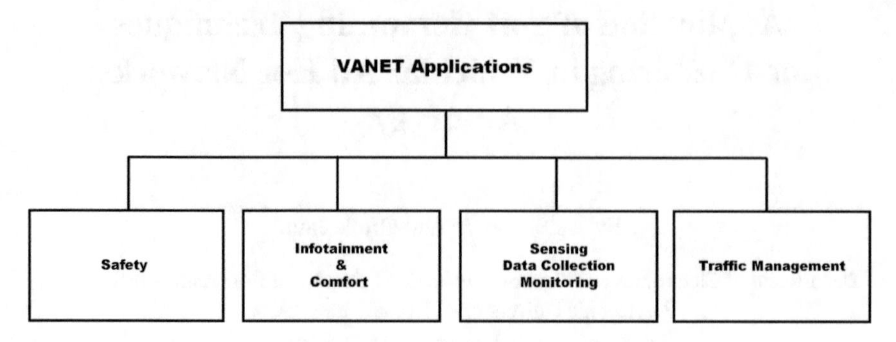


**Fig. 1.** VANETs Applications.

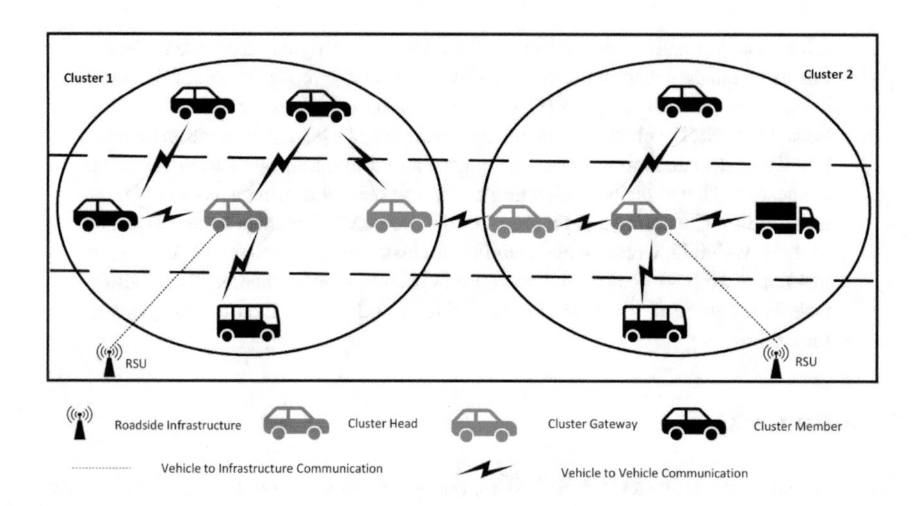


**Fig. 2.** Clustering in VANETs.

mobility of vehicular networks. These adaptive techniques are very important to be applied for clustering in frequently and dynamic changed VANET environment.

In Fig. 3 is presented a classification of soft computing techniques for clustering in VANETs. Clustering techniques can be based on Fuzzy Logic (FL), Metaheuristics or Machine Learning (ML).

FL [10] handles uncertainty, imprecision, and ambiguity to solve problems. FL takes into consideration different input parameters and the degree of membership for the clustering process. FL uses linguistic variables and fuzzy rules and can be easily adapted when network topology or condition changes.

Metaheuristics is a very powerful approach used for clustering in VANETs. Metaheuristics [11] are algorithms that mimic natural behavior and can be classified as evolutionary-based and swarm-based. Multi-objective optimization can be used for optimization of conflicting objectives. Genetic Algorithms (GAs), Ant Colony Optimization (ACO), PSO (Particle Swarm Optimization), Whale Optimization Algorithm

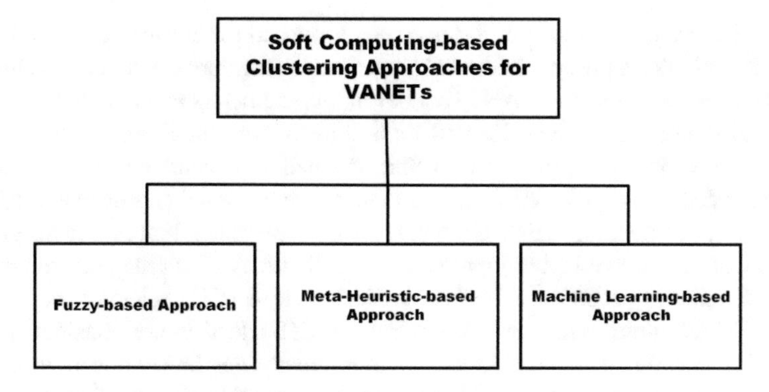


**Fig. 3.** Classification of Soft Computing approaches for clustering in VANETs.

(WOA), Grey Wolf Optimization Algorithm (GWOA) are some of the algorithms used for optimization in the cluster formation process in VANETs.

ML [12] uses statistical techniques and computational algorithms to learn from the data, make predictions and take actions. ML algorithms are suitable for VANETs because they can handle large amounts of data.

In this survey, different papers where soft computing-based clustering techniques are used, are reviewed, and advantages and limitations of these techniques are presented.

The contribution of this paper:

1. Detailed study of soft computing-based intelligence clustering techniques in VANETs
2. Categorization of these techniques in FL-based, Metaheuristics-based, and ML-based approaches.
3. Advantages and limitations for different research papers considered.
4. Discussion on advantages of using hybrid/integrated approaches for clustering in VANETs.

The rest of the paper is organized as follows: In Sect. 2, we present fuzzy-based clustering techniques. Metaheuristic-based clustering techniques are presented in Sect. 3. In Sect. 4 are presented machine learning-based clustering techniques. A discussion for Hybrid/Integrated techniques for clustering in VANETs is presented in Sect. 5. Finally, the conclusions are provided in Sect. 6.

## 2 Fuzzy Logic-Based Clustering Techniques for VANETs

FL is a computing approach based on the human way of making decisions. FL uses fuzzy sets and fuzzy rules to make decisions and makes calculations by using words. In a fuzzy controller there are three processes.

1) Fuzzification: During this process, membership functions are applied to the input.
2) Fuzzy rule-based inference system: Rules from the experts are applied to the fuzzified data.
3) Defuzzification: This process is done using the same membership functions used in the fuzzification process and the output is obtained.

Several approaches based on FL have been proposed for clustering in VANETs.

In [13] authors proposed a fuzzy-based system that evaluates whether a vehicle will remain or leave the cluster in VANETs. Four input parameters and a table with 81 rules are applied to control if a vehicle will remain in the cluster. From the results, if the distance of a vehicle from the cluster center is small and security is strong, there is a high chance that this vehicle will be in the cluster. The proposed approach is very simple and can easily decide for a vehicle if it will remain in the cluster, however in the proposed system are not considered other input parameters that may affect the performance.

In [14] authors proposed a FL-based Cluster Head (CH) selection approach for VANETs. Three input parameters to control the CH selection are considered for the fuzzy logic controller: average distance, relative velocity, and link connectivity duration. The proposed approach creates stable clusters and extended cluster lifetime compared with other approaches.

In [15] authors proposed a fuzzy logic with three input parameters: average distance, average compatibility and average velocity that controls the output of this system that is CH eligibility.

The mean number of clusters, mean cluster-head lifetime and mean cluster-head stability metrics are compared with two other approaches. Simulation results show that the proposed fuzzy logic-based approach performs better than the other approaches. However, the results are only for low and medium speed of vehicles and for 2 hop communication.

In [16] authors proposed a fuzzy logic system for cluster head selection. Four input parameters are considered (direction, speed, acceleration, and distance), and the output parameter is the best cluster head candidate. The system uses 27 fuzzy rules. Different from other works, authors considered two-way multilane highway scenario.

In [17], is proposed a fuzzy logic-based clustering scheme for VANETs that considers relative speed and distance from the cluster members. Compared with other works, this approach creates stabler clusters since it uses learning process to predict the driver's behavior.

In Table 1 are summarized advantages and limitations of the considered research works that uses FL-based approaches.

**Table 1.** FL-based approaches, advantages, and limitations.

| Reference | Advantages | Limitations |
|---|---|---|
| [13] | The proposed system uses simple linguistic expressions to model the problem and to find the solution | Only group speed, relative acceleration, security, and distance from the cluster center are considered to decide if the vehicle will remain or leave the cluster |
| [14] | The proposed approach creates Stable clusters, and their lifetime is extended compared with other works | Security and other parameters are not taken into consideration in the proposed approach |

(*continued*)

**Table 1.** (*continued*)

| Reference | Advantages | Limitations |
|---|---|---|
| [15] | The proposed approach creates Stable clusters, and their lifetime is extended compared with other works | The study is limited or low and medium speed (results are for speed 30 and 70 kmph) and for high speed there are no results. Also, only 2 hop scenario is considered |
| [16] | The advantage of this approach is that direction and multilane effect are considered | In this approach, other parameters that may affect the CH selection are not considered |
| [17] | This approach uses a learning mechanism to predict the speed of vehicles | In this approach only one directional 4 lanes are considered |

## 3 Metaheuristic-Based Clustering Techniques for VANETs

Metaheuristics clustering algorithms [18] are general-purposed optimization algorithms used in VANETs. Given the optimization objectives the metaheuristics algorithm will explore and exploit search space to find near optimal or good solutions.

Metaheuristics-based clustering techniques can be classified in swarm-based and evolutionary-based algorithms.

In swarm-based algorithms, the main advantage of using it for clustering in VANETs is the ability to adapt with the rapidly changing decentralized network topology.

GAs and other evolutionary techniques can be applied for optimizing the clustering process. They can dynamically adjust cluster head selection based on network performance metrics like communication quality, energy efficiency, and load balancing.

In [19] a clustering algorithm based on Moth Flame Optimization (MFO) is proposed. The proposed approach creates less clusters compared with other approaches. However, the number of nodes is small, 30–60 nodes are considered and details about the scenario and mobility model of the vehicles are not given, speed and direction are set arbitrary.

In [20] Whale Optimization Algorithm (WOA) is used to decide the optimal number of clusters for vehicular communications. It is bio-inspired from Bubble-net attacking Whales behavior. The simulation results show that the proposed approach uses a reduced number of clusters compared to Comprehensive Learning Particle Swarm Optimization (CLPSO).

In [21] a cluster optimization technique inspired by Harris Hawks behavior is proposed. This method in exploration phase initializes the vehicles position, speed, direction and in exploitation phase the CH is selected.

In [22] a reputation based weighted clustering protocol for VANETs is proposed. In this approach Firefly algorithm is used for tunning the parameters. Simulation scenarios are using real maps.

In [23] a hybrid clustering technique based on Genetic Algorithms and Ant Colony Optimization is proposed. GAs is utilized to explore the search space and identify appropriate cluster centers and ACO is used to attain the most optimal solution, thereby reducing the overall number of clusters.

**Table 2.** Metaheuristic-based approaches, advantages, and limitations.

| Reference | Advantages | Limitations |
|---|---|---|
| [19] | This approach creates a small number of clusters even when the density of nodes is low | Speed and node directions are arbitrary, and the number of nodes and grid size is very small |
| [20] | The proposed approach minimizes the number of clusters and end-to-end delay | Vehicles are randomly deployed within the search space and the number of vehicles and grid size considered is very small |
| [21] | Reduce the number of cluster heads and increase cluster stability | Number of nodes considered is small and the grid size and results are only for highway scenario |
| [22] | The average number of clusters is smaller compared with the other approaches in the paper. Lane direction, intersections and traffic lights are considered | The number of vehicles considered is small |
| [23] | Reduce the number of clusters and increased average cluster lifetime | Evaluations are done for a single lane |

In Table 2 are presented the advantages and limitations of all the considered research works for cluster selection in VANETs using metaheuristics.

## 4 Machine Learning-Based Clustering Techniques for VANETs

Clustering based in ML gained attention in the context of VANETs due to their ability to provide adaptive and data-driven cluster formation.

ML techniques can be applied to adapt clusters in real-time. This is very important in VANETs where network conditions can change rapidly due to vehicle mobility and varying traffic scenarios.

Unsupervised learning methods are often preferred compared to supervised learning because of the dynamic and unpredictable nature of VANETs. Unsupervised learning does not rely on labeled data. In unsupervised learning, algorithms try to find out structures or patterns in the data without specific instructions and these techniques are suitable with the nature of VANETs. Various unsupervised learning algorithms used in VANETs are presented in the following:

1) K-Means clustering: Is an algorithm that partitions vehicles into k clusters. These clusters are non-overlapping and are formed based on the similarity or distance between vehicles.
2) Hierarchial clustering: uses two approaches bottom up and top down to build a hierarchy of clusters. It is a promising technique in VANETs for organizing vehicles into clusters based on their geographical location or their directions of moving.
3) Affinity propagation: is used in VANETs for creating clusters based how similar are vehicles with each other. In this method clusters are created using a graph based technique by sending messages between pairs of samples.
4) Gaussian mixture models (GMM): uses a probalisitic model which assumes data in VANET come from several Gaussian distributions for example different directions of the vehicles or different speeds.
5) Density based spatial clustering of applications with noise: this method is used in VANET networks to indentify a large number of vehicles grouped together such as traffic jams. It groups together vehicles with nearby neighbours and identifies as outlier points which stand alone in low density areas.
6) Spectral clustering: is used when vehicles in VANET networks create a non-convex shape. It uses matrix values to reduce a cluster in smaller dimensions.
7) Ordering points to indentify the clustering structure: uses the same logic as density based spatial clustering of applications with noise just is more adaptable with varying densities. Since VANETs change densities because of moving vehicles it is a promising approach.
8) Mean shift clustering: a promising approach in VANETs since its not necessary to specify the number of clusters in advance. It uses a sliding window algorithm that finds the areas of dense data points.

In [24] authors proposed a clustering for VANETs using online machine learning. The proposed approach uses continuous learning and update in real time and can predict the behaviour in the next intersection. The cluster eficiency and stability is improved.

In [25] authors proposed a dynamic clustering technique for VANETs using Deep reinforcement learning. The dynamic policy applied extends the cluster lifetime.

In [26] authors proposed the weighted approach based in deep learning for selecting CH and creation of stable and efficient clusters. Trust metric is used as a new concept in the weighted approach.

In [27] authors used ML in 5G-Cellular V2X to create a multi-dimensional affinity propagation clustering. The proposed approach forms stable clusters.

In [28] author proposed a new validation index model, important for choosing and selecting the best machine learning clustering algorithm. M2I try to get minimum intra-cluster distance and maximum inter-cluster distance. It is useful for energy-efficient clustering.

In Table 3 are presented the advantages and limitations of all the considered research works for cluster selection in VANETs using ML-based approaches.

**Table 3.** ML-based approaches, advantages, and limitations.

| Reference | Advantages | Limitations |
|---|---|---|
| [24] | Efficient and stable clusters in intersection scenarios | Is tested only in highway scenario |
| [25] | Extended cluster lifetime | Is tested only in highway scenarios |
| [26] | An improvement in cluster head stability is seen when there is a high vehicular density. Improved throughput and network performance along with decreased packet delay and energy consumption | Using the trust value there is a misclassification of almost 14% for primary users' detection |
| [27] | This method enables dynamic cluster determination and reduction in cellular network load | For real world applications would be challenging because of limited hardware and because of significant computational complexity |
| [28] | High correct value clustering and increased accuracy of the index | It has a low efficiency when clusters overlap |

## 5 Discussion

Clustering in VANETs is a very complex problem to solve due to the dynamic and complex nature of these networks. Soft computing techniques can be used for clustering in VANETs due to their adaptability, fault tolerance, and ability to handle uncertainties.

Different soft computing techniques have unique advantages for clustering in VANETs. FL deals with imprecise information and employs linguistic terms, linguistic variables, and simple rules for clustering in VANETs.

Metaheuristics deals with optimization of complex problems. The main advantage in Metaheuristic-based algorithms is that they explore many solutions for cluster configuration, allowing for a more comprehensive search across the solution space and increased likelihood of finding optimal or near-optimal solutions in complex environments.

ML deals with pattern recognition and prediction. ML algorithms, through their ability to learn from data, can adapt to dynamic network conditions, handle complex relationships among variables, and provide insights into evolving traffic patterns and communication dynamics, contributing to more effective and adaptive clustering strategies in VANETs.

We propose a combination of different soft computing approaches and their strengths to create new innovative hybrid techniques for cluster formation in VANETs. Hybrid approaches can overcome the limitations that a single approach may have.

Integration of ML and FL can be used to create stable clusters. FL can be used for initial cluster formation and ML to enhance the system's adaptability by continuously learning and making decisions based on historical information.

The integration of Metaheuristics with Fuzzy Logic not only enhances the exploration of the solution space but also leverages the strengths of both approaches, combining the global search capabilities of Metaheuristics with the precise reasoning and adaptability of Fuzzy Logic.

Integration of FL, Metaheuristic and ML can result in highly effective clustering solutions for VANETs. This integration not only addresses the dynamic nature of VANETs but also ensures adaptability, robustness, and improved performance by combining the fuzzy reasoning of FL, global search capabilities of Metaheuristics, and learning and predictive capabilities of ML.

## 6 Conclusions

In this paper we have presented a survey of different soft computing techniques used for clustering in VANETs. Different papers where FL, Metaheuristics and ML approaches are used for clustering are reviewed and the advantages and limitations of these approaches are presented.

While the existing literature does not explicitly endorse it, our novel proposal is to use hybrid soft computing approaches. Hybrid approaches can be the best for clustering in VANETs because they tend to utilize the benefits of each technique and to overcome the limitations of individual techniques.

The development of ITS is moving beyond VANETs, transitioning towards the broader framework known as Internet of Vehicles (IoV). In the future, we will review the possibility of fitting the clustering techniques used in VANETs with the unique characteristics and requirements of IoV.

**Acknowledgments.** Authors would like to thank Polytechnic University of Tirana for supporting this research work.

## References

1. Qafzezi, E., Bylykbashi, K., Spaho, E., Barolli, L.: A new fuzzy-based resource management system for SDN-VANETs. Int. J. Mob. Comput. Multimedia Commun. (IJMCMC) **10**(4), 1–12 (2019). https://doi.org/10.4018/IJMCMC.2019100101
2. Spaho, E., Ikeda, M., Barolli, L., Xhafa, F.: Performance comparison of OLSR and AODV protocols in a VANET crossroad scenario. In: Park, J., Barolli, L., Xhafa, F., Jeong, H.Y. (eds.) Information Technology Convergence. Lecture Notes in Electrical Engineering, vol. 253, pp. 37–45. Springer, Dordrecht (2013). https://doi.org/10.1007/978-94-007-6996-0_5
3. Ono, R., Kulla, E., Spaho, E.: Performance evaluation of VANETs in different real map scenarios. In: Barolli, L., Xhafa, F., Yim, K. (eds.) Advances on Broad-Band Wireless Computing, Communication and Applications. BWCCA 2016. Lecture Notes on Data Engineering and Communications Technologies, vol. 2, pp. 639–647. Springer, Cham (2017). https://doi.org/10.1007/978-3-319-49106-6_63
4. Spaho, E., Ikeda, M., Barolli, L., Xhafa, F., Younas, M., Takizawa, M.: Performance of OLSR and DSDV protocols in a VANET scenario: evaluation using CAVENET and NS3. In: 2012 Seventh International Conference on Broadband, Wireless Computing, Communication and Applications, Victoria, BC, Canada, pp. 108–113 (2012). https://doi.org/10.1109/BWCCA.2012.28

5. Pumwa, J.: Time variant predictive control of autonomous vehicles. Int. J. Innov. Technol. Interdisc. Sci. **2**(1), 62–77 (2019). https://doi.org/10.15157/IJITIS.2019.2.1.62-77
6. Poulain, T.: Trajectory generation and control of autonomous vehicles. Int. J. Innov. Technol. Interdisc. Sci. **2**(3), 200–211 (2019). https://doi.org/10.15157/IJITIS.2019.2.3.200-211
7. Saha, S.: Automated traffic law enforcement system: a feasibility study for the congested cities of developing countries. Int. J. Innov. Technol. Interdisc. Sci. **3**(1), 346–363 (2020). https://doi.org/10.15157/IJITIS.2020.3.1.346-363
8. Opoku, D., Kommey, B.: FPGA-based intelligent traffic controller with remote operation mode. Int. J. Innov. Technol. Interdisc. Sci. **3**(3), 490–500 (2020). https://doi.org/10.15157/IJITIS.2020.3.3.490-500
9. Zadeh, L.A.: Fuzzy logic, neural networks, and soft computing. Commun. ACM **37**(3), 77–84 (1994). https://doi.org/10.1145/175247.175255
10. Krause, P.: Fuzzy logic for the management of uncertainty edited by Lotfi Zadeh and Janusz Kacprzyk, John Wiley & Sons, New York (1992), pp. 1–676 (1995). ISBN 0-471-54799-9. The Knowledge Engineering Review **10**(1), 101–101. https://doi.org/10.1017/S0269888900007347
11. Abdel-Basset, M., Abdel-Fatah, L., Sangaiah, A.K.: Metaheuristic Algorithms: A Comprehensive Review. In: Sangaiah, A.K., Sheng, M., Zhang, Z. (eds.) In Intelligent Data-Centric Systems, Computational Intelligence for Multimedia Big Data on the Cloud with Engineering Applications, pp. 185–231. Academic Press (2018). ISBN 9780128133149. https://doi.org/10.1016/B978-0-12-813314-9.00010-4
12. Ayyub, M., Oracevic, A., Hussain, R., Khan, A.A., Zhang, Z.: A comprehensive survey on clustering in vehicular networks: current solutions and future challenges. Ad Hoc Netw. 124, 102729 (2022). ISSN 1570-8705. https://doi.org/10.1016/j.adhoc.2021.102729
13. Ozera, K., Bylykbashi, K., Liu, Y., Ikeda, M., Barolli, L.: Clustering in VANETs: a fuzzy-based system for clustering of vehicles. In: Barolli, L., Kryvinska, N., Enokido, T., Takizawa, M. (eds.) Advances in Network-Based Information Systems. NBiS 2018. Lecture Notes on Data Engineering and Communications Technologies, vol. 22, pp. 810–821. Springer, Cham (2019). https://doi.org/10.1007/978-3-319-98530-5_72
14. Aissa, M., Bouhdid, B., Ben Mnaouer, A., Belghith, A., AlAhmadi, S.: SOFCluster: safety-oriented, fuzzy logic-based clustering scheme for vehicular ad hoc networks. Trans. Emerg. Telecommun. Technol. **33** (2022). https://doi.org/10.1002/ett.3951
15. Tal, I., Muntean, G.-M.: User-oriented fuzzy logic-based clustering scheme for vehicular ad-hoc networks. In: 2013 IEEE 77th Vehicular Technology Conference (VTC Spring), Dresden, Germany, pp. 1–5 (2013). https://doi.org/10.1109/VTCSpring.2013.6692801
16. Çalhan, A.: A fuzzy logic-based clustering strategy for improving vehicular ad-hoc network performance. Sadhana **40**, 351–367 (2015). https://doi.org/10.1007/s12046-014-0315-9
17. Hafeez, K.A., Zhao, L., Liao, Z., Ma, B.N.-W.: A fuzzy-logic-based cluster head selection algorithm in VANETs. In: 2012 IEEE International Conference on Communications (ICC), Ottawa, ON, Canada, pp. 203–207 (2012). https://doi.org/10.1109/ICC.2012.6363839
18. Ahsan, W., et al.: Optimized node clustering in VANETs by using meta-heuristic algorithms. Electronics **9**, 394 (2020). https://doi.org/10.3390/electronics9030394
19. Shah, Y.A., Habib, H.A., Aadil, F., Khan, M.F., Maqsood, M., Nawaz, T.: CAMONET: moth-flame optimization (MFO) based clustering algorithm for VANETs. IEEE Access **6**, 48611–48624 (2018). https://doi.org/10.1109/ACCESS.2018.2868118
20. Shah, Y.A., et al.: An evolutionary algorithm-based vehicular clustering technique for VANETs. IEEE Access **10**, 14368–14385 (2022). https://doi.org/10.1109/ACCESS.2022.3145905
21. Husnain, G., et al.: An intelligent harris hawks optimization based cluster optimization scheme for VANETs. J. Sens. **2022**, Article ID 6790082, 15 p (2022). https://doi.org/10.1155/2022/6790082

22. Joshua, C.J., Duraisamy, R., Varadarajan, V.: A reputation based weighted clustering protocol in VANET: a multi-objective firefly approach. Mobile Netw. Appl. **24**, 1199–1209 (2019). https://doi.org/10.1007/s11036-019-01257-z
23. Goswami, V., Verma, S.K., Singh, V.: A novel hybrid GA-ACO based clustering algorithm for VANET. In: 2017 3rd International Conference on Advances in Computing, Communication & Automation (ICACCA) (Fall), Dehradun, India, pp. 1–6 (2017). https://doi.org/10.1109/ICACCAF.2017.8344740
24. Alsuhli, G.H., Khattab, A., Fahmy, Y.A., Massoud, Y.: Enhanced urban clustering in VANETs using online machine learning. In: 2019 IEEE International Conference on Vehicular Electronics and Safety (ICVES), Cairo, Egypt, pp. 1–6 (2019). https://doi.org/10.1109/ICVES.2019.8906296
25. Zhang, R., Wu, K.: A new dynamic clustering scheme for VANETs driven by deep reinforcement learning. In: Proceedings of SPIE 12636, Third International Conference on Machine Learning and Computer Application (ICMLCA 2022), p. 1263611 (2023). https://doi.org/10.1117/12.2675347
26. Abdulrazzak, H.N., Hock, G.C., Radzi, N.A.M., Tan, N.M.L.: A new unsupervised validation index model suitable for energy-efficient clustering techniques in VANET. IEEE Access. https://doi.org/10.1109/ACCESS.2023.3281302
27. Koshimizu, T., Gengtian, S., Wang, H., Pan, Z., Liu, J., Shimamoto, S.: Multi-dimensional affinity propagation clustering applying a machine learning in 5G-cellular V2X. IEEE Access **8**, 94560–94574 (2020). https://doi.org/10.1109/ACCESS.2020.2994132
28. Saleem, M.A., et al.: Deep learning-based dynamic stable cluster head selection in VANET. J. Adv. Transp. **2021**, Article ID 9936299, 21 p (2021). https://doi.org/10.1155/2021/9936299

# EarlyStage Diabetes Risk Detection Using Comparison of Xgboost, Lightgbm, and Catboost Algorithms

Henny Febriana Harumy(✉), Sri Melvani Hardi, and Muhammad Fajri Al Banna

Universitas Sumatera Utara, Dr. Mansyur No. 9, Padang Bulan, North Sumatera, Indonesia
{hennyharumy,vani.hardi}@usu.ac.id,
mfajribanna@students.usu.ac.id

**Abstract.** Diabetes Mellitus is a chronic metabolic disorder that elevates blood glucose levels due to insufficient insulin production or insulin resistance. This disease has significant and increasing global impact. Numerous studies have investigated the effects of using classification algorithms to detect and prevent diabetes mellitus. However, existing research faces various challenges in achieving optimal performance and efficient detection time. This study selects the LightGBM, XGBoost, and CatBoost methods for classification. These three algorithms are trained using the "early-stage diabetes risk prediction dataset" and their results are compared to determine which algorithm is best for classifying early-stage diabetes. The testing results indicate that the model trained using the CatBoost algorithm demonstrates superior performance with higher accuracy, precision, recall, F1-Score, and ROC-AUC scores compared to models trained with XGBoost and LightGBM. Additionally, the LightGBM algorithm exhibits faster computational time compared to XGBoost and CatBoost.

**Keywords:** Diabetes · Classification · XGBoost · LightGBM · CatBoost

## 1 Introduction

Diabetes Mellitus, commonly known as diabetes, is a chronic metabolic disorder characterized by elevated blood glucose levels due to insufficient production of the hormone insulin or ineffective utilization of insulin by the body. Based on the disease's pathogenesis, two main mechanisms are involved in the development of diabetes: insufficient insulin production due to damage to insulin-producing β cells in the pancreas, known as Type 1 Diabetes Mellitus (T1DM), and insulin resistance, which occurs when body cells become less sensitive to insulin, making it difficult for insulin to transport glucose into cells for energy. There are three main types of diabetes: Type 1 Diabetes Mellitus (T1DM), Type 2 Diabetes Mellitus (T2DM), and gestational diabetes (GDM). Recent findings indicate a significant increase in the prevalence of diabetes over the past decade, making it an emerging epidemic. In 2021, it was estimated that 10.5% of all adults aged 20-79 worldwide were affected by diabetes, representing 537 million individuals [1]. It


L. Barolli (Ed.): AINA 2024, LNDECT 203, pp. 12–24, 2024.
https://doi.org/10.1007/978-3-031-57931-8_2

is projected that by 2030, there will be 643 million people with diabetes, and by 2045, this number will rise to 783 million adults aged 20-79 [1]. Almost half of the adults with diabetes are unaware of their condition, with an estimated 240 million people remaining undiagnosed [1]. Late diagnosis of diabetes leads to a higher likelihood of complications and increased healthcare utilization. It is crucial for individuals with diabetes and those without to be diagnosed as early as possible to prevent or delay complications, avoid premature death, and enhance quality of life. Therefore, a system capable of detecting the risk of diabetes is needed.

In the current era of modern technology, computer-based systems can assist in accurately detecting diseases, saving time and money through the use of data mining [2, 3]. Data mining involves extracting essential knowledge from large datasets to predict outcomes using techniques such as classification, clustering, and association, making it an effective method for diabetes risk detection. Data mining significantly aids in identifying diabetes risk and ultimately improves the quality of healthcare. Data mining methods in disease diagnosis employ various machine learning algorithms to uncover hidden patterns and enhance the accuracy of detection used for predicting future events [4–7]. Numerous studies have investigated the impact of using classification algorithms to detect diabetes. However, existing research still faces challenges in achieving optimal performance in detection systems and minimizing detection delays. Gradient Boosting Decision Tree is a popular method for prediction problems in the classification and regression domains. Gradient Boosting Decision Tree (GBDT) is an ensemble machine learning algorithm that combines multiple decision trees to make more accurate predictions. GBDT works by iteratively adding decision trees to the model and adjusting the weights on each tree to minimize prediction errors. XGBoost, LightGBM, and CatBoost are popular Gradient Boosting Decision Tree algorithms that have demonstrated superior accuracy and efficiency in various competitions. Each of these algorithms has its strengths and weaknesses. Based on the aforementioned explanations, the author designs an early-stage diabetes risk detection system by comparing the XGBoost, LightGBM, and CatBoost algorithms to determine which algorithm can better detect the risk of diabetes. The expected objectives of this research are as follows to Develop an application or system capable of detecting the risk of early-stage diabetes using the XGBoost, LightGBM, and CatBoost algorithms and Compare the performance metrics of Accuracy, Precision, Recall, F1-score, ROC- AUC, and running time between the XGBoost, LightGBM, and CatBoost algorithms trained on both CPU and GPU platforms.

## 2 Methodology

The dataset was analyzed using the following classification algorithms. The steps for data analysis are formulated as follows (Fig. 1).

### 2.1 XGBoost

XGBoost is one of the boosted tree algorithms that follows the gradient boosting rule, which is an ensemble learning method that combines predictions from multiple weak models to generate a stronger prediction [2, 8, 9]. In this algorithm, decision trees

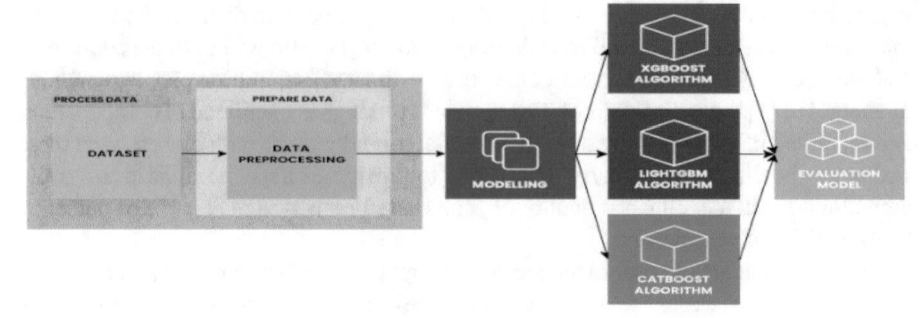


**Fig. 1.** Architecture Research. Gambar ini menunjukan tentang arsitektur model dimulai dari preprocesing data kemudian modeling hingg evaluation model.

are created sequentially. Weights (w) play a significant role in XGBoost. Weights are assigned to all independent variables, which are then inputted into the decision tree to predict the outcome. The weight of variables that are predicted incorrectly by the tree is increased, and these variables are then included in the second decision tree. These individual classifiers/predictors are then combined (ensemble) to generate a stronger and more accurate model.

$$Obj^{(t)} = \sum_{i=1}^{t} l(y_i, \hat{y}^{(t)}) + \sum_{i=1}^{t} \Omega(f_i) \tag{1}$$

Description:

$\hat{y}_i(t)$ = Prediction value
$y_i$ = Actual value
$l(y_i, \hat{y}_i^{(t)})$ = *Loss Function*
$\Omega(f_i)$ = Regularization

### 2.2 LightGBM

LightGBM (Light Gradient Boosting Machine) is one of the GBDT algorithms designed by [3]. The LightGBM model uses XGBoost as its baseline but approaches the classification problem differently [4], by introducing and applying a combination of two novel techniques: Gradient-based One-Side Sampling (GOSS) and Exclusive Feature Bundling (EFB) [3]. The GBDT model is defined as follows:

$$F(x; w) = \sum_{t=0}^{T} \alpha_t h_t(x; w) \tag{2}$$

where the function $h$ represents the $t$-th model decision tree, the function $F$ is interpreted as the predictive values of the GBDT model, x represents input samples, $w$ is the parameter of the decision tree, and α is the weight of each tree. By minimizing the loss function $L$ to map the $x$ space to the y space, the optimal model is solved as:

$$F* = \arg\min_{F} \sum_{i=0}^{N} L(y, F(x; w)) \tag{3}$$

The LightGBM algorithm utilizes GOSS as a sampling algorithm. Samples with large gradients are retained, while samples with small gradients are randomly selected and given a constant weight. This way, GOSS focuses more on poorly trained samples without altering the distribution of the raw data. The variance gain of splitting instances on feature subsets A and B is defined by the following equation:

$$\tilde{V}(d) = \frac{1}{n}(L_1 + L_2) \tag{4}$$

### 2.3 CatBoost

CatBoost, also known as Categorical Boosting, is a new gradient boosting algorithm proposed by [5] and [6]. This algorithm differs from other GBDT algorithms in several aspects. The CatBoost algorithm can handle categorical features during training without the need for preprocessing. CatBoost enables the use of the entire dataset for training. Target Statistic (TS) is an efficient method for handling categorical features. For each example, CatBoost randomly permutes the dataset and calculates the average label value for examples with the same category value placed before the given example in the permutation. If the permutation is $\Theta = [\sigma_1, \cdots \sigma_n]_n^T$ it is replaced with:

$$x_{\sigma_{p,k}} = \frac{\sum_{j=1}^{p-1} \left[x_{\sigma_{j,k}} = x_{\sigma_{p,k}}\right] \cdot Y_{\sigma_j} + \beta \cdot P}{\sum_{j=1}^{p-1} \left[x_{\sigma_{j,k}} = x_{\sigma_{p,k}}\right] + \beta} \tag{5}$$

where $P$ is the prior value or previous value and β is the weight of the prior. For regression tasks, the standard technique for calculating priors is to take the average value of the labels in the dataset [7, 11–14].

## 3 Experimental Analysis

Details regarding the dataset and results analysis are presented in this section.

### 3.1 Dataset

This dataset contains symptom reports from 520 people related to diabetes. It includes data about people including symptoms that could lead to diabetes. This dataset has been created from direct questionnaires to people who have recently developed diabetes, or who are still nondiabetic but have a few or more symptoms. The data has been collected from patients using a questionnaire directly from Sylhet Diabetes Hospital Sylhet, Bangladesh. Data preprocessing was done by handling missing values. The proposed methodology was evaluated on the "early-stage diabetes risk prediction dataset" retrieved from the UCI Repository. This *dataset* consists of medical details of 520 cases, with a detailed description in Table 1.

**Table 1.** Dataset Description

| Atribut | Values |
|---|---|
| Age | 0–91 |
| Gender | 1.Male, 2.Female |
| Polyuria | 1.Yes, 2.No |
| Polydipsia | 1.Yes, 2.No |
| Sudden weight loss | 1.Yes, 2.No |
| Weakness | 1.Yes, 2.No |
| Polyphagia | 1.Yes, 2.No |
| Genital thrush | 1.Yes, 2.No |
| Visual blurring | 1.Yes, 2.No |
| Itching | 1.Yes, 2.No |
| Irritability | 1.Yes, 2.No |
| Delayed healing | 1.Yes, 2.No |
| Partial paresis | 1.Yes, 2.No |
| Muscle stiffness | 1.Yes, 2.No |
| Alopecia | 1.Yes, 2.No |
| Obesity | 1.Yes, 2.No |
| Class | 1.Positif, 2.Negatif |

### 3.2 Preprocessing

Of the 520 cases from the dataset used, there are still duplicate data. Duplication of existing news needs to be removed because it can increase the sample weight and cause bias in the model to be generated. At this stage, duplicate news is removed based on the similarity of the value of the dataset features. Furthermore, data transformation will be carried out from categorical data to numerical so that it can be trained.

### 3.3 Result Analysis

In this step, the model is trained to take inputs that have passed the data preprocessing stage and predict the target data. The input used will be divided into three data separation formations, namely Training data split 0.5 and test data 0.5, Training data split 0.7 and test data 0.3 and Training data split 0.8 and test data 0.2.The three data separation formations were then trained with the three algorithms used in this study, namely the XGBoost, LightGBM and CatBoost algorithms on the CPU and GPU using several parameters as shown in Table 2. The performance of the XGBoost, LightGBM and CatBoost algorithms using the "early-stage diabetes risk prediction dataset" trained on CPU and GPU with detailed information as follows.

**Table 2.** Parameters Used in Training the Model

| Parameter | Value |
|---|---|
| depth | 5 |
| learning_rate | 0.001 |

### 3.4 XGBoost

The data that has undergone data preprocessing is then trained using the XGBoost algorithm. Training is performed using three data splitting configurations, namely Training data split 0.5 and test data 0.5, Training data split 0.7 and test data 0.3 and Training data split 0.8 and test data 0.2. XGBoost algorithm training is done on CPU and GPU, then the matrix evaluation of the resulting model will be measured. The model performance is shown as follows (Table 3):

**Table 3.** Score XGBoost Model

| Data Train | Data Test | Computation | Accuracy | Precision | Recall | F1-Score | AUC |
|---|---|---|---|---|---|---|---|
| 0.5 | 0.5 | CPU | 92 | 91 | 89 | 90 | 89 |
| 0.5 | 0.5 | GPU | 90 | 90 | 86 | 88 | 86 |
| 0.7 | 0.3 | CPU | 89 | 89 | 88 | 88 | 87 |
| 0.7 | 0.3 | GPU | 89 | 89 | 88 | 88 | 87 |
| 0.8 | 0.2 | CPU | 90 | 88 | 89 | 89 | 89 |
| 0.8 | 0.2 | GPU | 90 | 91 | 86 | 88 | 86 |

### 3.5 LightGBM

The data that has undergone data preprocessing is then trained using the LightGBM algorithm. Training is performed using three data splitting configurations, namely 50% training data and 50% testing data splitting, 70% training data and 30% testing data splitting and 80% training data and 20% testing data splitting. The LightGBM algorithm is trained on both CPU and GPU, and the evaluation metrics of the resulting model will be measured. The performance of the model is presented as follows (Table 4):

### 3.6 CatBoost

The data that has undergone data preprocessing is then trained using the CatBoost algorithm. Training is performed using three data splitting configurations, 50% training data and 50% testing data splitting, 70% training data and 30% testing data splitting and 80% training data and 20% testing data splitting. The CatBoost algorithm is trained on both CPU and GPU, and the evaluation metrics of the resulting model will be measured. The performance of the model is presented as follows (Table 5):

**Table 4.** Score Model LightGBM

| Data train | Data test | Computation | Accuracy | Precision | Recall | F1-Score | AUC |
|---|---|---|---|---|---|---|---|
| 0.5 | 0.5 | CPU | 84 | 82 | 77 | 79 | 77 |
| 0.5 | 0.5 | GPU | 84 | 82 | 77 | 79 | 77 |
| 0.7 | 0.3 | CPU | 89 | 90 | 87 | 88 | 87 |
| 0.7 | 0.3 | GPU | 89 | 90 | 87 | 88 | 87 |
| 0.8 | 0.2 | CPU | 88 | 86 | 88 | 87 | 88 |
| 0.8 | 0.2 | GPU | 88 | 86 | 88 | 87 | 88 |

**Table 5.** Score Model LightGBM

| Data Train | Data Test | Computation | Accuracy | Precision | Recall | F1-Score | AUC |
|---|---|---|---|---|---|---|---|
| 0.5 | 0.5 | CPU | 96 | 95 | 96 | 95 | 96 |
| 0.5 | 0.5 | GPU | 94 | 94 | 91 | 92 | 91 |
| 0.7 | 0.3 | CPU | 93 | 93 | 92 | 93 | 92 |
| 0.7 | 0.3 | GPU | 91 | 92 | 88 | 89 | 88 |
| 0.8 | 0.2 | CPU | 92 | 91 | 91 | 91 | 91 |
| 0.8 | 0.2 | GPU | 92 | 91 | 91 | 91 | 91 |

### 3.7 Model Evaluation

In the model evaluation stage, all the components required to train the model are re-examined and evaluated. Upon further investigation, it was found that the dataset suffers from imbalanced data. Imbalanced data is a problem in machine learning where there is an unproportional ratio in each class. Unbalanced data between the majority and minority classes can lead to errors in classification. Imbalanced data will heavily rely on the majority class in the classification process. The consequences of such misdiagnosis include obtaining incorrect classification results and potentially causing patients to receive incorrect medical treatments. Therefore, to address the issue of imbalanced data, the ADASYN data resampling technique is employed.

```
Before Upsampling Using Adasyn : Counter ({1:173,0:78})
Before Upsampling Using Adasyn : Counter ({1:173,0:168})
```

Additionally, in this stage, the K-fold cross-validation method is implemented with three folds to evaluate the model and obtain the best model. Furthermore, the performance metrics including accuracy, precision, recall, F1-Score, ROC-AUC, logging utilization, and running time are measured again when training the model using the XGBoost, LightGBM, and CatBoost algorithms on both CPU and GPU (Table 6).

**Table 6.** Evaluation Models

| Evaluation | Data Train | Data Test | GPU /CPU | Acc | Precision | Recall | F1-Score | AUC |
|---|---|---|---|---|---|---|---|---|
| XGBoost | 0.5 | 0.5 | CPU | 91 | 91 | 91 | 91 | 94 |
| XGBoost | 0.5 | 0.5 | GPU | 91 | 91 | 91 | 91 | 94 |
| XGBoost | 0.7 | 0.3 | CPU | 90 | 90 | 90 | 90 | 95 |
| XGBoost | 0.7 | 0.3 | GPU | 90 | 90 | 90 | 90 | 95 |
| XGBoost | 0.8 | 0.2 | CPU | 92 | 92 | 91 | 92 | 97 |
| XGBoost | 0.8 | 0.2 | GPU | 92 | 92 | 91 | 92 | 97 |
| LightGBM | 0.5 | 0.5 | CPU | 87 | 88 | 87 | 87 | 93 |
| LightGBM | 0.5 | 0.5 | GPU | 87 | 88 | 87 | 87 | 93 |
| LightGBM | 0.7 | 0.3 | CPU | 88 | 89 | 88 | 88 | 92 |
| LightGBM | 0.7 | 0.3 | GPU | 88 | 89 | 88 | 88 | 92 |
| LightGBM | 0.8 | 0.2 | CPU | 89 | 89 | 89 | 89 | 95 |
| LightGBM | 0.8 | 0.2 | GPU | 89 | 89 | 89 | 89 | 95 |
| CatBoost | 0.5 | 0.5 | CPU | 91 | 91 | 91 | 91 | 97 |
| CatBoost | 0.5 | 0.5 | GPU | 90 | 90 | 90 | 90 | 97 |
| CatBoost | 0.7 | 0.3 | CPU | 92 | 92 | 92 | 92 | 97 |
| CatBoost | 0.7 | 0.3 | GPU | 90 | 91 | 90 | 90 | 97 |
| CatBoost | 0.8 | 0.2 | CPU | 94 | 95 | 94 | 94 | 98 |
| CatBoost | 0.8 | 0.2 | GPU | 94 | 95 | 94 | 94 | 98 |

Based on the conducted tests, it was found that identification using the CatBoost algorithm has higher levels of accuracy, precision, recall, F1-score, and ROC-AUC compared to the XGBoost and LightGBM algorithms. The scores generated by the three models are shown in Table 7. The CatBoost algorithm also requires higher resources compared to the XGBoost and LightGBM algorithms for training on all data separation formations (Figs. 2, 3, 4, 5, 6 and 7).

However, models trained with the CatBoost algorithm have longer computation times for training and model prediction. Models trained with the LightGBM algorithm exhibit faster computation times compared to the XGBoost and CatBoost algorithms. The computation times generated by the three models are shown in Table 8.

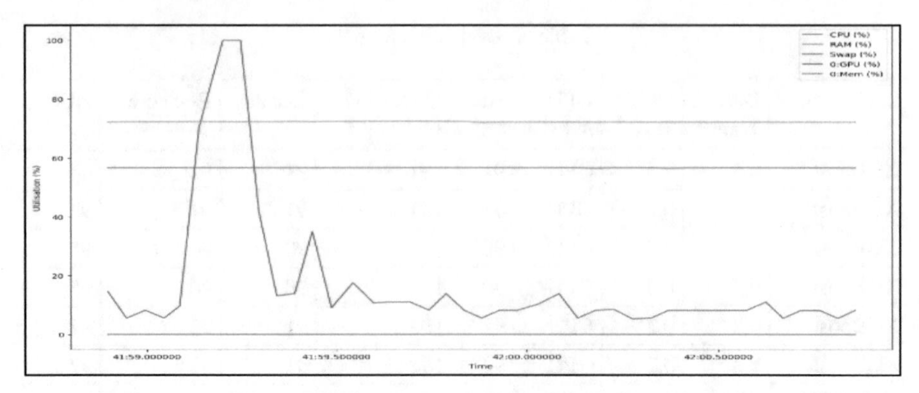


**Fig. 2.** Graph Logging Utilization Computing CatBoost Model Training on CPU with Training Data Separation 0.2 and Test Data 0.8

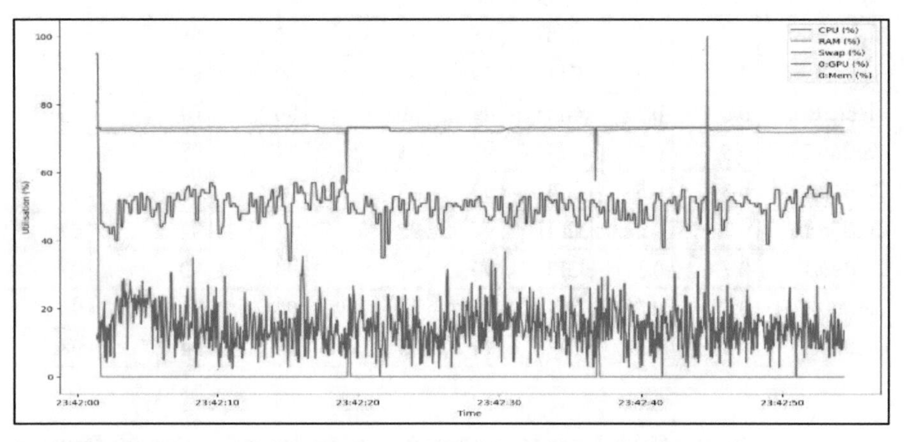


**Fig. 3.** Graph of Logging Utilization Computing CatBoost Model Training on GPU with Training Data Separation 0.2 and Test Data 0.8

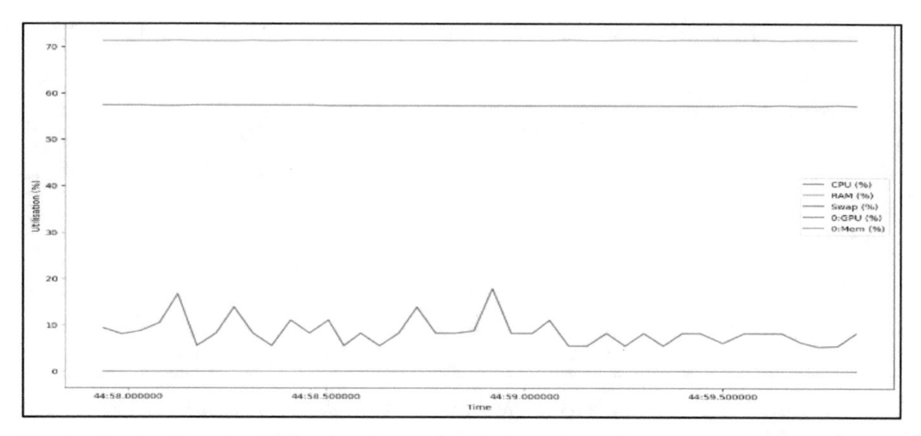


**Fig. 4.** Graph of Logging Utilization Computing CatBoost Model Training on CPU with Training Data Separation 0.3 and Test Data 0.7

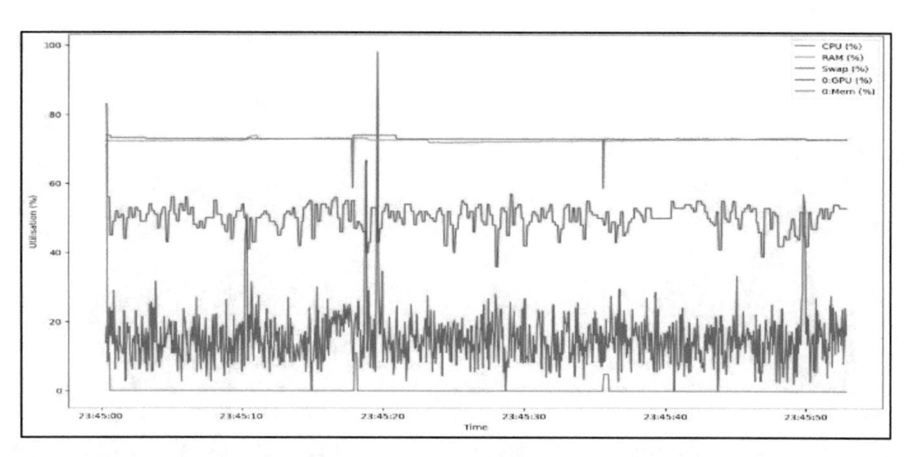


**Fig. 5.** Graph of Logging Utilization Computing CatBoost Model Training on GPU with Training Data Separation 0.3 and Test Data 0.7

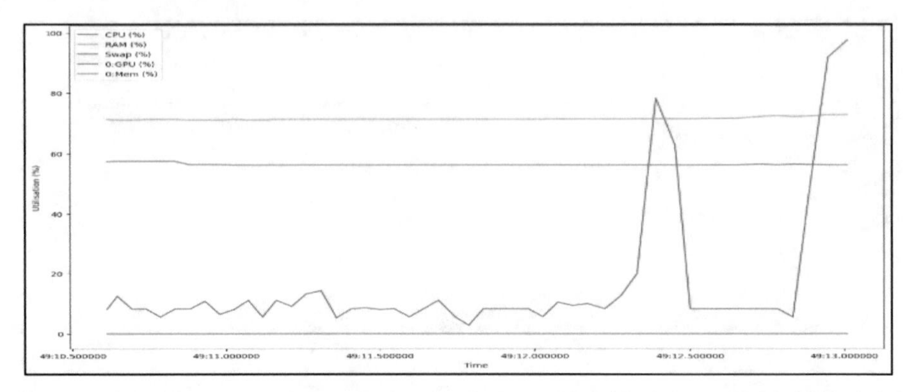


**Fig. 6.** Graph of Logging Utilization Computing CatBoost Model Training on CPU with Training Data Separation of 0.5 and Test Data 0.5.

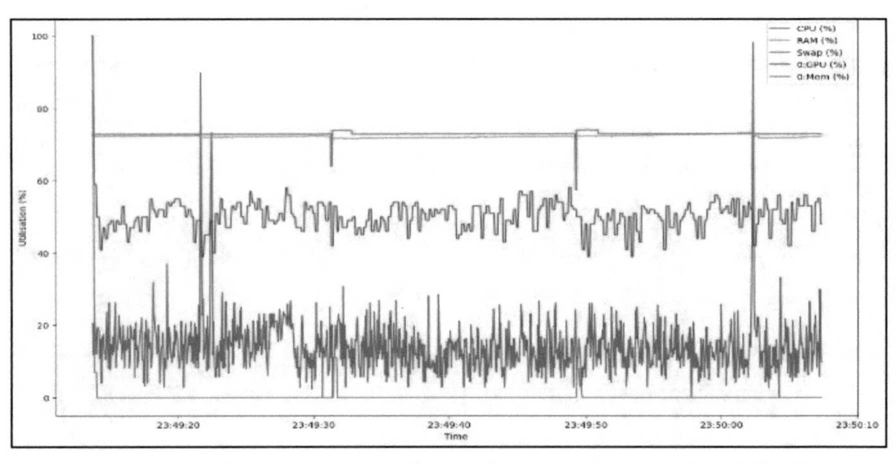


**Fig. 7.** Graph of Logging Utilization Computing CatBoost Model Training on GPU with Training Data Separation of 0.5 and Test Data 0.5

**Table 7.** Computation Time

| Evaluation | Data Train | Data Test | GPU/CPU | Computation Time |
|---|---|---|---|---|
| XGBoost | 0.5 | 0.5 | CPU | 0.1552448272705078 |
| XGBoost | 0.5 | 0.5 | GPU | 0.6926624774932861 |
| XGBoost | 0.7 | 0.3 | CPU | 0.15127205848693848 |
| XGBoost | 0.7 | 0.3 | GPU | 0.6484582424163818 |
| XGBoost | 0.8 | 0.2 | CPU | 0.32184720039367676 |
| XGBoost | 0.8 | 0.2 | GPU | 0.6732831001281738 |
| LightGBM | 0.5 | 0.5 | CPU | 0.15065789222717285 |
| LightGBM | 0.5 | 0.5 | GPU | 0.6794736385345459 |
| LightGBM | 0.7 | 0.3 | CPU | 0.1133260726928711 |
| LightGBM | 0.7 | 0.3 | GPU | 0.6180007457733154 |
| LightGBM | 0.8 | 0.2 | CPU | 0.1148526668548584 |
| LightGBM | 0.8 | 0.2 | GPU | 0.6537370681762695 |
| CatBoost | 0.5 | 0.5 | CPU | 2.654798746109009 |
| CatBoost | 0.5 | 0.5 | GPU | 58.77682685852051 |
| CatBoost | 0.7 | 0.3 | CPU | 2.3303232192993164 |
| CatBoost | 0.7 | 0.3 | GPU | 58.7476589679718 |
| CatBoost | 0.8 | 0.2 | CPU | 2.7965521812438965 |
| CatBoost | 0.8 | 0.2 | GPU | 59.93782114982605 |

## 4 Conclusion

Based on the testing and analysis conducted on the system as outlined in Chapter 4, the following conclusions can be drawn:

1. The research successfully produced a solution for detecting diabetes and demonstrated that the XGBoost, LightGBM, and CatBoost algorithms can be used to identify individuals who are likely to have diabetes or not.
2. Upsampling the data using ADASYN and employing K-Fold Cross Validation during the model evaluation stage proved to enhance the model's accuracy and stabilize the scores generated, thereby obtaining the best model.
3. Based on the testing conducted using the XGBoost, LightGBM, and CatBoost algorithms, models trained with the CatBoost algorithm achieved higher scores in terms of accuracy, precision, recall, F1-Score, and ROC-AUC when compared to models trained with the XGBoost and LightGBM algorithms. This was observed across the data splits of 0.5 training and 0.5 testing, 0.7 training and 0.3 testing, and 0.8 training and 0.2 testing.

4. The computation time of the LightGBM algorithm was faster across the data splits of 0.5 training and 0.5 testing, 0.7 training and 0.3 testing, and 0.8 training and 0.2 testing, compared to the XGBoost and CatBoost algorithms.
5. Training the models on a CPU was faster than training on a GPU.

## References


1. International Diabetes Federation (IDF): IDF Diabetes Atlas 10th edition (2021). www.diabetesatlas.org
2. Chen, T., Guestrin, C.: XGBoost: A Scalable Tree Boosting System (2016). https://doi.org/10.1145/2939672.2939785
3. Ke, G., et al.: LightGBM: A Highly Efficient Gradient Boosting Decision Tree. https://github.com/Microsoft/LightGBM.
4. Kinnander, M.: Predicting Profitability Of New Customers Using Gradient Boosting Tree Models
5. Prokhorenkova, L., Gusev, G., Vorobev, A., Dorogush, A.V., Gulin, A.: CatBoost: unbiased boosting with categorical features. https://github.com/catboost/catboost
6. Dorogush, A.V., Ershov, V., Gulin, A.: CatBoost: gradient boosting with categorical features support (2018). http://arxiv.org/abs/1810.11363
7. Huang, G., et al.: Evaluation of CatBoost method for prediction of reference evapotranspiration in humid regions. J. Hydrol. **574**, 1029–1041 (2019). https://doi.org/10.1016/j.jhydrol.2019.04.085
8. Balamurugan, P., Amudha, T., Satheeshkumar, J., Somam, M.: Optimizing neural network parameters for effective classification of benign and malicious websites. J. Phys. Conf. Ser. **1**, 2021 (1998). https://doi.org/10.1088/1742-6596/1998/1/012015
9. Riza, H., Santoso, E.W., Tejakusuma, I.G., Prawiradisastra, F., Prihartanto, P.: Utilization of artificial intelligence to improve flood disaster mitigation. J. Sains dan Teknol. Mitigasi Bencana **15**(1), 1–11 (2020). https://doi.org/10.29122/jstmb.v15i1.4145
10. Hidayat, M.A., Husni, N.L., Damsi, F.: Image processing based flood detector using convolutional neural network (CNN) within surveillance camera Pendeteksi Banjir dengan image processing berbasis convolutional neural network (CNN) pada Kamera Pengawas, vol. 2, no. October, pp. 10–18 (2022)
11. Alagoz, B.B., Simsek, O.I., Ari, D., Tepljakov, A., Petlenkov, E., Alimohammadi, H.: An evolutionary field theorem: evolutionary field optimization in training of power-weighted multiplicative neurons for nitrogen oxides-sensitive electronic nose applications. Sensors **22**(10) (2022). https://doi.org/10.3390/s22103836
12. Sadollah, A., Eskandar, H., Lee, H.M., Yoo, D.G., Kim, J.H.: Water cycle algorithm: a detailed standard code. SoftwareX **5**, 37–43 (2015). https://doi.org/10.1016/j.softx.2016.03.001
13. Vinayakumar, R., Alazab, M., Soman, K.P., Poornachandran, P., Al-Nemrat, A., Venkatraman, S.: Deep learning approach for intelligent intrusion detection system. IEEE Access **7**, 41525–41550 (2019). https://doi.org/10.1109/ACCESS.2019.2895334
14. Rathor, A.: A Review at Machine Learning Algorithms Targeting Big Data Challenges, pp. 753–759 (2017)

# Blockchain Based Artificial Intelligence of Things (AIoT) for Wildlife Monitoring

R. Madhusudhan(✉) and P. Pravisha

Department of Mathematical and Computational Sciences, National Institute of Technology, Karnataka, India
madhu@nitk.edu.in

**Abstract.** Climate change poses a significant threat to wild animals and their habitats, increasing the chance of human-wildlife conflict. Traditional camera-based imaging systems are centralized and require operators to install the camera and monitor the video recording constantly. However, manually processing the massive number of images and videos gathered from camera traps is expensive and time-consuming. In this article, we will develop a framework for wildlife monitoring systems that make use of Artificial Intelligence of Things (AIoT), the Interplanetary File System (IPFS), and blockchain. A wildlife camera that uses AIoT to detect wild animal movement in real-time gathers the dynamic properties of animals. Cloud computing solutions are impractical for critical data management in wildlife monitoring due to their high latency and constant internet connectivity requirements. IPFS is a distributed file system that offers efficient data storage, distribution, and persistence, enabling offline-centric paradigms. In our framework, IPFS is used for permanent data storage, and the hash value of data is stored on a private blockchain. The data from multiple forest zones is stored on a consortium blockchain. A simulation is carried out using CNN and a method to improve the scalability of the framework is presented.

**Keywords:** Artificial Intelligence of Things (AIoT) · Blockchain · InterPlanetary File System (IPFS) · Convolutional Neural Network(CNN) · Wildlife monitoring

## 1 Introduction

In a world where the human population keeps expanding, there is an increasing demand for land for agriculture and natural resources for industry, leading to people coming into contact with the area where wild animals are present and ensuing conflict [1]. Temperature and weather fluctuations might cause some species to change their migration routes or timing. These disturbances may cause animals to appear in unforeseen places and interact with human populations. Technology enables the real-time and accurate collection of data from a variety of sources, including sensors, cameras, and satellites. It offers researchers


L. Barolli (Ed.): AINA 2024, LNDECT 203, pp. 25–36, 2024.
https://doi.org/10.1007/978-3-031-57931-8_3

and conservationists valuable insights for informed decision-making and efficient conservation strategies. Existing wild animal tracking techniques include GPS tracking, radio tracking, vehicle tracking, and camera traps. One of the widely used geospatial data collection tools is GPS tracking, which involves attaching a GPS sensor to the animal and tracking the animal's location [2]. The data can be downloaded remotely, or the tracker must be retrieved from the animal, depending on the model. One of the primary challenges with GPS tracking is that the data might be lost if the tracker is lost. Very High Frequency (VHF) radio tracking is a frequently used form of radio telemetry. It uses radio transmitters that are attached to an animal in the same way that GPS devices are. The researcher uses an antenna to monitor signals from the animal's device if it is within range. When using radio transmitters, the distance an animal can travel from the antenna to collect data is limited. This is not optimal for animals that cover long distances.

The Internet of Things (IoT) is an emerging paradigm that enables communication between electronic devices and sensors through the Internet to facilitate our lives [3]. IoT facilitates the collection of data in inaccessible or remote areas, thereby facilitating a greater understanding of wildlife activities and their responses to environmental changes. One of the challenges of IoT-based wildlife monitoring is the size and durability of the embedded devices. The devices worn by animals must be lightweight, convenient, and durable. Wearable devices can cause direct injury to animals due to the harmful effects of exposure, hardware adaptation or wearing, and technical breakdowns [6]. Using environmental indicators such as temperature, humidity, and sound patterns, IoT sensors, and Artificial Intelligence(AI) algorithms can detect and analyze potential hazards or changes in wildlife habitats. AI refers to the simulation of human intelligence by a system or a machine [4]. AI is designed to develop machines that can imitate human behaviors and actions, such as perception, reasoning, learning, planning, prediction, etc. A combination of AI and IoT is referred to as AIoT. IoT ensures that real-time data is accessible anytime, and AI is beneficial for processing that data. AI-powered wildlife cameras that monitor wild animals' movements near populated areas may significantly improve the quality of life for humans.

Cloud-based data management is vulnerable to attack due to centralized administrative control. The main disadvantage of cloud storage is that it requires constant access to the internet. Due to internet connectivity concerns in the forest, the InterPlanetary File System (IPFS) can be used for permanent data storage. IPFS is a decentralized file system that uses Distributed Hash Table (DHT) technology [8]. IPFS makes it possible to store data across a network of storage providers rather than a centralized provider. Blockchain is a distributed ledger technology that records data through a peer-to-peer computer network [5]. A blockchain is a distributed ledger that consists of a chain of blocks, with each block containing transactions, nonce, hash value, and previous hash [18]. Each block in the blockchain contains a unique cryptographic hash that is formed based on the block's data and the hash of the preceding block, which makes it highly resistant to tampering[19]. There are three types of blockchain: public, private, and consortium [5]. In public blockchains, anyone can join the network, but in private blockchains, the network is controlled by a central authority. A consortium blockchain is a combination of public and private blockchains managed by a group of organizations. Blockchain can be used for various conservation projects, and it provides cyber-security to establish trust in data as it is shared between government, academia, and local environmental groups [7]. Blockchain can help improve transparency in the forestry sector, which is often plagued by corruption and illegal logging, by creating a tamper-proof record of timber production and sales [4]. For security and scalability reasons, a private blockchain is preferable in wildlife monitoring as it deals with critical data.

In this article, emerging technologies such as AIoT, IPFS, and blockchain are used to develop a framework for wildlife monitoring. The article makes the following scientific contributions:

1. An AIoT-blockchain framework for tracking the movement of wild animals near the boundary of the forest has been proposed. The proposed framework stores data on IPFS and the hash value of data on blockchain that can be shared with authorities to enhance coordination and collaboration for monitoring wild animals. The data from multiple forest zones is brought together on a consortium blockchain.
2. A simulation is carried out for the framework using the deep learning algorithm CNN to identify and classify different wild animals using images. An analysis is conducted to evaluate the performance of storing data in IPFS and blockchain, and methods to improve the framework's scalability are presented.

The rest of the paper is organized as follows: Sect. 2 describes the literature review. Section 2 presents preliminaries, and Sect. 3 describes the literature survey. The proposed framework is discussed in Sect. 4. The simulation results are presented in Sect. 5, and the conclusion and future work are presented in Sect. 6.

## 2 Literature Review

Real-time wildlife monitoring has become possible as a result of the proliferation of the Internet of Things, which is based on sensors that are connected to fast wireless technology such as 5G, 6G, and LTE (Long-Term Evolution). Andy et al. [9] proposed a cloud-edge, sensing multi-tier end-to-end, distributed IoT system for wildlife monitoring. The system makes use of image processing based on machine learning to classify animal images gathered by remote camera traps. Where's The Bear (WTB) is a system that uses Google TensorFlow and OpenCV programs to automatically classify and categorize images. The system uses edge devices to carry out classification near the IoT devices and cloud systems to train machine learning models. The system was able to identify the images of bears, deer, coyotes, and empty images while minimizing the time and bandwidth required for image transfer and end-user processing. While WTB and other systems have demonstrated the viability of using IoT and edge cloud systems for wildlife monitoring, there are still some challenges that must be addressed. For instance, IoT devices are frequently resource-constrained, making it challenging to deploy advanced machine-learning models on them. In addition, edge cloud systems must be designed to be robust and reliable, as they will be deployed in remote environments.

Eyuel et al. [10] proposed a dual radio IoT network architecture for wildlife monitoring systems that uses LPWAN IoT network infrastructure to minimize the consumption of energy. Designing a wildlife monitoring system with a high delivery ratio, low latency, and low power consumption is challenging. Because performing data pre-processing locally consumes less energy, the proposed system addresses these challenges by performing local processing and transmitting the concatenated data to the LoRa server with lower energy consumption. The authors emphasize that the combination of Bluetooth Low Energy (BLE) sensing and LoRa IoT infrastructure is more applicable for applications requiring a long network lifetime at high packet traffic intensities. The proposed architecture is ideal for building a wildlife monitoring system with low latency and low energy consumption.

Patel et al. [12] proposed a secure wildlife monitoring system that leverages blockchain and AIoT. The AIoT devices receive data from animal sensors, which are then analyzed in real-time using machine learning techniques. AIoT devices often collect and transmit sensitive information, such as the locations and images of wild animals, which must be protected against unauthorized access. Blockchain technology is a promising solution to these challenges. In the proposed system, AIoT devices encrypt and transmit collected data to a blockchain network. On the blockchain network, smart contracts store and manage encrypted data. On the blockchain network, only authorized users can access the data by decrypting it with their private keys. The major challenge with the proposed system is that blockchain networks can be slow and expensive to scale in the context of wildlife monitoring systems that collect and process large amounts of data.

Juan et al. [13] proposed an edge-computing Embedded Neural Networks (ENNs) based solution to identify three different horse gaits. Multiple ENNs were implemented and deployed in various microcontroller architectures to get the optimal balance of energy consumption and computational performance. The evaluation demonstrated that ENNs can be used to classify animal behavior with high accuracy on microcontroller-based devices. The ENNs they evaluated were more accurate at identifying three different horse gaits. In addition, the ENNs were found to be energy-efficient. Data collected from sensors can be noisy in the context of wildlife monitoring. One of the challenges of the system is the sensitivity of ENNs to noise in input data. In addition, ENNs are costly to implement and train computationally. Developing new ENN architectures that are more efficient could be the solution to these challenges.

Wild et al. [15] proposed the Sigfox IoT network, which is a real-time multi-sensor data retrieval and tag commanding system for analyzing wild animals. The system provides low-power, long-range remote digital data transfer and location estimates. Sigfox offers a long range and low power consumption, and it can be used to monitor a wide variety of species in a variety of habitats, including remote areas. Sigfox could also be used to monitor wildlife populations in response to challenges such as climate change and habitat loss. The main drawback of the Wild et al. study is that it did not evaluate the effect of Sigfox tracking devices on wild animals to make sure that devices do not harm them. Another limitation of the study is that it did not assess the use of Sigfox for real-time tracking of animals. Real-time monitoring could be used to monitor animal movements more closely and respond promptly to potential threats. Traditional wildlife monitoring techniques, such as manual observation and camera traps, are labor-intensive and ineffective. There is a significant lack of comprehensive studies focusing on the scalability, energy efficiency, and real-world implementation challenges of existing systems.

A blockchain-based coordination system could be ideal for wildlife data management for privacy, security, and access control because IoT devices at various locations in forests generate massive amounts of data. This article will contribute to developing a framework for wildlife monitoring that incorporates technologies such as AIoT, blockchain, and IPFS that contribute to future developments in this field.

# 3 Proposed Framework

The field of wildlife conservation is undergoing a rapid transformation as a result of technological improvements, which is opening up new potential for the development of frameworks that monitor wild animals. The limited storage capacity of the blockchain makes it impossible to implement a strategy that involves storing huge amounts of data on the blockchain. Since cloud computing requires a constant internet connection and is under the control of a single administrative body, it is not an effective option for data storage. In addition, the forest has challenges with connectivity to the internet, which makes cloud computing impractical for data storage. To overcome these limitations and take advantage of the tamper-resistant properties of blockchain, the proposed framework stores data in a distributed IPFS system, and its hash value can be stored in blockchain. By storing only data hashes in blockchain, one can reduce costs and increase energy efficiency, thereby making blockchain a more cost-effective solution. In the proposed framework, smart environmental sensors are installed in the forest as part of an AIoT-powered network, and these sensors can identify any wild animals that enter the boundary. Figure 2 shows the proposed framework. The proposed framework consists of six components: AIoT, wildlife monitoring officers, users (local people), a blockchain network, mobile applications, and IPFS. The three layers used in the proposed framework are AIoT, blockchain, and IPFS. The data collected may include animal migration patterns, behavior, and habitat changes. The following is a summary of the proposed architecture:

- AIoT devices are deployed in various forest regions, and these devices collect real-time data.

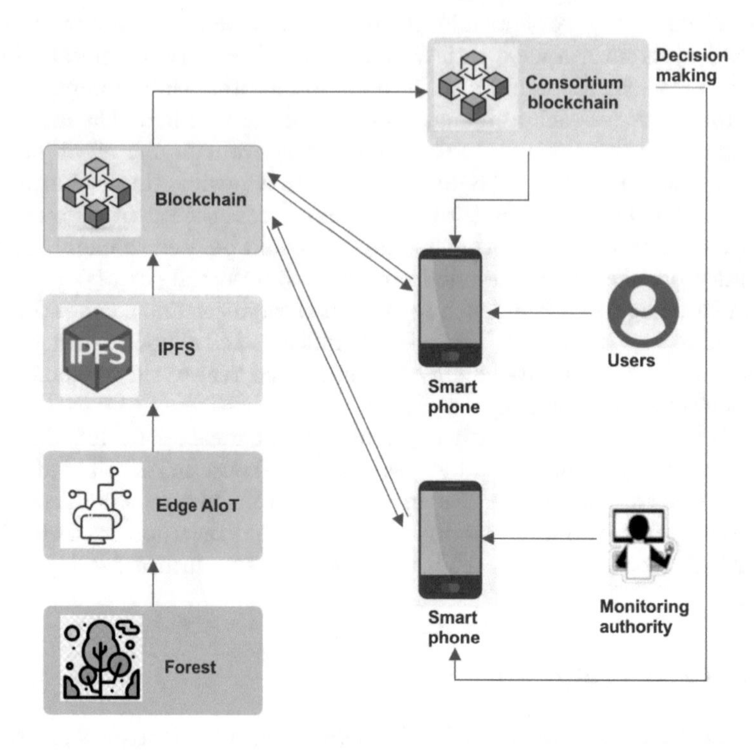


**Fig. 1.** Proposed Framework

- The data collected by AIoT device is stored on a private IPFS and its hash is stored on a private blockchain
- A consortium blockchain is implemented to interconnect data from the various regions of the forest.

The following subsections provide detailed information on each component used in the proposed framework.

### 3.1 AIoT for Collection of Data

AIoT devices equipped with various sensors, such as motion detectors, temperature sensors, cameras, and audio recorders, can gather data from forests. AI tools analyze the data that IoT devices collect to improve productivity and efficiency. These devices can collect data on animal movement, behavior, and habitat conditions, as well as recognize specific species through images or sound. The gathered information can be transmitted to IPFS for storage. Images and audio captured by IoT devices can be analyzed by AI algorithms to identify and categorize wild animals. AIoT devices can generate alerts in real time based on predefined boundaries or incidents. For instance, if an animal enters a restricted area or if there is an unexpected change in environmental conditions, wildlife officials can receive automated alerts to take immediate action. This enables conservation measures and reduces possible risks. AIoT improves human-machine interactions, IoT operations, data management, and analytics.

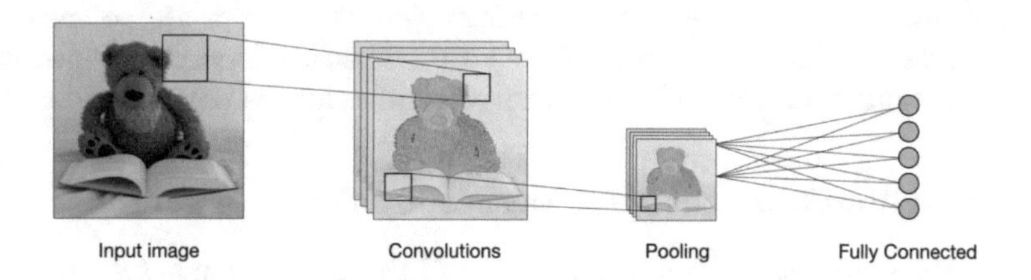


**Fig. 2.** The different layers used in CNN [11]

AIoT makes use of convolutional neural networks (CNNs), which automatically learn the image features and generate a classifier. In comparison to neural networks that have fully connected layers, CNNs have a deep feed-forward architecture, which enables generalization in an effective manner[20]. CNN is based on the concept of weight sharing, which significantly reduces the number of parameters required for training, leading to better generalization. Figure 2 shows the various layers of the CNN. A convolutional layer consists of various convolutional maps. In a convolution stage, feature maps are convoluted with different kernels, which are equivalent to filters in the field of image processing [21]. A pooling layer consists of different pooling maps, and the pooling process is performed on convolutional layers. CNN's fully connected layer learns complex data patterns to make classifications based on the features extracted by the network's earlier layers.

### 3.2 IPFS for Storage of Data

The data collected from the AIoT smart devices is permanently stored in IPFS. IPFS is a distributed file storage system that stores and shares data on the Internet. Since IPFS was created to function in a peer-to-peer manner, data can be accessed offline. If a file has been previously accessed on a local IPFS node, it can be retrieved and accessed even when disconnected from the IPFS network as a whole. By using IPFS for wildlife monitoring, forest authorities, and conservationists can take advantage of decentralized and secure data storage, efficient sharing, data integrity, and offline access. These features enhance collaboration, data integrity, and conservation in wildlife monitoring initiatives.

### 3.3 Blockchain for Storage of Hash of Data

In the proposed architecture, the hash value generated by IPFS for each piece of data is transferred to the blockchain for storage. The hash of the data is stored on a blockchain, which makes it resistant to tampering and can be verified by multiple parties. This ensures the integrity of the collected data. This can help monitoring and conservation endeavors by providing reliable and accessible information for decision-making. Wildlife monitoring involves gathering huge amounts of data, including high-resolution images or sensor readings in real-time. Private blockchains offer a greater degree of control over the network, which enhances scalability and performance that is appropriate to meet the requirements of wildlife monitoring. This ensures that the blockchain network is capable of storing and processing data without affecting its performance.

### 3.4 Consortium Blockchain for Decision-Making

In forests, there are many distinct sections, and the data belonging to these sections could be brought together on a consortium blockchain. If there are critical circumstances in which humans and wild animals come into confrontation with one another, suitable action is required. In such circumstances, various authorities, including wildlife conservationists, forest officers, and higher authorities, may engage in collaborative decision-making.

### 3.5 Mobile Application for Monitoring Wild Animal Movement

In the proposed framework, verified users can use mobile applications to keep track of wild animals. Verified users are wild animal monitoring center officers, local authorities, forestry administrators, researchers, and habitats near forests. The mobile application enables wildlife officers and local people to report animal sightings and movements. The application allows users to input relevant information, such as species, location, time, and behavior. This data can be stored locally on the user's device, enabling offline access and reducing the need for continuous internet connectivity.

## 4 Simulation Results

### 4.1 The Classification of Images Using CNN

A simulation is performed for the proposed framework using the CNN deep learning algorithm to identify and classify images of two species: a lion and a tiger. The model was simulated using a MacBook Air with an Apple M1 processor, an 8-core GPU, 16 GB of memory, and 512 GB of storage. The CNN model consists of three convolution layers, two max-pooling layers, and two output layers with fully connected output layers. The input images of two species, lion and tiger, were collected from a dataset of 400 images downloaded from Kaggle [14], with a test sample of 40 images. In the first convolution layer, there are 32 kernels with a size of 3*3 and a stride of 2. The activation function for all convolution layers is 'relu'. The input to the first convolution layer is 32*32*3. The output of the first convolution layer is 32*32*30, which is fed to the second layer, which is a max-pooling layer. The max-pooling layer consists of a 2*2 max-pooling kernel with a stride of 2. The output of the max-pooling layer is a 15*15*32 that is fed to the convolution layer, which is the third layer. This convolution layer has a kernel size of 3*3 and a stride value of 1. The output of this convolution layer, which is of size 13*13*64, is fed to the fourth layer, which is the max-pooling layer. This max-pooling layer has a max-pooling kernel of 2*2 with stride 2. The output from this max-pooling layer is of size 6*6*64, which is fed to the fifth layer, which is the convolution layer. This convolution layer has a 3*3 convolution kernel with a stride value of 1. The output of this convolution layer has a dimension of 4*4*64 and is passed to the flatten function. This is because the output of the convolution layer is 3D.

**Fig. 3.** Training and testing accuracy

The dense layer only admits input in one dimension; therefore, the three-dimensional vector needs to be flattened down to one dimension before it can be passed to the dense layer. The resulting one-dimensional vector is fed into a dense layer with 64 nodes, and its output is fed to an output layer with two nodes representing the number of classes in the data set. The performance analysis of the proposed framework is shown in Fig. 3 and Fig. 4, which represent the training and validation accuracy and training and validation loss, respectively. The evaluation is restricted to a maximum of a hundred epochs. An effective prediction model should have a high level of accuracy and a low rate of loss. We obtained a training accuracy of 95.93%, a validation accuracy of 97.36%, a training loss of 10.70%, and an 8.21% validation loss, which demonstrates that the model is efficient.

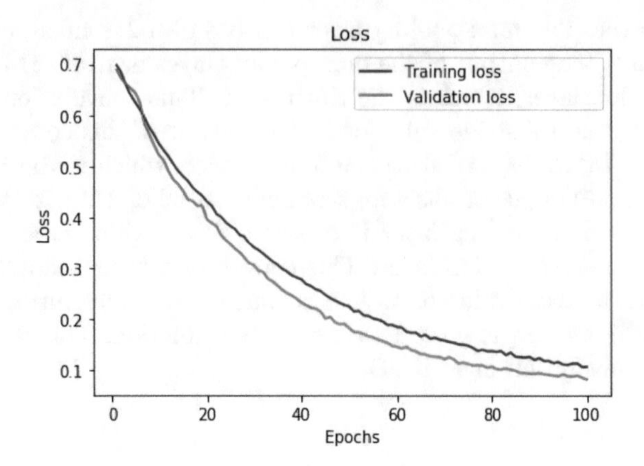


**Fig. 4.** Training and testing loss

## 4.2 Uploading Data to IPFS and Blockchain

The files are uploaded to the IPFS network using Web3 Storage. A unique CID value is generated for each file. A CID is a cryptographic hash representing the content of the file. The smart contract is written in Solidity language in Remix IDE. The smart contract is created to store the CID on the blockchain. This process typically involves sending a transaction to the smart contract, which adds the hash to the blockchain's storage. Users can later retrieve the data by querying the smart contract with a specific identifier or key.

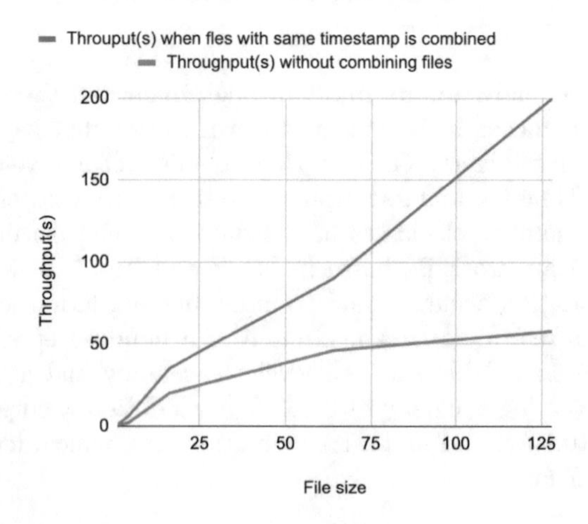


**Fig. 5.** Transaction throughput for storing data in blockchain

The smart contract will return the corresponding IPFS hash (CID), which users can use to retrieve the actual data from IPFS. We have deployed the smart contract on the Ganache blockchain for demonstration purposes. Two transactions were created; the first transaction stores the hash on the blockchain, while the second transaction retrieves it from the blockchain. Each transaction in a blockchain network uses a certain amount of gas. The amount of gas consumed depends on the computational resources required for executing the transaction. When uploading each image or video to the blockchain, the blockchain consumed more gas, so we combined similar data with a timestamp into a single file. Figure 5 shows the transaction throughput for storing data in the blockchain when files with the same timestamp are combined. It reduces the amount of energy and cost required to generate a hash for each file and results in better throughput. Scalability can be improved by storing data in this manner.

## 5 Conclusion and Future Work

The article investigated how blockchain-based AIoT can be used to monitor wild animals. The combination of AI algorithms and IoT devices enables the efficient collection and analysis of wildlife data. By combining the benefits of blockchain technology, artificial intelligence, and IoT, the proposed framework enhances wildlife conservation efforts. The integration of blockchain technology ensures the immutability, transparency, and security of data, thereby addressing the challenges associated with data integrity and reliability in wildlife monitoring. Moreover, real-time monitoring and alerts facilitate timely intervention in the event of threats or emergencies. We performed a simulation of the framework by using the CNN deep learning algorithm to identify and classify images of various wild animals. Simulation for storing data into IPFS and the hash of data into the blockchain is also carried out. Future research directions may involve improving the proposed framework by exploring the integration of emerging technologies such as 6G connectivity to enhance the efficiency and responsiveness of the framework. The implementation of sharding methods will also be a focus of future work to enhance the blockchain's scalability and transaction throughput.

## References


1. Sillero-Zubiri, C., Switzer, D.: Crop raiding primates: searching for alternative, humane ways to resolve conflict with farmers in Africa. Oxford University, People and Wildlife Initiative. Wildlife Conservation Research Unit. Oxford (2001)
2. Bernstein-Root M., Gooden J., Boyes A.: Rewilding in practice: Projects and policy. Geoforum, pp. 292–304 (2018)
3. Kumar, S., Tiwari, P., Zymbler, M.: Internet of Things is a revolutionary approach for future technology enhancement: a review. J. Big Data **6**, 111 (2019). https://doi.org/10.1186/s40537-019-0268-2
4. Yongjun et al.: Artificial intelligence: a powerful paradigm for scientific research. Innovation **2** (2021)

5. Ecological Impact of Blockchain, InsideEcology. https://insideecology.com/2023/04/17/ecological-impact-of-blockchain/
6. Tuyttens, F.A., Molento, C.F., Benaissa, S.: Twelve threats of precision livestock farming (PLF) for animal welfare. Front. Vet. Sci. Sec. **9**, 889623 2022
7. Why blockchain technology for conservation? living planet technology hub, WWF. https://techhub.wwf.ca/technology/blockchain-applications/
8. Politou, E., Alepis, E., Patsakis, C., Casino, F., Alazab, M.: Delegated content erasure in IPFS. Futur. Gener. Comput. Syst. **112**, 956–964 (2020)
9. Elias, A.R., Golubovic, N., Krintz, C., Wolski, R.: Wheres the bear?- automating wildlife image processing using IoT and edge cloud systems. In: 2017 IEEE/ACM Second International Conference on Internet-of-Things Design and Implementation (IoTDI), Pittsburgh, PA, USA, pp. 247–258 (2017)
10. Ayele, E.D., Meratnia, N., Havinga, P.J.M.: Towards a new opportunistic IoT network architecture for wildlife monitoring system. In: Proceedings of the 2018 9th IFIP International Conference on New Technologies, Mobility and Security (NTMS), pp. 1–6. IEEE (2018)
11. Amidi, A., Amidi, S.: Convolutional neural networks cheatsheet, CS 230 - Deep Learning. https://stanford.edu/~shervine/teaching/cs-230/cheatsheet-convolutional-neural-networks
12. Patel, R., Chaudhary, S., Goel, S., Goyal, P.: Secure wildlife monitoring using blockchain and AIoT. In: Proceedings of the IEEE International Conference on Artificial Intelligence and Internet of Things, pp. 89–94. IEEE (2020)
13. Pedro, J., et al.: Wildlife monitoring on the edge: a performance evaluation of embedded neural networks on microcontrollers for animal behavior classification. Sensors (Basel, Switzerland) **21**, 2975 (2021)
14. Lions-Tigers, Alexandr Krasnov. https://www.kaggle.com/datasets/akrsn-v/lions-and-tigers
15. Wild, T.A., van Schalkwyk, L., Viljoen, P., et al.: A multi-species evaluation of digital wildlife monitoring using the Sigfox IoT network. Anim Biotelemetry **11**, 13 (2023)
16. Castro, M., Liskov, B.: Practical byzantine fault tolerance. In: Proceedings of the Third Symposium on Operating Systems Design and Implementation, New Orleans, USA (1999)
17. Lamport, L., Shostak, R., Pease, M.: The Byzantine generals problem. Dr. Dobbs J. **33**, 30–36 (2008)
18. Nakamoto, S.: Bitcoin: a peer-to-peer electronic cash system (2008). https://bitcoin.org/bitcoin.pdf
19. Huo, R., et al.: A comprehensive survey on blockchain in industrial internet of things: motivations, research progresses, and future challenges IEEE Commun. Surv. Tutor. **24**(1), 88–122 (2022)
20. Nebauer, C.: Evaluation of convolutional neural networks for visual recognition. IEEE Trans. Neural Networks **9**(4), 685–696 (1998)
21. Chen, R., Jalal, M.A., Mihaylova, L., Moore, R.: Learning capsules for vehicle logo recognition In: Proceedings of the 21st International Conference on Information Fusion, Cambridge, UK, pp. 565–572 (2018)

# Research on Screw Detection Using Artificial Intelligence Technology

Wei-Chun Hsu[1(✉)], Hsing-Chung Chen[2,3(✉)], Pei-Yu Hsu[2], and Yong-Jian Siao[2]

[1] Department of Vehicle Technology and Entrepreneurship, WuFeng University, 117, Sec 2, Chiankuo Road, Minhsiung, Chiayi County 62153, Taiwan, R.O.C.
ingmar.hsu@wfu.edu.tw

[2] Computer Science and Information Engineering, Asia University, 500, Lioufeng Road, Wufeng, Taichung 41354, Taiwan, R.O.C.
cdma2000@asia.edu.tw, shin8409@ms6.hinet.net, hcliebe2019@gmail.com, 107121005@live.asia.edu.tw

[3] Research Consultant With Department of Medical Research, China Medical University Hospital, China Medical University, Taichung, Taiwan

**Abstract.** The purpose of this study is to develop a screw screening automation machine that can achieve a large number of rapid detection of screws. At present, in Taiwan's screw machinery industry, most of the inspection technology uses 2D photography technology. Therefore, in the process of testing, screws must be placed in each fixture to perform photography, and different screws need different fixtures. The screening machine, therefore combined with AI detection technology and robot application in the development of various types of screw screening automation machinery could achieve a large number of rapid detection and automated screening functions.

## 1 Introduction

The screw industry in Taiwan has a long history of development and is characterized by moderate capital intensity, labor intensity, and high levels of globalization, particularly as an export-driven sector. The largest cluster of screw industry in Taiwan is located in the southern regions of Tainan and Kaohsiung, with over 600 businesses concentrated in the Gangshan area of Kaohsiung, highlighting the strategic importance of the Kaohsiung region in industrial distribution. The screw industry primarily operates in the southern region of Taiwan, with the development of the Gangshan area facilitated by the presence of China Steel Corporation, making it a key hub for Taiwan's screw industry. Due to the high demand for production, automated inspection methods are necessary to achieve rapid shipping goals. To achieve rapid automated inspection, automated photography coupled with image recognition technology is employed to detect defective products.

From scholarly literature, Huang et al. (2018) invented a screw inspection data collector [1], which is installed on a screw production machine and includes a microcontroller and sensing unit connected electrically to the microcontroller to detect screw appearance and generate inspection data. In 2014, Guo and Hsu invented a screw straightening device [2] comprising a detection unit, straightening unit, and rotating unit. The straightening


L. Barolli (Ed.): AINA 2024, LNDECT 203, pp. 37–45, 2024.
https://doi.org/10.1007/978-3-031-57931-8_4

unit is placed on a lifting seat, while the rotating unit is positioned below the detection unit to rotate the screw. This device serves both screw inspection and straightening functions, improving product quality and minimizing screw damage, thereby enhancing product yield and production speed. In 2011, Chen developed a screw inspection device consisting of a machine, positioning unit [3], detection unit, and control unit, wherein the detection unit includes mounting base, emitter, receiver, and power source. By utilizing the receiver in conjunction with the emitter, images of screws placed in the positioning gap of the positioning unit's clamp are captured, processed by the control host, and displayed on the output interface, offering high applicability, accuracy, and ease of operation.

## 2 Current 2D Inspection Technology

**Fig. 1.** The screening process involves pouring screws into a vibratory bowl [4]

Based on the literature, most screw manufacturing industries in Taiwan currently employ 2D inspection technology using flat photography. Each screw needs to be individually guided into fixtures and photographed one by one using two 2D cameras. The images are then analyzed for defects. Sorting operations involve dumping screws into a vibratory bowl feeder, sequentially guiding screws into fixtures, and photographing them from above and the side using two 2D cameras, as shown in Figs. 1, 2 and 3. During transportation, separating good and defective products using guide plates is necessary, with each bucket of screws requiring several hours for inspection. This process is time-consuming as each screw must be sequentially loaded into fixtures and photographed.

**Fig. 2.** The screws are directed into fixtures for the 2D imaging process [4]

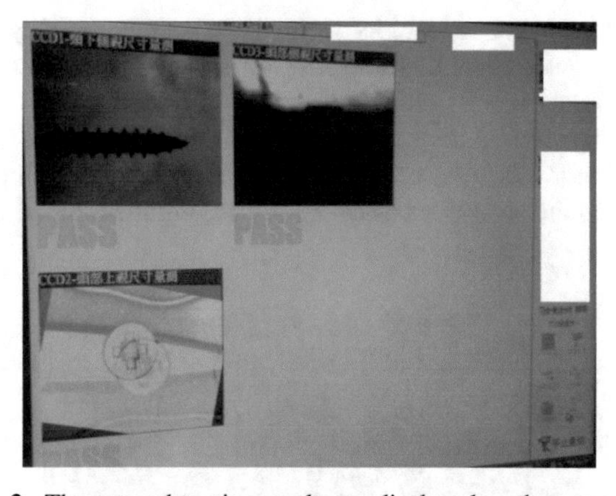


**Fig. 3.** The screw detection results are displayed on the screen [4]

Moreover, the wide variety of screw types produced by each factory, differing in size and shape, necessitates the use of different fixtures for inspection, resulting in the need for different inspection machines, increasing inspection machine costs for factories.

## 3 AI Inspection Technology

Given the multitude of screw types, the current inspection methods require the procurement of numerous inspection machines. To address this issue, it is crucial to resolve the use of fixtures for photography. With advancements in machine vision recognition technology, artificial intelligence can assist in enhancing detection effectiveness. Therefore, developing lower-cost detection methods by leveraging this technology is essential. This involves scattering numerous screws on a conveyor belt, transporting them under cameras, photographing all screws on the plane, utilizing machine vision recognition technology to quickly identify screw types and screen anomalies and defects, and employing robotic arms to automate anomaly and defect screening, as shown in Fig. 4.

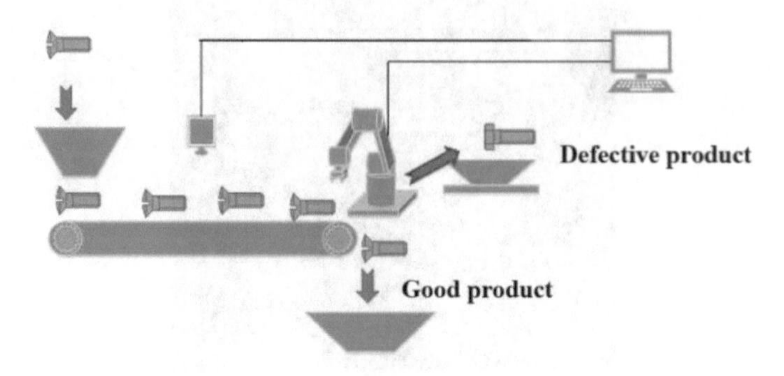


**Fig. 4.** The operation of artificial intelligence machine vision for screw inspection

To utilize machine vision recognition technology, we must employ machine learning to train with numerous correct and defective screw photos. Correct screw photos, as depicted in Figs. 5 and 6, and defective screw photos, as shown in Figs. 7 and 8, are necessary for training. Thus, during actual inspection, the system can distinguish between good and defective products, relay the positions of each screw to the robotic arm, and have the robotic arm grip the screws accordingly, classifying them based on the inspection results.

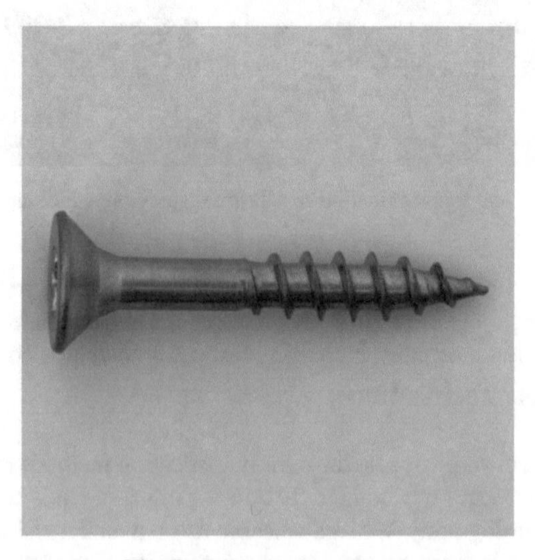

**Fig. 5.** Correct screw photo1

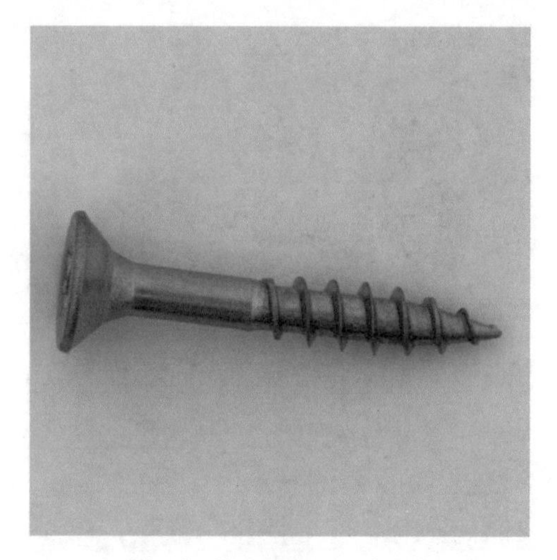

**Fig. 6.** Correct screw photo2

**Fig. 7.** Defective screw photo1

Using the machine vision recognition technology described above, we can also classify different types of screws on a flat surface. By training the system with photos of various types of screws and nuts, it can recognize the positions of identical screws or nuts, as depicted in Fig. 9. During inspection, the system can identify the same screws or nuts, relay their positions to the robotic arm, which then grips them accordingly, and classify them based on their types.

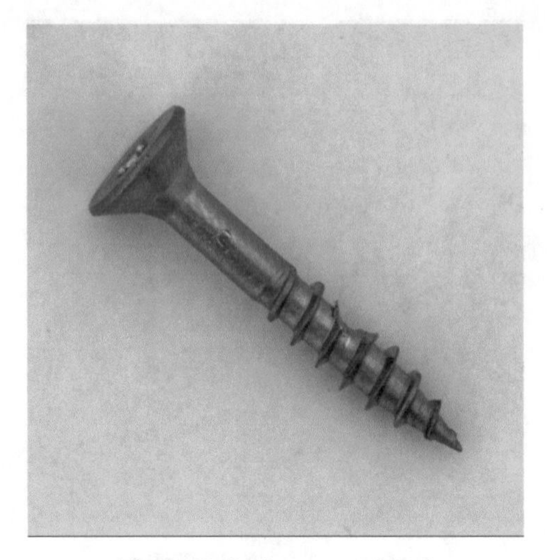

**Fig. 8.** Defective screw photo2

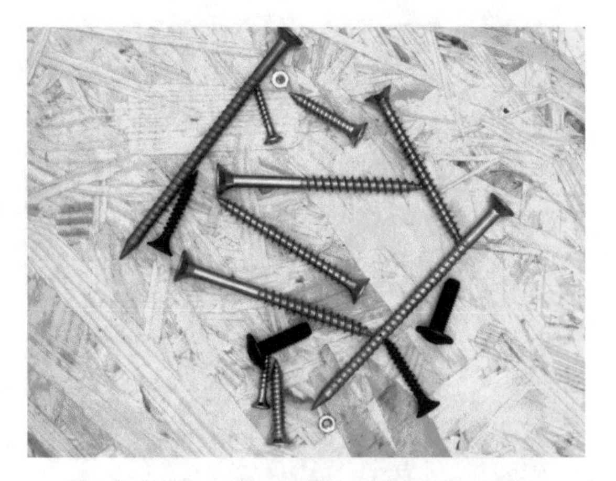

**Fig. 9.** Different types of screws on a flat surface

## 4 The Referred CNN Model for Screw Dataset in Kaggle

The anomaly images detection challenge for screw dataset where the screw dataset consists of normal and anomaly screw images as the open access dataset are shown in Ref. [5]. In addition, its simple sample code written in Python language and shown in Fig. 10 is accessed in Ref. [6]. However, there are some mistakes in this proposed scheme and the corresponding codes for his Convolutional Neural Network (CNN) model. Theses mistakes are described and then corrected below.

First, the input_shape = (256, 256, 3) in the proposed architecture of CNN model [6] for the first Conv2D layer specifies the dimensions of the input image, with 256 × 256 being the height and width, and 3 representing the number of color channels (typical for RGB images). Within the architecture of this CNN model [6], when it refers to the

```
model = Sequential()

model.add(Conv2D(filters=10, kernel_size=(4,4), padding='same', input_shape=(256, 256, 3), activation='relu'))
model.add(Conv2D(filters=10, kernel_size=(3,3), padding='same', input_shape=(64, 64, 8), activation='relu'))
model.add(Conv2D(filters=10, kernel_size=(2,2), padding='same', input_shape=(16, 16, 16), activation='relu'))

model.add(Flatten())
model.add(Dense(2, activation='softmax'))

model.compile(loss='categorical_crossentropy', optimizer=Adam(lr=0.0003), metrics=['accuracy'])
```

**Fig. 10.** The referred CNN Model for anomaly images detection [6]

shape of each layer, such as (256, 256, 3), (64, 64, 8), or (16, 16, 16), some mistakes are discussed below.

a. The input_shape = (64, 64, 8) for the second Conv2D layer is not accurate as it is unnecessary and indeed should not be to specify the input_shape beyond the initial layer, since Keras automatically infers the input dimensions from the output of the preceding layer. However, should this shape represent the output dimensions of the second layer, then 64 × 64 would refer to the spatial dimensions of the feature maps, and 8 to the number of feature maps.
b. Similarly, the input_shape = (16, 16, 16) for the third Conv2D layer is incorrect for the same reasons. If this shape were indicative of the output dimensions of the third layer, then 16 × 16 would represent the spatial dimensions of the feature maps, with 16 being the count of feature maps.

# 5 The CNN Model for Anomaly Images Detection and Experiments Results

The CNN Model for Anomaly Images Detection which is modified from the proposed CNN model. The details are described below. First, a sequential model instance is still deployed in this modified CNN model, which is a linear stack of layers in the Keras library.

a. The first layer, Conv2D, is added to the model. This is a 2-dimensional convolutional layer appropriate for processing 2-dimensional data such as images. It is configured with 10 filters, indicating the layer will produce 10 feature maps. The kernel size for each filter is set to 4 × 4. Padding is specified as 'same', which ensures the spatial dimensions of the output feature maps remain unchanged relative to the input. The activation function is set to the Rectified Linear Unit (ReLU). This layer is unique in that it specifies the input shape of the data it will receive - in this case, images of dimensions 256 × 256 pixels with 3 color channels, which is typical for RGB images.
b. Subsequent Conv2D layers are added, also with 10 filters each, but with kernel sizes of 3 × 3 and 2 × 2, respectively. In these layers, the input shape parameter is omitted as Keras auto-calculates the shape of the input based on the output of the preceding layer.

c. A Flatten layer is introduced following the convolutional layers to transform the output, which is a multi-dimensional tensor, into a one-dimensional tensor (vector) that can be fed into the subsequent densely connected layer.
d. A Dense layer follows, which is a fully connected neural network layer. It comprises 2 units corresponding to the number of output classes for the classification task at hand. The SoftMax activation function is applied, which is standard for the output layer of a multi-class classification problem.

In addition, the model is compiled, specifying the loss function, optimizer, and performance metrics. 'Categorical_crossentropy' is still chosen as the loss function, which is appropriate for multi-class classification problems. The Adam optimizer is employed for adjusting the weights, with the learning rate set to 0.0003. 'Accuracy' is selected as the performance evaluation metric (Fig. 11).

```
model = Sequential()

model.add(Conv2D(filters=10, kernel_size=(4,4), padding='same', input_shape=(256, 256, 3), activation='relu'))
model.add(Conv2D(filters=10, kernel_size=(3,3), padding='same', activation='relu'))
model.add(Conv2D(filters=10, kernel_size=(2,2), padding='same', activation='relu'))

model.add(Flatten())
model.add(Dense(2, activation='softmax'))

model.compile(loss='categorical_crossentropy', optimizer=Adam(learning_rate=0.0003), metrics=['accuracy'])
```

**Fig. 11.** The modified CNN Model for anomaly images detection

The experimental results are presented in Fig. 12, illustrating the performance comparison between the referenced CNN model [6] and our enhanced CNN model for anomaly image detection, depicted in Figs. 12a and 12b, respectively. Our modified CNN model demonstrates superior performance in anomaly image detection, achieving a test accuracy of 0.85, an improvement over the previous model's 0.81. Additionally, our model exhibits a reduced test loss of 0.434, surpassing the earlier model's 0.81.

| Train Loss: 0.003<br>Train accuracy: 1.0<br>Test Loss: 0.527<br>Test accuracy: 0.81 | Train Loss: 0.005<br>Train accuracy: 1.0<br>Test Loss: 0.434<br>Test accuracy: 0.85 |
|---|---|
| a. The referred CNN model [6] | b. Our modified CNN Model |

**Fig. 12.** The experiment results

## 6 Conclusions

The performance comparison between the referenced CNN model [6] and our enhanced CNN model for anomaly image detection reveals significant improvements. Specifically, our modified CNN model achieves a test accuracy of 0.85, surpassing the previous model's accuracy of 0.81. Furthermore, it demonstrates a test loss of 0.434, an improvement over the earlier model's loss of 0.81. These advancements in anomaly detection technology facilitate the rapid identification and classification of screw types. By enabling the integration of manufacturing data with computer systems, our approach lays the groundwork for future incorporation with Industry 4.0 technologies, thereby contributing to industrial modernization efforts.

## References

1. Huang, H.C., et al.: Screw Inspection Data Collector. Republic of China Patent, Certificate No.: M566117 (2018)
2. Kuo, L.Y., Hsu, H.Y.: Screw Straightening Device. Republic of China Patent, Certificate No.: M491530 (2014)
3. Chen, Y.L.: Screw Inspection. Republic of China Patent, Certificate No.: M404739 (2011)
4. Hsu, W.C., Huang, B.Z., Jiang, Q.H.: Screw Inspection Equipment Planning. ICSSMET A3-808 (2018)
5. Ruru, R.: Screw Dataset: Normal and anomaly screw Images. Kaggle (2021). https://www.kaggle.com/datasets/ruruamour/screw-dataset/code. Accessed 05 Feb 2024
6. Ruru, R.: Screw Dataset: Simple Sample Code. Kaggle (2021). https://www.kaggle.com/code/ruruamour/simple-sample-code. Accessed 05 Feb 2024

# An Efficient Method for Lung Cancer Image Segmentation and Nodule Type Classification Using Deep Learning Algorithms

Dorsaf Hrizi[1,3](✉), Khaoula Tbarki[2,3], and Sadok Elasmi[1]

[1] UCAR COSIM Lab, Higher School of Communications of Tunis, Tunis, Tunisia
{dorsaf.hrizi,elasmi}@supcom.tn
[2] Researcher in the LR-SITI research laboratory ENIT, Tunis, Tunisia
[3] Private Higher School of Technology and Engineering, Tek-Up University, Ariana, Tunisia

**Abstract.** Lung cancer is one of the types of cancer in the world. It is a type of disease that develops from control and forms abnormal cells in the lungs. These cells do not function like other normal cells due to deoxyribonucleic acid (DNA) mutation by various genetic factors. However, early detection and treatment of cases can reduce the risk of cancer mortality. In turn, the use of models based on convolutive neural network (CNN) architecture in the field of medical imaging diagnostics is widespread, but these architectures have different results in terms of diagnostic accuracy. In this paper, we propose an efficient method for CT lungs images segmentation and nodule type classification using deep learning algorithms based on convolutional neural network architecture with two fundamentally distinct deep learning algorithms, U-NET for segmentation and U-Net for lung nodules classification type.

**Keywords:** segmentation · Convolutional Neural Network · Deep Learning · U-Net · Segnet · VGG16 · classification

## 1 Introduction

The most dangerous type of cancer is lung cancer which has had the highest mortality rate. Lung cancer is the second most common cancer, with prostate cancer in men and breast cancer in women. An estimated 9.74 million people died from cancer in 2022 and 19.96 million new lung cancer cases were recorded [1]. To accelerate the healing process, lung cancer must be detected effectively. However, there are four basic processes involved in diagnosing lung cancer. An X-ray of the chest is the initial step in identifying any lesions. It is unable to identify the specific type of nodule, but it can detect their presence. The second tool is the fiberscope, an extensible tube with an illumination device that is put into the trachea through the nostril to look for anomalies. The biopsy,


L. Barolli (Ed.): AINA 2024, LNDECT 203, pp. 46–56, 2024.
https://doi.org/10.1007/978-3-031-57931-8_5

which is the third stage of diagnosis, entails obtaining samples of aberrant tissue and classifying the type of cells in those samples by analysis. From computed tomography (CT) lung imagery, we are able to detect cancer using image processing and deep learning techniques. By taking away with manual processes and enhancing outcomes throughout all phases of chemotherapy, radiation, surgery, and immunotherapy, this technology saves time. Much work has been put into developing automated methods that are more effective. An adaptive boosting technique and a densely linked convolutional neural network were used to classify the lung picture as either cancerous or benign, R. Dhiaa and O. Awad in[4] sought to predict and identify lung cancer diagnoses as well as predict relapse following surgery utilizing the eXtreme Gradient Boosting (XGBoost) model. Song et al. [2] developed three types of deep neural networks for lung cancer classification which were applied to a computed tomography image classification task with slight modifications for benign and malignant lung nodules. The CNN network showed accuracy, sensitivity, and specificity of 84.15%, 83.96%, and 84.32%, respectively. A method of classifying colon cancer into two groups, such as polyps and adenocarcinomas, was developed in [3]. The CNN technique combined with the MobileNet structure was used to construct this solution. To categorize lung nodules, Mahmood and Ahmed [4] developed an automatic CAD system based on the AlexNet architecture. To obtain better performance, the suggested AlexNet architecture was adjusted with a number of layers and hyperparameters. The model attained 98.9% specificity and 98.7% accuracy in the lung cancer screening experiment. In order to automatically segregate pulmonary nodules in CT 2D images, Singadkar et al. [5] used a residual network de déconvolution profunde (DDRN). This model was formed piecemeal with fully captured resolution characteristics. An inverted residual block was proposed by Bruntha et al. [6] and used by the decoder and encoder to segment lung nodules. They used a pyramid attention network in their suggested Lung PAYNet architecture to extract dense features from the encoder and decoder. Using 13,292 pictures, Taylor et al. [7] proposed the use of a deep convolutional neural network architecture for automatic pneumothorax identification. There are 3,107 positive classes in their dataset. They evaluated the sensitivity, specificity, and area under the curve of the deep neural networks. After analysis, they received an AUC of 0.94, a sensitivity score of 0.84, and a specificity score of 0.90. They had a sensitivity of 0.80, an AUC-ROC of 0.96, and a specificity of 0.97 for a high-specificity model. They also ran experiments on the NIH dataset, however the performance measurements did not yield satisfying results. Lindsey et al. [8] proposed a tChexNet with 122 layers and trained this architecture from scratch. They used deep transfer learning architectures to identify collapsed lungs. Utilizing the weight transfer method from CheXNet to tCheXNet, they outperformed CheXNet by 10% in terms of AUC on the test set. PneumoNet is an ensemble deep learning model for pneumothorax detection proposed by Kumar et al. [9]. PneumoNet employs a segmentation algorithm to find dark spots and data augmentation to overcome class difference. The accuracy of the model is 98.41 Kalaivani et al. [10] created a deep neural network for identifying lung cancer from CT scans. This process produced accurate results of 90.85%. In addition to a binary classifier for the classification approach.

In 2020, H. Yu et al. [11] proposed the "Adaptive Hierarchical Heuristic Mathematical Model (AHHMM)" for the automatic diagnosis of lung cancer. This method consists of five steps. The first step is to get the image. The second stage is Pre-processing. The third step is Binarization. Next, Thresholding and segmentation. Finally, feature extraction and detection by a Deep Neural Network (DNN).

H. Guo et al. [12] proposed a Knowledge-based Analysis of Mortality Prediction Network (KAMP-Net) to predict mortality risk in lung cancer. In this method, data augmentation is used to train Convolutional Neural Network (CNN). They assumed that data augmentation is effective to improve the performance of CNN. Clinical measurements are combined to train a support vector machine (SVM) classifier, and CNN and SVM results are combined to generate mortality risk. Clinical measurements were obtained manually. The method includes the following steps: multi-channel images coding, network design and implementation, integration of deep learning and clinical knowledge. The National Lung Screening Trial (NLST) dataset is used in this method.

Kumar et al. [13] have focused on the problem of co-morbidity recognition from patients' clinical records. The author aimed to employ both classical machine learning and deep learning approaches. The goal of this work was to develop a classification system to identify whether a certain health condition occurs for a patient by studying his/her past clinical records. Finally, they have compared the outcomes of classical machine learning and deep learning approaches with the employed feature representation methods and feature selection methods.

Kaucha et al. [13] have introduced a CNN-HHO integrated automated identification model, which makes use of SeaLion optimization methods for improving overall network optimization.

S. Pang et al. [14] used deep learning to identify the type of lung cancer in the CT images of patients in a hospital in Shandong province. To solve the problem of low data collected by the patient, they used image pre-processing methods such as "rotate, translate and transform" in order to expand the training data. The authors trained "dense-connected convolutional networks (DenseNet)" to classify lung cancer images.

In this paper, we suggest a method for CT image preprocessing, segmentation, and classification for lung cancer. This article's reminders are arranged as follows. We outline our suggested method's structure in Sect. 2. The datasets used in our work are examined in Subsect. 2.1. The pre-processing and lung mask creation processes are described in Subsects. 2.2 and 2.3. The segmentation section is covered in Subsect. 2.4, where a pair deep learning algorithms based on convolution neural network architectures are compared. The classification portion, wherein we implemented a convolutional neural network architecture model for nodule subtypes categorization, is described in depth in part

II.E. Our study's review results are presented in Sect. 3. The conclusion is finally presented in Sect. 4.

## 2 Lung Cancer Segmentation and Discrimination Proposed System

Figure 1 provides an illustration of the suggested technique. After obtaining it, the lung CT imaging data set is first preprocessed. Following that, the mask-making procedure is followed by the lung segmentation process and, ultimately, the deep learning algorithms are used to classify lung nodules as cancerous or non-cancerous.

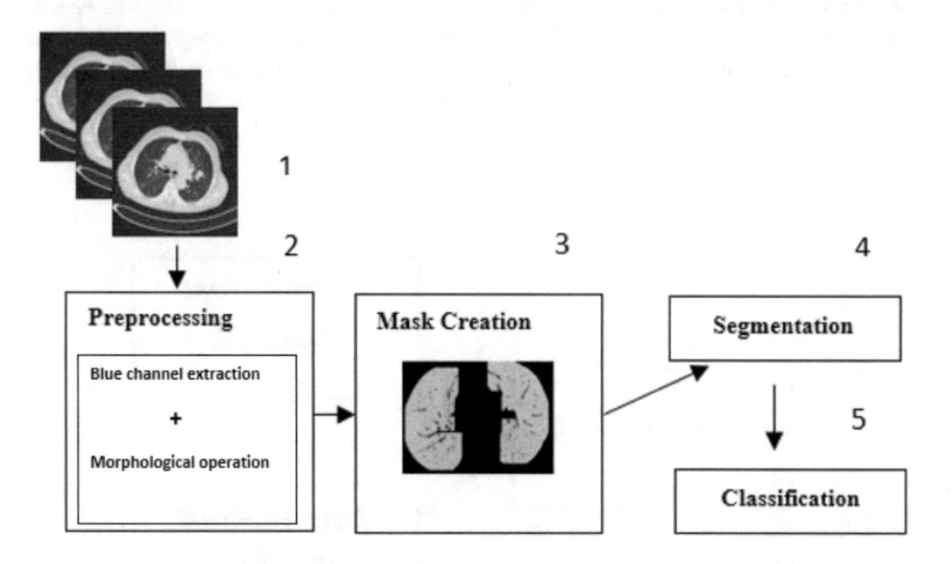


**Fig. 1.** WORKFLOW OF OUR PROPOSED METHOD

### 2.1 Data Exploration

This study uses a set of lung scintigraphy pictures, including computed tomography (CT) scans from patients with various cancer stages. Furthermore, the dataset includes 1190 individuals from 110 CT sections that are divided into three categories.

Benign, malignant, and normal(Out of them, 40 are considered malignant, 15 benign, and 55 normal.) [15].

In fall 2019, across three months, the aforementioned specialist hospitals hosted the lung cancer dataset for the Iraq-Oncology Teaching Hospital/National Center for Cancer Diseases (IQ-OTH/NCCD). It includes CT images of lung cancer patients in various stages of the disease as well as volunteers in good health. In these two centers, IQ-OTH/NCCD slides were marked by radiologists and oncologists. The initial CT image collection was completed in DICOM format, holding the breath until fully inspired, a window center range of 50 to 600 HU, a window width range of 350 to 1200 HU (Hounsfield), and a slice thickness of 1 mm are all required by the CT procedure.

## 2.2 Image Preprocessing

The purpose of image preprocessing is to improve the image quality for subsequent analysis steps. This improvement is based on distorsions elemination, or on image enhanced characteristics. In our work, the preprocessing step aims to prepare the image for segmentation, and to make features extraction easier. The preprocessing architecture is shown in Fig. 2. channel extraction: The scan image is coded on Red, Green and Blue (RGB) mode, where each image pixel consists of 3 channels and each channel represents a color. In order to perform the cell count, we extract only the blue channel to get nuclear staining.

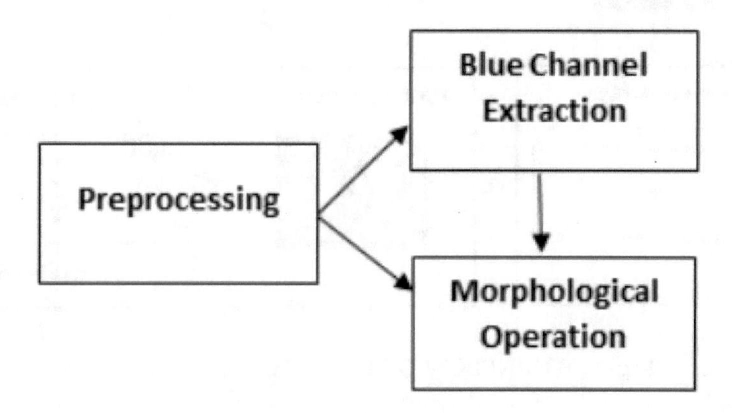


**Fig. 2.** Preprocessing architecture

Morphological operation: We used the opening morphological operation to remove unnecessary noise in the image, then the closure morphological operation that is used for filling the holes. Aiming to facilitate features extraction step.

## 2.3 Mask Creation

For precise and meaningful segmentation findings in image analysis, mask production has to come before segmentation. In order to help locate and isolate areas of interest in an image, masks are a crucial component of the preprocessing process. The mask creation process took five steps: The first step is converting the image mode from RGB(Red Green Blue) [16] to grayscale [17]. Then the resulting image is simply linked by the Otsu thresholding method that turns the image from grayscale into a binary image. After that the morphological closure approach is used to fill the holes. The resulting binary mask contains pixels, the majority of which have a value of **0** in the lung lobes and a value of **255** outside, while the goal is to multiply the CT image by a mask that recovers only the inner part of the lungs. Therefore, to solve this problem, the mask is inverted and the result mask at this point has **255** pixel values inside the lobes pulmonary regions and **0** outside. After that the **255** pixels values are replaced by **1**. Finally the image is multiplyed by the mask to obtain the region inside.

### 2.4 Segmentation

In order to produce an output segmentation map, we employed the semantic segmentation of lung pictures methodology in our work, which comprises building a framework for features extraction through successive convolutions. We put into practice two convolutive neural network-based deep learning methods.

**U-NET** is an architecture for a convolutional neural network designed specifically for image segmentation [18]. The U-Net architecture's U-shaped form has two primary elements.

There are two main components to the U-shaped form of the U-Net architecture. The network can track both high-level contextual data and precise geographic information thanks to the contract path and the expansion path. The contract path's duties include acquiring contextual data and extracting features from the input image. It consists of several convolution and clustering layers that increase the number of feature channels while progressively decreasing the spatial dimensions of the input. The expanding path makes an effort to reconstruct the segmentation map using the features found in the contracting path. The feature maps are progressively oversampled through the use of upward convolution layers. Subsequently, the required feature mappings from the contract path are concatenated with the oversampled features via hop connections.

**SegNet** is a convolutional neural network architecture designed for semantic segmentation. It comprises of an encoder-decoder structure with skip connections [19]. The encoder part is responsible for extracting the high-level feature representations from the input image. The decoder uses the previously learned features and upsampling to rebuild the segmentation map. The skip connections are used by the encoder and decoder to enable the decoder to obtain comprehensive spatial data from the encoder. Attach the corresponding encoder and decoder layers at the same spatial resolution. The softmax activation function layer is employed last to assign probability to each pixel across many classes. This produces a segmentation map that is pixel-by-pixel and identifies the predicted class probabilities for every pixel.

### 2.5 Classification

At this stage, the type of cancer must be classified as either benign or malignant. In this last stage, we classified the type of lung nodules using the VGG16 deep learning algorithm.

**VGG16** [20] has only sixteen weight layers, but is made up of thirteen convolutional layers, five Max Pooling layers, and three Dense layers, for a total of twenty-one layers. The input tensor size that VGG16 accepts is 512 by 512 with three RGB channels, which matches the data that the U-Net and SegNet algorithms were previously used to segment. The most distinctive feature of VGG16 is that, rather than using a large number of hyper-parameters, they concentrated on $3 \times 3$ filter convolution layers with stride 1 while maintaining the use of the $2 \times 2$ filter's maxpool layer and padding from stride 2. Throughout the architecture, the convolution and maximum pool layers are arranged uniformly. There are 64 filters in Conv-1 Layer, 128 filters in Conv-2, 256 filters in Conv-3, 512 filters in Conv-4, and 512 filters in Conv-5. A stack of convolutional layers is followed by three fully connected (FC) layers. The first two of these layers each have 4096 channels, while the third does 1000-way classification and so has 1000 channels (one for each class). Soft-max layer is the last layer.

## 3 Results and Discussion

There are two subsections in this section. The pre-processing and mask construction findings utilizing the dataset mentioned in part II.A are presented in the first subsection. The outcomes of the segmentation and classification procedure are described in depth in the second subsection.

### 3.1 Pre-Processing and Mask Creation Results

Figures 3 and 4 presents the result after the blue channel extraction and morphological operation.

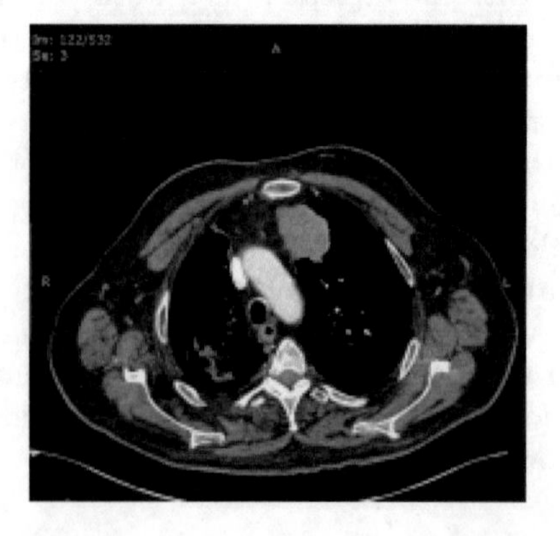

**Fig. 3.** image after blue channel extraction

**Fig. 4.** image after opening application

Five processes were taken to create the masks. Grayscale rendering of RGB (red, green, and blue) images. The Otsu thresholding technique [21], which converts the grayscale image into a binary image, and then the morphological closure technique, which fills in the gaps. The Fig. 5 presents an example of the Non inverted binary mask for normal, benign and malignant case respectively.

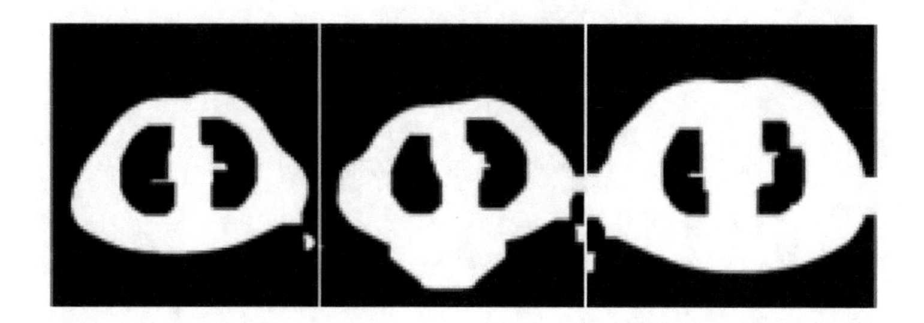

**Fig. 5.** Non inverted binary mask for normal, benign and malignant case respectively

The resulting mask must be inverted to get a **1** pixels values in the lob regions and **0** out the pulmonary lobs. Figure 6 illustrates the binary mask inverted for the typical scenario.

In Fig. 7, We illustrate the result image of the multiplication between the preprocessing result image and the binary mask. [22] to extract the inner region.

**Fig. 6.** Inverted binary mask for normal, benign and malignant case respectively

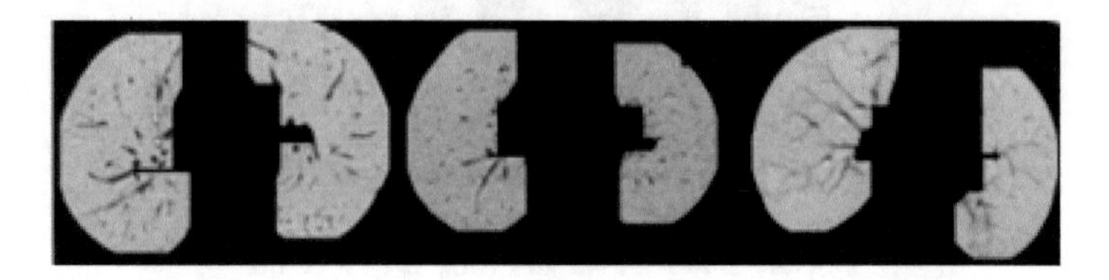

**Fig. 7.** The final mask for normal, benign and malignant case respectively

### 3.2 Segmentation and Classification Results

The suggested model's performance metrics are shown in this subsection. To do that, we apply the corresponding equations for accuracy, recall, precision, and f1score., (1)-(4).

$$Accuracy = \frac{TP + TN}{TP + FP + TN + FN} \tag{1}$$

$$Recall = \frac{TP}{TP + FN} \tag{2}$$

$$Precision = \frac{TP}{TP + FP} \tag{3}$$

$$f_1Score = 2 \times \frac{Recall \times Precision}{Recall + Precision} \tag{4}$$

The results are given in Table 1.

**Table 1.** VGG16 classification's result

| Segmentation algorithm | Accuracy | f1score | Recall | Precision |
|---|---|---|---|---|
| SegNet | 95.1% | 97.41% | 97.33% | 97.5% |
| U-Net | 98.6% | 99.26% | 99.1% | 99.43% |

From these tables we can clearly see that U-Net gives better results compared to SegNet algorithm explained by the efficient of the U-Net in segmentation and VGG16 in classification.

## 4 Conclusion

The purpose of this planned study is to address the obstacles to early detection of lung cancer nodules before conditions worsen. To this end, this work develops an effective computer-aided diagnosis approach for the early identification of this fatal cancer. Here, data from chest tomography scans have been fed into the suggested model. Finding an effective combination of CNN architecture models that could reliably identify and categorize lung cancer nodules was the aim of this study. The recommended approach diagnoses 1190 digital tomography images from the human image datasets with a 99.26% diagnostic accuracy.

We want to present a reinforcement learning-based cancer status prediction approach in a later study.

## References

1. International Agency for Research on Cancer (IARC): Five Key Facts About Cancer. Technical report, International Agency for Research on Cancer (2024)
2. Song, Q., Zhao, L., Luo, X., Dou, X., et al.: Using deep learning for classification of lung nodules on computed tomography images. J. Healthc. Eng. **2017**, 8314740 (2017)
3. Kelana, S.e.S.S., Yusuf, G., Cahya et, R.: Classification des images histopathologiques du cancer du côlon 'a l'aide de la méthode des réseaux neuronaux convolutifs. In: Conférence Internationale 2023 sur L'informatique, les Technologies de L'information et L'ingénierie (ICCoSITE), pp. 821–826 (2023)
4. Mahmood, S.A., Ahmed, H.A.: An improved CNN-based architecture for automatic lung nodule classification. Med. Biol. Eng. Comput. **60**(7), 1977–1986 (2022)
5. Singadkar, G., Mahajan, A., Thakur, M., Talbar, S.: Deep deconvolutional residual network based automatic lung nodule segmentation. J. Digit. Imaging **33**, 678–684 (2020)
6. Bruntha, P.M., Pandian, S.I.A., Sagayam, K.M., Bandopadhyay, S., Pomplun, M., Dang, H.: Lung paynet?: un réseau pyramidal d'apprentissage profond basé sur l'attention pour la segmentation des nodules pulmonaires. Rapports scientifiques **12**, 20330 (2022)
7. Taylor, A.G., Mielke, C., Mongan, J.: Automated detection of moderate and large pneumothorax on frontal chest x-rays using deep convolutional neural networks: a retrospective study. PLoS Med. **15**(11), 1002697 (2018)
8. Lindsey, T., Lee, R., Grisell, R., Vega, S., Veazey, S.: Automated pneumothorax diagnosis using deep neural networks. In: Vera-Rodriguez, R., Fierrez, J., Morales, A. (eds.) CIARP 2018. LNCS, vol. 11401, pp. 723–731. Springer, Cham (2019). https://doi.org/10.1007/978-3-030-13469-3_84
9. Kumar, V.D., Rajesh, P., Geman, O., Craciun, M.D., Arif, M., Filip, R.: diagnostic quo vadis?: Application de l'informatique 'a la détection précoce du pneumothorax. Diagnostics **13**, 1305 (2023)
10. Kalaivani, N., Manimaran, N., Sophia, D., Devi, D.: Deep learning based lung cancer detection and classification. IOP Conf. Ser. Mater. Sci. Eng. **994**, 012026 (2020). https://doi.org/10.1088/1757-899X/994/1/012026
11. Yu, H., Zhou, Z., Wang, Q.: Deep learning assisted predict of lung cancer on computed tomography images using the adaptive hierarchical heuristic mathematical model. IEEE Access **8**, 86400–86410 (2020)

12. Guo, H., Kruger, U., Wang, G., Kalra, M.K., Yan, P.: Knowledge-based analysis for mortality prediction from CT images. IEEE J. Biomed. Health Inform. **24**(2), 457–464 (2019)
13. Sukumaran, A., Abraham, A.: Automated detection and classification of meningioma tumor from MR images using sea lion optimization and deep learning models. Axioms **11**(1), 15 (2021)
14. Pang, S., Zhang, Y., Ding, M., Wang, X., Xie, X.: A deep model for lung cancer 10 type identification by densely connected convolutional networks and adaptive boosting. IEEE Access **8**, 4799–4805 (2019)
15. F. Al-Yasriy, H.: The IQ-OTH/NCCD Lung Cancer Dataset (2021). https://doi.org/10.17632/bhmdr45bh2.2
16. Zhou, T., Fan, D.-P., Cheng, M.-M., Shen, J., Shao, L.: RGB-D salient object detection: a survey. Comput. Vis. Media **7**, 37–69 (2021)
17. Kanan, C., Cottrell, G.W.: Color-to-grayscale: does the method matter in image recognition? PLoS ONE **7**(1), 29740 (2012)
18. Ronneberger, O., Fischer, P., Brox, T.: U-Net: convolutional networks for biomedical image segmentation. In: Navab, N., Hornegger, J., Wells, W.M., Frangi, A.F. (eds.) MICCAI 2015. LNCS, vol. 9351, pp. 234–241. Springer, Cham (2015). https://doi.org/10.1007/978-3-319-24574-4_28
19. Badrinarayanan, V., Kendall, A., Cipolla, R.: SegNet: a deep convolutional encoder-decoder architecture for image segmentation. IEEE Trans. Pattern Anal. Mach. Intell. **39**(12), 2481–2495 (2017). https://doi.org/10.1109/TPAMI.2016.2644615
20. Sugata, T., Yang, C.: Leaf app: leaf recognition with deep convolutional neural networks. IOP Conf. Ser. Mater. Sci. Eng. **273**, 012004 (2017). https://doi.org/10.1088/1757-899X/273/1/012004
21. Yousefi, J.: Image Binarization using OTSU Thresholding Algorithm. Ontario, Canada: University of Guelph 10 (2011)
22. Image, G.-s., Parallel, S., Cohen, H., Duong, C.: Harvey a. cohen and cung h. duong, gray-scale image segmentation using a parallel (2002)

# Targeted and Automatic Deep Neural Networks Optimization for Edge Computing

Luca Giovannesi, Gabriele Proietti Mattia(✉), and Roberto Beraldi

Department of Computer, Control and Management Engineering "Antonio Ruberti", Sapienza University of Rome, Rome, Italy
{giovannesi,proiettimattia,beraldi}@diag.uniroma1.it
https://www.diag.uniroma1.it

**Abstract.** DNNs, commonly employed for complex tasks such as image and language processing, are increasingly sought for deployment on Internet of Things (IoT) devices. These devices operate with constrained resources, including limited computational power, memory, slower processors, and restricted energy requirements. Consequently, optimizing DNN models becomes crucial to minimize memory usage and computational time. However, traditional optimization methods require skilled professionals to manually fine-tune hyperparameters, striking a balance between efficiency and accuracy. This paper introduces an innovative solution for identifying optimal hyperparameters, focusing on the application of pruning, clusterization, and quantization.

Initial empirical analyses were conducted to understand the relationships between model size, accuracy, pruning rate, and the number of clusters. Building upon these findings, we developed a framework that proposes two algorithms: one for discovering optimal pruning and the second for determining the optimal number of clusters. Through the adoption of efficient algorithms and the best quantization configuration, our tool integrates an optimization procedure that successfully reduces model size and inference time. The optimized models generated exhibit results comparable to, and in some cases surpass, those of more complex state-of-the-art approaches.

The framework successfully optimized ResNet50, reducing the model size by 6.35x with a speedup of 2.91x, while only sacrificing 0.87% of the original accuracy.

**Keywords:** Deep Neural Networks · DNN Acceleration · DNN Compression · Edge Computing

## 1 Introduction

Deep Neural Networks (DNNs) have garnered increasing popularity over the past decade due to their adeptness in addressing intricate problems such as image


L. Barolli (Ed.): AINA 2024, LNDECT 203, pp. 57–68, 2024.
https://doi.org/10.1007/978-3-031-57931-8_6

recognition, language processing, signal processing, and more. Simultaneously, the rapid proliferation of the Internet of Things (IoT) has led to an increasing demand for deploying DNN models on IoT devices.

However, while these devices offer the advantage of easy installation in numerous contexts where traditional servers may be impractical, this convenience comes at a price. IoT devices typically feature constrained computational power, limited memory capacity, and comparatively slower processors in comparison to conventional desktops and laptops. Furthermore, a substantial number of these devices operate on battery power, introducing additional constraints related to energy consumption. Usually, DNNs require a powerful GPU to run effectively, a component that is generally not available in most IoT devices, and therefore, here emerges the need to optimize neural network models to reduce memory usage and the computational time required for the inference operation.

Nowadays, the most common approaches for neural network optimization require heuristic hand-tuning of some fundamental hyper-parameters by an expert in the field, whose goal is to find the optimal balance point between model optimization and model accuracy degradation. In this paper, we present a tool that is able to automatically find these hyper-parameters and use them to apply the optimizations, which are pruning, weight clustering and quantization.

The rest of the paper is organized as follows. In Sect. 2 we give some background to the reader about the techniques that are used for compressing DNN models, in Sect. 3 we present different related work to our proposed solution, in Sect. 4 we describe the algorithms used in the tool and, finally, in Sect. 6 we draw the conclusions.

## 2 DNN Compression Techniques

### 2.1 Network Pruning

DNNs exhibit notable redundancy in their parameterization, containing portions that are not truly essential [1]. Recognizing this phenomenon, we can eliminate some of these redundant parts to create smaller and simpler models. However, when removing a portion of the DNN, the accuracy typically decreases. The key challenge lies in identifying the most suitable parts of the network to be pruned while minimizing the decrease in accuracy.

Network pruning can be categorized into three main types: Channel pruning, Filter pruning, and Connection pruning.

**Channel Pruning.** The idea behind channel pruning is to decrease the number of input and/or output channels in convolutional layers without compromising performance. In this context, the authors of [2] achieved a 5x speedup on VGG-16 [3]. However, on modern networks such as ResNet [4] and Xception [3], the acceleration is only 1.4%.

**Filter Pruning.** The Convolutional Neural Network (CNN) employs a substantial number of filters to enhance precision. However, each filter adds the overall number of floating-point operations, thereby increasing the latency of the network. The goal of filter pruning is to eliminate less crucial filters. In [5], the authors achieved a reduction in floating-point operations by approximately 3.2 and 1.58 times for VGG-16 and ResNet-50, respectively, while maintaining nearly the same level of accuracy.

**Connection Pruning.** The size of a DNN is dictated by the number of connections between its layers. A larger number of parameters result in a bigger model, requiring more operations. The concept behind this pruning technique is to eliminate unimportant connections.

A straightforward yet effective approach is Global Magnitude Pruning (MP) [6], where all weights with absolute values below a predefined threshold are removed.

### 2.2 Weight Sharing

In weight sharing, also called weight clustering [7], only a small subset of values are used to represent all the weights of the model.

For the creation of the clusters different algorithms can be used, in [8] the weights distribution is based on a low-cost hash function, in which all connections with the same hash share a single parameter value.

Another approach is the one proposed by Han et el. [9], where k-means algorithm is used for the clusterization, in each iteration it assigns the values of the weights to the clusters with the nearest centroid, it converges when no addition assignments are needed. The time required for the convergence depends by the initial value of the centroids, the most used method is using the k-means++ algorithm [10].

### 2.3 Quantization

Quantization [11] reduces the bit width of weights and activation functions, typically from 32 bits (FP32) to 8 bits (INT8/UINT8). This compression yields advantages such as a 4x reduction in memory overhead and a quadratic decrease in computational cost for matrix multiplication by a factor of 16.

Quantization can be applied to weights only or both activation functions and weights. While NNs are generally robust to quantization, when a low bit-width quantization ($<$8 bits) is employed, noise is introduced, potentially decreasing accuracy. Robustness varies across networks, necessitating additional efforts to leverage quantization benefits.

The two methods for quantizing neural networks are Post Training Quantization (PTQ) and Quantization Aware Training (QAT).

**Post Training Quantization (PTQ).** In PTQ a pre-trained FP32 neural network is converted directly to an integer model without additional tuning. If some sample data are provided, they may help the calibration of the parameters. PTQ can be applied on both weights and activation functions using 8 bits integers. It can be applied only weights keeping the original floating-point activation functions. Otherwise a combination such as int8 weights and int16 activation functions. Alternatively, the entire model can be quantized to 16-bit floating point. Our experiments (Table 1) demonstrate the impact of various quantization approaches on model size and inference time.

**Table 1.** MobileNet V2 quantized with different configurations

| Quantized elements | Accuracy | Model size (MB) | Inference time (ms) |
|---|---|---|---|
| Weights fp32 (Original)<br>Activation functions fp32 | 0.832 | 12.28 | 9.240 |
| Weights int8<br>Activation functions int8 | 0.821 | 3.18 | 4.055 |
| Weights int8<br>Activation functions fp32 | 0.820 | 3.18 | 4.200 |
| Weights int8<br>Activation functions int16 | 0.820 | 3.32 | 277.52 |
| Weights fp16<br>Activation functions fp16 | 0.832 | 6.26 | 9.240 |

Using int8 for weights and int16 for activation functions (4th row in Table 1) results in slower inference due to incomplete support by the kernel, as noted in TensorFlow documentation. The FP16 model maintains the original inference time while reducing the model size by half.

**Quantization Aware Training (QAT).** PTQ is fast to implement, it does not require labeled data, and in most of the cases the quantized model has an accuracy close to the original model, however, when the bit width is very low, like 4 bits, the PTQ introduces too many errors.

In QAT some nodes are added to the network in order to simulate the error introduced by the quantization, and then the model is trained with the presence of the added nodes improving the robustness to the quantization noise. In general, QAT provides better performance but requires a labeled dataset and fine-tuning which takes computational time.

Figure 1 illustrates the weight distribution of the mentioned optimization approaches.

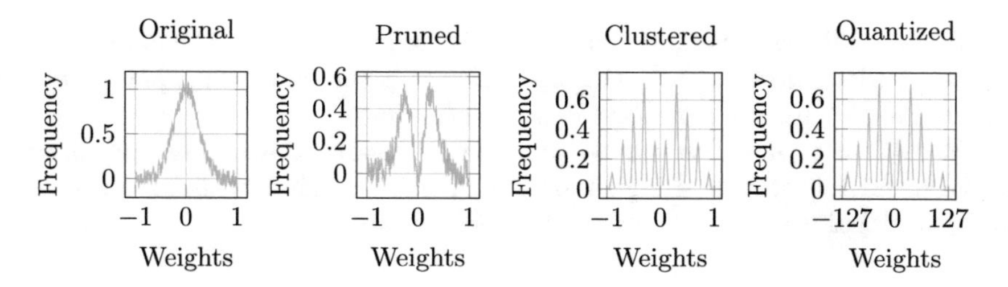


**Fig. 1.** Weight distribution in the different steps of the described optimization methods

## 3 Related Work

In the literature, a lot of works use the illustrated techniques for model compression, but only a few proposals find automatically the hyper-parameters necessary to apply them. Regarding the pruning, the necessary hyper-parameter is the pruning rate, for the clustering is the number of clusters, while for the quantization the searching space is about the optimal configuration between the ones presented in Table 1.

### 3.1 Pruning Rate Search

In [12] the authors claim that if the absolute values of the weights of each layer are placed in order, they are very similar to each other, approximating a function in which in the first part the values of the weights increase linearly, while in the second part, the values increase exponentially, these two parts are divided by a point called "demarcation point". They find a correlation between the threshold used for Global Magnitude Pruning manually chosen in [9] and the demarcation point, claiming that usually the optimal threshold is near this point. However they have reported this correlation only for Lenet-5 [13], AlexNet [14], VGG [15], additional tests are necessary to check if the correlation is always present.

In Optimal Brain Surgeon (OBS) [16], a method determines which weights to remove based on their impact on loss function $L$. This often involves inverting the Hessian matrix $\mathbf{H} = \partial^2 L / \partial^2 W$, a very hard task with modern neural networks that have millions of parameters $W$.

L-OBS [17] addresses this by applying OBS to individual layers and proposing an algorithm to reduce the Hessian matrix for feasible inversion. NAP [18] uses an efficient approximation of the Hessian, utilizing a Kronecker-factored Approximate Curvature method. However, these methods provide a lower compression rate than methods based on weights magnitude pruning.

The authors of [19] propose a reinforcement learning approach, in which the agent receives layer characteristics and returns the pruning rates for each layer, which will be used to prune the model. Results are similar or better to manual methods on VGG-16[15] and ResNet 50 [4].

Runtime Neural Pruning [20] (RL based approach), dynamically optimizes models during run-time adjusting the balance point based on input data, the original model is always preserved for restoration.

The solutions in which the hyper-parameter search is solved as optimization problem [21,22] provide a higher precision, but this kind of method has large computational costs slowing the convergence time.

### 3.2 Number of Clusters Search

In [23] LetNet-5 [13] is clustered using K-Means with a greedy algorithm for cluster number selection, However, this approach is slow for larger models. Instead, in [24] authors solve the problem as a minimization task, but the computational cost tends to be high despite optimal solutions.

To the best of our knowledge, the only relevant work in which all the previous optimizations are used together is the one by Song Han et al. [9], but there is not evidence of methods to automatically assign the hyperparameters.

## 4 Proposed Method

The proposed framework receives a trained deep neural network model, a dataset, and additional parameters as input. It then autonomously identifies the optimal pruning rate, followed by determining the optimal number of clusters, and ultimately, finding the best configuration for quantization.

### 4.1 Weight Pruning

Our idea uses Global Magnitude Pruning (MP), despite the simplicity of this method, it is possible to get results comparable to the state-of-the-art [25,26], with respect also to complicated pruning algorithms present in the literature.

Global Magnitude Pruning (MP) is a method where weights with an absolute magnitude value below a specified threshold are eliminated. The threshold is set to achieve a predetermined sparsity rate after the pruning process. Through experiments, an inversely proportional relationship between the pruning rate and accuracy was observed, as depicted in Fig. 2. This finding enables the use of binary search to determine the pruning rate that yields the desired accuracy.

In the searching algorithm, it is allowed an error of $\epsilon$ between the desired accuracy and the reached accuracy. In our case, the desired accuracy is slightly less than the one initially measured on the original model. Looking for a smaller accuracy guarantees that the returned value is the one before the quick accuracy decreases.

Binary search converges since both the pruning rate and the measured accuracy are linear, this implies that changing the pruning rate is always possible to reach the desired accuracy.

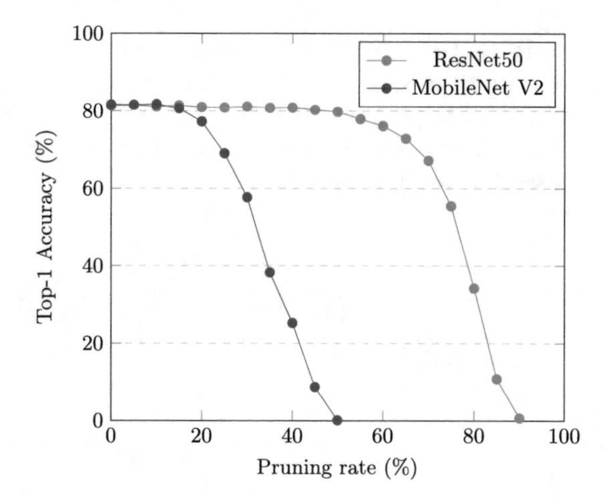


**Fig. 2.** Accuracy variation with different pruning rates for ResNet50 and MobileNet V2

### 4.2 Number of Clusters

In order to minimize the model size, we have to minimize the number of clusters, however a small number of clusters cannot be sufficient to reach the original model accuracy.

In our experiments, MobileNetV2 was clustered with different numbers of clusters. The results are visible in Fig. 3 in which there is an evident a direct correlation between the model size and the number of clusters, while there isn't any evident relation between the number of clusters and the inference time. Therefore the optimal amount of clusters is the smaller one in which is possible to reach the desired accuracy, in other words, given an original model $M_{original}$, the pruned model $M_{pr}$, a function $CL(M, n_{cl})$ which clusters a model $M$ with $n_{cl}$ clusters and a $\epsilon$ defined as the maximum allowed difference between the original model accuracy and the clusterized model accuracy. The optimal number of clusters $cl_{opt}$ are:

$$cl_{opt} = \underset{cl}{\operatorname{argmin}} |Acc(M_{original}) - Acc(CL(M_{pr}, cl)| < \epsilon \tag{1}$$

Then giving the function $f$ defined as:

$$f(x) = \begin{cases} 1, & \text{if } |Acc(M_{original}) - Acc(CL(M_{pr}, cl)| < \epsilon \\ 0, & \text{otherwise} \end{cases} \tag{2}$$

Which returns 1 if the given number of clusters $x$ returns a model that matches the required accuracy constraints, 0 otherwise, from the analysis we know that $f(x) \leq f(x+1)$, then to find the optimal number of clusters we have to find a $x$ such that $f(x) < f(x+1)$. The problem of finding the optimal value of $x$ can be solved using an algorithm based on ternary search, so given a range of clusters we have to find the step (the transaction between 0 and 1). Then for each iteration the range is split in 3 intervals by two points, and computing the values of these points is possible to determine in which segment the step belongs, then the algorithm is applied again in the new segment, it ends when the difference between the two points is 1.

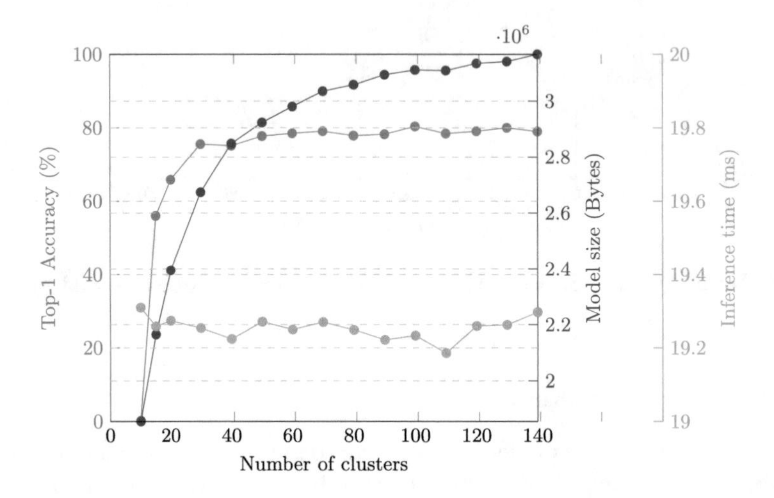


**Fig. 3.** Metrics affected with different number of clusters on MobileNet V2

### 4.3 Quantization

The used technique for quantization is the PTQ, because, differently from QAT, it does not require fine-tuning and the accuracy loss is negligible until we don't go below 8 bits. Quantization affects models differently; what works for one may not for another.

The efficiency of PTQ allows testing various methods outlined in Table 1 without significant time overhead, methods are tested from the more aggressive quantization for faster but less accurate models to softer quantization for slower but more accurate models, using this rank, the first method that matches the original desired accuracy is selected.

From the proposed methods (ones used in Table 1) the one that uses int8 for weights and int16 for activation functions is excluded, since as reported in the documentation of TensorFlow Lite (the DNN framework used for our tests) the lack of implementation brings slow inference time.

### 4.4 Overall Procedure

This section describes the full procedure, including all of the previous techniques. Figure 4 illustrates a flowchart depiction. Initially, the program examines the input model and establishes an accuracy baseline for the next operation. Then

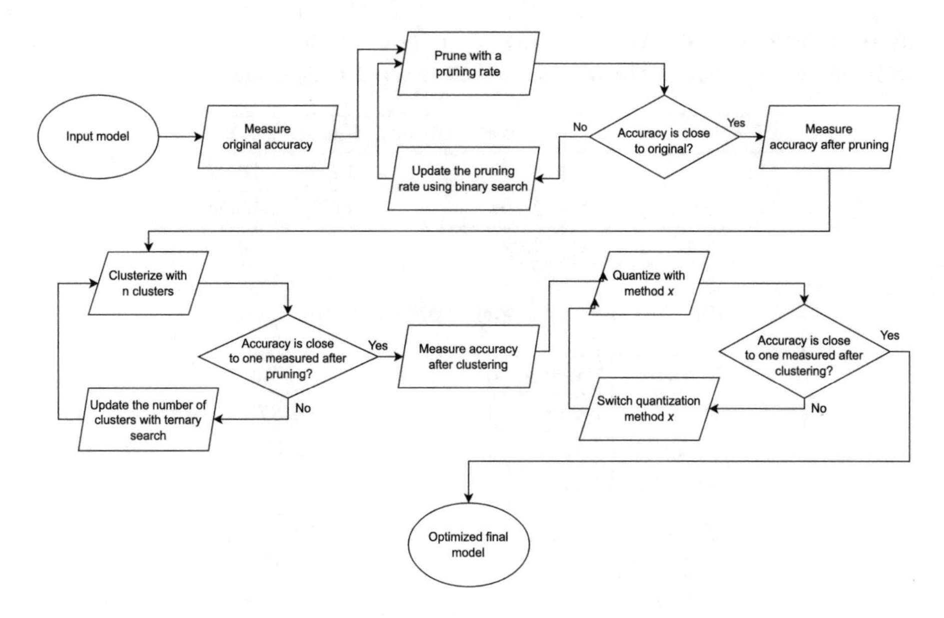


**Fig. 4.** Flowchart representing the overall optimization procedure

it starts to prune the network with its methods. After experimenting with several hyper-parameters, it returns the best pruning configuration. Next, the pruned model is evaluated to establish a new baseline, thus weight clustering is used, and the program runs the algorithm designed specifically to calculate the ideal number of clusters of the pruned model, returning the optimal one. Finally, after establishing a new baseline from the clustered model, it tries multiple quantization algorithms, ranking them from more aggressive to kinder, and returns the first that meets the required target accuracy. At this stage, the software builds the optimized model based on all of the previously determined optimal parameters.

## 5 Results

Making a comparison with the literature is not easy, since the proposed software can optimize different models used to solve different problems. Furthermore, the literature employs diverse metrics: some use inference times for speedup calculation, and others focus on FLOPs or MAC operations reduction. Regarding model compression, metrics range from parameter count reduction to actual model size reduction.

Our evaluation metrics are the speedup factor, calculated as the ratio of the inference times between original and optimized models, and the compression rate, calculated as the ratio of the original model's size to the optimized one.

We remark by adjusting $\epsilon$ values in pruning and clustering algorithms, models with varying accuracies can be achieved.

**Table 2.** VGG16 on CIFAR10 comparison, the model is optimized with our tool also with different $\epsilon$ values for the pruning and clusterization algorithms.

| Model | Compression rate | Speed up | $\Delta$ Top1 |
|---|---|---|---|
| Our | 12.95x | 4.77x | 0.9% |
| Our (lower $\epsilon$) | 12.19x | 3.40x | 0.26% |
| Play and Prune [21] | 13.0x | – | 0.3% |

**Table 3.** ResNet50 model optimization comparison.

| Method | Speedup | Compression rate | $\Delta$ Top1 |
|---|---|---|---|
| **Our** | **2.91x** | **6.35x** | **0.87%** |
| NAP[18] | 2.3x | – | 2% |
| Thinet [5] | 2.3x | – | 4.3% |
| CPI [2] | 2x | – | 3.3% |
| AMC [19] | – | 5x | 0.02% |
| LWC [6] | – | 2.7x | 0.01% |

The listed results (Tables 2 and 3) are executed on different hardware and as reported in other works, the speedup changes with different hardware, it decreases when the device parallelism is higher. Since most of the competitor projects have the source code not available, we are not able to run the proposed frameworks in order to run the optimized models on the same hardware, so the data reported is the one written in the related articles.

Regarding the compression rate, we remark that it does not depend on the hardware and it can be comparable with the ones reported by the authors in their papers. The convergence times of the optimized models on an nVidia GTX 1080 are 125 min for ResNet50 and 40 min for VGG16-CIFAR10.

## 6 Conclusion

We propose a framework to optimize NNs efficiently, it does not require a deep knowledge by the user in the optimization procedures. The framework simply takes as input a model, a dataset, and some information like batch size and the data format expected by the networks, then it returns an optimized model ready to be deployed on an edge device. The proposed framework relies on simple algorithms to solve intricate problems. The design of the framework allows future integration of new pruning and clustering algorithms keeping the original search algorithms, in the way the speedup and the compression rate can be improved. The experiments show results comparable to the state of the art on ResNet and VGG architecture.

However, the suggested solution's simplicity allows for future improvement, indeed, by using the same search algorithms, we can employ superior pruning strategies, such as filter pruning, to significantly boost inference time. The current version of the framework can handle only models for classification tasks, thus our next goal is to implement the frequently employed object detection and classification tasks. Actually, because a large number of fine-tuning phases are required, the optimizer can take time to converge, particularly on large models; therefore, it would be desirable to improve the used algorithms with more targeted solutions to reduce converge times.

**Acknowledgements.** This work was supported by project no. 202277 WMAE CUP B53D23012820006, "EdgeVision against Varroa (EV2): Edge computing in defence of bees" funded by the Italian's MUR PRIN 2022 (ERC PE6) research program and by the European Union – Next Generation EU.

## References

1. Denil, M., Shakibi, B., Dinh, L., Ranzato, M.A., de Freitas, N.: Predicting parameters in deep learning. In: Burges, C., Bottou, L., Welling, M., Ghahramani, Z., Weinberger, K. (eds.) Advances in Neural Information Processing Systems, vol. 26. Curran Associates, Inc. (2013)
2. He, Y., Zhang, X., Sun, J.: Channel pruning for accelerating very deep neural networks. In: 2017 IEEE International Conference on Computer Vision (ICCV), pp. 1398–1406 (2017)
3. Simonyan, K., Zisserman, A.: Very deep convolutional networks for large-scale image recognition (2015)
4. He, K., Zhang, X., Ren, S., Sun, J.: Deep residual learning for image recognition (2015)
5. Luo, J.-H., Wu, J., Lin, W.: ThiNet: a filter level pruning method for deep neural network compression (2017)
6. Han, S., Pool, J., Tran, J., Dally, W.J.: Learning both weights and connections for efficient neural networks (2015)
7. Larsen, J.S., Clemmensen, L.: Weight sharing and deep learning for spectral data. In: ICASSP 2020 - 2020 IEEE International Conference on Acoustics, Speech and Signal Processing (ICASSP), pp. 4227–4231 (2020)
8. Chen, W., Wilson, J.T., Tyree, S., Weinberger, K.Q., Chen, Y.: Compressing neural networks with the hashing trick (2015)
9. Han, S., Mao, H., Dally, W.J.: Deep compression: Compressing deep neural networks with pruning, trained quantization and Huffman coding (2016)
10. Arthur, D., Vassilvitskii, S.: K-means++: the advantages of careful seeding. In: Proceedings of the Eighteenth Annual ACM-SIAM Symposium on Discrete Algorithms, SODA 2007, (USA), p. 1027-1035. Society for Industrial and Applied Mathematics (2007)
11. Nagel, M., Fournarakis, M., Amjad, R.A., Bondarenko, Y., van Baalen, M., Blankevoort, T.: A white paper on neural network quantization (2021)
12. Liu, C., Liu, Q.: Improvement of pruning method for convolution neural network compression. In: Proceedings of the 2018 2nd International Conference on Deep Learning Technologies, ICDLT 2018, (New York, NY, USA), pp. 57–60. Association for Computing Machinery (2018)
13. Lecun, Y., Bottou, L., Bengio, Y., Haffner, P.: Gradient-based learning applied to document recognition. Proc. IEEE **86**(11), 2278–2324 (1998)

14. Krizhevsky, A., Sutskever, I., Hinton, G.E.: ImageNet classification with deep convolutional neural networks. In: Pereira, F., Burges, C., Bottou, L., Weinberger, K. (eds.) Advances in Neural Information Processing Systems, vol. 25. Curran Associates, Inc., (2012)
15. Simonyan, K., Zisserman, A.: Very deep convolutional networks for large-scale image recognition. arXiv:1409.1556 (2015)
16. Hassibi, B., Stork, D., Wolff, G.: Optimal brain surgeon and general network pruning. In: IEEE International Conference on Neural Networks, vol. 1, pp. 293–299 (1993)
17. Dong, X., Chen, S., Pan, S.J.: Learning to prune deep neural networks via layer-wise optimal brain surgeon. arXiv:1705.07565 (2017)
18. Zeng, W., Xiong, Y., Urtasun, R.: Network automatic pruning: start nap and take a nap. arXiv:2101.06608 (2021)
19. He, Y., Lin, J., Liu, Z., Wang, H., Li, L.-J., Han, S.: AMC: automl for model compression and acceleration on mobile devices. arXiv:1802.03494 (2019)
20. Lin, J., Rao, Y., Lu, J., Zhou, J.: Runtime neural pruning. In: Guyon, I., Luxburg, U.V., Bengio, S., Wallach, H., Fergus, R., Vishwanathan, S., Garnett, R. (eds.) Advances in Neural Information Processing Systems, vol. 30. Curran Associates, Inc. (2017)
21. Singh, P., Verma, V.K., Rai, P., Namboodiri, V.P.: Play and prune: adaptive filter pruning for deep model compression. arXiv:1905.04446 (2019)
22. Sun, Q., Cao, S., Chen, Z.: Filter pruning via automatic pruning rate search. In: Proceedings of the Asian Conference on Computer Vision (ACCV), pp. 4293–4309 (2022)
23. Dupuis, E., Novo, D., O'Connor, I., Bosio, A.: On the automatic exploration of weight sharing for deep neural network compression. In: 2020 Design, Automation & Test in Europe Conference & Exhibition (DATE), pp. 1319–1322 (2020)
24. Wu, J., Wang, Y., Wu, Z., Wang, Z., Veeraraghavan, A., Lin, Y.: Deep $k$-means: Re-training and parameter sharing with harder cluster assignments for compressing deep convolutions. arXiv:1806.09228 (2018)
25. Gupta, M., et al.: Is complexity required for neural network pruning? A case study on global magnitude pruning. arXiv:2209.14624 (2023)
26. Gale, T., Elsen, E., Hooker, S.: The state of sparsity in deep neural networks. arXiv:1902.09574 (2019)

# Edge to Cloud Network Function Offloading in the ADAPTO Framework

Alessio Botta[1], Roberto Canonico[1(✉)], Annalisa Navarro[1], Giovanni Stanco[1], Giorgio Ventre[1], Antonio Buonocunto[2], Antonio Fresa[2], Vincenzo Gentile[2], Leonardo Scommegna[3], and Enrico Vicario[3]

[1] Department of Electrical Engineering and Information Technology (DIETI), University of Naples Federico II, Naples, Italy
{a.botta,roberto.canonico,annalisa.navarro,giovanni.stanco, giorgio.ventre}@unina.it

[2] Ericsson Telecomunicazioni S.p.A., Pagani, Italy
{antonio.buonocunto,antonio.fresa,enzo.gentile}@ericsson.com

[3] Department of Information Engineering, University of Florence, Florence, Italy
{leonardo.scommegna,enrico.vicario}@unifi.it

**Abstract.** As telcos increasingly adopt cloud-native solutions, classic resource management problems within cloud environments have surfaced. While considerable attention has been directed toward the conventional challenges of dynamically scaling resources to adapt to variable workloads, the 5G promises of Ultra-Reliable Low Latency Communication (URLLC) remain far from being realized. To address this challenge, the current trend leans toward relocating network functions closer to the edge, following the paradigm of Mobile Edge Computing (MEC), or exploring hybrid approaches. The adoption of a hybrid cloud architecture emerges as a solution to alleviate the problem of the lack of resources at the edge by offloading network functions and workload from the Edge Cloud (EC) to the Central Cloud (CC) when edge resources reach their capacity limits. This paper focuses on the dynamic task offloading of network functions from ECs to CCs within cloud architectures in the ADAPTO framework.

**Keywords:** Resource management · Virtual Network Function · Edge-to-cloud offloading

## 1 Introduction

Telco Networks have always been subject to a continuous architectural evolution to satisfy the demanding requirements associated with each network generation. One of the most important evolutions introduced with the 5G networks is the massive adoption of the cloud-native principles that provide a flexible way to design and orchestrate Network Functions. A cloud native network function is designed to better sustain all mutable network characteristics thanks to a finer granularity achieved by decomposing a monolithic SW component in several


L. Barolli (Ed.): AINA 2024, LNDECT 203, pp. 69–78, 2024.
https://doi.org/10.1007/978-3-031-57931-8_7

microservices. The advantages gained in breaking a Network Function (NF) come with the risk of increasing resource utilization if the NF is operated without a clear and automated optimization strategy to keep the application dimension in line with the network demands.

The advent of 5G Mobile Edge Computing (MEC) represented a significant stride towards materializing the promise of Ultra-Reliable Low Latency Communication (URLLC) within the domain of 5G networks. MEC extends the cloud-computing capabilities near mobile users, providing computational and data processing capabilities and enabling time-sensitive services such as driverless vehicles, augmented reality, robotics, and immersive media [1]. However, the scarcity of resources in the Edge Clouds (ECs) limits the delivery of URLLC, Enhanced Mobile Broadband (eMBB), and Massive Machine Type Communication (mMTC) services. Consequently, hybrid methodologies have emerged, proposing the integration of Edge and the Central Cloud (CC) to provide the proximity advantages offered by edge computing and the extensive computational resources housed within the central cloud, catering to the multifaceted and stringent demands of URLLC, eMBB, and mMTC use cases in 5G networks.

In this context and with these objectives, the ADAPTO project was established. ADAPTO is part of the larger RESTART initiative, funded by the Italian government in the context of the *National Recovery and Resilience Plan* (NRRP) as part of the Next Generation EU (NGEU) programme. Leveraging recent advancements in Software-Defined Networking (SDN) and Network Function Virtualization (NFV), ADAPTO seeks to craft a framework to orchestrate the utilization of computational and network resources across distributed cloud infrastructures to adapt to variable workloads and network conditions. The overarching goals of the ADAPTO framework are threefold: first, the autonomous scaling and orchestration of Virtual Network Functions (VNFs) based on the operational load; second, the intelligent offloading of the VNFs between edges and the central cloud, based on 5G backhaul conditions and edge resource status; and finally, the selection of the most suitable 5G backhaul between the edge and the central cloud to ensure a seamless handoff and guarantee URLLC requirements.

This paper specifically focuses on a critical aspect within the framework - i.e. the seamless handoff mechanism - presenting the examination of the role of an intelligent agent employing Reinforcement Learning (RL) positioned at the network's edge. This agent is tasked with decisions on the offloading of the workload between EC and CC, as well as the optimal selection of the most suitable 5G backhaul for data and signaling transfer between EC and CC. Through continuous assimilation of real-time data encompassing network conditions, edge resource utilization, and workload variations, this RL-based agent acquires the capacity to make informed decisions. The rest of the paper is organized as follows: the technological background of the application context investigated by the ADAPTO project is provided in Sect. 2. In Sect. 3, we describe in detail how the ADAPTO project can help telco operators fulfill 5G promises, describing the design goals and the architectural view (Sect. 3). Finally, we summarize our paper and draw conclusions in Sect. 4.

## 2 Technological Context

In this section, we briefly highlight the technological context in which the ADAPTO project provides its research contributions.

### 2.1 5G Architecture

The 5G system, standardized by 3GPP [2], comprises three primary components: User Equipments (UEs) comprising mobile devices such as smartphones or tablets, the Radio Access Network (RAN), that connects UEs to the fixed infrastructure through wireless connections, and the Core Network (CN), that performs essential functionalities such as authentication, session management, and data routing.

The 5G Core Network (depicted in Fig. 1) consists of various 5G Core network functions engineered as cloud-native SW components that, with their modular architecture, can better adapt to the high and various demands of an evolved and dynamic telco network. The Access and Mobility Function (AMF) handles communication between the UEs and the RAN, controlling access, registration, and mobility. The Session Management Function (SMF) manages the User Plane Functions (UPFs) for routing decisions and Quality of Service (QoS) settings. The Policy and Charging Function (PCF) is used for enforcing subscriber policies and charging the users. A centralized state is kept in the Unified Data Management (UDM), which synchronizes the states in the different network functions. All of the 5G Core components have been designed according to the Service-based Architecture (SBA) framework, implementing an HTTP/2-based Service Bus Interface (SBI). The main characteristic of this architecture is a decomposition of function in multiple independent entities (microservices) that interact via Application programming Interface (API) to deliver network functionalities.

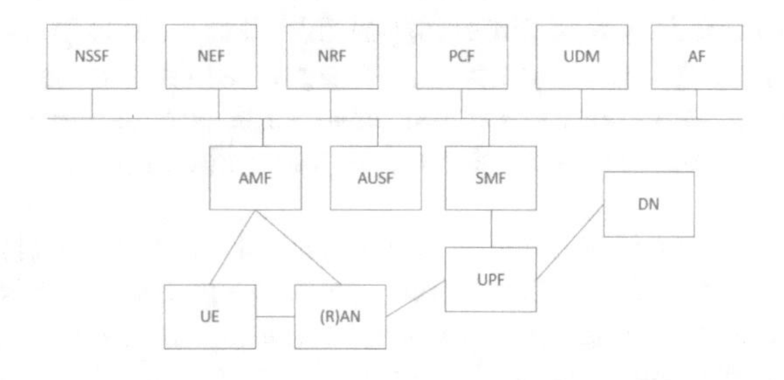


**Fig. 1.** 5G network functions

### 2.2 Network Function Virtualization Scaling and Placement

Network Function Virtualization (NFV) technology stands as a cornerstone in the evolution of 5G and beyond networks by detaching the traditional network functions from specialized hardware and transforming them into versatile, cross-platform, Virtualized Network Functions (VNFs).

**VNF Scaling.** NFV enables flexible Network Function (NF) deployments, in which they can scale dynamically according to the rapid fluctuations in mobile data traffic demands and can be placed at different locations, e.g. at the edge or at a central position [3]. Two options are available for VNF scaling: Vertical and Horizontal. *Vertical scaling* involves resizing existing VNFs by adjusting computational, memory, or storage resources as needed. Conversely, *Horizontal scaling* creates additional instances of the same VNF or terminates redundant instances. While horizontal scaling enhances scalability and service reliability, it intensifies resource consumption and poses challenges related to state migration. Vertical scaling optimizes resource utilization but falls short in scalability and lacks flexibility in changing VNF hosts, impacting practical implementation.

**VNF Placement.** The promises of 5G to provide low-latency and ultra-reliable connectivity have led to the need for reconsidering the fully centralized architecture typical of 4G networks, where all core functions were confined to a single central location connected to the RAN through high-capacity backhaul.

Reenvisioning cloud computing, edge computing moves resources from centralized data centers to the network's edge, bringing them closer to users and application-produced data. Edge computing stands out as one of the vital elements essential for meeting the rigorous Key Performance Indicators (KPIs) of 5G, driving the development of *5G Mobile Edge Computing (MEC)* [4]. MEC offers the flexibility to position both the essential 5G core VNFs and application VNFs at the mobile network's edge. Notably, certain 5G core VNFs, such as UPF, maintain close ties with application VNFs. Consequently, it becomes imperative to locate the UPF, and not only services, at MEC servers closer to users [3]. Strategically placing VNFs in both edge and core cloud servers enables the accommodation of different possible demands. Edge servers cater to applications requiring immediate responsiveness, while core servers accommodate VNFs needing more computational power or serving larger user bases. This dual placement ensures each service receives tailored computational capabilities and responsiveness [5].

Several possibilities have been so far explored for the so-called *Edge-Core split Option* [6]: the Local Offload Split, where only the UPF is at the edge, causing significant signaling through the backhaul; the Locally Administered Edge Split, involving the AMF and SMF at the edge, reducing backhaul signaling but potentially leading to handovers due to mobility or overload; and the Autonomous Edge Split, which operates autonomously at the edge, handling all functions locally but limiting its scope to a local area and eliminating handover possibilities in case of overload or mobility. Two specific connectivity setups have been examined experimentally [7]: direct connectivity, where the UPF is positioned within the central core, and a local offloading scenario, in which the UPF is placed near the gNodeB. This analysis revealed that deploying a local UPF notably reduces latency. This effect was especially evident when utilizing a satellite backhaul between the edge and the central core.

### 2.3 SD-WAN for 5G Backhauls

The 5G backhaul between the edges and the core network, as well as between the edges and a possible central cloud, is critical for meeting strict latency and throughput requirements. Backhaul options are usually based on fiber or millimeter-wave wireless connections. However, employing these technologies alone could not be feasible, especially in cases when there is the need to cover very distant areas [8]. For this reason, the edge and the central core can be interconnected through a joint combination of backhaul networks based on different technologies, such as fibers, wireless, satellite, each one bringing its own benefits in terms of guaranteed throughput or latency [6].

The advancements in the Software Defined Networking (SDN) paradigm have led to extending its applicability across time and scope, towards increasingly broader areas and with a finer temporal granularity. SD-WAN (Software Defined Wide Area Network) comes into play when catering to the real-time and adaptable demands of new 5G services flexibly and cost-effectively [9]. SD-WAN orchestrates traffic routing between multiple distant geographical sites. The distributed sites keep a connection with the Internet and with each other through one or more Edge Routers (ERs) leveraging a combination of different transport technologies (e.g., mobile, fiber). Each ER is linked to an SD-WAN controller that enforces network policies by sending rules to the ERs, exerting specific actions. These actions enable leveraging the multiple transport networks available in order to perform different kinds of optimization such through bandwidth aggregation, duplicating packets across different transports to ensure successful delivery or error correction of packets, or adaptively switching the used transport if the current one fails to guarantee desired performance or becomes unavailable. SD-WAN has also been successfully employed to enable communications between Service VNFs spread across the globe to form SFCs [10].

The application of SD-WAN to monitor and control traffic traversing the 5G backhauls involves placing providers' ERs at the edges and at the central site (e.g. between the access network and the core network [11]). The ER function in the network serves as an SD-WAN router, making decisions on available backhauls and directing data traffic based on specific characteristics of these backhauls.

## 3 The ADAPTO Project

Virtualization of network functions, along with service function chaining, allows network infrastructure providers to achieve both agility and cost reductions by replacing traditional dedicated hardware devices with flexible software modules running in virtualized components. To satisfy the expectations of operators, however, sophisticated algorithms are required to properly map virtualized network functions onto the physical infrastructure that is composed of distributed datacenters located both in the core and at the edge of the network. This requires management and control plane functions that are able to take optimal resource allocation decisions within a fair time constraint and that properly take into account the different end-to-end requirements of diverse classes of applications. The ADAPTO project aims to define a management framework for 5G and beyond networks that allows infrastructure providers to properly manage the computational resources required to accommodate the dynamic instantiation of network functions by combining QoS requirements with at-scale energy-saving objectives.

In a telco network, an optimization function requires a comprehensive view of all entities involved, from network-level metrics to the characteristics of individual microservices composing a network function. Once a proper data strategy is in place, it becomes possible to monitor the behavior of each functionality active in the network by also tracking the resources consumed to deliver it.

In the following, we give a brief overview of the design goals of the ADAPTO project and an overview of the edge agent's architecture.

- **VNFs autoscaling**: The ADAPTO framework dynamically adjusts the allocated resources for Virtual Network Functions (VNFs) based on operational load. The primary aim is to increase resources when the workload is high, avoid overburdening available resources, and reduce them during low load periods, optimizing the overall energy consumption. This requires monitoring of VNF resource utilization and the development of intelligent approaches seeking to adapt the allocated resource to the variable workload dynamically. The ADAPTO framework investigates the two possible VNF scaling possibilities mentioned earlier in this paper: vertical and horizontal.
- **VNFs placement:** The placement and scaling of functions and applications in edge environments are constrained by limited computational resources. Consequently, apart from deploying these functions and applications at the edge, their deployment in a central cloud with an assumed pool of nearly limitless resources is also anticipated. An offloading agent is envisioned to detect when the edge resources are inadequate for accommodating the actual workload and utilize resources present in the central cloud.
- **Seameless Offloading:** Another objective is to mask the geographical dispersion of resources, aiming to access central cloud resources as if they were local, ensuring low latency and consistently meeting the Quality of Service (QoS) requirements for various applications. This involves employing multiple backhaul connections between the edge and the central cloud and utilizing SD-WAN Edge Routers to select the backhaul that facilitates achieving the required low latency.

Figure 2 illustrates the reference architecture. The UEs establish connections to the edge cloud via the RAN. Within the edge deployment, the UPF manages

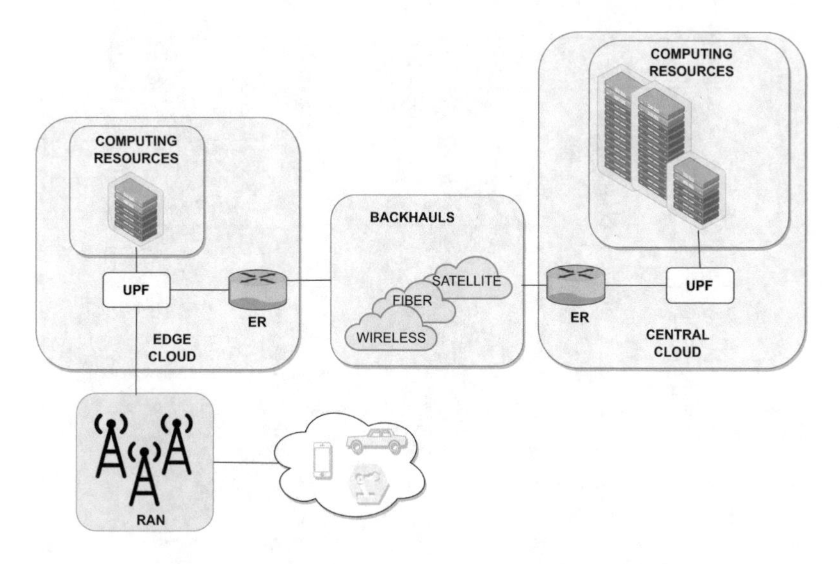


**Fig. 2.** Reference Architecture

traffic direction, either directing it to the edge cloud or the central cloud through the ER. Different ERs are employed - one situated at the edge and another at the central cloud - as in the SD-WAN paradigm. These ERs hold the responsibility of selecting the optimal backhaul among available options (such as Satellite, Fiber, or Wireless) to connect the edge and central cloud, aiming for the lowest latency possible. In Fig. 3, the architecture of the ADAPTO agent devised for the Edge Cloud is shown. In the following, the main components are described:

- **NFV Controller**: The NFV controller is tasked with continuous monitoring of resources at the edge, specifically tracking the percentage usage of memory, CPU, and storage across various virtual network functions (VNFs). It also oversees the overall utilization of computing resources to assess available computation capacity.
- **NFV Scaling Decision Engine**: This component holds the responsibility of determining whether a particular instance of an NFV should scale in or out based on the potential overloading of edge resources. Additionally, it assesses and decides to offload services to the central cloud in case the edge resources become insufficient.
- **SD-WAN Controller**: The SD-WAN controller's primary function is to monitor latency by employing probing mechanisms on the available 5G backhauls. It configures the ERs to select the 5G backhaul that aligns with the Quality of Service (QoS) requirements.
- **Backhaul Selection Decision Engine**: Activated when the decision is made to offload services to the central cloud, this engine determines the backhaul to utilize for reaching the central cloud based on backhaul monitoring. It employs an intelligent Reinforcement Learning (RL) agent, which will be further discussed in the subsequent section.

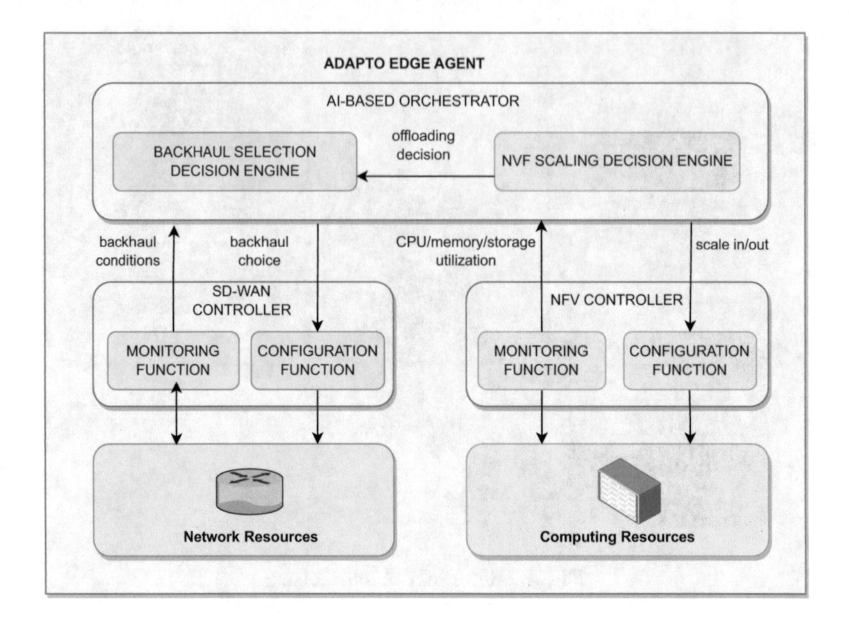


**Fig. 3.** Architecture of the ADAPTO agent deployed at the Edge.

### 3.1 Seamless Offloading Based on Reinforcement Learning

We propose an Edge-Cloud offloading mechanism that considers the possible overload of resources at the edge and the near real-time edge-to-cloud latency. It is assumed that cloud resources are unlimited, yet the edge-to-cloud latency varies and is generally higher than the time required to complete a task at the edge, in case the edge resources are not overloaded [12]. The proposed mechanism entails real-time monitoring of edge resources (CPU or memory) and end-to-end latency between the cloud and the edge. Specifically, an agent situated at the edge collects network device status and computational resource data from edge servers (e.g., CPU, memory, and storage utilization). These collected data, together with the WAN latency, serve as inputs for intelligent algorithms to determine whether a service should be processed at the edge or in the cloud.

It is crucial that when offloading the task to the cloud, the WAN transport chosen ensures the QoS required by the application. Specifically, our focus is on reducing edge-to-cloud latency to guarantee a seamless handoff between edge services and services in the central cloud. The solution proposed is based on RL, employing an agent and an environment as depicted in Fig. 4. The agent interacts with the environment, altering its state based on actions and receiving rewards or penalties for its behavior. This continuous trial-and-error process allows the agent to learn the best strategies for achieving objectives without prior knowledge of the network's details. The environment comprises the edge and central site that needs to communicate, the 5G backhauls, and the ERs. Reward functions are designed to reflect adherence or violation of network policies, where positive rewards are given to the agent when the policy is met and negative rewards are given when the policy is violated. Policies enable establishing a maximum allowed latency for the 5G backhaul connection to provide certain QoS levels for 5G applications. The control action involves instructing ERs to select the 5G backhaul suitable for forwarding traffic between the edge and the cloud. Thanks to the reward, the agent can learn which backhaul to use based on the current backhaul conditions, providing, in this way, a method for enhancing the QoS for 5G applications and enabling a seamless handoff.

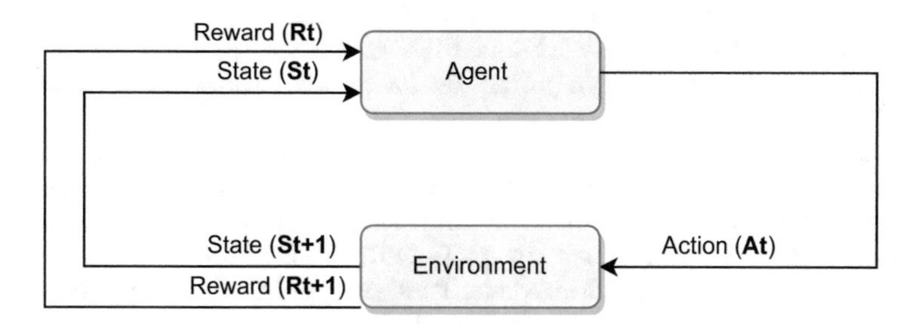


**Fig. 4.** Reinforcement Learning Loop

## 4 Conclusion

The design goals of the ADAPTO project revolve around the realization of 5G use cases such as URLLC, eMBB, and MTC while emphasizing energy efficiency. By synergizing the strengths of 5G Core Network SBA, advancements in NFV, 5G MEC, and SD-WAN, this project aims to establish a network framework tailored for telcos reliant on cloud-native architectures and next-generation networks. This paper has provided an overview of the foundational principles and technologies underpinning the ADAPTO framework, namely NFVs, SD-WAN, SBA, and 5G MEC. Additionally, it offers a high-level insight into the ADAPTO EDGE agent, focusing on the control and orchestration of computing resources and network elements to facilitate intelligent scaling, edge-to-cloud transitions, and optimal selection of 5G backhauls.

Furthermore, a specific aspect of the ADAPTO framework regards the strategic integration of a Reinforcement Learning (RL) agent at the edge. This intelligent agent assumes a pivotal role in enabling seamless edge-to-cloud transitions by proficiently selecting the most suitable 5G backhaul. Leveraging its decision-making capabilities, the RL agent can ensure the desired Quality of Service (QoS) for applications, thereby contributing to the realization of 5G promises for URLLC.

**Acknowledgment.** This work was partially supported by the European Union through the ADAPTO project, part of the RESTART program, NextGenerationEU PNRR, CUP E63C2 2002040007, CP PE0000001.

## References

1. Pham, Q.-V., et al.: A survey of multi-access edge computing in 5G and beyond: fundamentals, technology integration, and state-of-the-art. IEEE Access **8**, 116974–117017 (2020)
2. 3GPP: System architecture for the 5G System (5GS) (2022). v16.12.0
3. Harutyunyan, D., Behravesh, R., Slamnik-Kriještorac, N.: Cost-efficient placement and scaling of 5G core network and MEC-enabled application VNFs. In: 2021 IFIP/IEEE International Symposium on Integrated Network Management (IM), pp. 241–249. IEEE (2021)
4. Kekki, S., et al.: MEC in 5G networks. ETSI White Paper **28**(2018), 1–28 (2018)
5. Zhang, Q., Liu, F., Zeng, C.: Adaptive interference-aware VNF placement for service-customized 5G network slices. In: IEEE INFOCOM 2019-IEEE Conference on Computer Communications, pp. 2449–2457. IEEE (2019)
6. Corici, M., Chakraborty, P., Magedanz, T.: A study of 5G edge-central core network split options. Network **1**(3), 354–368 (2021)
7. Agbo-Adelowo, P., Weitkemper, P.: Analysis of different MEC offloading scenarios with LEO satellite in 5G networks. In: 2023 IEEE International Conference on Omni-layer Intelligent Systems (COINS), pp. 1–6. IEEE (2023)
8. Ahamed, Md.M., Faruque, S.: 5G backhaul: requirements, challenges, and emerging technologies. Broadband Commun. Netw. Recent Adv. Lessons Pract. **43**, 2018 (2018)
9. Hung, M.-H., Teng, C.-C., Chuang, C.-P., Hsu, C.-S., Gong, J.-W., Chen, M.-C.: A SDN controller monitoring architecture for 5G backhaul networks. In: 2022 23rd Asia-Pacific Network Operations and Management Symposium (APNOMS), pp. 1–4. IEEE (2022)
10. Leivadeas, A., Pitaev, N., Falkner, M.: Analyzing the performance of SD-WAN enabled service function chains across the globe with AWS. In: Proceedings of the 2023 ACM/SPEC International Conference on Performance Engineering, pp. 125–135 (2023)
11. Corici, M., et al.: SATis5 solution: a comprehensive practical validation of the satellite use cases in 5G. In: Proceedings of the 24th Ka and Broadband Communications Conference, Niagara Falls, ON, Canada, pp. 15–18 (2018)
12. Zhang, Y., Xu, C., Muntean, G.-M.: Revenue-oriented service offloading through fog-cloud collaboration in SD-WAN. In: GLOBECOM 2022-2022 IEEE Global Communications Conference, pp. 5753–5758. IEEE (2022)

# DEFEDGE: Threat-Driven Security Testing and Proactive Defense Identification for Edge-Cloud Systems

Valentina Casola[1], Marta Catillo[2], Alessandra De Benedictis[1(✉)], Felice Moretta[3], Antonio Pecchia[2], Massimiliano Rak[3], and Umberto Villano[2]

[1] Department of Electrical Engineering and Information Technology, University of Naples Federico II, Naples, Italy
{casolav,alessandra.debenedictis}@unina.it
[2] Department of Engineering, University of Sannio, Benevento, Italy
{marta.catillo,antonio.pecchia,villano}@unisannio.it
[3] Department of Computer Engineering, University of Campania Luigi Vanvitelli, Aversa, Italy
{felice.moretta,massimiliano.rak}@unicampania.it

**Abstract.** Nowadays, the edge-cloud (EC) paradigm is adopted in several domains, including manufacturing, health, and critical infrastructure management. Identifying existing threats and vulnerabilities of an EC system and determining appropriate countermeasures is a costly and time-consuming process due to the inherent system complexity and to the heterogeneity of involved assets. Moreover, even when appropriate security measures are enforced, attacks may still succeed because of the natural degradation of security mechanisms' effectiveness due to attackers' reconnaissance efforts and/or to unknown vulnerabilities coming into play. This paper describes the objectives of the DEFEDGE project, which aims to define a set of techniques for the development of secure and resilient edge-cloud systems and for their assessment based on a threat-driven approach. The main idea is to leverage the results of a guided threat modeling process to derive both the security controls and the mechanisms to be enforced, as well as the security tests to perform in order to verify the effectiveness of controls in place. Security controls selection and enforcement will follow Moving Target Defense principles. Security testing will exploit existing threat intelligence and attack patterns knowledge bases to derive a set of general-purpose attack procedures that can be suitably customized to test a target system. For the generation of attack procedures and their customization, the project will also explore machine learning techniques to infer new attack patterns and scenarios, in order to improve the overall testing effectiveness.

## 1 Introduction

The edge-cloud (EC) paradigm distributes part of the processing and storage burden of an IoT system among edge devices, highly-heterogeneous appliances that are closer to the data sources and are able to collect and pre-process data locally in order to reduce latency and bandwidth consumption. Nowadays, the EC paradigm is adopted in several domains, including manufacturing, health, and critical infrastructure management and,


L. Barolli (Ed.): AINA 2024, LNDECT 203, pp. 79–88, 2024.
https://doi.org/10.1007/978-3-031-57931-8_8

therefore, suitable security measures must be devised in the design of EC systems not only to prevent but also to withstand, recover from, and adapt to adverse conditions and attacks, in order to improve their overall resilience.

Unfortunately, identifying existing threats and vulnerabilities of an EC system and determining appropriate countermeasures is a costly and time-consuming process due to the inherent system complexity and to the heterogeneity of involved assets. Moreover, even when appropriate security measures are enforced that are able to address identified security requirements, attacks may still succeed because of the natural degradation of security mechanisms' effectiveness due to attackers' reconnaissance efforts and/or to unknown vulnerabilities coming into play [1].

Independently of the security measures implemented in a system, their assessment by means of suitable security testing techniques is fundamental to evaluate the extent to which they are able to thwart possible attacks. Among existing security testing techniques, penetration testing is particularly suited to verify the effectiveness of implemented enforced security controls and the impact of identified vulnerabilities as it emulates the behavior of an attacker by means of targeted attacks. Unfortunately, penetration testing is not always viable or convenient since it is mainly a human-driven procedure, whose validity heavily depends on the pentester's experience and skill.

In this paper, we introduce the DEFEDGE project, aimed at defining a set of techniques for the development of secure and resilient edge-cloud systems and for their assessment based on a threat-driven approach. The main idea is to leverage on the results of a guided threat modeling process to derive both the security controls and mechanisms to enforce as a mitigation for existing threats and the security tests to perform in order to verify the effectiveness of controls in place. In particular, security controls selection and enforcement will follow Moving Target Defense (MTD) principles [1,2], according to which the attack surface of a system is continually and proactively changed to reduce attack success probability. Security testing will exploit existing threat intelligence and attack patterns knowledge bases to derive a set of general-purpose attack procedures that can be suitably customized to test a target system. For the generation of attack procedures and their customization, the project will also explore machine learning techniques to infer new attack patterns and scenarios, in order to improve overall testing effectiveness.

The remainder of the paper is organized as follows. Section 2 discusses the project motivations and introduces the main challenges that the project intends to face related to (i) ensuring security and resilience via moving target defense methodologies and techniques, (ii) assessing the level of security and resilience of a CE system via a cost-effective penetration testing and (iii) dynamically generating sound and targeted attacks via Machine Learning. Section 3 illustrates the main project objectives and methodologies, while Sect. 4 summarizes the expected project impact and draws our conclusions.

## 2 DEFEDGE Research Challenges

As anticipated in the Introduction, DEFEDGE aims at developing a set of methods and techniques to design, implement and test secure and resilient edge-cloud systems. The project will explore proactive defense techniques to effectively enforce security and resilience and will leverage on security testing (with a special focus on penetration testing) to stress an EC system and the enforced countermeasures. Both defense and attack will strongly rely upon the results of threat modeling activities to identify more focused results. Achieving the project objectives entails addressing several scientific challenges, summarized in the following.

**Challenge 1 - Asset-Centric EC Threat modeling:** The first challenge addressed by the project is the definition of a methodology to automatically identify threats applicable to a specific EC system based on its architecture and asset properties.

An EC system consists of multiple heterogeneous interacting devices and sub-systems. The heterogeneity of involved devices, their distributed management and the inherent sensitivity of exchanged data make EC systems an attractive target for attackers, which can exploit vulnerabilities exposed by protocols, devices and applications [3]. At state, several industrial and scientific initiatives exist that aim to provide a taxonomy of threats related to the edge-cloud world[1] [2] [4,5]. Moreover, threat modeling automation is currently offered by several open and commercial tools, including Microsoft Threat Modeling Tool, OWASP Threat Dragon, IriusRisk and ThreatModeler. Despite the availability of taxonomies and tools, threat modeling is not a solved problem. In fact, most of the time, threat classifications are generic for a given technology (e.g., cloud computing, IoT, etc.) and are not directly linked to specific assets and related properties while this would be very useful to spot actual existing weaknesses and determine appropriate countermeasures. On the other hand, available tools are often tied to specific vendors and technologies and adopt proprietary threat taxonomies that are hardly comparable.

**Challenge 2 - Threat-Driven Resilience Enforcement:** A complete and fine grained threat model is a need for secure system design, but it is only the first step towards a secure architecture. The second challenge addressed by the project relates to identifying appropriate controls to enforce based on existing threats, along with suitable strategies to preserve the efficacy of controls in place over time by proactively changing the attack surface. It is quite clear indeed that, even when appropriate security controls are enforced within a system, security incidents may still happen, and in many cases it is just a matter of time, as attackers are able to gain more and more information on the system thanks to reconnaissance efforts. Resilience can be achieved by means of proactive approaches aimed at anticipating attackers' moves in order to thwart reconnaissance attempts, adapt to adverse conditions, preserve availability and recover from attacks. This is the goal of MTD [1,6], a proactive cyber-defense paradigm according to which a system's attack surface is continually changed over time by means of reconfiguration performed at different levels, to increase complexity and cost for attackers during reconnaissance activities, limit the exposure of vulnerabilities, and increase overall system resilience. Existing MTD approaches basically rely upon the principles of *diversity*, *redundancy* and *shuffling* (system settings at various layers are rearranged

[1] Microsoft Inc. IoT Security. https://azure.microsoft.com/en-gb/resources/cloud-computing-dictionary/what-is-iot/security/.

[2] Cloud Security Alliance. Top Threats to Cloud Computing: Egregious Eleven. https://cloudsecurityalliance.org/artifacts /top-threats-to-cloud-computing-egregious-eleven/.

during system operation), and are characterized by different impacts on system functionality and by different costs (in terms of both implementation cost and operation overhead) and benefits. Several MTD approaches have been proposed in the literature in the last decade, targeting cloud-based systems as well as IoT and edge systems [7]. They devise reconfiguration at different architectural layers including network, runtime environment, software, data format, platform and firmware. Despite the availability of MTD techniques targeting IoT and edge devices, which show that reconfiguration is actually feasible in the IoT and edge context even if generally harder to implement due to existing resource constraints, existing approaches essentially lack a rationale for the identification of the appropriate reconfiguration level and technique based on involved assets, on their capabilities, and on the actual threats that affect a system. Moreover, most existing approaches do not mention when reconfiguration should be triggered, depending for example on the current and forecast operating conditions. Finally, a few MTD solutions concretely address the problem of preserving service continuity during reconfiguration, which is one of the main factors preventing their actual adoption in real scenarios.

**Challenge 3 - Threat-Driven Penetration Testing (PT):** Independently of the security measures implemented in a system, their assessment by means of suitable security testing techniques is fundamental to evaluate the extent to which they are able to thwart possible attacks. The NIST SP-800-115 special publication [8] describes various security testing measures and provides guidelines for organizations on planning and conducting different types of security assessments. Among all, it defines a PT methodology model based on four phases: planning, discovery, attack, reporting. Despite the recent growing adoption of PT in several contexts and the availability of several PT methodologies and frameworks (e.g., PTES[3] , OWASP Testing Guide[4], OSSTMM[5], PTS[6], ISSAF), there is no state-of-the-art standard dedicated to the description of PT activities [9], while it is currently heavily dependent on testers' skills and experience. The main problem with PT is the long time necessary to test even a medium-size system: in fact, nowadays security consultants (pentesters) typically perform time-boxed assessments, which represents a limited and, often, very small attempt to address security at the end of the development cycle. In practice, PT can only identify a small representative set of possible security risks in a system. More often than not, it becomes prohibitive to fix any critical design security issues discovered with PT late in the life cycle, that is a point where both time and budget severely limit the remediation options [10]. Automation is a natural solution to obtain efficient penetration testing and it may also help increase the testing coverage, discovering further possible system vulnerabilities [11].

[3] Penetration Testing Execution Standard (PTES). http://www.pentest-standard.org/index.php/PTES_Technical_Guidelines.

[4] OWASP Testing project page. https://owasp.org/www-project-web-security-testing-guide/.

[5] Pete Herzog. Osstmm 3: The open source security testing methodology manual-contemporary security testing and analysis. https://www.isecom.org/OSSTMM.3.pdf. 2010.

[6] Penetration Testing Framework (PTS). http://www.vulnerabilityassessment.co.uk/Penetration%20Test.html.

An interesting example of automation in literature is given by MITRE Adversary Emulation Plans[7].

Our approach assumes that penetration testing should rely (i) on a synthetic model of the System under Attack (SuA) outlining all involved assets, (ii) on a detailed threat model of the SuA highlighting attack targets and (iii) on a knowledge base of attack patterns, techniques, tactics and tools (like CAPEC or ATT&CK). In this regard the open problems are: how to select, from a high level description of a threat, the possible attacks? How to automatically perform such attacks, relying on the information collected in the model? How to measure the completeness of the process? And, consequently, to which extent are the testing plan and its implementation complete?

**Challenge 4 - Machine-Learning (ML)-Based Attack Generation:** Successful automation of PT tasks requires not only an automated mechanism to reproduce previously-planned, static, attack graphs, but also to generate in an intelligent way variants of known attacks to discover new vulnerabilities. This is the point where AI and ML come into play, helping the generation of targeted dynamic attack trees and so providing a solution that goes beyond the simple automation of static penetration tests. The use of ML techniques is an active field of research in the contest of fuzzing techniques, customarily used to stress software so as to detect coding errors and security vulnerabilities [12, 13]. In the last few years, the use of ML-guided generation testing is spreading also in the PT context [14–16], relying -mainly but not exclusively - on Reinforcement Learning methodologies [17]. Nowadays, the use of machine learning (ML) techniques in cyberattack analysis and identification is a common practice; however, modern malwares and attacks can adapt to the detection systems and conceal their malicious behavior. This challenge relates to boosting the usability of ML and to define its role in threat-driven security testing and proactive defense. In a continuous battle among self-adaptable attacking systems and self-adaptable defensive systems, there is still a key role in humans that operate the system and define the attacking strategies. A key challenge addressed by the project is the automated generation of new attacks by means of ML applied to data coming from direct measurement, Cyber Threat Intelligence (CTI)-oriented knowledge bases (like ATT&CK) and indicators of compromise (IoC) from open communities through platforms such as MISP. It is a fact that ML can be used to address massive data: however, there are several challenges that prevent from fruitfully applying ML in production networks for attack generation, such as imperfection of the training data, representativeness and uncertainty of normal baselines, and tuning of ML algorithms.

## 3 DEFEDGE Objectives and Methodologies

In this section, the main objectives posed by the project with respect to the challenges discussed in Sect. 2 are illustrated, and the methodologies that will be followed are sketched. The main project concepts are sketched in Fig. 1.

[7] Mitre Corporation. Adversary Emulation Plans. https://attack.mitre.org/resources/adversary-emulation-plans/.

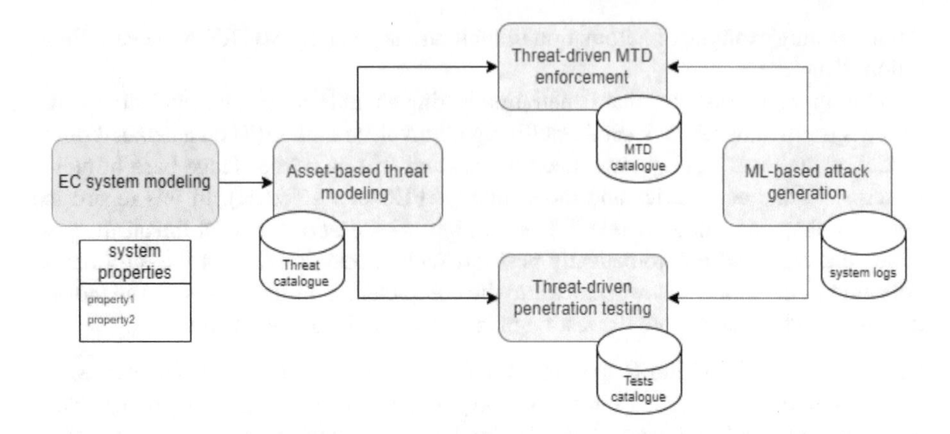


**Fig. 1.** DEFEDGE project main concepts

**Addressing Challenge 1 - Asset-Centric EC Threat Modeling:** Regarding the challenges posed by threat modeling in EC systems, two main objectives have been identified in the project:

- Objective 1.1: to define a comprehensive yet simple model of edge-cloud systems, able to identify and describe those assets and asset properties that are relevant to threat modeling.
- Objective 1.2: to build a catalogue of technical threats associated with specific assets and asset properties in order to systematize existing disjointed knowledge bases on threats.

To pursue the first objective, the project will leverage and extend the MACM (Multi-purpose Application Composition Model) formalism [5] to define a modeling framework for edge-cloud systems, taking into account both architectural and technological aspects based on recent technology advancements. With regard to the second objective, the project will first carry out a thorough analysis of existing initiatives (both by Industry and Academia) on threat classifications and taxonomies for edge, IoT and cloud systems. Based on the analysis results and taking into account the system model identified as the main outcome of Objective 1.1, a list of unique and disjoint threats explicitly mapped to assets and asset properties will be derived. To collect such threats, the project will define a threat catalogue structure specification and provide an open threat catalogue implementation, and will deliver a threat modeling tool to automatically retrieve applicable threats based on the system model.

**Addressing Challenge 2 - Threat-Driven Resilience Enforcement via MTD:** As anticipated in the previous section, challenges identified in the design of secure and resilient EC systems are addressed by DEFEDGE by leveraging on the MTD approach. Two complementary objectives have been established:

- Objective 2.1: to define a set of MTD mechanisms and techniques applicable to different assets based on existing threats and asset properties.

- Objective 2.2: to define a reconfiguration strategy for the triggering of appropriate reconfigurations based on relevant conditions.

In order to identify suitable MTD mechanisms and techniques for EC systems, the project will analyze the state of the art on available MTD solutions identifying the possible architectural levels at which reconfiguration can be applied and the general mechanisms that can be adopted. For each high-level mechanism (e.g., address shuffling), specific techniques will be identified (e.g., address randomization with the use of hash functions) tailored to involved assets. Mechanisms will be associated to specific security controls and mapped to threats as possible means to mitigate them. Finally, a set of metrics will be derived to evaluate the footprint of each technique, its effectiveness and the associated reconfiguration overhead. The main results of this activity will be the definition of a taxonomy of MTD techniques applicable to the various assets of an edge-cloud system, the development of a catalogue of MTD techniques associated with assets and threats, and the identification of set of metrics for MTD technique evaluation.

Regarding Objective 2.2, the project will take into account several factors to build a reconfiguration strategy: (i) the natural degradation of the security level due to the exposure of the attack surface for a given period of time during which an attacker can gain knowledge on it and exploit some exposed vulnerability; (ii) the possible occurrence of a security incident; (iii) system-specific risk indicators such as the mission, the type of enforced security policies, the expected damage, the skills required from the attacker; (iv) the cost of reconfiguration in term of the overhead introduced and of possible outage periods. Based on these factors, a set of rules and algorithms for reconfiguration planning and execution will be released.

**Addressing Challenge 3 - Threat-Driven PT:** Regarding the challenges related to security testing, two main objectives have been established:

- Objective 3.1: to define a process to automatically generate a list of concrete attacks from system and threat models.
- Objective 3.2: to automatically implement an attack plan and validate the process.

Towards the first objective, the project will perform a systematic review of existing knowledge bases of attacks, attack patterns, attack techniques, attack tactics and will identify and/or develop a synthetic way to describe a specific attack and, accordingly, an attack plan. The manual process required for generating the relationship among threats and attacks (described in [11]) will be automated, in case adopting IA-based techniques, in order to create a continuously growing mapping of threats-assets-attacks. Finally, an expert system will be designed in order to automatically generate attack plans from system and threat models. As a result of these activities, the following outcomes are expected: (i) a knowledge base that collects a mapping among threats, assets and attacks, (ii) an automated procedure that enriches the knowledge base in an automated way, (iii) an expert system that automatically produces testing plans. In addition to the main results, the project may produce additional side results: (i) a language for detailed attack description, (ii) a language for penetration testing plan description and (iii) a set of metrics for security testing plan completeness evaluation.

Regarding the automated execution of attack plans, the project will design and develop a system able to receive as an input a penetration testing plan and a detailed system model and automatically perform the planned attacks, generating an automated report. The project will identify a set of target systems, in the context of the edge-cloud domains, to be used as case studies. The case studies will be of three categories: (i) Toy systems, (ii) real small systems, (iii) real complex systems. The proposed solution will be developed in an agile way, focusing on the toy systems, and validated by a subsequent application to small and complex real systems. A tool able to implement automated tests on the basis of a predefined penetration testing plan and of a system model will be developed, capable of producing a set of penetration testing reports on real environments. In addition to the main results, the project may produce additional side results: (i) Discover new vulnerabilities, (ii) enrich existing attacking tools and (iii) develop new tools for penetration testing (attacking tools).

**Addressing Challenge 4 - ML-Based Attack Generation:** Challenges related to the generation of attacks via ML will be addressed by the project by pursuing three objectives:

- Objective 4.1: to transform raw data into suitable representations and to address the gap between ML development and operation for security testing.
- Objective 4.2: to improve the detection of anomalies by the convergence of different learning approaches and explainability.
- Objective 4.3: to support security testing by means of ML-based attacks in order to validate defense and countermeasures.

Regarding the first objective, the project will develop a uniform framework for handling and transforming structured and unstructured data into representations suitable for automated learning. The framework will also include novel methodologies to improve the generalization of the models obtained to support security testing. An ecosystem of transformation techniques and supplementary tools will be provided to boost the usability of ML.

The accurate detection of anomalies from data (both from the attacker and defender perspectives) is crucial to gain insight on the runtime and to drive the generation of effective attacks on the one side and the triggering of reconfigurations on the other. The project aims to improve anomaly detection capabilities by leveraging the complementarity of different learning paradigms, with a focus on supervised and semi-supervised approaches in order to come up with the generation of zero-days attacks. This will be achieved through the combination of concepts such as outlier detection, deep autoencoding and reconstruction error. The project will search for novel approaches to reduce the gap between the outcome of anomaly detection and its interpretation through human-understandable explanations on the conditions that make an attack successful. As a result of these activities, anomaly-based attack detection techniques and methods to support explainability will be developed.

Finally, related to the Objective 4.3, the project will develop novel ML-based approaches to generate attack cases for security testing. In particular, ML algorithms will be used to infer the parameters that make an attack successful against the defense in hand. The core of the methodology is to feed ML with various data sources (e.g.,

CTI, IoC and runtime monitors outputs), in order to form a control loop aiming to tune the attack based on the observed outcome of the system. The project will capitalize on recent advancements on network traffic replay, generative adversarial networks (GAN) and attack tools to develop methods and tools for ML-based attack generation.

## 4 Conclusions

In this paper the main research challenges and objectives of the DEFEDGE project (December 2023 - October 2025) have been illustrated. The project will advance the state of art related to the enforcement and assessment of security and resilience of edge-cloud systems by pursuing a threat-driven approach, by which suitable defense mechanisms - based on MTD principles - and testing techniques - based on automated PT processes leveraging also ML-aided attack generation - are selected, planned and (automatically) implemented in an EC system based on existing threats. Project main results include a security-oriented modeling framework for EC systems, a set of catalogues and tools for security enforcement and testing automation (i.e., a threat catalogue and threat modeling tool, an MTD techniques catalogue and associated metrics, a knowledge base mapping attacks onto threats and assets and a framework to automatically build and execute penetration tests), and a set of methods and tools for data transformation, anomaly detection and attack generation based on ML. These outcomes will help (i) prevent/mitigate incidents thanks to novel approaches able to continually shift the system attack surface, (ii) dynamically assess the system security posture by means of effective security testing procedures, and (iii) reduce the impact of unavoidable incidents by providing administrators with early and accurate detection and event correlation capabilities. This will result in faster and more targeted responses to incidents, which -in turn- decreases latency and costs of resilience.

The outcomes generated by the DEFEDGE project will have two major groups of international stakeholders, namely the scientific community interested in topics as EC security and resiliency enforcement and evaluation, and the commercial community including financial service providers, CPS-IoT companies and Cloud providers. Finally, the project aims to contribute to international initiatives on industrial cybersecurity sponsored by the European Commission, such as the European Cyber-Security Organization (ECSO), and is perfectly in line with the directions endorsed by the European programs. Just to mention the *New EU Cybersecurity Strategy*,[8] aiming to develop the industrial and technological resources to make digital entities more resilient and the Digital Single Market initiative pushed by the European Commission, in which data protection regulation plays a big role (GDPR) as well as the Cybersecurity Act approved in June 2019, which is expected to boost the adoption of advanced security solutions.

**Acknowledgements.** This work has been partially funded by the European Union - NextGenerationEU - National Recovery and Resilience Plan (NRRP) - MISSION 4 COMPONENT 2, INVESTIMENT N. 1.1, CALL PRIN 2022 D.D. 1409 14-09-2022 - (Threat-driven security testing and proactive defense identification for edge-cloud systems). PROJECT CODE: P2022TT7A7. CUP. E53D23016380001.

[8] https://ec.europa.eu/commission/presscorner/detail/en/IP_20_2391.

## References

1. Jajodia, S., et al.: Moving Target Defense: Creating Asymmetric Uncertainty for Cyber Threats, vol. 54. Springer, Cham (2011)
2. Casola, V., De Benedictis, A., Albanese, M.: A multi-layer moving target defense approach for protecting resource-constrained distributed devices. In: Bouabana-Tebibel, T., Rubin, S. (eds.) Integration of Reusable Systems. Advances in Intelligent Systems and Computing, vol. 263, pp. 299–324. Springer, Cham (2014). https://doi.org/10.1007/978-3-319-04717-1_14
3. Ometov, A., et al.: A survey of security in cloud, edge, and fog computing. Sensors **22**(3), 927 (2022)
4. Xiao, Y., et al.: Edge computing security: state of the art and challenges. Proc. IEEE **107**(8), 1608–1631 (2019)
5. Ficco, M., Granata, D., Rak, M., Salzillo, G.: Threat modeling of edge-based IoT applications. In: Paiva, A.C.R., Cavalli, A.R., Ventura Martins, P., Perez-Castillo, R. (eds.) Quality of Information and Communications Technology. Communications in Computer and Information Science, vol. 1439, pp. 282-296. Springer, Cham (2021). https://doi.org/10.1007/978-3-030-85347-1_21
6. Cho, J.-H., et al.: Toward proactive, adaptive defense: a survey on moving target defense. IEEE Commun. Surv. Tutorials **22**(1), 709–745 (2020)
7. Navas, R.E., et al.: MTD, where art thou? A systematic review of moving target defense techniques for IoT. IEEE Internet Things J. **8**(10), 7818–7832 (2020)
8. Scarfone, K., et al.: Technical guide to information security testing and assessment. NIST Spec. Publ. **800**(115), 2–25 (2008)
9. Knowles, W., Baron, A., McGarr, T.: The simulated security assessment ecosystem: does penetration testing need standardisation? Comput. Secur. **62**, 296–316 (2016)
10. Arkin, B., Stender, S., McGraw, G.: Software penetration testing. IEEE Secur. Priv. **3**(1), 84–87 (2005)
11. Rak, M., Salzillo, G., Granata, D.: ESSecA: an automated expert system for threat modelling and penetration testing for IoT ecosystems. Comput. Electr. Eng. **99**, 107721 (2022)
12. Lin, P.-Y., et al.: ICPFuzzer: proprietary communication protocol fuzzing by using machine learning and feedback strategies. Cybersecurity **4**(1), 1–15 (2021)
13. Wang, Y., et al.: A systematic review of fuzzing based on machine learning techniques. PLoS ONE **15**(8), e0237749 (2020)
14. McKinnel, D.R., et al.: A systematic literature review and meta-analysis on artificial intelligence in penetration testing and vulnerability assessment. Comput. Electr. Eng. **75**, 175–188 (2019)
15. Confido, A., Ntagiou, E.V., Wallum, M.: Reinforcing penetration testing using AI. In: 2022 IEEE Aerospace Conference (AERO), pp. 1–15. IEEE (2022)
16. Jiao, J., Zhao, H., Cao, H.: Using deep learning to construct auto web penetration test. In: 2021 13th International Conference on Machine Learning and Computing, pp. 59-66 (2021)
17. Ghanem, M.C., Chen, T.M.: Reinforcement learning for efficient network penetration testing. Information **11**(1), 6 (2019)

# Swarmchestrate: Towards a Fully Decentralised Framework for Orchestrating Applications in the Cloud-to-Edge Continuum

Tamas Kiss[1(✉)], Amjad Ullah[2], Gabor Terstyanszky[1], Odej Kao[3], Soren Becker[3], Yiannis Verginadis[4,5], Antonis Michalas[6], Vlado Stankovski[7], Attila Kertesz[8], Elisa Ricci[9], Jörn Altmann[10], Bernhard Egger[10], Francesco Tusa[1], Jozsef Kovacs[1,11], and Robert Lovas[11,12]

[1] University of Westminster, London, UK
{kisst,terstyg,f.tusa,j.kovacs}@westminster.ac.uk,
jozsef.kovacs@sztaki.hu

[2] Edinburgh Napier University, Edinburgh, UK
a.ullah@napier.ac.uk

[3] Technische Universitaet Berlin, Berlin, Germany
{odej.kao,soeren.becker}@tu-berlin.de

[4] Institute of Communications and Computer Systems, National Technical University of Athens, Athens, Greece
jverg@aueb.gr

[5] School of Business, Department of Business Administration, Athens University of Economics and Business, Athens, Greece

[6] Tampere University, Tampere, Finland
antonios.michalas@tuni.fi

[7] University of Ljubljana, Ljubljana, Slovenia
vlado.stankovski@fri.uni-lj.si

[8] FrontEndArt Software Ltd, Szeged, Hungary
attila.kertesz@frontendart.com

[9] Fondazione Bruno Kessler, Trento, Italy
eliricci@fbk.eu

[10] Seoul National University, Seoul, South Korea
jorn.altmann@acm.org, bernhard@csap.snu.ac.kr

[11] Institute for Computer Science and Control (SZTAKI), Hungarian Research Network (HUN-REN), Budapest, Hungary
robert.lovas@sztaki.hu

[12] John Von Neumann Faculty of Informatics, Óbuda University, Budapest, Hungary

**Abstract.** Collecting and analysing large amounts of data in the Cloud-to-Edge computing continuum raises novel challenges that traditional centralised orchestration solutions cannot handle efficiently. To overcome the limitations of current centralised application management approaches, this paper presents a fully decentralised application-level orchestrator, based on the notion of self-organised interdependent Swarms. Application microservices are managed in a dynamic Orchestration Space by decentralised Orchestration Agents, governed by distributed intelligence that provides matchmaking between application requirements


L. Barolli (Ed.): AINA 2024, LNDECT 203, pp. 89–100, 2024.
https://doi.org/10.1007/978-3-031-57931-8_9

and resources, and supports the dynamic self-organisation of Swarms. Knowledge and trust, essential for the operation of the Orchestration Space, are managed through blockchain-based trusted solutions and the utilisation of emerging methods such as Self-Sovereign Identities (SSI) and Distributed Identifiers (DID). End-to-end security of the overall system is assured by utilising state-of-the-art cryptographic and privacy-preserving data analytics algorithms. A digital twin, that runs in parallel to the physical system, further improves its behaviour with predictive feedback. The presented concept is going to be implemented in the EU-funded Swarmchestrate project that starts in 2024.

## 1 Introduction

Orchestration is referred to as the coordination and management processes of physical computational resources of an infrastructure environment to serve the application requirements, as defined by Jiang et al. [1]. The definition of infrastructure environment here is contextual. For example, Tomarchio et al. [2] discussed cloud orchestration and, therefore, referred to it as coordination and management processes of cloud resources, whereas Costa et al. [3] discussed this within the context of the fog paradigm. Irrespective of the underlying infrastructure environment, the overall goal of the orchestration system in terms of users' business requests is to ensure the meeting of Quality of Service (QoS) goals of applications. Therefore, in the absence of a universally agreed definition, a Cloud-to-Edge orchestration system is responsible for providing simultaneous access to the heterogeneous resource landscape of the continuum for the automation of application deployment and management over the resource landscape. It guarantees QoS goals, by handling the required complex tasks of resource selection, allocation, deployment and monitoring, and the run-time reconfiguration control of the resources and applications.

Although there are several research efforts and even relatively mature solutions providing orchestration capabilities, none of these can fulfil completely the highly dynamic and complex requirements imposed by the Cloud-to-Edge continuum.

The efficient and effective management and processing of the large amounts of data generated at the edges of the network must deal with versatile requirements. Some data need to be processed locally due to regulations, privacy issues and/or performance constraints, while some others may require access to long-term sophisticated computational cloud resources. Applications running in such systems have a wide range of requirements, including the execution of low-latency analytics closer to the data source, privacy sensitivity, context awareness, time and location awareness, as well as the need for simultaneous access to geographically distributed arrays of sensors, remote localised heterogeneous computational resources, and large-scale on-the-fly allocated multi-cloud infrastructure. Therefore, a new generation of orchestration tools and solutions is needed to handle this complexity efficiently and to take into consideration this dynamically changing set of complex requirements in an intelligent way.

All currently available orchestration tools, responsible for deploying and managing data processing applications in the Cloud-to-Edge computing continuum, are based on a certain level of centralisation [4]. Such centralisation, while relatively easy to implement, carries several disadvantages. The central component can become a single point of failure, can be easily overloaded as the system scales, and provides a good target for security attacks [5]. Additionally, such a centralised approach does not fit well with the highly distributed and dynamically changing nature of the computing environment. A centralised management approach cannot react fast enough to some changes in local environments (e.g. volatility of resources) and cannot support fast adaptation of resources and application requirements (e.g. due to the movement of certain computing elements).

An alternative is a decentralised self-adaptive system that can be aware of its surroundings and can organise and reorganise itself without any central control and management. While the implementation of such a system is complex, recent advances in various fields of computing, including Swarm computing, distributed AI, distributed ledger systems and decentralised identity management, now enable the efficient realisation of such an approach.

To address these challenges Swarmchestrate, a new EU-funded research project kicking off in 2024, aims to combine and extend the above-mentioned emerging technologies and create a completely decentralised autonomous and self-organised application management system. The approach applied by Swarmchestrate is fundamentally new to application orchestration and suitable to manage hyper-distributed applications that span across large distances and the different layers of the dynamic compute continuum.

The rest of this paper overviews the current efforts towards decentralised orchestration, introduces the high-level concepts and fundamental building blocks of the Swarmchestrate framework, and outlines its main components that need to be implemented.

## 2 From Centralised to Decentralised Orchestration

Existing Cloud-to-Edge orchestration solutions can be divided into three categories: static [6], rule-based [7] and machine learning (ML) based [8] intelligent approaches. Static approaches place the burden on system engineers for structuring the system, as they statically map the different parts of an application to different resources of the compute continuum. Rule-based approaches embed some predefined threshold-based rules that help in determining the selection of resources. Lastly, ML-based approaches make informed decisions based on the collected data at runtime. For a more detailed list of different orchestration solutions from these categories, refer to Ullah et al. [9]. Most of these existing solutions, irrespective of their individual characteristic and underlying implementation techniques, follow a centralised model where a central entity, usually running in the cloud, collects data for further decision-making from the entire compute continuum.

In contrast to the centralised model, there are only a few solutions that follow the decentralised approach. Some examples include HYDRA [10] and Caravela [11] that provide a peer-to-peer (P2P) network of nodes where each node is both orchestrator and resource; mF2C [7] that follows a hierarchical architecture, where different agents work at different layers of the ecosystem and facilitate proactive decision making; Ozyar et al. [12] that utilises Blockchain to ensure security, however, its scope is limited to container placement at the edge layer; and finally, EPOS Fog [13], a multi-agent system where each node has its own software agent that defines which service is deployed on which host in the neighbourhood of the agent.

The Swarmchestrate project has the ambition to evolve the concept of decentralised Cloud-to-Edge orchestration in dimensions: (1) Elaborate the concept of decentralised orchestration from an application centric approach, in contrast to the currently available resource-oriented orchestration solutions that target and optimise resource provider objectives. (2) Develop novel standards and protocols for the collaboration of the decentralised orchestrators based on Swarm computing principles and the collaboration of multiple Swarms for the fulfilment of the overall objectives of the target applications. (3) Develop novel deployment and reconfiguration strategies with the aim of optimising application centric objectives based on certain requirements, e.g. application topology, fault tolerance, performance goals and various contextual constraints, such as resources, energy utilisation and geographical location.

The developed algorithms and solutions will be utilised to transform a centralised open-source Cloud-to-Edge orchestrator, called MiCADO-Edge [14], into a fully decentralised solution based on Swarm computing principles. The new decentralised orchestrator will then be applied in several use cases to improve wastewater manhole management, create a metaverse digital twin of natural habitat, provide more efficient parking space management, and improve video scene analytics in cities.

## 3 The Proposed Swarmchestrate Architecture

The Swarmchestrate project aims to develop a novel and innovative decentralised application-level orchestration solution with the potential to change the fundamentals of how applications are managed and executed in the highly dynamic Cloud-to-Edge compute continuum. Within the decentralised orchestration spectrum, Swarmchestrate's approach follows an application-centric view in contrast to the currently available resource-oriented orchestration solutions that target and optimise resource provider objectives.

The overall methodology of Swarmchestrate has been developed to address the inherent highly decentralised nature of the compute continuum, which is formed of several heterogeneous resources, spanning multiple administrative domains, that exhibit highly dynamic behaviour in terms of capacity and availability. Therefore, the development of intelligent mechanisms for the orchestration of applications deployed within this continuum becomes essential. Application owners do not have to be aware of the complexity of the underlying resource infrastructure. In fact, they can use a single and uniform high-level and interoperable descriptor that incorporates both the topology and the constraints of an application, as well as the related optimisation goals and performance requirements. Swarmchestrate abstracts the low-level details of the Cloud-to-Edge continuum and allows users to run their applications' microservices in a multi-domain infrastructure while taking advantage of its built-in trust and guaranteed security features.

Based on these essential aspects, Fig. 1 presents a high-level architecture, depicting the overall vision of Swarmchestrate and how it works. From a structural viewpoint, the overall system is divided into four distinctive parts, comprising of (1) Application view – dealing with application specification, (2) Orchestration space – handling all core functions of orchestration, (3) Trusted knowledge management – secure handling and management of all system-wide knowledge and interface required to enable trust and

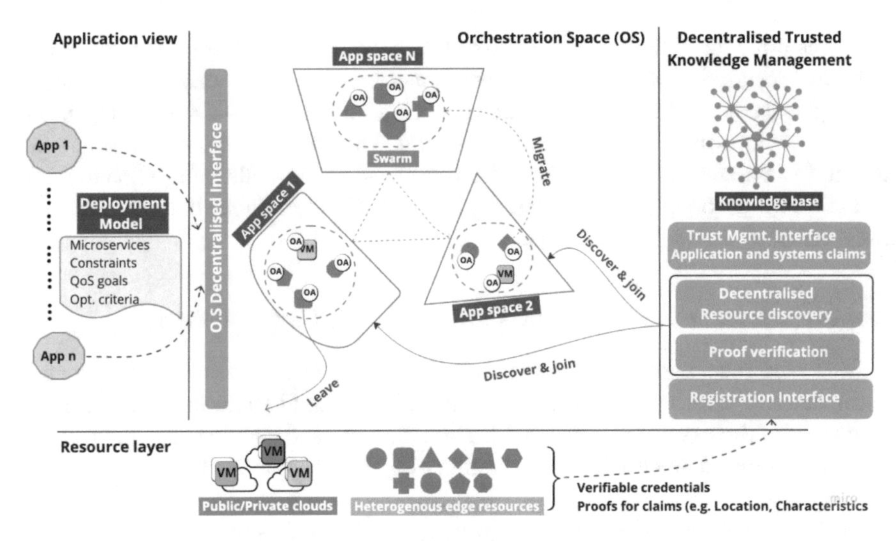


**Fig. 1.** High-level Architecture of Decentralised Orchestration in Swarmchestrate

transparency, and (4) Resource layer – representing resources of the continuum that span across multiple cloud environments and non-cloud layers. The technical implementation of these parts is being realised using the following fundamental concepts described below.

### 3.1 Application View

Within the Application view, DevOps can describe their applications, including resource needs, governing policies and QoS requirements. There are several such specification formats, especially for cloud computing environments, and these are essential to ensure portability and avoid vendor lock-in. However, the emergence of the Cloud-to-Edge model raises the quintessential need for the extension of such cloud description languages. There are clear shortcomings when capturing deployment, monitoring, contextualisation, and reconfiguration aspects of Cloud-to-Edge applications. Current approaches cannot sufficiently cope with the concept of decentralisation required in application management and orchestration, especially when distributed Cloud-to-Edge continuum resources are involved.

To improve this, Swarmchestrate introduces modelling artefacts that can extend a standards-based specification, such as TOSCA [15], to cover all the necessary specification details, allowing the decentralised and automated DevOps platform to manage and orchestrate microservices-based applications in the heterogeneous and dynamic Cloud-to-Edge continuum. The project concentrates on enhancing the specification for the entire application lifecycle, allowing application and resource specification across the Cloud-to-Edge continuum, which is still missing at large.

In Swarmchestrate, an application can be of different types (e.g., batch-based, or long-lasting services) and can consist of multiple container-based microservices, whose blueprints are possibly hosted in different repositories. As represented in Fig. 1, by using the above-mentioned enhanced modelling features, application owners can describe a deployment model of their applications consisting of the application topology and the high-level description of its contextual requirements in terms of resources (e.g., security, geographical constraints), the application optimisation criteria and the related QoS goals in terms of performance (e.g., latency, trust). Such an approach takes care of automating the instantiation of the required microservices on the underlying Cloud-to-Edge infrastructure according to the given deployment model, without the need of further involvement of the application owner.

### 3.2 Orchestration Space - Decentralisation, Swarms and Intelligence

In Swarmchestrate, the applications are submitted to the Orchestration Space, which is a distributed entity with no central access point. The notion of Orchestration Space is the confluence of three concepts – decentralisation, Swarms and intelligence – for achieving efficient, optimised and trusted orchestration of applications in the Cloud-to-Edge ecosystem.

As it was analysed in Sect. 2, the majority of existing orchestration solutions are typically based on centralised architectures. Whilst such centralised solutions have a number of benefits, they do not fit well with the distributed nature of the Cloud-to-Edge continuum, which requires a more decentralised orchestration approach. Resources closer to the edge of the network are typically volatile, their network connection may be lost from time to time, and their processing and data storage capabilities are limited. Therefore, more emphasis must be put onto local decision-making and to the collaboration of multiple interacting entities every time global decisions related to the behaviour of applications need to be made. To address these issues, Swarmchestrate follows a decentralised approach towards orchestration where multiple players are responsible for executing the core functions of orchestration in the Cloud-to-Edge ecosystem in place of a centralised entity. This notion of decentralisation will be realised through the implementation of Swarms.

The key characteristic of Swarm computing [16] is the emergence of the collective behaviour and intelligence of individual agents as a result of interactions between them, rather than being explicitly controlled by a central entity. The usage of distributed agents enables a self-organised, highly scalable, and adaptable orchestration approach, which fits perfectly with the highly dynamic and distributed nature of Cloud-to-Edge systems. The common goal that the distributed agents are aiming to achieve in this scenario is the execution of orchestration functions for the applications submitted to the Orchestration Space.

Swarms are commonly associated with the concept of close proximity [17] meaning that agents geographically located near to each other can come together to form a Swarm dynamically, in order to cooperate to the completion of a task. However, in the Cloud-to-Edge compute continuum, an application – often consisting of multiple interconnected microservices – requires simultaneous access to resources distributed across the different layers of the compute continuum. Based on this aspect, Swarmchestrate extends the concept of Swarm-formation from close proximity to logical proximity [18]. This is determined based on the application's requirements and characteristics, i.e., resource requirements (such as CPU, memory, storage), security requirements, locality, performance, availability, energy constraints, trust factors, etc., instead of considering the resources' geographical distribution only. In this regard, we formalise the semantics for logical proximity based on application requirements and characteristics of involved resources and further define the mechanism of self-organisation based on logical proximity. We are also defining protocols for intra- and inter-swarm coordination.

In Orchestration Space, Swarms can be formed based on the above-mentioned concept of logical proximity. They are self-organised and fully dynamic, as resources can join or leave a Swarm based on their changing requirements/availability (or preferences) and/or dynamically changing application characteristics. Furthermore, based on the notion of logical proximity, we also envision that a single resource may be part of more than one Swarm. Each Swarm aims to fulfil the requirements of a particular application within the Orchestration Space. Hence, a Swarm provides this notion of Application Space (see Fig. 1) that is potentially changing at any point in time, based on application requirements and resource behaviours. Although Swarms can change, the overall lifetime of the Application Space is directly correlated with the application's lifetime. After completing/terminating the application, the Swarm dissolves as a result of collective self-organising decisions.

Swarms within Orchestration space are aware of each other, hence, they can also influence each other's behaviour. For example, a resource that is part of multiple Swarms can become overloaded as a result of the load in one particular Swarm, ultimately affecting the performance of other Swarms too. Therefore, strategies for inter-Swarm coordination are further formalised to achieve the overall objectives at global level.

From a technical viewpoint, Orchestration Agents (OA) in Swarms are responsible for picking up submitted applications and their microservices, and carrying out the tasks. OAs are attached to microservices and responsible for the self-organisation of the Swarms and for the inter-Swarm communication, as described above. In the absence of a central entity, these agents interact with each other based on certain simple principles and are able to share information within and across Swarms. Through the holistic intelligence based on the interaction and cooperation amongst OAs, the application-level objectives can be achieved.

Swarmchestrate exploits the use of distributed AI techniques to establish Swarm intelligence systems with an aim to optimise the overall dynamics of Swarms. This includes aspects such as the dynamic formation of Swarms, the individualistic behaviour of OAs, principles for intra- and inter-Swarm coordination, interaction with the environment, information sharing, and adaptation in case of changes in the operating conditions.

### 3.3 Trusted Knowledge Management

Trust in a decentralised environment is difficult to achieve. Swarmchestrate intends to generate various verifiable credentials/presentations and proofs that can be used in the context of obtaining and providing trust (e.g. proof of presence, proof of location, proof of computing capabilities). The possibility to generate various Zero-Knowledge proofs is considered across all levels of the applications. Swarmchestrate deals with trust as the most essential and fundamental pillar of the platform.

For this purpose, and to support the overall philosophy of decentralisation, Swarmchestrate develops a Blockchain-based decentralised knowledge and trust engine/infrastructure, which is shown on the right-hand side of Fig. 1. The role of this infrastructure is twofold: it is responsible for the global handling and management of knowledge, as well as for facilitating overall transparency and assuring trust amongst the system, the distributed resource layer and the users' applications running on the Swarmchestrate platform. The functionalities of this persistent trusted knowledge management infrastructure are available and used by both the Resource and the Orchestration Space layers.

Relevant sets of trust attributes, essential for the transparent and trustworthy interactions amongst the entities of the system within the context of orchestration in the Cloud-to-Edge compute continuum, are the subject of an initial investigation carried out by the project. This aims at the development of formal models of trustworthiness that help us in guaranteeing the dependable and trusted interaction between entities, stakeholders and services in a decentralised environment. These formal models provide foundation to the development of evidence- and blockchain-based trust management solutions using methods of SSI (Self-Sovereign Identity) [19] and DID (Decentralised Identifier) [20]. Using such a solution, identities are created and associated with various functionalities when resources join Swarmchestrate, so that these can only operate under circumstances where proper rights are given to them. Hence, it will be demonstrated that full transparency, traceability and privacy-preserving identity and role management can be achieved based on the above mechanisms.

Furthermore, the Blockchain-based knowledge base manages the overall information related to resource descriptions, system interactions and decision making, as well as applications using smart contracts and decentralised oracles. Therefore, at any point in time, the resources can be discoverable based on various contextual attributes and trust factors for Swarms' formation, as well as for the verification of system and application-level claims through external entities.

### 3.4 Resources Layer

In the Swarmchestrate concept, a resource is referred to as any computational resource ranging from a dynamically created virtual machine in the cloud, or a physical node that exists at any layer of the compute continuum, to an intelligent sensor with processing capability. Furthermore, a resource can also be a pre-deployed software service running on some dedicated hardware. These resources, shown at the bottom of Fig. 1, can be heterogeneous and can belong to different administrative domains. A resource can be characterised by various contextual attributes such as hardware characteristics, supported operating system, geographical location, mobility and battery power. Such characteristics are used to identify the suitability of a resource for a particular task at any given time.

In Swarmchestrate, a resource can be considered as a trusted resource, once it gets registered using DID-based identities and becomes able to produce proofs for each of the claims, i.e., the assertions made regarding its characteristics. These proofs are verified before the formation of the Swarm and/or when the resource joins a particular Swarm. The verification aims to establish the truth of the above claims to ensure the suitability of that resource for a particular Swarm.

## 4 Implementation of the Swarmchestrate Concept

The implementation of the above-mentioned fundamental concepts of Swarmchestrate is being achieved through a set of independent technical components that interwork to realise the overall vision of the project. Figure 2 shows these components as pluggable blocks from which the overall framework is built. A short description of each component is provided below.

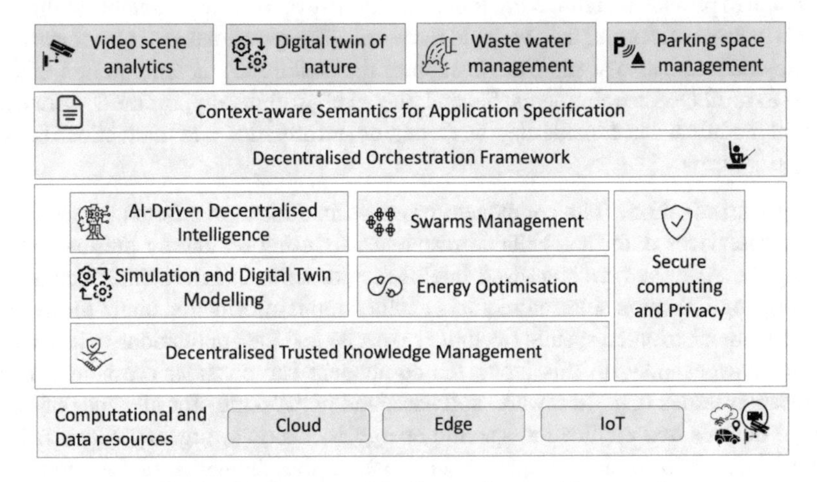


**Fig. 2.** Components of the Swarmchestrate Framework

**Context-Aware Semantics for Application Specification.** This component supports the modelling of fog and edge nodes in addition to cloud resources. Moreover, it focuses on modelling entities that are necessary from the viewpoint of decentralisation and AI-enabled operations, such as the specification of context-aware attributes and aspects of reconfiguration. The modelling approach is being enhanced with suitable constructs related to the decentralised concept of the Swarmchestrate platform to allow the parallel deployment of topology by decentralised orchestrators in the application space.

**Decentralised Orchestration Framework.** The component is responsible for producing the overall integrated orchestration solution, in line with the decentralised vision of Swarmchestrate. The implementation of the component substantially transforms and extends an existing open-source Cloud-to-Edge orchestrator, called MiCADO [21], into a fully decentralised solution by utilising the developed algorithms and solutions implemented in other technical components of the framework.

**Swarms Management.** It is responsible for managing the ad-hoc formation of Swarms from the available resources, which are then used for the deployment and execution of an application. In contrast to traditional Swarm computing management tasks, the utilised resources are located in different dimensions of the Cloud-to-Edge continuum, which requires a peer-to-peer overlay network protocol to enable communication between them. Subsequently, this component implements such a protocol in each of the participating resources to enable a discovery and connection establishment process that automates the formation of a common overlay network, independently from the underlying network architectures.

**AI-Driven Decentralised Intelligence.** This component comprises of solutions for enabling effective and flexible utilisation of AI algorithms, in support of decentralised orchestration decisions. The AI algorithms serve two different but related purposes. First, the distributed algorithms are executed throughout the Cloud-to-Edge computing continuum to provide matchmaking functionality between the requirements of the application's microservices and the available resources. The matchmaking is performed both at deployment and also at run-time to support the continuous reconfiguration of applications to fulfil QoS requirements. Second, the AI algorithms support the Swarms Management module when considering the formation, reformation and interactions between multiple Swarms.

**Energy Optimisation.** The component provides an allocation schema for all application microservices onto the available resources, focusing on energy preservation and utilising the AI-driven decentralised intelligence described above. The calculation of the energy optimisation solution requires detailed input information about the available devices in the distributed system (technical aspects) and the applications to be executed (economic aspects). With this input, the component can consider cross-layer energy optimisation issues (e.g., hardware, software, and networking) for all types of Cloud-to-Edge devices and application-specific energy issues (e.g., urgency expressed by a high willingness to pay for computing services). As the calculation of the optimisation solution (even with the help of heuristics for multi-objective optimisation algorithms) is computationally too expensive, time-consuming, and not accounting for the highly dynamic environment of the envisioned Cloud-to-Edge ecosystem, continuous learning approaches will be applied.

**Simulation and Digital Twin Modelling.** This component is responsible for modelling decentralised self-organising orchestration services using simulation. The component is based on the open-source DISSECT-CF-Fog simulator [22], which is able to utilise a multi-layered Cloud-to-Edge infrastructure. However, new modelling constructs are being defined to incorporate decentralised resource management and decision-making by introducing an orchestration layer and a Swarm manager component to the simulation architecture. Later in the project, the simulator will be further enhanced into a digital twin solution that runs in parallel to the real system, evaluates its behaviour in real-time based on possible alternative scenarios, and initiates certain reconfiguration decisions, if required.

**Decentralised Trusted Knowledge Management.** The component provides a persistent and trusted knowledge base for the global management of application-descriptors and market-tradable resources and services in the Cloud-to-Edge continuum. It is based on standard compatible and blockchain-based self-sovereign identities that describe all participating entities in the fully distributed environment. The component is being implemented by following the principles of Decentralised Identities (DID) and Verifiable Credentials as defined by the W3C [23].

**Secure Computing and Privacy.** The secure computing and privacy layer provides end-to-end security for the orchestrator and the targeted and managed applications. It utilises several modern encryption techniques, e.g. Functional Encryption (FE) and Hybrid Homomorphic Encryption (HHE), to analyse encrypted data stored in distributed locations as if they were unencrypted (i.e., in a privacy-preserving way). It also provides a decentralised trust management solution using blockchain-based FE mechanism to facilitate overall transparency and assure trust amongst the system, the distributed computing infrastructure, and the storage resources. Finally, it aims to provide an anonymous Sybil-resistant DID solution so that each entity can only get one ID.

## 5 Conclusions and Further Steps

Due to the increasing adaptation of the Cloud-to-Edge continuum by applications with complex and changing requirements, it is crucial to develop novel mechanisms for the management of large microservices-based applications in such environments. Traditional centralised application management and orchestration approaches are quickly becoming bottlenecks in these scenarios. To tackle this challenge, the Swarmchestrate project developed the concept of a novel fully decentralised application-focused orchestration framework that is based on Swarm computing principles and utilises distributed AI and self-sovereign identities for application life-cycle management.

The project starts in 2024, and after conducting a detailed analysis of currently available technologies, it will develop its framework using an incremental and iterative methodology. Swarmchestrate will demonstrate its results by reengineering an existing centralised orchestrator and implementing four real-life use cases.

## References


1. Jiang, Y., Huang, Z., Tsang, D.H.: Challenges and solutions in fog computing orchestration. IEEE Netw. **32**(3), 122–129 (2017). https://doi.org/10.1109/MNET.2017.1700271
2. Tomarchio, O., Calcaterra, D., Modica, G.D.: Cloud resource orchestration in the multi-cloud landscape: a systematic review of existing frameworks. J. Cloud Comput. **9**(1), 49 (2020). https://doi.org/10.1186/s13677-020-00194-7
3. Costa, B., Bachiega, J., Jr., de Carvalho, L.R., Araujo, A.P.: Orchestration in fog computing: a comprehensive survey. ACM Comput. Surv. **55**(2), 1–34 (2022)
4. Svorobej, S., Bendechache, M., Griesinger, F., Domaschka, J.: Orchestration from the cloud to the edge. The cloud-to-thing continuum: opportunities and challenges in cloud. Fog Edge Comput. 61–77 (2020)
5. Hong, C.H., Varghese, B.: Resource management in fog/edge computing: a survey on architectures, infrastructure, and algorithms. ACM Comput. Surv. **52**(5), 1–37 (2019)
6. Kumara, I., et al.: SODALITE@RT: orchestrating applications on cloud-edge infrastructures. J. Grid Comput. **19**, 1–23 (2021)
7. Masip-Bruin, X., et al.: Managing the cloud continuum: lessons learnt from a real fog-to-cloud deployment. Sensors **21**(9), 2974 (2021)
8. Verginadis, Y., et al.: Prestocloud: a novel framework for data-intensive multi-cloud, fog, and edge function-as-a-service applications. Inf. Resour. Manage. J. **34**(1), 66–85 (2021)

9. Ullah, A., et al.: Orchestration in the Cloud-to-Things compute continuum: taxonomy, survey and future directions. J. Cloud Comput. **12**(1), 135 (2023)
10. Jimenez, L.L., Schelen, O.: HYDRA: decentralised location-aware orchestration of containerized applications. IEEE Trans. Cloud Comput. **10**(4), 2664–2678 (2020)
11. Pires, A., Simão, J., Veiga, L.: Distributed and decentralised orchestration of containers on edge clouds. J. Grid Comput. **19**, 1–20 (2021)
12. Özyar, U.C., Yurdakul, A.: A Decentralised framework with dynamic and event-driven container orchestration at the edge. In: Espoo, Finland, 2022, pp. 33–40 (2022). https://doi.org/10.1109/iThings-GreenCom-CPSCom-SmartData-Cybermatics55523.2022.00017
13. Nezami, Z., Zamanifar, K., Djemame, K., Pournaras, E.: Decentralised edge-to-cloud load balancing: service placement for the Internet of Things. IEEE Access **9**, 64983–65000 (2021)
14. Ullah, A., Dagdeviren, H., Ariyattu, R., DesLauriers, J., Kiss, T., Bowden, J.: MiCADO-edge: towards an application-level orchestrator for the cloud-to-edge computing continuum. J. Grid Comput. **19**(4), 47 (2021). https://doi.org/10.1007/s10723-021-09589-5
15. Tsagkaropoulos, A., Verginadis, Y., Compastié, M., Apostolou, D., Mentzas, G.: Extending TOSCA for edge and fog deployment support. Electronics **10**(6), 737 (2021)
16. Kaur, K., Kumar, Y.: Swarm intelligence and its applications towards various computing: a systematic review. In: 2020 International Conference on Intelligent Engineering and Management (ICIEM), London, UK, pp. 57–62 (2020). https://doi.org/10.1109/ICIEM48762.2020.9160177
17. Lera, I., Guerrero, C., Juiz, C.: Availability-aware service placement policy in fog computing based on graph partitions. IEEE Internet Things J. **6**(2), 3641–3651 (2018)
18. Sharma, V., Kumar, R., Rathore, N.: Topological broadcasting using parameter sensitivity-based logical proximity graphs in coordinated ground-flying ad hoc networks. J. Wirel. Mob. Netw. Ubiquitous Comput. Dependable Appl. **6**(3), 54–72 (2015)
19. Preukschat, A., Reed, D.: Self-Sovereign Identity. Manning Publications (2021)
20. Decentralised Identifiers (DIDs) v1.0. https://www.w3.org/TR/did-core/. Accessed 10 Dec 2023
21. Kiss, T., et al.: MiCADO—microservice-based cloud application-level dynamic orchestrator. Future Gener. Comput. Syst. **94**, 937–946 (2019)
22. Markus, A., Kertesz, A.: Investigating IoT application behaviour in simulated fog environments. In: Ferguson, D., Pahl, C., Helfert, M. (eds.) Cloud Computing and Services Science (CLOSER 2020). CCIS, vol. 1399, pp. 258–276. Springer, Cham (2021). https://doi.org/10.1007/978-3-030-72369-9_11
23. Sporny, M., Longley, D., Chadwick, D.: Verifiable credentials data model V1.1. W3C (2022). Retrieved 18 March 2023. https://www.w3.org/TR/vc-data-model

# REDRAW: fedeRatED leaRning for humAn Wellbeing

Rocco Aversa[1], Mario Bochicchio[2], Dario Branco[1], Mario Magliulo[3], Albina Orlando[4], Anna Pristner[3], Adriano Tramontano[1,2,3,4,5], Erika Schirinzi[5], Gabriele Siciliano[5], and Salvatore Venticinque[1](✉)

[1] University of Campania Luigi Vanvitelli, Aversa, Italy
{rocco.aversa,dario.branco,salvatore.venticinque}@unicampania.it
[2] Università degli Studi di Bari Aldo Moro, Bari, Italy
mario.bochicchio@unibari.it
[3] Institute of Biostructures and Bioimaging IBB-CNR, Napoli, Italy
mario.magliulo@ibb.cnr.it, anna.pristner@cnr.it
[4] Istituto per le applicazioni sul calcolo "Mauro Picone" IAC-CNR, Napoli, Italy
albina.orlando@cnr.it
[5] Università di Pisa, Pisa, Italy
gabriele.siciliano@unipi.it

**Abstract.** The REDRAW project investigates the exploitation of the federated learning computing paradigm to improve the technologies adopted for the monitoring, diagnosis and treatment management of specific health conditions, developing approaches more respectful of the constraints of privacy, confidentiality and cybersecurity, which are still largely absent from the market. REDRAW proposes the study and fine-tuning of dynamic cloud-edge deployment techniques, which exploits Federated Learning (FL) models, in three real-world contexts, to improve the technological features of existing solutions, while respecting the strategic and non-functional constraints that characterize the Italian and European scenarios.

## 1 Introduction

The health care sector has been characterized by profound changes and large-scale adoption of heterogeneous digital solutions. An increasing number of citizens across Europe expect to access care quickly and easily choosing remote health smart devices to share with physicians own health data to generate information flows aimed at health care providers, to improve diagnosis, monitoring and treating processes of a growing number of conditions, mainly in frail patients. Public Health Administrations are performing increasingly relevant agreement to establish which technologies are most effective and which IT infrastructure can deliver safe, secure and interoperable access for collecting, processing, and distributing information flows from patients, but also healthy general population, to provide a sustainable predictive, preventive, personalized and participatory system. Individual European Countries has launched plans for the digitization


L. Barolli (Ed.): AINA 2024, LNDECT 203, pp. 101–109, 2024.
https://doi.org/10.1007/978-3-031-57931-8_10

of cross-border health services and the creation of a European health data space. A rational approach to the management of such a complex system cannot disregard the use of computational systems based on cloud resources and edge devices through heterogeneous and multi-level systems (cloud-edge continuum).

Development of research activity in this area is particularly relevant for at least two reasons: a) improving the technologies adopted for the monitoring, diagnosis and treatment management of specific health conditions; b) the need to develop approaches more respectful of the constraints of privacy, confidentiality, cybersecurity, accountability, and European digital sovereignty, which are still largely absent from the market.

REDRAW (fedeRatED leaRning for humAn Wellbeing) is a research project funded by MIUR (Italian Ministry of University and Research) that proposes the study and fine-tuning of dynamic cloud-edge deployment techniques, which exploits Federated Learning (FL) models, in three real-world contexts, to improve the technological features of existing solutions, while respecting the strategic and non-functional constraints that characterize the Italian and European scenarios. By the application of the FL paradigm into the cloud-edge continuum, REDRAW will enable the development of smart monitoring tools, which train locally a personalized model that learns how to evaluate information about the user's well-being. The centralized collection and elaboration of trained models will be exploited to learn how such conditions are evolving in the overall population without infringing the users' privacy but allowing for deploying new models that consider global trends. The three case-studies investigate (1) maternal-fetal health care in pregnancy, conducted in collaboration with the Obstetrics and Gynecology Clinic of the Bari Polyclinic, (2) treatment of neuromuscular disorders that characterize many genetic and degenerative diseases related to trauma or aging, in collaboration with the Neurological Clinic of the University of Pisa, and (3) monitoring of sleep disorders at IBB CNR of Naples.

## 2 Related Work

As of today, most E-health applications that involve the use of smart devices focus on two separate types of tools [2,7]. *Clinical grade devices* requires complex and onerous approval to be used in medical practice or clinical research. *Non-clinical devices* that are commercially distributed by private organizations for nonclinical purposes (e.g., wellness, sports, etc.). These devices usually collect on their respective manufacturers' cloud systems huge volumes of personal data over long periods. Although theoretically bound to comply with stringent privacy protection rules (GDPR), manufacturers often impose user licenses that protect their business. The REDRAW project will advance the state of art providing clinical researchers with the ability to monitor and analyze a wide range of patients' parameters, through clinical and non-clinical devices that permit obtaining qualitative measures useful for Patient Quality of Life (QoL) information and this at the edge rather than centrally, in compliance with non-functional requirements, but at the same time maintaining full control over applied algorithms, methodologies, and potential customizations (e.g., based on the patient history). The intent to keep the data close to their source, is based on recent experiences of transition from centralized programming paradigms to the Cloud-Continuum [4]. The extension of centralized approaches to the Edge Computing paradigm is still under development, especially as regards requirements expressed by distributed applications. The lack of a standard methodology and a reference framework prevents the possibility to compare alternative solutions and different scientific contributions [1]. It needs to develop techniques for the efficient deployment and management of algorithms, to be run on devices used by patients during day-by-day monitoring and treatment. Machine and Deep learning techniques are currently applied in clinical trials to analyze patients' data, but the distribution of computations over several edge devices represents an important challenge for researchers. We will exploit Privacy Preserving Federated Learning techniques (PPFL) [9], which guarantee the absence of personal or identifiable data in the parameters exchanged over networks to train Machine and Deep learning algorithms. PPFL approaches are useful for the efficient exploitation of computational nodes in distributed environments, and most importantly for the positive impact they have on privacy. Since several PPFL approaches are possible, being guided by architectural and computational patterns becomes fundamental to reducing risks. FL Patterns and reference architectures have already been defined [3] and their applications are available [5], but support for the application of such Patterns and the development of PPFL algorithms is still to be explored. To better understand the specific requirements and constraints characterizing the healthcare field, the proposal is focused on three medical areas in which cloud-edge solutions can have a very positive impact: maternal-fetal well-being, monitoring of muscular diseases progression and sleep monitoring. Advanced digital techniques have been developed [6,8], to detect early signs of possible damaging conditions, but their widespread adoption is very limited, also because of the problems investigated in this proposal.

## 3 The REDRAW Project

The REDRAW project is motivated by the following four research questions:

- *RQ1:* May the exploitation of privacy-preserving federated learning techniques help to maximize the benefits of low-cost sensing technologies and the ubiquitous interconnected embedded devices in the e-Health domain?
- *RQ2:* Is it possible to demonstrate the effectiveness of privacy-preserving federated learning techniques in terms of technical and technological choices in a reproducible and objective way, so to allow for a comparison of different solutions and in different real-world scenarios?
- *RQ3:* To what extent can the exploitation of privacy preserving federate learning techniques extract valuable knowledge about human well-being and Quality of Life, on a large scale, that can be correlated with a small amount of data collected from real-world scenarios to extract new insights?
- *RQ4:* To what extent can, the adoption of a proper set of architectural primitives, drive and ease the decomposition of health-related algorithms in a

distributed environment helping to meet the nonfunctional requirements of real-world healthcare scenarios?

Driven by these research questions the project will pursue three main research objectives:

- *OR1.* The definition of a methodology that allows for effective exploitation of privacy-preserving federated learning techniques and technologies in the field of e-health, with a focus on human well-being and Quality of Life monitoring. This objective will be pursued by investigating and extending existing models, developing original models, by the efficient decomposition of the code and finding optimal deployment strategies in the Cloud-Edge continuum, looking for architectural and computational patterns for the e-health application domain.
- *OR2.* The definition of a theoretical framework that allows for designing, implementing, and testing privacy preserving federated learning solutions to permit a systematic and measurable evaluation of the key characteristics of the solution, the reproducibility of experiments and the comparison with other solutions based on different techniques and technologies. This objective will be pursued focusing on the formalization of the requirements, coming from the analysis of three selected real-world scenarios contributed by two project partners, and their generalization, to pave the way toward a reference standard and a state of art framework for the Cloud-Edge research community working on privacy preserving federated. This OR also includes the definition of the main KPIs of our proposal.
- *OR3.* The definition and implementation of an evaluation strategy that allows estimating the impact of federated learning solutions developed in the project through the new methodology in three selected real-world scenarios which aim at exploiting health technologies, new tools and digital solutions for human well-being monitoring.

As shown in Fig. 1, the project development is based on four pillars: *Applications*, that are represented by three selected application scenarios. The requirement analysis will provide the input necessary to develop the PPFL algorithms, define the architecture of the edge-cloud computing infrastructure, simulate or generate the related data, process it and verify, together with the involved health professionals, the effectiveness of the proposed solution.

*Data*, that is the most important asset to test and evaluate the privacy preserving federated learning models adopted by the project. For these purposes we will a) generate real data from human subjects, where appropriate, or retrieve real data from online archives (e.g. Kaggle, physionet.org, etc.); b) generate synthetic data purposely built to test the models investigated by this proposal.

*Privacy Preserving Federated Learning deployment model*, which will be developed according to the results of the "requirement analysis" phase, and to the architecture of the cloud-edge computing infrastructure. A virtualized solution, designed to comply with the constraints and requirements observed in hospitals and home healthcare providers, will be used to test de developed PPFL models and to adapt them to the different application scenarios.

*Privacy Preserving Federated Learning reference model*, with related KPIs measurement and evaluation procedures, which represent the outcome of the conducted experiments ad aim at assessing the impact of PPFL algorithms in the cloud-edge continuum and the e-health domain. KPIs of interest will be defined to measure the scalability, the consumed computational resources, the quality of results and the compliance to non functional requirements in comparison to the other solutions available in the literature.

The proposed framework will allow us to describe the design and development process till the assessment of results following the methodological step. The expected results, for each experiment, will be a machine readable research report that guarantees reproducibility and a transparent comparison of the research outcome with solutions investigated by other researchers in the future. In particular, when they can be made publicly available, the report will include all the links to the software repository and open data archive where the code, the data and other software artifacts can be retrieved for reproducing the experiment.

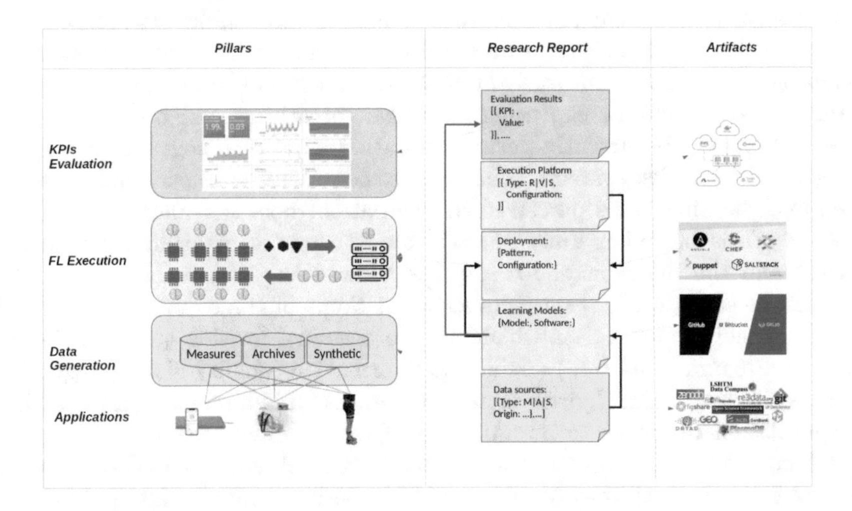


**Fig. 1.** Conceptual framework

# 4 Application Scenarios

## 4.1 Analysis of Sleep Disorders

Sleep disorders, very widespread in the general population but often scotomized nevertheless their medical relevance, as also recently underlined by EU health regulation[1] which consider management and prevention of their detrimental effects mandatory. In particular, emphasis on OSAs was given also because of poor and bad sleeping related to this condition remains underrepresented in clinical practice. Subjects with sleep disorders complain of excessive daytime drowsiness, impaired cognition, mood changes, anxiety, fatigue, or reduced tolerance to pain more often than good sleepers. They seek medical assistance or anticipate retirement more frequently and use more prescribed medication according to surveys and population-based studies. Indeed, quality of life is a major outcome variable in choosing and evaluating treatment alternatives for sleep disorders. Only minimal estimates of the economic impact of sleep disorders and their derivative consequences are possible because of under-recognition and underreporting. This is why protocols and procedures for identifying and monitoring particularly vulnerable individuals or categories of sleepers are absolutely necessary. The investigation of models to study the correlation between sleep disorders and serious pathologies like stroke, diabetes, heart disease, neurodegenerative disorders, and even potentially cancer, which are frequently overlooked are the main concerns in this scenario.

In this context, the utilization of a federated learning solution would allow the utilization of low-cost smart technologies which are accessible, discreet, contactless, and suitable for long-term monitoring at a large scale, under physiological conditions and in a non-subjective way. It will use, among others, ballistocardiogram (BCG) sensors for capturing heartbeat, breathing, and body movement. Sleep monitoring is currently performed with the polysomnography technique, using highly invasive systems. The use of these devices considerably disturbs sleep and the annoyance perceived by the patient does not allow for a correct analysis of his behavior. Only recently has the miniaturization technology of devices and the diffusion of IoT devices allowed the creation of new sensors that can be easily used by patients at home. Some of these devices can provide high-frequency sampled data which, after processing, provide quantitative (respiration rate and heart rate) qualitative information (quality of sleep, agitation, tremors). Many of the algorithms used to process this kind of signal need high computational resources to obtain some of the information about the patient health conditions. The use of mixed processing techniques based on cloud or edge architectures can open new scenarios for in-home monitoring and in medical diagnosis.

### 4.2 Monitoring Maternal and Fetal Health Status During Pregnancy

Maternal and fetal well-being during pregnancy is strictly linked and based on homeostatic maternal-fetal tolerance along which fetal responses vary in response to different types of physiological adaptations or pathological stimulous occurring in the mother. However, mother and fetus although biologically jointed

[1] http://eur-lex.europa.eu/legal-content/EN/TXT/?uri=uriserv:OJ.L_.2014.194.01.0010.01.ENG.

are two defined entities and, if apparently a maternal-fetal conflict does not exist, sometimes there are situations in which apparently their clinical interests diverge. Current medical technologies have great potential but they do not have in fact an absolute value in determining with a reasonable threshold of certainty and uncertainty the construction of the decision-making process in this dualism. Early recognition and monitoring of maternal and fetal clinical signs can help in closing the gap to build effective decisional clinical management routes.

Monitoring maternal and fetal health status during pregnancy based on a programmable Asclepyus device to detect vital parameters (temperature, heart rate, blood oxygenation level, and blood pressure variances) of the mother and to report any abnormalities can have an impact on more than one million mothers/year in Europe, and 35M mothers/years worldwide. These figures triple when considering that during the COVID-19 pandemic, remote monitoring proved essential to limit the risk of infection for pregnant women and health professionals. Non-invasive monitoring of fetal health rate through electrocardiographic sensors placed on the maternal abdomen is based on sophisticated denoising techniques and subsequent separation of signals originating from the maternal body (heartbeat, uterine contractions, muscle activity) from those originating from the fetus (mainly cardiac signal). To solve this problem, the most advanced and performing solutions make extensive use of machine learning techniques that require a significant computational load and a high energy expenditure, so that the possibility to distribute the computational load between the Asclepyus device and a connected smartphone (or other edge resources) could extend the energy autonomy and the monitoring time of the device while minimizing/avoiding the exchange of sensitive data with the cloud. To avoid the complexities of involving real patients, all tests will be performed on a real or simulated swarm of devices produced by Asclepyus, in a realistic operating environment, working on the anonymized "Fetal ECG" dataset available on Physionet.

### 4.3 Monitoring of Neuromuscular Disorders

Muscle diseases are rare and while persons affected by muscle or, more in general, neuromuscular disorders have seen improvements in the last years in symptoms management and have a longer life perspective, in many cases remain still potentially lethal with the lifelong debilitating course. In particular progressive motor impairment, muscle fatigue, and respiratory function decline due to respiratory muscle compromission are dominant and worsen over time often in an unpredictable manner, changing the disease trajectory and shifting the care priorities with a remarkable impact on global management [16]. Thus, the burden of those diseases is considerable from a personal, social and economic perspective. The intersection between health digital applications and the therapeutic pipelines rapidly evolving, but not fully exploited for these diseases, offers broad new opportunities to customize and improve the effectiveness of care solutions.

Long-term continuous monitoring of functional parameters of patients affected by neuromuscular diseases and is based on the same device described for the second scenario for cardiopulmonary vital signs registration and on inertial sensor systems for both hospital and home monitoring of strategic motor functions, such as deambulation, trunk axial posture, harm and hand movements and detection of surface electromyography data. Providing measurements that describe disease evolution, both in relation to the natural history course of the diseases and in response to tailored therapeutic interventions, the aim is to early recognize detrimental events as descriptors of clinical progression, to elaborate a predictive model to generate interventional decisional trees, useful for patients in preserving and compensate the loss of functions. In doing that, considering the extreme phenotypic variability of several neuromuscular diseases, we choose as cases study to test the hypothesis two muscle diseases, Pompe Disease and Spinal Muscle Atrophy (SMA), as models respectively of mild-medium and high clinical and relative management complexity.

## 5 Conclusion

Digital transformation represents a not more postponable process to optimize the management chain of healthcare, closing the gap between demand and supply in a system that involves multiple stakeholders, including health and assistance personnel, technical IT staff, data scientists, industry, national and local health care providers, policy makers at any levels. In a critical appraisal of several needs of care which can be addressed to digital transformations to find acceptable solutions, the investigated solutions proposed by REDRAW innovate the way according to which the data are currently collected and managed by low-power wearable electronic devices in general, and in particular in the context of monitoring of human well-being. The Cloud-Edge paradigm is being largely adopted in several domains, especially where the possibility to keep the computations close to the data source is deemed necessary to boost efficiency, and privacy and reduce energy consumption. Improving the current methodologies applied to the development and deployment of Cloud-Edge software will have a strong impact on the exploitation of existing Cloud-Edge environments, whose adoption is still hindered by the lacking standard, open, and easy-to-apply Programming Models and Tools. In particular, the scientific results of the REDRAW project contribute to changing the design methodology of a large class of applications, even where the Security Driven approach, absolutely necessary in the HealthCare field in order to ensure patient privacy, is required. The results of both theoretical and applied research will leverage the spread of such technology preventing issues related to the user's acceptance and mitigating the complexity to be addressed by service providers to comply with legal and privacy requirements. Application fields include monitoring of users' activities, monitoring of appliances and utilities, such as energy and water consumption, smart cities and smart transport applications and others. Moreover, the utilization of personal archived data, or part of that, can be made available by users' hosting on their edge device the training of software downloaded on request, because the mechanism for authentication and integrity check of code already exists. In this context the REDRAW research activities will also contribute to understanding the level of scalability and performance which could be achieved. In fact, the project focuses mainly on those topics which are relevant to those researchers who work either on edge-cloud programming paradigms or the community interested in artificial intelligence and software engineering. REDRAW creates also new connections, and reinforces the existing ones, with the biomedical community that finds an opportunity to experiment with such new techniques and technologies in their field. Clearly, patients and medical researchers who deal with the analysis of the collected data will benefit from the innovations we intend to bring. Carrying out preventive actions or a timely diagnosis following the analysis of the Soft-Real Time data collected can lead to sure benefits in terms of patient health but also economic benefits to the community as well as reducing the potential stress of patients.

**Acknowledgements.** This work has been supported by the REDRAW research project (P2022MWE3S - Prin 2022 PNRR, DR n. 1409 of 14-09-2022) funded by the Italian Ministry of Research and by the European Union.

## References

1. Di Martino, B., Venticinque, S., Esposito, A., D'Angelo, S.: A methodology based on computational patterns for offloading of big data applications on cloud-edge platforms. Future Internet **12**(2), 28 (2020)
2. Izmailova, E.S., Wagner, J.A., Perakslis, E.D.: Wearable devices in clinical trials: hype and hypothesis. Clin. Pharmacol. Ther. **104**(1), 42–52 (2018)
3. Lo, S.K., Lu, Q., Zhu, L., Paik, H., Xu, X., Wang, C.: Architectural patterns for the design of federated learning systems. CoRR **abs/2101.02373** (2021)
4. Martino, B.D., Esposito, A.: Applying patterns to support deployment in cloud-edge environments: A case study. In: Barolli, L., Woungang, I., Enokido, T. (eds.) Advanced Information Networking and Applications - Proceedings of the 35th International Conference on Advanced Information Networking and Applications (AINA-2021), Toronto, ON, Canada, 12-14 May, 2021, Volume 3, Lecture Notes in Networks and Systems, vol. 227, pp. 139–148. Springer, Cham (2021). https://doi.org/10.1007/978-3-030-75078-7_15
5. Martino, B.D., Graziano, M., Cante, L.C., Cascone, D.: Analysis of techniques for mapping convolutional neural networks onto cloud edge architectures using splitfed learning method. In: Barolli, L., Hussain, F., Enokido, T. (eds.) Advanced Information Networking and Applications - Proceedings of the 36th International Conference on Advanced Information Networking and Applications (AINA-2022), Volume 3, Lecture Notes in Networks and Systems,, vol. 451, pp. 163–172. Springer, Cham (2022). https://doi.org/10.1007/978-3-030-99619-2_16
6. Mohebbian, M.R., Vedaei, S.S., Wahid, K.A., Dinh, A.T., Marateb, H.R., Tavakolian, K.: Fetal ECG extraction from maternal ECG using attention-based Cycle-GAN. IEEE J. Biomed. Health Inform. **26**, 515–526 (2021)
7. Polat, E.O.: Seamlessly integrable optoelectronics for clinical grade wearables. Adv. Mater. Technol. **6**(3), 2000853 (2021)
8. Sadek, I., Biswas, J., Abdulrazak, B.: Ballistocardiogram signal processing: a review. Health Inf. Sci. Syst. **7**(1), 10 (2019)
9. Xu, J., Glicksberg, B., Su, C., Walker, P., Bian, J., Wang, F.: Federated learning for healthcare informatics. J. Healthc. Inform. Res. **5**, 1–19 (2021)

# Greening AI: A Framework for Energy-Aware Resource Allocation of ML Training Jobs with Performance Guarantees

Roberto Sala[1(✉)], Federica Filippini[1], Danilo Ardagna[1], Daniele Lezzi[2], Francesc Lordan[2], and Patrick Thiem[3]

[1] Politecnico di Milano, Milan, Italy
{roberto.sala,federica.filippini,danilo.ardagna}@polimi.it
[2] Barcelona Supercomputing Center, Barcelona, Spain
{daniele.lezzi,francesc.lordan}@bsc.es
[3] Cloud and Heat, Dresden, Germany
patrick.thiem@cloudandheat.com

**Abstract.** The rapid expansion of Machine Learning (ML) and Artificial Intelligence (AI) has profoundly influenced the technological landscape, reshaping various industries and applications. This surge in computational demands has led to the widespread adoption of Cloud data centers, crucial for supporting the storage and processing requirements of these advanced technologies. However, this expansion poses significant challenges, particularly in terms of energy consumption and associated carbon emissions. As the reliance on cloud data centers intensifies, there is a growing concern about the environmental impact, necessitating innovative solutions to enhance energy efficiency and reduce the ecological footprint of these computational infrastructures. This paper focuses on addressing the challenges linked to training ML and AI applications, emphasizing the importance of energy-efficient solutions. The proposed framework integrates components from the AI-SPRINT project toolchain, such as Krake, Space4AI-R, and PyCOMPSs. Our reference application involves training a Random Forest model for electrocardiogram classification, profiling available resources to obtain a performance model able to predict the training time, and dynamically migrating the workload to sites with cleaner energy sources providing guarantees on the training process due date. Results demonstrate the framework's capacity to estimate execution time and resource requirements with low error, highlighting its potential for establishing an environmentally sustainable AI ecosystem.

## 1 Introduction

Training ML and, in particular, AI applications is challenging. It demands large computational resources and powerful accelerators, e.g., GPUs and TPUs, resulting in large investment for organizations, especially those with large-scale ML infrastructure [1]. On the other hand, the AI training process is energy-intensive and associated with carbon emissions that significantly contribute to environmental degradation [2].

As ML and AI become more pervasive, there is a pressing need to adopt energy-efficient solutions both for algorithm implementation and hardware configuration [3].


L. Barolli (Ed.): AINA 2024, LNDECT 203, pp. 110–121, 2024.
https://doi.org/10.1007/978-3-031-57931-8_11

The environmental impact of Cloud data centers has indeed become a significant concern, with global electricity usage surging to 200 Tera-Watt hours (TWh) in 2019 and 240–340 TWh in 2022 [4]. Cloud datacenters significantly contribute to carbon emissions, prompting global policymakers to establish ambitious targets, such as Net-Zero emissions by 2050 in countries like the UK, France, Denmark, and New Zealand. The Top500 list [5], a benchmark for supercomputing performance, is indeed complemented by the Green500 [6], emphasizing energy efficiency.

A strategic allocation of computational tasks to clusters exploiting sustainable energy, such as solar or wind power, is crucial, although challenges arise from the intermittent nature of renewable energy. Migrating AI workloads during periods of increased clean energy availability optimizes costs and fosters a more ecologically sustainable AI ecosystem, emphasizing the integration of environmental considerations into the ML infrastructure deployment.

The framework presented in this paper showcases results from integrating various components within the AI-SPRINT project toolchain [7]: Krake, SPACE4AI-R, and PyCOMPSs. Krake is an orchestrator engine that streamlines and manages containerized and virtualized workloads across diverse cloud platforms. SPACE4AI-R [8] is a runtime tool designed to tackle the Resource Selection and Component Placement problem efficiently. It aids in executing AI applications across a spectrum of Computing Continuum resources, spanning Edge devices, Cloud Virtual Machines, and Function as a Service configurations. PyCOMPSs [9] is a parallel programming framework for Python applications. It streamlines the development of computational workflows for distributed infrastructures with a programming model based on sequential development.

In our study, we focus on a PyCOMPSs application that trains a Random Forest model tailored for classifying electrocardiograms. This application is profiled on the existing resource infrastructure, collecting data to construct ML-based performance models used by SPACE4AI-R to identify the optimal configuration in terms of cost and maximum execution time. We consider a reference scenario where the application is initially deployed at site $A$ and later migrated to a different site $B$ when new resources powered with a cleaner energy source become available; the checkpointing features of PyCOMPSs allow transparently re-starting the computation from the last saved point on the new resources. The interaction between Krake and SPACE4AI-R allows determining the computational resources required on the new site to guarantee a user-specified maximum execution time, considering the setup time at site B and the consequent migration. Additionally, the application runs periodically, allowing us to access logs from previous executions. We aim to ensure an upper bound on the execution time, which may vary for each run. This assumption is reasonable in the case of AI applications, given that models are regularly retrained to prevent model drift. The results show that our model can estimate both the execution time on the new cluster and the number of cores needed to satisfy the time constraint, with an error lower than 20%.

## 2 Related Work

In recent years, there has been widespread attention on energy consumption and efficiency in cloud data centers due to the exponential growth in the number and power of machines. For instance, [10] conducts a comprehensive analysis of power consumption characteristics in HPC workloads, examining them from the perspectives of systems, jobs, and users. Conversely, [11] surveys power conservation techniques at both hardware and software levels, emphasizing the importance of developing energy-efficient solutions for green IT.

The scheduling of jobs and orchestration of resources play a crucial role in reducing energy consumption and implementing green policies in HPC. In [12], the authors introduce the Energy-Aware Multi-Cluster scheduling policy (EAMC-policy), automating job placement and optimal clock frequency selection. This prioritizes a delicate balance among performance, energy consumption, and response time within heterogeneous environments. Some works propose Deep Reinforcement Learning (DRL) models for resource planning. [13] presents a partition-based task scheduling approach for efficiently managing heterogeneous resources. It pre-assigns tasks to partitions based on current conditions and allocates them to suitable servers, utilizing nonlinear regression for a precise energy consumption model without network fitting. On the other hand, [14] develops a scheduling policy incorporating workload shifting and cloud bursting within a geographically distributed hybrid multi-cloud environment, aiming to maximize renewable energy utilization and prevent deadline constraint violations. Supervised learning models are also employed to optimize CPU frequency selection during job execution, aiming to minimize energy consumption in HPC systems [15].

Finally, dynamic resource allocation methods are developed, as in [16–18]. [16] focuses on autonomously managing converged edge platforms to enable resource-efficient workload orchestration and promote environmental sustainability. The proposed solution emphasizes intelligent dynamic resource configuration in multi-tenant edge computing platforms, aiming to ensure Service Level Objectives (SLO) for each service while encouraging eco-friendly communication practices. Moreover, in [17], it is shown that the effectiveness of prevalent static power allocation strategies in value-based algorithms depends on the applied power constraint. The study reveals shortcomings in these static approaches, resulting in underutilized resources despite system oversubscription. To address this, the paper suggests a dynamic power management strategy for value-based algorithms, utilizing application power-performance models to reallocate power and optimize system productivity, resource utilization, and job completion rates. Finally, [18] focuses on the management of Deep Learning training applications executed in GPU-accelerated clusters. It proposes a stochastic scheduler that automatically selects the optimal type and amount of resources to be assigned to each training job over time, minimizing the energy execution costs and the penalties for due dates violations. The developed method considers the stochasticity in the applications training times and enables GPU space sharing to maximize the resources utilization for less demanding applications.

## 3 AI Training Job Energy Management Framework

In this section, we provide an overview of the AI-SPRINT tools we used for the creation and the training of our application. The PyCOMPSs application consists of a Random Forest model designed for the classification of Electrocardiograms (ECGs) into categories such as normal ECG, Atrial Fibrillation (AF), inconclusive, or noise. Specifically, the training process utilises a set of 15 distinct functions, identified by their IDs in PyCOMPSs log files. Each function is invoked a specific number of times during the training process. The coordination of the distributed execution of these interdependent functions is managed by the COMPSs framework. In the following we present PyCOMPSs (Sect. 3.1), the performance models (Sect. 3.2), Krake (Sect. 3.3) and SPACE4AI-R (Sect. 3.4).

### 3.1 PyCOMPSs

PyCOMPSs is a Python based programming framework aiming at easing the development of general-purpose applications targeting the Cloud-Edge-IoT Continuum. It is composed of a task-based programming model and of a runtime that orchestrates the execution of such tasks in a serverless manner on top of any distributed platform. PyCOMPSs applications are described as workflows composed of many tasks with data dependencies and, at runtime, the engine determines the best host where to run a task and handles all the necessary data movements to offload the task execution.

The PyCOMPSs runtime provide an automatic checkpointing mechanism. As the application runs, the runtime engine decides to copy some of the data values of the application. In the case of an application error or an abrupt end of the execution (the user decides to kill all the VMs), a future execution could resume the application execution from that point. Being able to save the state of the application at any point allows the migration of stateful executions to improve the execution time or the energy consumption of the execution. Besides the default mechanisms, the application developer can request checkpointing all the values on demand adding a specific API call in the application or develop a customized mechanism implementing a simple interface with three methods.

### 3.2 Performance Models

During the training of the application, PyCOMPSs produces a log which provides a snapshot of the training status, i.e. the number of executed, running and completed functions, to be extracted at each instant. Analysing the aforementioned logs, we developed the following model for estimating the residual execution time, which is also valid for estimating the total training time:

$$\tilde{T} = \sum_{f \in F} \left\lceil \frac{n_f}{cores} \right\rceil t_f \tag{1}$$

where $n_f$ and $t_f$ are, respectively, the number of times the single function (executed in a task) $f$ have to be executed until the end of the training of the application and its average execution time, and $F$ is the set of functions called during the entire execution. Here by function we mean the code executed by a COMPSs task. This formula is based on the fact that functions are executed in parallel, when multiple workers are available. The ratio $\lceil \frac{n_f}{cores} \rceil$ indicates the number of waves with which the single function is executed. This approximate model is accurate when the number of function calls is greater than the number of workers, which happens in general in HPC systems and for PyCOMPSs [19]. Since our objective is to estimate the residual execution time in unobserved configurations, the treatment of $t_f$ largely influences this estimate. Therefore, we developed four different scenarios, which differ in the way the "history", i.e. the previous runs, and the current run-times are taken into account:

- *Naive* approach: if the individual function was used at least once in the current simulation, $t_f$ is calculated on the current log. Otherwise, the one calculated on the history is used.
- *Mixed* approach: $t_f$ is a weighted average between the history and the current execution, where the weights are given by the number of executions of the single functions versus their total number of executions.
- *Balanced* approach: $t_f$ is a weighted average between the history and the current execution, where the weights are given by the step at which any migration occurs versus the total number of steps.
- *Full post-processing* approach: $t_f$ is only computed on the history.

The results for the estimation of residual training times are shown in Sect. 4.2.

### 3.3 Krake

Krake[1] is an orchestrator engine for containerized and virtualized workloads across distributed and heterogeneous cloud platforms. It creates a thin layer of aggregation on top of the different platforms (such as OpenStack, Kubernetes) and presents them through a single interface to the cloud user. The user's workloads are scheduled depending on both user requirements (hardware, latencies, cost) and platform characteristics (energy efficiency, load). Krake can be leveraged for a wide range of application scenarios such as central management of distributed compute capacities and application management in Edge Cloud infrastructures. In this paper settings, Krake performs a PyCOMPSs training application migration across Kubernetes clusters that reside in different sites characterised, possibly, by different energy profiles. Furthermore, it ascertains the "best" level of resource usage based on user-defined parameters and re-evaluates the deployment periodically.

### 3.4 SPACE4AI-R

SPACE4AI-R[2] (System PerformAnce and Cost Evaluation on Cloud for AI applications Runtime) [8] is an optimization framework that deals with the runtime management of AI applications executed in the Computing Continuum. The SPACE4AI-R optimizer implements several heuristic algorithms to effectively tackle the Resource Selection and Component Placement problem. It determines the minimum-cost configuration that guarantees user-imposed Quality of Service constraints, in response to load variations or whenever a component migration determines the need to re-consider the current deployment.

[1] https://gitlab.com/rak-n-rok/krake.

[2] https://github.com/ai-sprint-eu-project/space4ai-r-optimizer.

In this work, among the available optimization algorithms, we used a dichotomous search function to determine the resources needed on the new site, following the migration. The motivations behind this choice are manifold. Firstly, in the regime of profiled configurations, the execution time decreases monotonically as the number of workers increases. In addition, the dichotomous function allows us to balance our two objectives, i.e., to minimise the cost, and therefore the number of used cores, while guaranteeing a global constraint on execution time.

## 4 Experimental Results

In this section, we present the experimental results of our PyCOMPSs application. In particular, Sect. 4.1 describes the setting from which we extracted the data used in our simulations which constitutes the application "history", Sect. 4.2 shows the related results, while in Sect. 4.3 we show the results of a real migration of our application from a site $A$ to a different site $B$, to show the accuracy of our models. The data and scripts to replicate all experimental results are available on Zenodo[3].

### 4.1 Experimental Setting

The PyCOMPSs application has been tested on a machine Intel(R) Xeon(R) CPU, 2.20 GHz, utilizing 1 to 10 VMs, each equipped with 4 VCPUs, 15 GB of memory, and 50 GB of storage size, operating on an Ubuntu OS. This involves considering configurations with 4, 8, 12, ..., 40 cores, by varying the number of active workers. To ensure robustness, three independent runs were conducted for each setting. It is noteworthy that while the profiling campaign was executed on a single machine, the models are easily extendable to machines of different types, as indicated by the following results.

PyCOMPSs generates a log file that records scheduling, start, and finish times of the execution of each function, enabling the extrapolation of a snapshot of the training status at any moment. To simulate the partial execution, the log files were truncated after a fixed number of steps, from the total number of 11,655.

With regard to estimating the number of cores on the new site, following the migration, as discussed previously we perform a dichotomous search on all available configurations on the new machine, selecting the smallest number of workers that guarantees satisfaction of the global time constraint.

As introduced before, the PyCOMPSs runtime has the ability to save the state of the application. In this case, the application is started on a cluster and prepares a dataset, trains a Random Forest model with it and evaluates the accuracy of the resulting model. Checkpointing sets a tradeoff between the amount of data values saved and the amount of computation lost in the case of an error. To that end, several checkpointing mechanisms able to work with different granularity have been defined. The first mechanism, Instantiated Tasks (IT), decides to save the output values of a set of N tasks according to their creation order. Finished Tasks (FT), does the same as IT but considering the order in which tasks finish; the checkpointing will save all the values every N tasks finished.

[3] https://doi.org/10.5281/zenodo.10592846.

Finally, the third mechanism, Periodic Time (PT), triggers the saving of all the values periodically after N seconds.

Table 1 shows the different impact of the different checkpointing mechanisms on a PyCOMPSs execution and compares their execution time to an execution without checkpointing.

**Table 1.** Execution times of running a complete execution of COMPSs when applying different checkpointing policies

| No checkpoint | Instantiation Tasks (10 tasks) | Finished Task (10 tasks) | Periodic Time (15 s) |
|---|---|---|---|
| 112.97 s | 144.58 s | 153.06 s | 145.58 s |

### 4.2 Simulation Results

First, we analyse the accuracy of our model (1) in estimating the residual training time of our application, in the four different scenarios described in Sect. 3.2, every 100 steps of the execution. The error between the estimate and the actual residual time is calculated as follows:

$$\%\ error = \frac{residual_{estimated} - residual_{real}}{total_{real}} \tag{2}$$

This gives an non-biased evaluation of the error on the total execution time.

In order to validate our model, we first observed the predictive ability of the model with the same configuration between history and actual setting. Figures 1 and 2 show, respectively, the residual time estimated by our model in the four different scenarios and the relative error, calculated as in (2), with a 20-cores configuration.

The results show that all four models are able to estimate a residual time within 20% error. Only the naive model deviates more, especially in the early stages of training, due to the inaccurate estimation of the average execution time of those functions that have been performed a few times.

In a more general scenario, where a larger number of machines should be profiled, it is not possible to imagine performing such a large number of simulations for each machine. For this reason, we tried to estimate the residual time for all settings considered, using data from the profiling of a run with 40 cores as history. This configuration takes into account the resource contention of the functions to be executed. The results for the 20-core configuration are shown in Figs. 3 and 4.

Also, the model is sufficiently accurate for all the four scenarios but for the *Naive* one, observing an error of less than 20% with respect to the total run-time.

The second part of our analysis involves computing the number of cores on the new VM cluster required to meet a global time constraint, following the migration of the application. Building upon previous analyses, we have chosen the *Full post-processing* model to estimate the number of cores after migration, and the 40-core configuration as history. This choice for the treatment of $t_f$ in (1) is motivated by the fact that the

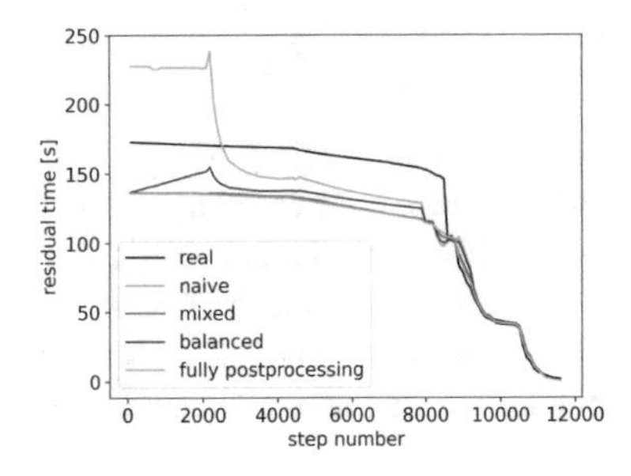


Fig. 1. Residual time for a 20-core configuration, using the same configuration history.

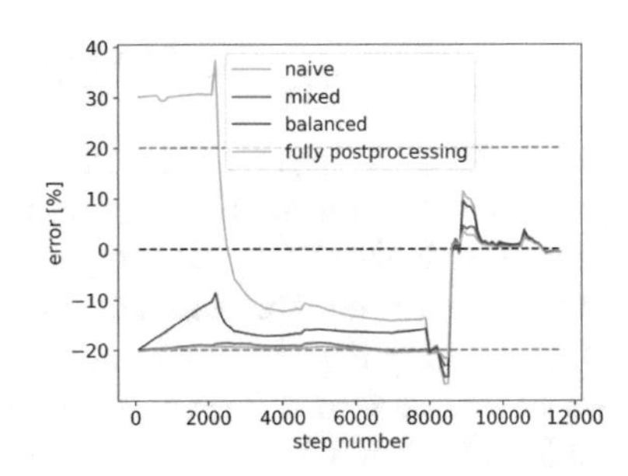


Fig. 2. Error for a 20-core configuration, using history with the same configuration.

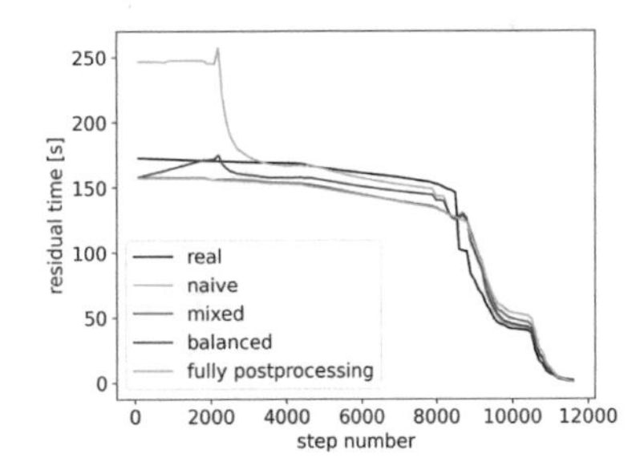


Fig. 3. Time remaining for a 20-core configuration, using the 40-core profiling data as history.

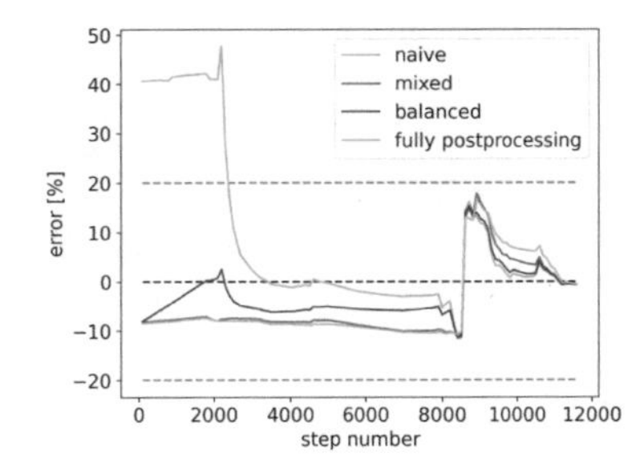


Fig. 4. Error for a 20-core configuration, using the 40-core profiling data as history.

"Fully post-processing" model best estimates the average time remaining over the entire training process. Moreover, in case of migration, the number of cores usually needs to increase to recover the migration delay.

The estimation of the minimum number of cores required to satisfy the global time constraint is carried out by means of a dichotomous search between 1 and 40 cores, based on the model chosen for residual time estimation. The global time was set, for each simulation, equal to the real execution time on the given profiled configuration, which is considered as initial deployment, plus 60 s of migration. In this way, we expect an optimal number of cores following migration usually equal to the one pre-migration. Furthermore, the estimation of the number of cores post-migration was carried out by partitioning the logs into intervals corresponding to 5% of the steps and randomly extracting the point of migration. This estimation was performed 20 times for each interval, for each trial and for each setting for which we have profiling, for a total of 60 values for each interval and for each number of cores. The percentage of extra cores through the overall execution time is computed as follows:

$$\% \ cores_{extra} = \frac{cores_{estimated} - cores_{real}}{cores_{real}} \cdot \frac{T_{total} - T_{elapsed}}{T_{total}} \tag{3}$$

where $T_{total}$ is the real total execution time in the starting configuration, while $T_{elapsed}$ is the elapsed time before migration. In this formula, the percentage of additional cores calculated for execution on the new VM cluster is normalized with respect to the remaining time. This normalization yields an estimate of the number of additional cores used during the entire training, and consequently, the actual energy costs due to model estimation errors. Figure 5 shows the percentage of extra cores estimated by our model for all the initial configurations.

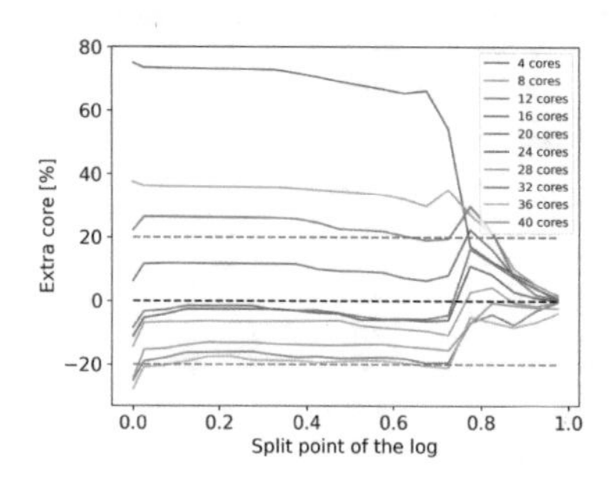


**Fig. 5.** Percentage of extra cores estimated.

Results show that for settings with more than 16 cores the estimate is sufficiently accurate, since the error remains below 20%. Note that for settings with only a few cores, the error is large in percentage terms, but not too large when we consider that an estimate of 75% extra cores on four cores results in an estimate of three additional cores.

In the final analysis, we look at the percentage of the number of violations of the overall time constraint and the percentage of time violation. In this case, the application migration point was randomly selected with respect to the entire application, the migration time was set at one minute and the global time constraint was randomly drawn between the average training execution times for the 32- and 4-core configurations. This experiment was carried out 1,000 times, for each available profiling run, and the results were averaged considering the initial number of cores. As for the estimation of the possible violation, the number of post-migration cores was calculated considering the 40-core profiling data, while the "real" post-migration residual time is calculated through our model on the actual number of cores we are using. This means that if initially the machine is using 40 cores and the dichotomic search function predicts that 33 cores are needed, the estimate of the "real" residual time is made on the 36 cores data, i.e., the first available data approximating by excess.

The percentage violation time for the individual experiment is calculated as follows:

$$\% \ violation \ time = \begin{cases} 0.0, \text{ if } T_{max} \geq T_{elapsed} + T_{residual} + T_{migration} \\ (T_{elapsed} + T_{residual} + T_{migration} - T_{max})/T_{max}, \text{ otherwise} \end{cases} \tag{4}$$

where $T_{max}$ is the global time constraint and $T_{migration}$ is the migration time. Figure 6 illustrates the percentage of violations in relation to the total number of migrations, while Fig. 7 depicts the percentage of time violations with respect to the overall time constraint, averaged over all experiments.

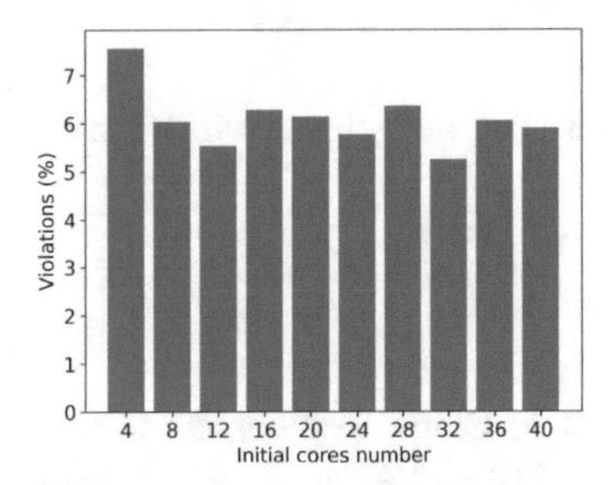


**Fig. 6.** Percentage number of violations.

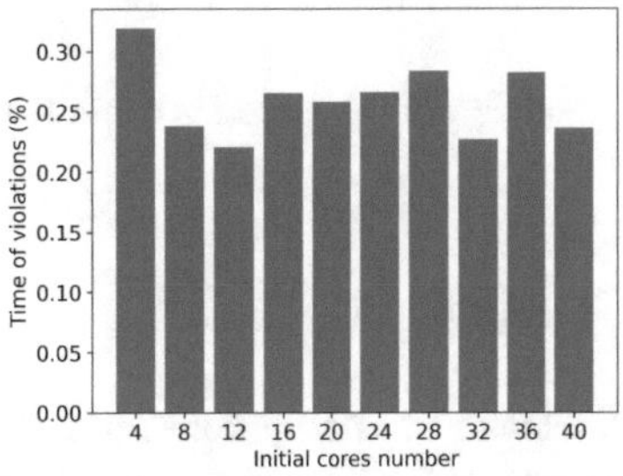

**Fig. 7.** Percentage violation time.

Results show that on average, about 6% of migrations involve a violation of the global time constraint, but the average time violation remains within 0.35%.

### 4.3 Analysis on a Prototype Environment

In this section, we present the results related to the real execution of our PyCOMPSs application, showcasing the migration from site *A* to a different site *B*. The cluster at site *A* has 2 cores and 15 GB of memory, while the one at site *B* has 30 GB of memory and can run up to 8 cores. The estimated migration time, as well as the checkpointing time at site *A*, ranges from 10 to 15 min. In addition, we profiled our application on site *B* with 2, 4, 6, and 8 cores.

We set the maximum execution time of the application to 45 min, and we started the execution of our application on site *A* considering 2 cores, which are enough to fulfill the global time constraint. Subsequently, after 12.5 min we attempted to migrate the application. It took about 12 min to save the checkpoints, and a few seconds for our dichotomous search function to compute the number of cores needed on site *B*. In particular, we determined that the minimum number of cores required to complete the execution in the specified time is 6, considering an estimate of 12.5 min to instantiate the machine on site *B* and download the checkpoints, and using as history the data coming from the initial profiling with 4 cores.

The actual time to instantiate the machine at site *B* was of 11.5 min, after which the execution was resumed and completed in additional 8.5 min. Overall, including migration overheads, the execution was completed in 44.5 min. The entire execution process with migration is shown in Fig. 8.

All the relevant files for our analyses are available on Zenodo[4].

[4] https://doi.org/10.5281/zenodo.10592846.

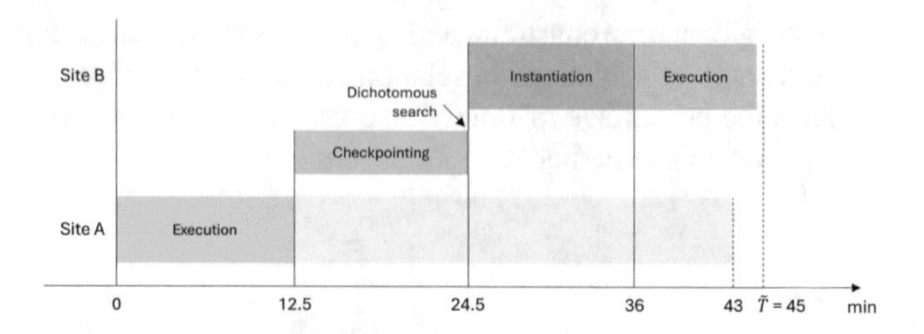


**Fig. 8.** Overview of the execution of our PyCOMPSs application with migration from site $A$ to $B$.

# 5 Conclusions

The challenge of energy consumption management in Cloud data centers is gaining significance as the computational power demands of ML and AI applications continue to rise. In this paper, we address a PyCOMPSs application, specifically focusing on the process of migrating computational resources from a site $A$ to an alternative site $B$ powered by cleaner energy sources with the help of Krake. We have developed an integrated model within SPACE4AI-R, capable of predicting both the remaining execution time of the application, given the system configuration, and the number of cores required to meet a time constraint, using profiling data from the maximum number of cores configuration alone.

Experimental results indicate that our model can predict the remaining runtime with an error of less than 20% compared to the actual runtime. Furthermore, in estimating the resources needed at the post-migration site through dichotomous search, we observed that only 6% of migrations violate the global time constraint, with an average delay of 0.35% relative to the specified threshold.

Future works will concentrate on estimating the energy overhead resulting from migration to enhance the accuracy of our energy model.

**Acknowledgments.** This work has been funded by the European Commission under the H2020 grant N. 101016577 AI-SPRINT: AI in Secure Privacy pReserving computINg conTinuum.

# References

1. Ilager, S., Toosi, A.N., Jha, M.R., Brandic, I., Buyya, R.: A data-driven analysis of a cloud data center: statistical characterization of workload, energy and temperature. In: 2023 IEEE/ACM 16th International Conference on Utility and Cloud Computing (UCC 2023), pp. 1–10 (2023). to appear
2. Lacoste, A., Luccioni, A., Schmidt, V., Dandres, T.: Quantifying the carbon emissions of machine learning. arXiv preprint: arXiv:1910.09700 (2019)
3. Filippini, F., et al.: ANDREAS: artificial intelligence traiNing scheDuler foR accElerAted resource clusterS. In: 2021 8th International Conference on Future Internet of Things and Cloud (FiCloud), pp. 388–393. IEEE (2021). https://doi.org/10.1109/FiCloud49777.2021.00063

4. https://www.iea.org/energy-system/buildings/data-centres-and-data-transmission-networks
5. https://www.top500.org
6. https://www.top500.org/lists/green500/
7. Sedghani, H., et al.: Advancing design and runtime management of AI applications with AI-SPRINT (position paper). In: 2021 IEEE 45th Annual Computers, Software, and Applications Conference (COMPSAC), pp. 1455–1462. IEEE (2021)
8. Filippini, F., Sedghani, H., Ardagna, D.: SPACE4AI-R: a runtime management tool for AI applications component placement and resource scaling in computing continua. In: 2023 IEEE/ACM 16th International Conference on Utility and Cloud Computing (UCC 2023), pp. 1–7. to appear, ISBN: 979-8-4007-0234-1/23/12. https://doi.org/10.1145/3603166.3632560.
9. Badia, R.M., Conejero, J., Ejarque, J., Lezzi, D., Lordan, F.: PyCOMPSs as an instrument for translational computer science. Comput. Sci. Eng. **24**(2), 66–82 (2022). https://doi.org/10.1109/MCSE.2022.3152945
10. Patel, T., Wagenhäuser, A., Eibel, C., Hönig, T., Zeiser, T., Tiwari, D.: What does power consumption behavior of HPC jobs reveal?: Demystifying, quantifying, and predicting power consumption characteristics. In: 2020 IEEE International Parallel and Distributed Processing Symposium (IPDPS), pp. 799–809. IEEE (2020)
11. Thakkar, A., Chaudhari, K., Shah, M.: A comprehensive survey on energy-efficient power management techniques. Procedia Comput. Sci. **167**, 1189–1199 (2020)
12. D'Amico, M., Gonzalez, J.C.: Energy hardware and workload aware job scheduling towards interconnected HPC environments. IEEE Trans. Parallel Distrib. Syst. (2021)
13. Li, J., Zhang, X., Wei, Z., Wei, J., Ji, Z.: Energy-aware task scheduling optimization with deep reinforcement learning for large-scale heterogeneous systems. CCF Trans. High Perform. Comput. **3**, 383–392 (2021)
14. Zhao, J., Rodríguez, M.A., Buyya, R.: A deep reinforcement learning approach to resource management in hybrid clouds harnessing renewable energy and task scheduling. In: 2021 IEEE 14th International Conference on Cloud Computing (CLOUD), pp. 240–249. IEEE (2021)
15. Ozer, G., et al.: Towards a predictive energy model for HPC runtime systems using supervised learning. In: Schwardmann, U., et al. (eds.) Euro-Par 2019: Parallel Processing Workshops. LNCS, vol. 11997, pp. 626–638. Springer, Cham (2020). https://doi.org/10.1007/978-3-030-48340-1_48
16. Guim, F., et al.: Autonomous lifecycle management for resource-efficient workload orchestration for green edge computing. IEEE Trans. Green Commun. Netw. **6**(1), 571–582 (2021)
17. Kumbhare, N., Akoglu, A., Marathe, A., Hariri, S., Abdulla, G.: Dynamic power management for value-oriented schedulers in power-constrained HPC system. Parallel Comput. **99**, 102686 (2020)
18. Filippini, F., Anselmi, J., Ardagna, D., Gaujal, B.: A stochastic approach for scheduling AI training jobs in GPU-based systems. IEEE Trans. Cloud Comput. (01), 1–17, 5555. https://doi.org/10.1109/TCC.2023.3336540
19. Ataie, E., Evangelinou, A., Gianniti, E., Ardagna, D.: A hybrid machine learning approach for performance modeling of cloud-based big data applications. Comput. J. **65**(12), 3123–3140 (2022)

# Optimization of University Services Through Cloud-Edge Integration: Study of a Use Case at the University of Naples "Federico II"

Barone Giovanni Battista and Sabella Gianluca(✉)

University of Naples "Federico II", Naples, Italy
{gbbarone,gianluca.sabella}@unina.it

**Abstract.** The University of Naples "Federico II," through its University Center for Information Services (CSI), enhances services through a strategic shift to the cloud. This transformation targets the dismantling of data silos, significantly enhancing service efficiency and accessibility. The hybrid exam reservation system, designed for an improved student experience, seamlessly integrates with the cloud, leveraging advanced APIs and Oracle's Autonomous Database. This strategic adoption empowers the university to address challenges promptly and autonomously like resource shortages. As we strive to balance our commitment to student services with innovative exploration, the cloud emerges as a pivotal compromise, ensuring our university remains technologically advanced and adaptable to evolving needs.

## 1 Introduction

In this era of rapid technological evolution, the University of Naples "Federico II" has distinguished itself through the proactive adoption of cutting-edge [16] solutions in the IT and digital services sector [1–3]. The University Center for Information Services (CSI) emerges as a crucial hub for innovation and optimization of IT infrastructures within the academic institution. The CSI not only acts as a catalyst for emerging technologies but also serves as a reference point for the integration and development of innovative computing solutions in the academic environment.

The increasing complexity of university services, coupled with the increasingly sophisticated needs of students and academic staff, has necessitated a transition towards new technological frontiers. A series of APIs have been developed and implemented to facilitate communication between an existing application [11], provided by the Interuniversity Consortium for Automatic Calculation of Northeast Italy (CINECA) and adopted by several universities, and the data management system of the University of Naples "Federico II." The innovative approach of the CSI focuses on the integration of a hybrid data management architecture: a combination of a cloud database with a local one. The adoption of the cloud environment represents a significant leap in optimizing academic infrastructures [4]. The Federico II University stands at the forefront of this transformation, aiming for a more agile and efficient experience for both students and academic staff.


L. Barolli (Ed.): AINA 2024, LNDECT 203, pp. 122–131, 2024.
https://doi.org/10.1007/978-3-031-57931-8_12

The very nature of our institution and the fragmented structure of pre-existing systems have given rise to a situation where essential data for academic management, student services, and research are confined within separate silos. These silos, because of previous technological developments and disjointed infrastructures, have hindered the flow of information and collaboration among different sectors of the university.

The need to overcome these limitations imposed by data silos becomes imperative. The inefficiency stemming from data fragmentation undermines responsiveness, slows down decision-making processes, and restricts the potential for synergies among various university sectors. Additionally, the absence of a holistic view of data can impede the development of innovative solutions and the creation of more personalized and optimized experiences for end users.

This article aims to thoroughly examine the experience of the University of Naples "Federico II" in optimizing university services through the adoption of the cloud environment. In Chapter 1, we will scrutinize the architecture of the system for exam booking management, while in Chapter 2, we will delve into the implementation of APIs for integration with the cloud, exploring the pivotal role of the Oracle Autonomous Database and the configurations we chose to implement. Finally, in the conclusion, we will present future perspectives and implications of the transition to cloud-edge in optimizing academic services.

## 2 Architecture of the Exam Booking Management System

Data management is a fundamental aspect of the efficiency and accuracy of the exam booking process. In the university context, the vast amount of information related to courses, schedules, instructors, and classroom availabilities is crucial to ensure students can book exams accurately and promptly.

Our system architecture is designed to handle this heterogeneous and dynamic data efficiently and securely. The centralization of information, coupled with real-time updating mechanisms, ensures that students have access to updated and relevant information.

Moreover, proper data management enables user experience personalization, allowing platforms to dynamically adapt to the specific booking needs of everyone. The ability to check exam availability, view schedules, and access course information from a smartphone enables students to plan their academic paths more efficiently.

In the university ecosystem, agility in the student experience plays a crucial role, especially in the context of exam bookings. The capacity to respond promptly to the needs and requests of students is a key element in enhancing their interaction with the booking system and, consequently, their overall satisfaction.

Agility in the application manifests through various aspects. Foremost, instant access to critical information, such as exam availability and course schedules, empowers students to make quick and informed decisions about their academic choices. This speed and immediacy in accessing data translate into better study planning.

Furthermore, the use of the application plays a fundamental role in the agility of the student experience by minimizing the time required to perform operations such as exam booking or cancellation.

Agility, therefore, is not merely a technical feature of the application but a central pillar to ensure that students can navigate the university ecosystem flexibly and swiftly, contributing significantly to their academic success.

## 2.1 Architecture of the System: Data Synchronization and Management

The approach adopted by the University of Naples Federico II (UniNa) to enhance university services through the implementation of Cineca's mobile app focuses on the creation of an advanced API layer. This layer plays a crucial role in interfacing the mobile app with the student's educational and administrative management system, known as GEDAS.

At the core of this architecture are several components working in synergy, as can be seen from the Fig. 1. Students use the mobile device as a portal to access the app, while Cineca's API Gateway, hosted in the cloud, serves as an intermediary between the app and the services of GEDAS and UniNa. The entire system relies on a structure composed of GEDAS's local database, UniNa's Cloud Database, and a series of APIs and services facilitating access to information. The APIs [7], developed in Java, play a pivotal role in orchestrating communication between the mobile app and the local database. They interface directly with Oracle's SQL procedures and packages, facilitating both read and write operations efficiently and coherently.

The UniNa Cloud Database, in its synchronization with the local GEDAS, employs advanced strategies to optimize this critical operation. Tables are implemented as materialized views, crucial for ensuring a coherent data perspective, leveraging Oracle's atomic refresh [9] option. The atomic refresh feature allows updating a materialized view without compromising data integrity in the event of interruptions or errors during the update process, a vital aspect in our context.

Additionally, to efficiently manage substantial-sized tables, the decision was made to adopt the fast refresh option [8, 9]. In this context, materialized views utilize local logs [10] to efficiently record changes, such as insertions, updates, and deletions. The fast refresh option enables the synchronization of materialized views by involving only the modified data, eliminating the need for a complete regeneration of tables, and ensuring optimal performance in large-scale scenarios.

This approach reflects UniNa's commitment to ensuring smooth and coherent access to information, both locally and through the cloud. Intelligent synchronization, advanced resource management, and the integration between local and cloud databases represent significant steps toward optimizing the academic experience through cutting-edge technological solutions.

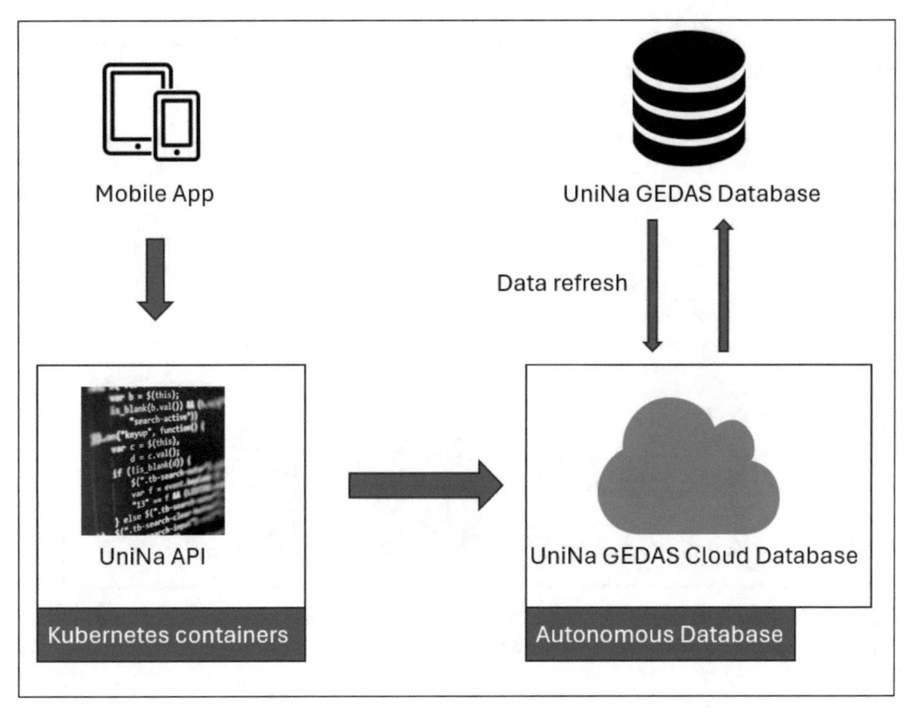


**Fig. 1.** High-level diagram of UniNa's exam reservation system: integration between local and cloud databases, APIs, and orchestration through Kubernetes. Implements Oracle's Autonomous Database for scalable and autonomous resource management.

## 3 Implementation of APIs for Integration with the Cloud

During the implementation of the APIs for integration with the cloud at the University of Naples Federico II, advanced strategies have been adopted to maximize the efficiency and flexibility of the university system.

The APIs, developed in Java, serve as a crucial interface between the students' mobile application and the complex data management system, ensuring efficient read and write operations through direct interaction with Oracle's SQL procedures and packages. The process involves 7 SQL packages, but the core of the entire process lies in the S3_APP_PRENOTAZIONI package, characterized by the following Oracle SQL functions:

- "prenota_esame": responsible for booking an exam instance, creating the reservation.
- "cancella_prenotazione": responsible for canceling the reservation instance when a user decides to cancel a previously booked exam reservation.

To test the functionality of the APIs, integration with the cloud, and study response times, a series of tests were conducted. In particular, a sample of 17,000 users over 60 min was executed. It's important to note that the times do not account for mobile device latency; indeed, the reported times correspond solely to backend calls and do not align with the actual user perception on the mobile device. Below are two graphs corresponding to the response times required to retrieve information about students and bookable exams, respectively (Figs. 2 and 3).

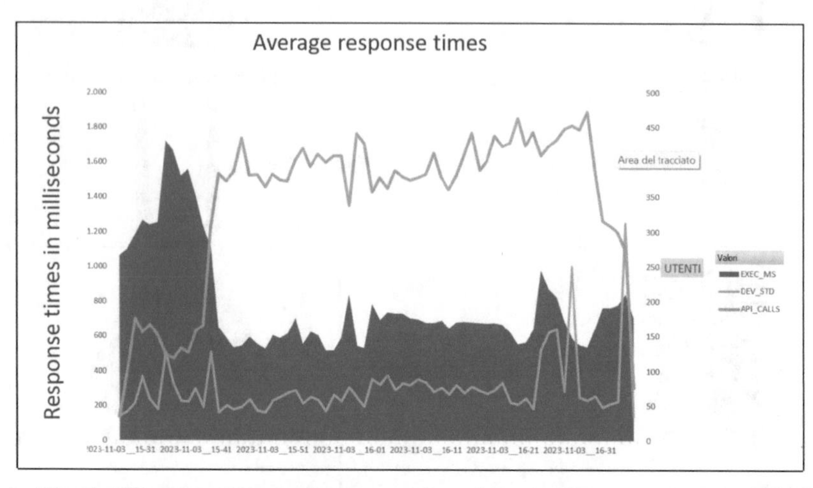


**Fig. 2.** The chart illustrates the average response times for accessing personal academic information. The times are consistently below 2 s on average, demonstrating the cloud's efficient response times.

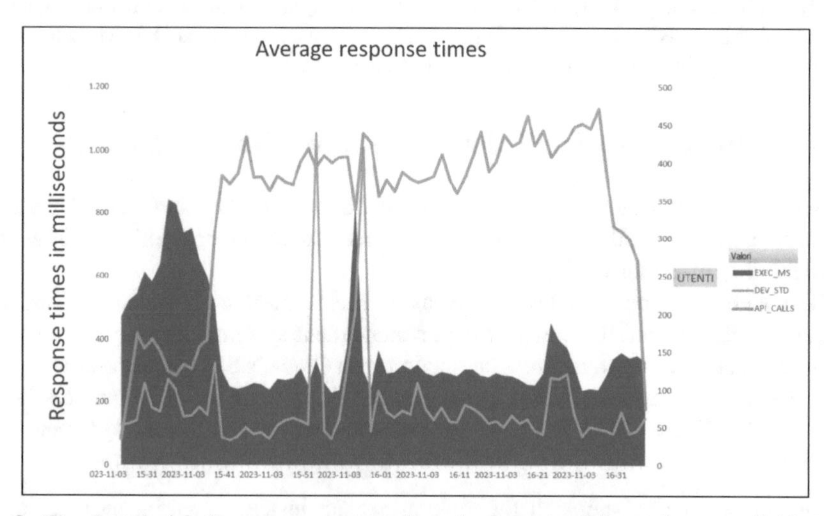


**Fig. 3.** The chart depicts the average response times for accessing information on available exam sessions. The times hover around one second on average, showcasing the cloud's swift response times.

**Kubernetes and Containers.** In the implementation process, orchestrating the APIs through Kubernetes [6] has proven to be a crucial element for dynamic and efficient resource management. Kubernetes is an open-source system designed for the scalability and reliable management of containerized applications. Containers [5] are lightweight and portable virtual environments containing applications and all their dependencies. This technology allows packaging the application and necessary resources into an isolated environment, ensuring consistency between the development and production environments. Below is a graphical representation of the Kubernetes architecture in Fig. 4.

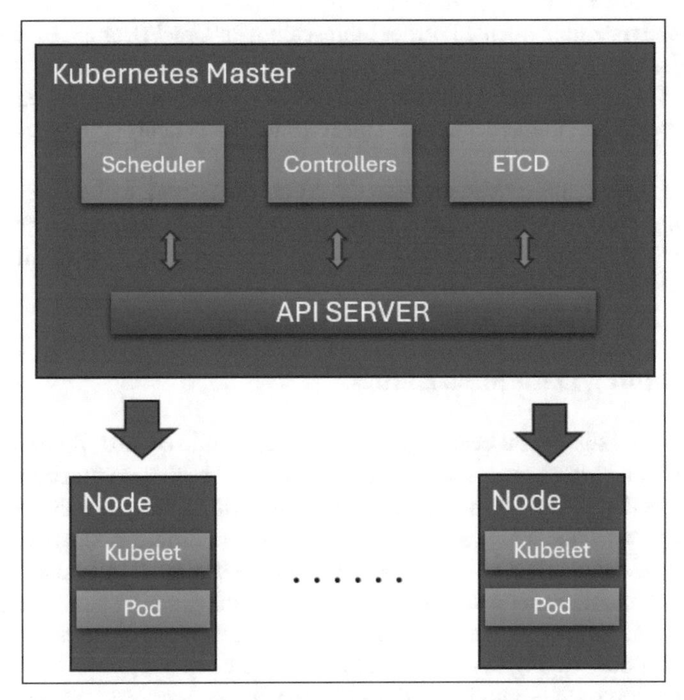


**Fig. 4.** Kubernetes architecture comprises a Master node, responsible for management and coordination, and n Worker nodes that execute containerized applications. The Master oversees application deployment and monitors the cluster's status.

The use of Kubernetes has allowed for the automation of the deployment process of containerized applications, managing their execution on a cluster of machines efficiently. This has led to a consistent and reliable distribution of resources while providing the necessary flexibility to adapt to dynamic changes in academic demand.

In practical terms, during the testing and initial implementation phase, the use of containers enabled rapid modification of resources, testing different configurations, and evaluating system performance. The portability and lightweight nature of containers facilitated a cohesive and consistent development environment, streamlining the process of managing and maintaining applications in various scenarios.

In summary, Kubernetes and containers have played a fundamental role in facilitating the integration of APIs into the university ecosystem, addressing the challenges and dynamic changes in the academic environment.

**Oracle Autonomous Database.** Another essential aspect was the choice to integrate the Oracle Autonomous Database. This tool, with its functionalities, automatically manages resources based on demand. During regular operations, the Autonomous Database, enabled with autoscaling functionality, optimizes itself by automatically scaling resources up or down without requiring manual intervention. This self-management directly impacts costs, reducing expenses during periods of inactivity and enhancing operational efficiency during peak times.

Specifically, we implemented vertical autoscaling for Oracle Compute Units (OCPU) [12, 14, 15], ranging from 2 to 6 OCPU[1] depending on demand, and horizontal autoscaling of pods [13], each affecting the total number of connections that can be established to the database: each container can establish up to 100 concurrent connections to the database.

Looking ahead, the implementation of APIs, orchestration through Kubernetes, and the adoption of the Autonomous Database form a solid foundation for a cutting-edge university system characterized by operational efficiency, flexibility, and cost optimization.

## 4 Overcoming Data Silos Limits

Data silos pose a significant challenge within the academic context, directly impacting the efficiency and effectiveness of university services. These silos function as stagnant compartments that isolate and segregate information within the university environment, restricting cross-sectional data sharing and access across different departments.

The most evident effect of data silos is the fragmentation of information. Essential data for academic management, such as exam scheduling, course timetables, resource availability, and student information, often resides in separate systems, making them inaccessible or challenging to integrate into a holistic view.

This isolation of data implies a series of negative consequences for university services. Difficulty in accessing data and the lack of a comprehensive view can hinder academic decision readiness, slow down administrative processes, and limit the adaptability of the academic institution to continuously evolving needs.

Furthermore, data silos can lead to errors or duplication of information. The lack of consistency and synchronized updates can create discrepancies between information available in different silos, compromising the accuracy and reliability of services provided to students and academic staff.

Overcoming these data silo limits requires a clear strategy and a robust data management architecture that promotes information integration and centralization. The adoption of cloud-edge technologies and the implementation of solutions that facilitate cross-sectional data access can significantly contribute to mitigating these problems, allowing for greater efficiency and flexibility in university services.

[1] 1 OCPU on x86 CPU Architecture (AMD and Intel) = 2 vCPUs [14].

The University of Naples Federico II, recognizing the constraints imposed by data silos, has embraced a strategy focused on cloud-edge integration to overcome these barriers and enhance data accessibility and management.

The cloud-edge approach is based on the idea of combining the strengths of local resources (on-premises) with the scalable and flexible capabilities of cloud computing [1]. This integration aims to create a synergistic environment where data is accessible, shared, and efficiently managed, regardless of its physical location.

The adoption of cloud-edge technologies enables the university to integrate and synchronize data from different systems, thereby reducing fragmentation and duplication of information. By employing a hybrid approach, where a portion of the data resides in on-premises infrastructure and another part is moved to the cloud environment, the limitations of space and scalability of local systems are overcome.

The ability to store, process, and analyze large volumes of data in the cloud optimizes the management of critical information for the university, enhancing data accessibility and security. Moreover, orchestrating cloud-edge resources, utilizing tools such as Kubernetes, facilitates dynamic resource management, ensuring an optimal user experience for both students and academic staff.

This cloud-edge integration strategy stands as a fundamental pillar in the university's digital transformation, aiming to create an environment where data can be unified and made accessible. The stated goal is to overcome the limitations of data silos and improve the quality and efficiency of the services offered.

## 5 Conclusion and Future Perspectives

The implementation of a hybrid [17] data management system at the University of Naples Federico II represents a significant step forward in optimizing university services. The integration of a local database with a cloud environment has proven to enhance efficiency and accessibility, especially in the context of exam reservation management.

This path of research and implementation has sparked a series of reflections. On one hand, there's the added value that a hybrid approach provides in terms of flexibility and local control. On the other hand, the managerial complexity of a dual system also emerges, with challenges related to the synchronization and maintenance of two distinct environments.

Looking to the future, an interesting consideration could be to evaluate a complete transition to the cloud environment, abandoning the local infrastructure altogether. This move could bring numerous advantages, including increased scalability, simplified resource management, and potential cost savings associated with local hardware maintenance.

However, it must be acknowledged that a complete transition to the cloud is not without challenges. Data security and reliance on external vendors are critical aspects that require careful consideration [18]. Additionally, migration could entail significant time and costs. Therefore, questions arise about how to balance potential benefits with challenges and the best strategies for a gradual and smooth transition.

Looking ahead, it might be useful to further explore the adoption of emerging technologies, such as blockchain-based storage or the integration of artificial intelligence, to further enhance data management and optimize academic services.

In conclusion, today's research represents a starting point rather than a definitive conclusion. The world of technology is constantly evolving, and each step taken offers new opportunities for learning and adaptation. We look to the future with enthusiasm and awareness of challenges, ready to contribute to progress in the field of university data management.

**Acknowledgments.** This work has been funded by the NextGenerationEU project (HPC National Center, Big Data and Quantum Computing, Italian Center for Super Computing (ICSC), Mission 4, Component 2, Investment 1.4) with project code CN_00000013 - CUP UNINA: E63C22000980007.

## References

1. Abbas, A., Khan, K.: Edge computing: extending the cloud to the edge of the network (2023)
2. Gannon, D., Barga, R., Sundaresan, N.: Cloud-native applications. IEEE Cloud Comput. **4**(5), 16–21 (2017). https://doi.org/10.1109/MCC.2017.4250939
3. Pahl, C., Xiong, H., Walshe, R.: A comparison of on-premise to cloud migration approaches. In: Lau, K.K., Lamersdorf, W., Pimentel, E. (eds.) Service-Oriented and Cloud Computing (ESOCC 2013). LNCS, vol. 8135, pp. 212–226. Springer, Heidelberg (2013). https://doi.org/10.1007/978-3-642-40651-5_18
4. Weinman, J.: Hybrid cloud economics. IEEE Cloud Comput. **3**(1), 18–22 (2016). https://doi.org/10.1109/MCC.2016.27
5. Bernstein, D.: Containers and cloud: from LXC to Docker to Kubernetes. IEEE Cloud Comput. **1**(3), 81–84 (2014). https://doi.org/10.1109/MCC.2014.51
6. https://kubernetes.io/docs/home/
7. Qiu, D., Li, B., Leung, H.: Understanding the API usage in Java. Inf. Softw. Technol. **73**, 81–100 (2016). https://doi.org/10.1016/j.infsof.2016.01.011
8. Oracle Corporation: Oracle database SQL language reference, v. 19c. https://docs.oracle.com/en/database/oracle/oracle-database/19/sqlrf/sql-language-reference.pdf
9. Oracle Corporation: Refreshing materialized views. Oracle Database Data Warehousing Guide. v. 12.2. Oracle. https://docs.oracle.com/en/database/oracle/oracle-database/12.2/dwhsg/refreshing-materialized-views.html#GUID-6EEA28AC-503B-4526-AD56-85378B547971
10. Oracle Corporation: Create materialized view log. Oracle Database SQL Language Reference, v. 19, Oracle. https://docs.oracle.com/en/database/oracle/oracle-database/19/sqlrf/CREATE-MATERIALIZED-VIEW-LOG.html#GUID-13902019-D044-4B79-9EB4-1F60652D037B
11. Cineca: Applicazione Mobile. Cineca - Sistemi Informativi Universitari. https://www.cineca.it/sistemi-informativi-universita/comunicazione/app-mobile
12. https://www.oracle.com/cloud/compute/
13. Nguyen, T.-T., Yeom, Y.-J., Kim, T., Park, D.-H., Kim, S.: Horizontal pod autoscaling in kubernetes for elastic container orchestration. Sensors **20**(16), 4621 (2020). https://doi.org/10.3390/s20164621
14. Oracle: Learn how to choose the best compute shape for machine learning. Document ID: F18367-04 (2019)
15. Jang, J., Jung, J., Hong, J.: An efficient virtual CPU scheduling in cloud computing. Soft. Comput. **24**, 5987–5997 (2020). https://doi.org/10.1007/s00500-019-04551-w
16. Atieh, A.T.: The next generation cloud technologies: a review on distributed cloud, fog and edge computing and their opportunities and challenges. ResearchBerg Rev. Sci. Technol. **1**(1), 1–15 (2021). https://www.researchberg.com/index.php/rrst/article/view/18

17. Dittakavi, R.S.S.: Evaluating the efficiency and limitations of configuration strategies in hybrid cloud environments. Int. J. Intell. Auto. Comput. **5**(2), 29–45 (2022). https://research.tensorgate.org/index.php/IJIAC/article/view/65
18. Turner, M., Williams, D.: Ethical considerations in cloud data management: a framework for best practices. Ethics Data Manage. **19**(2), 55–71 (2019)

# Distributed Collaborative AR on Cloud Continuum: A Case Study for Cultural Heritage

Beniamino Di Martino[1,2,3], Gennaro Junior Pezzullo[1,4], Dario Branco[1(✉)], Vincenzo Bombace[1], and Simonetta Ceglie[5]

[1] Department of Engineering, University of Campania "Luigi Vanvitelli", Caserta, Italy
{beniamino.dimartino,dario.branco}@unicampania.it,
gennaro.pezzullo@unicampus.it, vincenzo.bombace@studenti.unicampania.it
[2] Department of Computer Science and Information Engineering, Asia University, Taichung City, Taiwan
[3] Department of Computer Science, University of Vienna, Vienna, Austria
[4] Department of Engineering, University of Rome "Campus Bio-Medico", Rome, Italy
[5] Documentary Heritage Enhancement Service at "Archivio Centrale dello Stato" of Rome, Rome, Italy
simonetta.ceglie@cultura.gov.it

**Abstract.** In this work we explore distributed and collaborative augmented reality (AR) and motion capture technologies in a Cloud Continuum context through a case study focused on cultural heritage. In particular, the main objectives of the project are the definition of different logical and physical architectures through the creation of architectural, sequence and component diagrams, the implementation of these applications through reference platforms and programming languages such as Unity and Python, and finally the identification of limits from a computational and network speed point of view. This aspect is crucial to evaluate the feasibility and effectiveness of the proposed solutions. The actual testbed was realized thanks to the support of a documentary exhibition, held at the Central State Archives in Rome, which provided the ideal environment for testing the developed applications.

## 1 Introduction

On the occasion of the 80th anniversary of the round-up, at the Central Archives of the Italian State, in collaboration with the Cultural Assets and Activities Department of the Jewish Community of Rome, an exhibition entitled "28 Chariots of Jews" was inaugurated from October 16, 2023 until January 30, 2024, commemorating the tragic events with an exhibition of archival documents, bibliographic materials and artwork by the artist Georges de Canino. This exhibition, in addition to its significant cultural and awareness-raising impact, was also an opportunity to experiment with advanced Gesture Capture and Augmented [1]/Virtual [2] Reality "mixed" technologies for immersive enjoyment of the exhibition's digital content in a Cloud Continuum context [3]. In particular, this was an opportunity to implement a real testbed, allowing to collect concrete performance data and evaluate the effectiveness of the defined architectures.


L. Barolli (Ed.): AINA 2024, LNDECT 203, pp. 132–140, 2024.
https://doi.org/10.1007/978-3-031-57931-8_13

Through this event, one was able to test the systems in a real and dynamic environment by gathering information such as resilience, effectiveness, and scalability. The data collected during the exhibition provided crucial insights to further optimize the architectures and ensure that they can hold up effectively under various loads and operating conditions. The paper therefore is organized as follows: after showing the state of the art of the technologies used the proposed architectures in a Cloud Continuum environment will be defined in the 3th chapter, and then the 4th chapter will focus on how these architectures have been implemented and tested in real-world contexts and in particular in two case studies with applications usable in the exhibition. Finally, the data collected during the event will be discussed and emphasis is brought to how these will be used in future developments.

## 2 State of the Art

In the field of cultural heritage, more and more researchers and scholars are exploring the potential of the joint use of Cloud-Edge Continuum [4] and augmented reality (AR) techniques, offering a new perspective on the dissemination and exploration of cultural heritage. As evidence of the growing interest in these technologies, the review carried out in [5] that demonstrates the growing interest in the application of augmented reality techniques in the field of cultural heritage is interesting. Indeed, the use of these technologies has allowed the integration of digital content with physical artifacts, creating immersive and engaging experiences for visitors [6]. Moreover, in exhibitions where archival content is present, as in the use case presented in this paper, it is of fundamental importance not only to digitize (which allows the preservation of documents from the inevitable passage of time and their consequent deterioration) but also to enjoy the digitized content. Often, in fact, within exhibitions we find ancient documents placed under display cases of which we can only read the first page through the case, digitization and fruition in augmented reality allow, however, to access even the pages that otherwise would not be accessible to the user. Despite this premise, as demonstrated in [7] there are still few researchers who place their interest in the enjoyment of otherwise inaccessible assets. Contrary to what has been said so far, the possibility of integrating Cloud-Edge architectures into these domains remains little explored in the literature. In fact, existing studies have mostly focused on using the Cloud only to improve parameters such as scalability and accessibility of applications. The goal of this work, on the other hand, is to use the Cloud as a computing power for processing and storing the largest part of the data and to exploit the benefits of edge computing integration to meet the constraints of privacy, latency, and bandwidth.

## 3 Architecture

This chapter will show the design of two architectures for different purposes that will then be implemented. Among the commonalities we certainly have: **scalability:** since both architectures are designed to be deployed in an environment with a varying number of users, **cloud environment:** both architectures use Cloud Continuum technology although for different purposes. What instead differentiates them is the **privacy:** the second architecture unlike the first has been developed with privacy by design approach [8], in fact in the cloud it possesses only the trained model, while the data remains locally. While this choice requires more capacity on the serving side and results in less optimization, it also provides more privacy over the data Fig. 1.

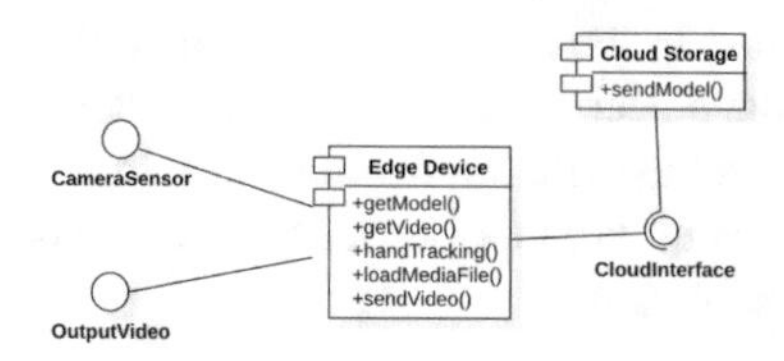


**Fig. 1.** Component Diagram of the Two Proposed Architectures

Below, we describe the two architectures separately in detail so as to motivate the respective design choices for each:

- **First Architectures (Augmented Reality):** In the first architecture, in which the goal is to enjoy objects in augmented reality, we distinguish mainly 3 components with their communication interfaces that are: mobile device, edge component and cloud. **Mobile Device** deals with what concerns in the rendering of assets in AR, the fruition of these by the end user and save performance measures such as processor type, ram capacity used, speed of download and others; **Edge Component** [9], which works as a bridge: in particular, the main purpose of this component is to optimize **Cloud** resources to avoid making numerous requests to storage. In particular, upon request by a device to receive an asset before requesting it from the cloud storage it checks if it has already been downloaded and, in case it has, it sends it directly to the device. If too much time passes and the asset in question is not requested it proceeds to delete it so as not to saturate the storage. On the other hand, about performance measures to avoid numerous writes and reads, the server takes care of receiving these measurements from the devices and, as soon as it reaches a fixed number of requests, it sends everything to the cloud database by doing a single write.
- **Second Architecture (Motion Capture):** The second architecture, on the other hand, is designed to use motion capture through the use of a trained model [10]. Here, we distinguish two components with a communication interface and a sensor to capture video images. In this case, since these media files are potentially sensitive because they could capture faces or other privacy-protected content, a local approach was chosen with regard to the media content. Specifically, this time the cloud has only the trained model that upon request of the server machine is downloaded. Once downloaded and after verifying the correct functioning, the server takes care of applying the latter on the video images that are transmitted in real time and managing the movements by mapping them in relation to the fruition of the multimedia files it possesses locally.

Both architectures described for clarity have also been made explicit through the use of the sequence diagrams in Fig. 2.

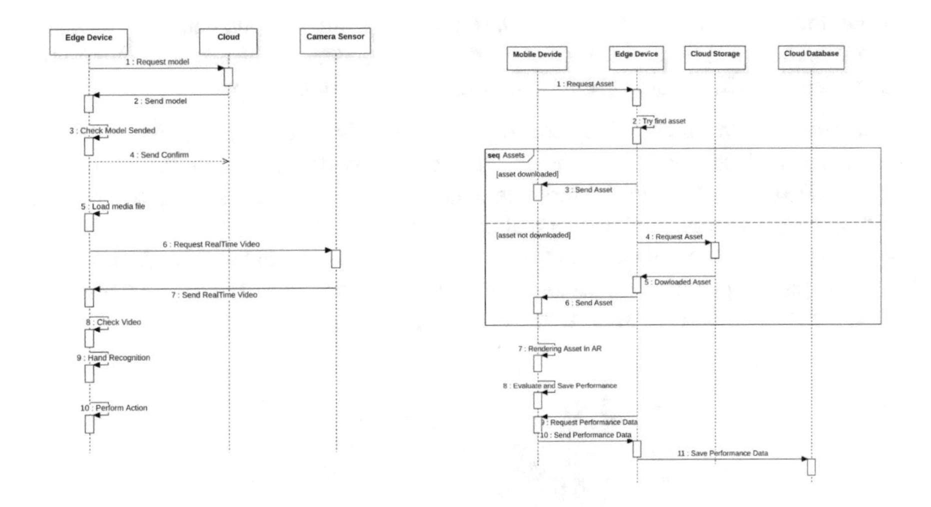


**Fig. 2.** Sequence Diagram of the Two Proposed Architectures

Once the design of the two architectures is finished, the next step is implementation, which requires not only the management of a backend that can comply with the design but also strong development both to easily deploy the application and attention to the frontend to make it as intuitive as possible for the exhibition participants in order to engage as many users as possible and consequently collect as many measurements as possible.

## 4 Implementation and Testing

With reference to the architectures mentioned above, two applications were developed. The first of these was developed using Unity. Known for its versatility in creating augmented reality experiences, the key functionality of the application is activated by scanning a QR code using the visitor's smartphone camera. Once the code is detected, AR is activated and the user can view digital documents in three-dimensional format. The use of the continuum cloud allows the full exploitation of AR so as to ensure high performance and a pleasant experience for the visitor. The second application blends machine learning and motion capture to allow users to browse the documents in the exhibition virtually. The software thanks to motion capture and a camera detects hand movements that will go into input to an artificial intelligence model previously trained in swipe recognition, showing on a projector the typed document.

### 4.1 First Case Study: Fruition of Digital Documents in Augmented Reality Through Smartphone

In this section, we are going to explore the implementation of the fruition of exhibition content through augmented reality in the cloud continuum environment. The first phase of development was the digitization and creation of documents and artworks into usable 3D models for AR. Specifically, the assets created were of two types:

- **Documents:** We used a free asset of a browsable book available online. This was adapted and customized to include the exhibition documents of the exhibition. Specifically, we developed a script associated with this element that allows for page flipping as close as possible to a real document (Fig. 3).
- **Pictures:** We used a 3D object of type quad, which customized itself with the texture of the paintings to be visualized. This approach made it possible to transform two-dimensional artworks into interactive 3D models (Fig. 4).

**Fig. 3.** Example of use with browsable book interaction

**Fig. 4.** Example of enjoying the virtual exhibition top photo with a normal camera, bottom photo with the application

Since the main purpose of this project is to best optimize the user experience and performance, taking into account the designed architecture, the assets were all uploaded within the Firebase cloud storage [11]. The second phase is the purely client part developed on unity. To use augmented reality, the project needs the ARFoundation library. [12]. This is needed because it simplifies the integration of augmented reality (AR) features into applications developed with the Unity game engine [13]. It provides a set of consistent APIs that facilitate floor detection, object tracking, user interaction, and other key AR features. This allows developers to focus more on AR application logic without having to handle complex platform-specific details. After configuring the project, we moved on to implementing the script associated with the AR camera. The default image tracking component allowed to effect of an association of an asset to multiple images, which meant having multiple QR codes to which the same asset was always appended. For this reason, it was decided to write an ad hoc script that would extend the functionality of the component. We used a list of tracked objects instead of a single object for each image, we used a dictionary to keep track of the tracked objects based on the name of the reference image. This allows multiple images to be handled simultaneously. In addition, we added a check to make sure that the object associated with the tracked image is present in the dictionary before modifying it. Queued to the logic of virtual object gendering was the call to the local server that is responsible for fetching the objects from Firebase cloud storage and sending them to the application. This was accomplished by replacing the string of code that was responsible for generating the object with a call to a function that downloads the right object. The presence of this near-edge component represents a bridge with which the smartphone communicates both to retrieve assets and especially also to retrieve data related to device specifications. The near-edge component is a server written in Python that allows the smartphone to be able to request via special APIs the assets it needs from cloud storage. The requested asset will be downloaded to the machine as well as sent and will reside on it for the duration of the server's operation. Another important functionality implemented within this component is that of writing data related to smartphones for testing. Unity gives us a way to know different information related to the device on which the application is run, among the many pieces of information useful for our research purposes are:

- **Device name**
- **Ram capacity:** This parameter represents the available RAM excluding the amount used by the operating system
- **Device battery quantity**.
- **Type of processor:** Here we have the model of the processor present inside the device
- **Server Response Time**
- **Network latency**
- **Network latency**
- **DB Write Latency**

These values are taken within the client, which constantly sends to the near component, which in turn saves to the Firebase database [14]. Data privacy in a context such as state storage is paramount, especially when it comes to information about exposed objects. Firebase Storage uses path-based security rules (path-based rules) to define who has access to which files. To ensure that these constraints are met, rules have been written that allow a device to access assets only if a unique ID is associated with the device that uniquely distinguishes it from others. This rule which may seem very trivial protects us from being able to read any other device not specified within the rules.

## 4.2 Second Case Study: Gesture Navigation of Multimedia Content

In this section, we will address the discussion of a mode of multimedia content enjoyment that allows the user to interact with the platform using only hand gestures. The architecture consists of a Web application that integrates a content gallery and a Real-time image recognition mechanism that uses the user's webcam. Thanks to this approach, the user can in fact select with a hand motion a collection of documents to view and confirm it by closing the palm of the hand toward the webcam and browsing through the collection that will open as if it were a real book and perform zoom in and zoom out operations through other predefined gestures Fig. 5.

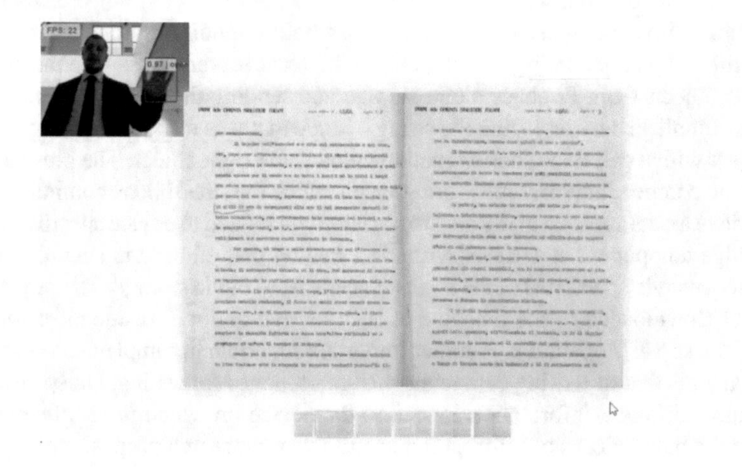


**Fig. 5.** Motion capture application interface for fruition of multimedia content

Such a system proves useful, especially in public environments such as exhibitions, installations, and museums where the number of multimedia content users is high and interaction through standard mechanisms produces potential hygiene risks. In fact, a fruition mechanism that does not involve contact with a surface not only makes the fruition experience more immersive and engaging but certainly drastically reduces the chances of transmission of viruses and bacteria. The software system has been designed from a Cloud-Edge perspective taking into account potential privacy limitations by having to process images that depict users. In order to solve problems related to the processing of user data, we chose not to process the images in the Cloud but rather to send the trained recognition model into Edge. This architecture allows us to process user data in the edge and ensures that no image is transmitted over the network but remains confined to the user device [15]. The model sent from the cloud to the edge is based on the popular Computer Vision OpenCV library [16] and allows us to recognize different gestures in static images. In the work performed, camera frames are sent to the computer vision mechanism, which in turn processes the frames and returns a confidence level for gesture prediction for each frame. The application in turn processes the output of the algorithm to calculate the position of the hand relative to the webcam's field of view and tracks the movements to make decisions in accordance with them (a hand moving from right to left means that the user intends to change the page of the book and flick to the next page).

## 5 Conclusion and Future Development

In this paper after a state of the art analysis, two distinct architectures were defined, implemented, and tested in a real-world environment. The former architecture focuses on the fruition of digital documents in augmented reality while the latter on motion capture for interactive navigation of multimedia content. Both leverage Cloud Continuum technology and in particular Cloud Edge to process data at the edge rather than directly in the cloud and in particular one of the two, through a privacy by-design approach, places special emphasis on privacy. During the testing phase, all the measurements needed to simulate the architectures later in controlled environments were properly salted. The next step now will be, on the basis of measurements taken, with the help of simulators such as Cloud Sim [17, 18] to stress these infrastructures as much as possible to understand what improvements will be made. On the other hand, as far as the implementation part is concerned, significant improvements could be made both from the user experience point of view, such as the use of even more immersive features, e.g. voice commands or gesture controls, and from the backend point of view, in particular using the emerging 5g technology to speed up the transfer of data from the cloud to the server.

**Acknowledgments.** The work has been supported by ARS01_00540 - RASTA project, funded by the Italian Ministry of Research PNR 2015–2020.

## References

1. Carmigniani, J., Furht, B.: Augmented reality: an overview. Handb. Augmented Reality 3–46 (2011)
2. Boas, Y.: Overview of virtual reality technologies. In: Interactive Multimedia Conference, vol. 2013 (2013)
3. Moreschini, S., Pecorelli, F., Li, X., Naz, S., Hästbacka, D., Taibi, D.: Cloud continuum: the definition. IEEE Access **10**, 131 876–131 886 (2022)
4. Di Martino, B.: Applications portability and services interoperability among multiple clouds. IEEE Cloud Comput. **1**(1), 74–77 (2014)
5. Silva, C., Zagalo, N., Vairinhos, M.: Towards participatory activities with augmented reality for cultural heritage: a literature review. Comput. Educ. X Reality **3**, 100044 (2023)
6. Ambrisi, A., Aversa, R., Branco, D., Ficco, M., Venticinque, S., Renda, G., Mataluna, S.: Intelligent agents for diffused cyber-physical museums. In: Camacho, D., Rosaci, D., Sarné, G.M.L., Versaci, M. (eds.) IDC 2021. SCI, vol. 1026, pp. 285–295. Springer, Cham (2022). https://doi.org/10.1007/978-3-030-96627-0_26

7. Boboc, R., Bautu, E., Florin, G., Popovici, N., Popovici, D.: Augmented reality in cultural heritage: an overview of the last decade of applications. Appl. Sci. **12**, 9859 (2022)
8. Kalloniatis, C.: Designing privacy-aware systems in the cloud. In: Fischer-Hübner, S., Lambrinoudakis, C., Lopez, J. (eds.) TrustBus 2015. LNCS, vol. 9264, pp. 113–123. Springer, Cham (2015). https://doi.org/10.1007/978-3-319-22906-5_9
9. Gezer, V., Um, J., Ruskowski, M.: An introduction to edge computing and a real-time capable server architecture. Int. J. Adv. Intell. Syst. (IARIA) **11**(7), 105–114 (2018)
10. Mishra, S., Verma, V., Akhtar, N., Chaturvedi, S., Perwej, Y.: An intelligent motion detection using OpenCV. J. Sci. Res. Sci. Eng. Technol. (2022)
11. Chatterjee, N., Chakraborty, S., Decosta, A., Nath, A.: Real-time communication application based on android using google firebase. Int. J. Adv. Res. Comput. Sci. Manag. Stud. **6**(4) (2018)
12. Linowes, J.: Augmented Reality with Unity AR Foundation. Packt Publishing (2021)
13. Nicoll, B., Keogh, B., Nicoll, B., Keogh, B.: The Unity Game Engine and the Circuits of Cultural Software. Springer, Cham (2019). https://doi.org/10.1007/978-3-030-25012-6
14. Goswami, L., Agrawal, P.: IoT based diagnosing of fault detection in power line transmission through google firebase database. In: 2020 4th International Conference on Trends in Electronics and Informatics (ICOEI) (48184), pp. 415–420. IEEE (2020)
15. Branco, D., Di Martino, B., Venticinque, S.: A big data analysis and visualization pipeline for green and sustainable mobility. In: Barolli, L., Woungang, I., Enokido, T. (eds.) AINA 2021. LNNS, vol. 227, pp. 701–710. Springer, Cham (2021). https://doi.org/10.1007/978-3-030-75078-7_69
16. Parveen, S., Shah, J.: A motion detection system in python and OpenCV. In: 2021 Third International Conference on Intelligent Communication Technologies and Virtual Mobile Networks (ICICV), pp. 1378–1382. IEEE (2021)
17. Kumar, R., Sahoo, G.: Cloud computing simulation using CloudSim. arXiv preprint arXiv:1403.3253 (2014)
18. Calheiros, R.N., Ranjan, R., De Rose, C.A., Buyya, R.: CloudSim: a novel framework for modeling and simulation of cloud computing infrastructures and services. arXiv preprint arXiv:0903.2525 (2009)

# Survey on Reference Architecture for Cloud Continuum and Multi-access Edge Computing (MEC) in 5G Networks

Beniamino di Martino[1,2], Gennaro Junior Pezzullo[1,3(✉)], Warren Low[4], Per Ljungberg[5], and Souradip Saha[6]

[1] Department of Engineering, University of Campania "Luigi Vanvitelli", Caserta, Italy
beniamino.dimartino@unicampania.it
[2] Department of Computer Science, University of Vienna, Vienna, Austria
[3] Department of Engineering, University of Rome "Campus Bio-Medico", Rome, Italy
gennaro.pezzullo@unicampus.it
[4] Capabilities Division, Allied Command Transformation, Norfolk, USA
warren.low@act.nato.int
[5] ERICSSON Group Function Technology, Stockholm, Sweden
per.ljungberg@ericsson.com
[6] Fraunhofer Institute for Communication, Information Processing and Ergonomics FKIE, Wachtberg, Germany
souradip.saha@fkie.fraunhofer.de

**Abstract.** This study aims to conduct an in depth analysis of Cloud Continuum and Multi Access Edge Computing architectures, with a focus on the importance of reference architectures even in a 5G context by highlighting how these technologies can synergistically enhance processing at the edge of the network. In particular, after identifying the largest number of reference architectures through a comparative analysis, the strengths, contexts and motivations in which one can benefit from its use and the challenges associated with each architecture will be shown for each of them, taking into account several parameters including privacy, modularity, orchestration or integration. Finally, some key use case examples will be explored in order to validate the importance of these technologies.

## 1 Edge Computing and 5G

Edge computing is a general concept in computing which has multiple definitions. Gartner defines edge computing as "a part of distributed computing topology where information processing is located close to the edge, where things and people produce or consume that information". To understand the concept of edge it is necessary to take a step back and introduce the concept of "Cloud Continuum" [1]. The first definitions of cloud continuum were both presented in 2016: by Harshit Gupta [2] that defined cloud continuum as "a continuum of resources available from the network edge to the cloud/datacenter" and Mung Chiang that defined cloud continuum explicitly mentioning computational, related aspects, for instance, where and how the computation is performed. Subsequently, over the following years up to the present date, various definitions of cloud continuum have been proposed, gathered in numerous publications. We


L. Barolli (Ed.): AINA 2024, LNDECT 203, pp. 141–150, 2024.
https://doi.org/10.1007/978-3-031-57931-8_14

will not dwell on citing them all, but it is important to focus on the fact that there are generally two main types of definitions. The former sees the continuum as the distribution of resources [3] across different network elements, including IoT, Fog, Edge, and even HPC, while the latter considers the continuum as an extension of the processing power to various nodes, often also mentioning the possibility of executing AI [4,5]. Edge computing can be seen in contrast to cloud computing where information processing takes place in a central hyperscale datacenter often placed geographically far from the users and devices. Edge computing mitigates latency and enhance response times for the users [6,19]. The advent of 5G has further amplified the value of edge computing. With significantly higher data transmission speeds and more stable connections provided by 5G it's possible to process a larger amount of data at the edge in real time. This synergy between 5G and edge computing improves operational efficiency and ensures a responsive user experience. In the context of 5G, when discussing edge, there are primarily two interpretations. The former sees the edge as "Computing Nodes" within the 5G network, emphasising processing close to the source of the data. The latter, on the other hand, characterises the edge as a manifestation of the "Cloud Edge", a virtualized and cloudified extension of the network, often leveraging technologies like NFV. Both these perspectives aim to enhance the efficiency and responsiveness of services in the 5G network: while one underscores the significance of physical proximity, the other highlights virtualization and flexible resource management. According to 3GPP, the motivation for introducing edge computing in mobile networks is to reduce latency and response times experienced by the users and to offload the communication networks [7]. Also, edge computing can reduce issues related to limited or unstable connectivity between the edge and a central cloud.

## 2 MultiAcess Edge Computing

The edge computing architecture known as Multi-access Edge Computing (MEC) was developed by the ETSI telecommunications community [8]. Even if the term "MEC" was introduced by ETSI, in this paper we refer to any edge compute platform that may be utilized with 5G networks and 3GPP's edge computing support. This is referred to as the "ETSI MEC" when discussing the architecture of ETSI. ETSI GS MEC 003 V3.1.1 [9] defines both a MEC Framework and Generic Reference Architecture 1 that has evolved since 2016.

The MEC System entities at the system, host, and network levels are described in the Framework. 3GPP networks are amongst the possible networking technologies mentioned in the framework. Prominent in the Reference Architecture are the functional elements and their associated external, management and platform reference points. The operator Operational Support System provides interface between the MEC system and the applications requesting to run applications in the MEC system. The multi-access nature of the reference architecture allows for a number of different access technologies (e.g. 3GPP) to interface with the MEC system setting up the potential for other unique vertical communications technologies to interface with the MEC system. ETSI MEC also allow for additional variant reference architectures such as the federation of MEC system from different providers.

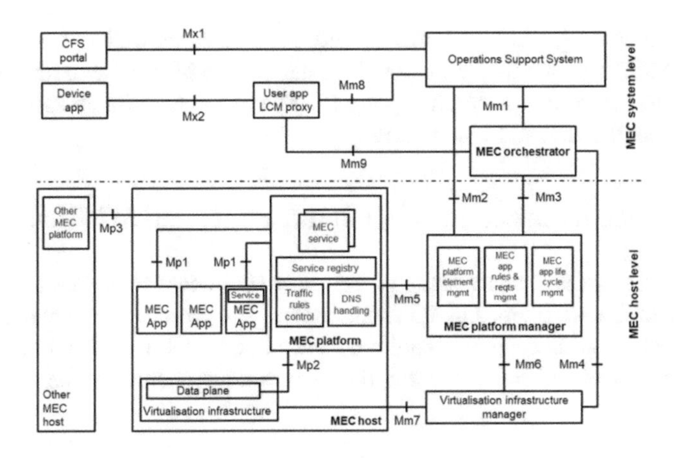


**Fig. 1.** ETSI MEC Generic Reference Architecture

Applications are usually managed in cloud data centers. These centers use distributed cloud architectures, content caching and high-bandwidth links between the access point to the mobile network and cloud providers. In specific situations, where rapid communication is needed and data is to be kept local, local management of User-Plane Functions (UPFs) can be opted for through a private cloud infrastructure, which can also include local or at-the-edge UPF deployments, as shown in the Fig. 2. MEC technology is applicable both in large mobile operator networks (PLMNs), which typically cover entire countries, and in private mobile networks (NPNs) with much more limited geographic coverage, such as a single base station (gNB).

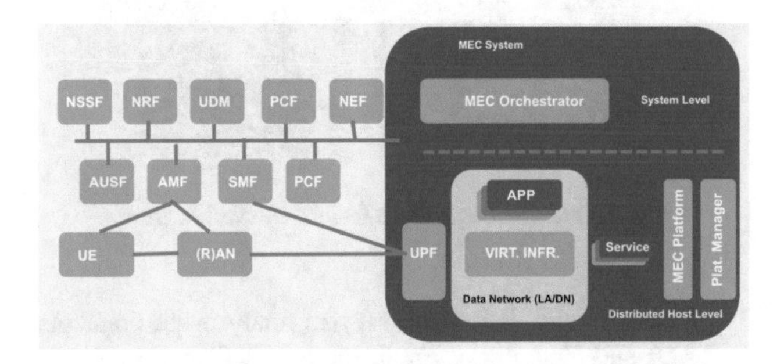


**Fig. 2.** 5G Service Based Architecture - MEC System Integration

In PLMNs, currently, the deployment of mobile networks is concentrated in a few central areas. However, with edge computing, this deployment can extend to regional or even local levels. For NPNs, edge serveristics can be integrated with processing systems for CN and RAN, making edge computing a fundamental component of the NPN concept. Several frameworks have been developed to explain how edge computing and cloud computing relate and interact with each other, as well as how they operate. The ETSI MEC has established a method to integrate the MEC system into 5G networks following the telecommunications approach. The 5G system interacts with the MEC system at both the host and system level.

# 3 Reference Architectures in Cloud, Edge, and 5G Environments

In this chapter, we will show some of the most significant reference architectures in the context of Cloud, Edge, and 5G technologies. The growing interoperability [10] of these technologies increasingly require advanced and well defined architectural solutions capable of effectively managing the various components and performance optimization.

## 3.1 National Institute of Standards and Technology (NIST)

National Institute of Standards and Technology (NIST) reference architecture for cloud continuum [11] defines the key actors, their activities and functions within cloud computing 3 [12].

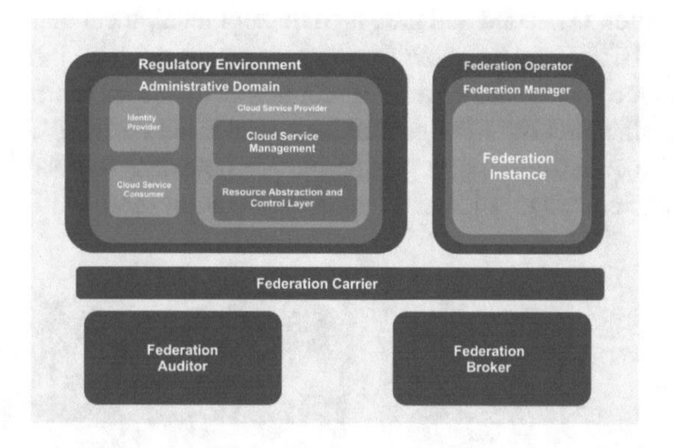


**Fig. 3.** Overview of NIST's Reference Architecture

Before delving into details, it is necessary to analyse the concepts of Administrative Domains (AD) and Regulatory Environments (RE) and explain how these are fundamental in this cloud federation model. In general, the basic authentication and authorization process exists within an Administrative Domain (DA) which is essentially composed of Identity Provider (IdP), Cloud Service Provider (SP) and Cloud Service Consumer (or simply a user). An IdP issues identity credentials to a Cloud Service Consumer, or User. When a User makes a service request, the SP validates the User's credentials with the IdP and then makes an access decision. All administrative domains exist within a Regulatory Environment. Cloud Service Consumer (CSC) and the Cloud Service Provider (CSP) both represent a person or organisation, but the former has a business relationship and uses the services of a Cloud Service Provider, while the latter is responsible for making cloud services available to interested parties. Lastly, the other actors shown in Fig. 3 are: **Federation Manager** that provides essential management functions for the duration of a federation; **Federation Operator** that facilitates the operations of one or more Federation Manager; **Federation Auditor**: an independent third party that can evaluate the compliance of any policy associated with a federation; **Federation Broker**: an intermediary between different federations that provides additional capabilities such as brokering, aggregating, and arbitrating services; **Federation Carrier**: it provides connectivity and transport of cloud services between consumers and providers. This architecture gives enhanced interoperability [13] for shared military operations. Security through authentication and authorization ensure compliance with strict defence standards.

## 3.2 FL Edge

Fl Edge, an open source industrial IoT system, is based on an extensible microservices architecture. The reference architecture shown in Fig. 4 is divided into:

- **Plugins** (components with a blue background) that are modules that allow for the extension of Fledge, written in Python or C++.
- **Microservices** (components with a light blue background) that can be distributed across single or multiple environments

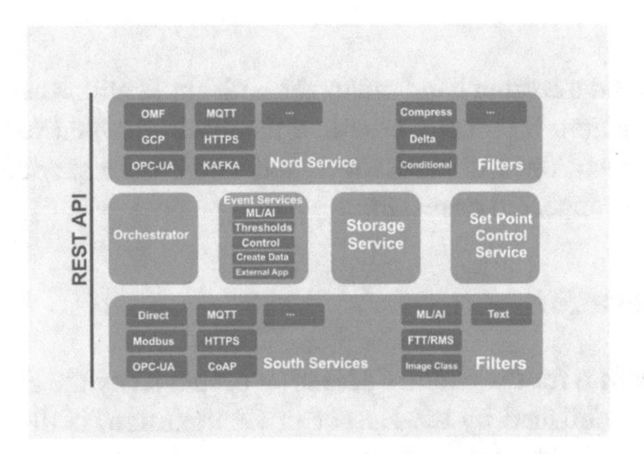


**Fig. 4.** Fl Edge Architecture

Going into a bit more detail, the key components of this architecture are: the **Fledge Core** which coordinate fundamental function such as: scheduling, configuration management, and others; **Storage Layer** which offers flexibility and scalability; **South and North Microservices** that connect Edge devices and broader systems and facilitate data transmission; **Filters and Pipelines** for the customization of data flows; **Event Service** a monitoring that responds based on predeterminated rules. Furthermore, the architecture has REST APIs and a User Graphic Interface and allows customizations through plugins and prewritten filters in Python. Fledge is a modular platform that offers advanced solutions for data management and processing, connecting edge devices and larger systems with effectiveness and flexibility.

### 3.3 KubeEgde

The entire architecture of KubeEdge [14], shown in Fig. 5, is divided into three layers: cloud, edge, and device. The control center is located in the cloud where on the left there is a Kubernetes master (virgin Kubernetes control center), while on the right there is CloudCore which includes EdgeController and DeviceController; these process information from the control centre, and, subsequently, the Cloud Hub transmits this information to the EdgeHub located at the edge section.

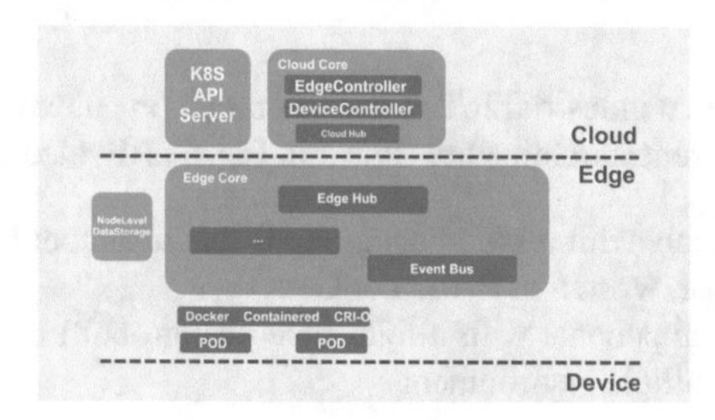


**Fig. 5.** KubeEdge Architecture

The edge layer, on the other hand, manages both applications and devices. EdgeHub receives the information, it is first stored in a database and then redirected to the edge or device. This allows Edged to access metadata and ensure service continuity even if the edge node is disconnected or restarted.

## 4 Cloud Federation

Reference Architecture for the Cloud Federation proposed by the Future Cloud Cluster, a working group established by the European Commission, is divided into three levels Fig. 6: The Edge Computing layer is responsible for the computational and storage capabilities made available and offers very low latency allowing for near real time data processing. Furthermore, this architecture stands out because it allows you to create clusters of Edge devices, including nested ones The Cloud layer, as can be seen from the name, provides interoperability between the actual cloud components. Federation Management mainly deals with the orchestration of cloud and edge resources and services.

The proposed architecture gives the possibility of creating clusters of Edge devices and provides a structure for management, device discovery, workload allocation, data management, networking, compliance, interoperability and security. This approach emphasises the importance of holistic management of the federation.

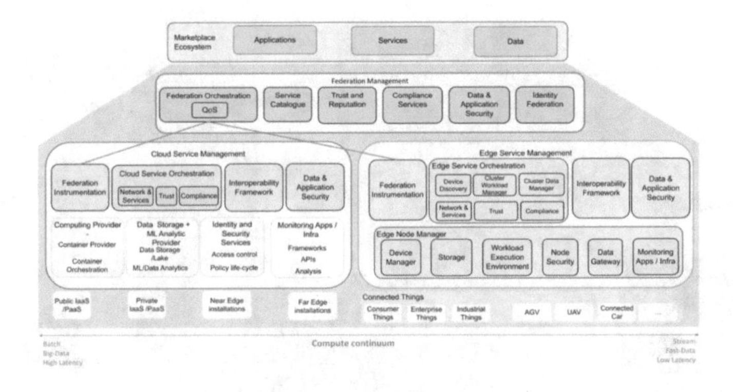


**Fig. 6.** Reference Architecture for Cloud Federation from the Future Cloud Cluster

## 5 Use Case Scenario: Sustainment of Military Operations

Military equipment can a diverse range of terminals that are mounted on land vehicles, aircrafts, maritime vessels as well unmanned units.

Additionally, such equipment deployed in military operations need to be sufficiently resilient to withstand extreme temperatures, operating at peak capacities, operate in harsh environmental conditions and including severe weather conditions over a long period of time. Under these circumstances, the operational effectiveness of military forces, depends on the conduct of preventative and restorative maintenance [15]. The conduct of that equipment maintenance is described as a logistics function within NATO Doctrine [16]. Within the NATO doctrine, a national or multinational logistics unit could be responsible equipment maintenance and at higher echelons. The military commander is responsible to maintain visibility on maintenance equipment and their status across multiple logistics units. The use of "intelligent dispatching and health monitoring" of equipment using embedded hardware, analytics software, and communications networks to provide commercial benefits [17] would thus be useful. Military operations can exploit existing vehicle telematics, additional internet of things sensors (or sensor networks), security mechanisms that exploit communications networks (e.g. 5G) utilizing multi-access edge computing. This is achieved by offloading the computational capabilities in proximity to the area of operations and processing a heterogeneous portfolio of data throughput, thus enhancing sustainment of military operations. As there is an expectation of hostile action and the potential inability to support maintenance close to combat areas, the use of virtual reality and augmented reality maintenance system using edge and cloud computing applications could provide training content and virtual assistance for maintenance in the field [18]. This is particularly beneficial for small communication networks close to theater, where there is a scarcity/absence of experts/specialists who can perform the required system maintenance tasks onsite. Combined vehicle maintenance with near real-time analytics and assistance through virtual and augmented reality could be implemented to provide evolutionary sustainment support to troops who need to maintain military equipment at geographically remote locations. On the other hand, to achieve a high degree of operational readiness,

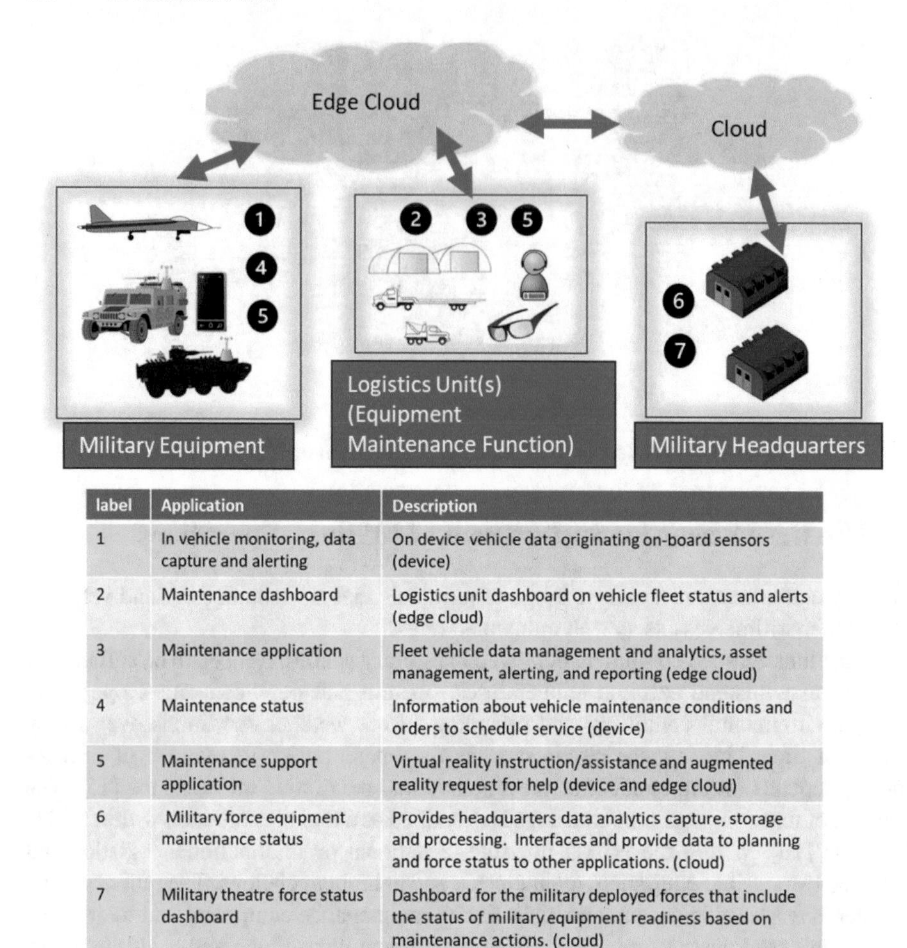


| label | Application | Description |
|---|---|---|
| 1 | In vehicle monitoring, data capture and alerting | On device vehicle data originating on-board sensors (device) |
| 2 | Maintenance dashboard | Logistics unit dashboard on vehicle fleet status and alerts (edge cloud) |
| 3 | Maintenance application | Fleet vehicle data management and analytics, asset management, alerting, and reporting (edge cloud) |
| 4 | Maintenance status | Information about vehicle maintenance conditions and orders to schedule service (device) |
| 5 | Maintenance support application | Virtual reality instruction/assistance and augmented reality request for help (device and edge cloud) |
| 6 | Military force equipment maintenance status | Provides headquarters data analytics capture, storage and processing. Interfaces and provide data to planning and force status to other applications. (cloud) |
| 7 | Military theatre force status dashboard | Dashboard for the military deployed forces that include the status of military equipment readiness based on maintenance actions. (cloud) |

**Fig. 7.** Military Maintenance Applications within edge and cloud computing infrastructure

summarized maintenance data from the national or multi-national units could be sent to higher echelons through cloud computing applications to satisfy wide area situational awareness of equipment and maintenance requirement. This could potentially avoid major supply-chain bottlenecks for sourcing equipment as well as improving the mission readiness level of deployed military units. Figure 7 provides an illustration of such a use case with the associated data and applications of interest.

## 6 Conclusion

In conclusion in this paper an in-depth analysis on Multi-access Edge Computing (MEC) and reference architectures in a Cloud Continuum environment has been carried out emphasizing their importance in a 5G context. In addition, in order to validate the importance of these technologies some case studies were shown including military ones that require the importance of factors such as: interoperability, modularity, offloading challenges or privacy. The analysis of these cases demonstrates the effectiveness of these solutions even in more critical contexts and paves the way for new implementation opportunities in different sectors.

**Acknowledgements.** This work has been conducted within the context of the Nato-STO (Science and Technology Organization) working group IST-187 on 5G.

Gennaro Junior Pezzullo is a PhD student enrolled in the National PhD in Artificial Intelligence, XXXVII cycle, course on Health and life sciences, organized by "Università Campus Bio-Medico di Roma".

## References

1. Di Martino, B., Esposito, A., D'Angelo, S., Maisto, S.A., Nacchia, S.: A compiler for agnostic programming and deployment of big data analytics on multiple platforms. IEEE Trans. Parallel Distrib. Syst. **30**(9), 1920–1931 (2019)
2. Gupta, H., Nath, S.B., Chakraborty, S., Ghosh, S.K.: SDFog: a software defined computing architecture for QoS aware service orchestration over edge devices. arXiv preprint arXiv:1609.01190 (2016)
3. Amato, A., Di Martino, B., Venticinque, S.: A distributed cloud brokering service. Informatica **26**(1), 1–15 (2015)
4. Di Martino, B., Esposito, A., Damiani, E.: Towards AI-powered multiple cloud management. IEEE Internet Comput. **23**(1), 64–71 (2019)
5. Pezzullo, G.J., Esposito, A., di Martino, B.: Federated learning of predictive models from real data on diabetic patients. In: Barolli, L. (ed.) AINA 2023. LNNS, vol. 655, pp. 80–89. Springer, Cham (2023). https://doi.org/10.1007/978-3-031-28694-0_8
6. Varghese, B., Wang, N., Barbhuiya, S., Kilpatrick, P., Nikolopoulos, D.S.: Challenges and opportunities in edge computing. In: 2016 IEEE international conference on smart cloud (SmartCloud), pp. 20–26. IEEE (2016)
7. Shi, W., Cao, J., Zhang, Q., Li, Y., Xu, L.: Edge computing: vision and challenges. IEEE Internet Things J. **3**(5), 637–646 (2016)
8. ETSI, M.: Mobile edge computing-introductory technical white paper. In: ETSI2014mobile, no. Issue (2014)
9. ETSI. Multi-access edge computing (MEC); framework and reference architecture. ETSI GS MEC 003 V3.1.1 (2022)
10. Di Martino, B.: Applications portability and services interoperability among multiple clouds. IEEE Cloud Comput. **1**(1), 74–77 (2014)
11. Plan, N.P.A.: National institute of standards and technology (NIST)
12. Liu, F., et al.: NIST cloud computing reference architecture. NIST SP 500-292. Consultado el **6**(08), 2013 (2011)
13. Di Martino, B., Cretella, G., Esposito, A.: Advances in applications portability and services interoperability among multiple clouds. IEEE Cloud Comput. **2**(2), 22–28 (2015)
14. Wang, S., Hu, Y., Wu, J.: Kubeedge. AI: AI platform for edge devices. arXiv preprint arXiv:2007.09227 (2020)
15. Nato logistics handbook (2012)
16. Allied joint doctrine for logistics, edition b, version 1 (2018)

17. Farahpoor, M., Esparza, O., Soriano, M.: Comprehensive IoT-driven fleet management system for industrial vehicles. IEEE Access (2023). https://doi.org/10.1109/ACCESS.2023.3343920.
18. Saha, S., Low, W., Di Martino, B.: Sustainment of military operations by 5G and cloud/edge technologies. In: Barolli, L. (ed.) AINA 2023. LNNS, vol. 655, pp. 70–79. Springer, Cham (2023). https://doi.org/10.1007/978-3-031-28694-0_7
19. Di Martino, B., Venticinque, S., Esposito, A., D'Angelo, S.: A methodology based on computational patterns for offloading of big data applications on cloud-edge platforms. Future Internet **12**(2), 28 (2020)

# Elastic Autoscaling for Distributed Workflows in MEC Networks

Benedetta Picano, Riccardo Reali(✉), Leonardo Scommegna, and Enrico Vicario

Department of Information Engineering, University of Florence, Florence, Italy
{benedetta.picano,riccardo.reali,leonardo.scommegna, enrico.vicario}@unifi.it

**Abstract.** With the recent advancements in computing technologies, new paradigms have emerged enabling users to access a large variety of distributed resources, overcoming several limitations of localized applications and information storage. Among these paradigms, Mobile Edge Computing (MEC) places storage and computing capabilities at the edge of the network, significantly decreasing congestion and service response times, at the cost of limited capacities. Within this context, the emergence of novel computationally intensive services has triggered the necessity to design algorithms that adaptively scale resources, achieving solutions tailored to traffic demand. In this paper, we present a preliminary scaling method to determine the resource provisioning of complex workflows of web services that are distributed on a MEC infrastructure, with the intent of improving the distribution of the end-to-end response time of the workflow. The method is designed to run compositionally, exploiting a structured hierarchical workflow representation, enabling efficient top-down determination of the resource provisioning. The method is also formalized to act considering the inherent limitations and complexities of an MEC network landscape. In so doing, we demonstrate the applicability of the approach on two synthetic application scenarios, confirming the validity of the proposed elastic scheme in optimizing resource management within a resource-constrained MEC network.

## 1 Introduction

Recent advancements in networking and computing technologies have sparked a growing interest in considering computation and communication in a collaborative and distributed manner, aligning with the network-computing paradigm [4]. Network computing introduces a novel computing paradigm wherein all information, data, and software applications exist on a network accessed by users on demand, whose infrastructure is typically deployed within the edge-to-cloud continuum. This computing approach promises to enable users to access a comprehensive range of resources from any location, eliminating the limitations associated with localized storage of information and applications [5]. This is due to the presence of Mobile Edge Computing (MEC) nodes which move storage and computing facilities at the edge of the network, close to end-users, resulting in significant performance enhancements, especially in terms of latency and response times. At the same time, the upcoming next-generation networks will


L. Barolli (Ed.): AINA 2024, LNDECT 203, pp. 151–160, 2024.
https://doi.org/10.1007/978-3-031-57931-8_15

enable the new era of computational intensive service classes, characterized by heterogeneous quality-of-service (QoS) and quality-of-experience requirements. To ensure effective service provisioning, the network resources need to be optimized and handled properly. Due to the computational-hungry nature of next-generation applications, and because of the high-velocity links expected to serve the new-generation networks, the usage of computational resources, if not optimized, risks becoming the bottleneck of novel applications. When resource exploitation is optimized, service providers only incur costs for the resources they utilize at a given moment. When managed effectively, this approach can lead to lower costs and a higher quality of service compared to hosting on traditional hardware [3]. Due to the dynamic and unpredictable nature of shared resources, autoscaling mechanisms must be designed to handle the complexity and time-varying nature of resource demand. The goal is to achieve a runtime scaling system that is self-aware, self-adaptive, and reliable in the face of changing demands [3].

Recently, autoscaling, originally designed and confined to cloud-based solutions, is gaining attention for its extension into distributed MEC contexts. For example, the paper [11] proposes a latency-optimal scheme to solve the monitoring function placement problem, considering an edge-to-cloud landscape. In order to favor the system scalability, authors design a hierarchical monitoring system topology, where a meta-heuristic algorithm is exploited to adaptively scale of resource pool of the MEC infrastructure. Then, an online scaling scheme was developed to perform real-time and on-demand resource allocation in such a monitoring system. In [9], the authors introduced deep learning models that encompass both centralized and federated strategies. These models are designed to execute both horizontal and vertical autoscaling across multi-domain networks. Authors in [6], propose an innovative auto-scaling method using a deep reinforcement learning-based algorithm. This method aims to dynamically adjust the number of instances assigned to an atomic microservice that composes a service, thereby optimizing resource allocation and potentially enhancing service performance.

An innovative approach to meet the dynamically changing network service demands in 5G networks was developed in [8]. The authors applied machine learning models for auto-scaling and predicting the required number of virtual network function instances based on traffic demand. Furthermore, Integer Linear Programming techniques were exploited to solve a joint user association and Service Function Chain placement problem. To address the scalability concern of the ILP model, they proposed a heuristic algorithm. However, existing solutions do not consider the necessity to realize coordinated scaling among distributed MEC resources for service workflows, i.e., services composed of atomic elementary tasks structured as directed acyclic graphs. In [1], a coordinated compositional approach is proposed to scale the resource provisioning of a workflow of services. The approach minimizes the workflow e2e response time, by considering topological information and exploiting a stochastic characterization of service durations, for an environment where any transmission costs occur. In this reference, the main contributions of this paper can be summarized as follows:

- The design and development of an effective coordinated vertical scaling scheme of MEC node resources to execute stochastic workflows within the distributed edge network. The objective is to meet QoS constraints by improving the end-to-end (e2e) response time distribution of the workflow.

- Definition of a heuristic that, through performing an efficient compositional analysis, deduces the resource provisioning of each sub-workflow. The proposed algorithm employs a structured workflow model to scale resources in a top-down fashion, including inter-MEC node communication costs.
- Application scenarios aimed at demonstrating the applicability of the method to workflow of services that are deployed on a MEC network, at illustrating a methodology to explore the solution space of the problem, and at highlighting the complexity of the problem which is opened to many relevant challenges.

## 2 Workflow Modelling

We model workflows by combining Stochastic Time Petri Net (STPN) [10] blocks. An STPN block is a single-entry/single-exit model, which receives a token when workflow execution starts, and eventually ends with probability 1 (w.p.1). Blocks are recursively combined with sequential, split/join, and choice/merge workflow patterns [7]. We reference workflows according to the following EBNF syntax:

$$
\begin{aligned}
< block > ::= \\
&\mathrm{SEQ}\,(< block > \{, < block >\})\;| \\
&\mathrm{AND}\,(< block > \{, < block >\})\;| \\
&\mathrm{XOR}\,(\langle < block >, prob\rangle \{, \langle < block >, prob\rangle\})\;| \\
&\mathrm{ACT}
\end{aligned}
$$

where ACT is an elementary activity (e.g., activity B in Fig. 1b), SEQ models sequential behaviors (e.g.,activity S1 in Fig. 1), AND models concurrent behaviors (e.g., activity A1 in Fig. 1), and XOR models alternative behaviors that occur with different probabilities (e.g., activity X1 in Fig. 1). A workflow is thus modeled as a *structure tree* [2] $S = \langle N, n_0 \rangle$, where $N$ is the set of nodes (i.e., blocks) and $n_0 \in N$ is the root node (i.e., the entire workflow). Each structure tree node $n_i \in N$ is characterized by the tuple $\langle n_i := R_i, Z_i, T_i, X_i \rangle$, where $R_i$ is the amount of provisioned resources, $Z_i$ is the generally distributed random variable characterizing the node e2e response time that arises from the given provisioning, $T_i = E[Z_i]$ is the expected value of the response time, and $X_i = R_i T_i$ is the *job size*, representing the amount of work required to complete the node with the assigned resources. As workflows are recursive compositions of STPN blocks, topological complexity can notably increases. Complexity is further exacerbated by the general characterization of the activity durations, which lead to the unfeasibility of many effective analysis methods. In such cases, to evaluate the e2e response time distribution of a node, we leverage on a compositional technique [2]. In particular, workflows are evaluated by performing a top-down visit of the structure tree to estimate the analysis complexity of blocks, evaluating the response time distribution of the identified sub-workflows in isolation, and finally performing a bottom-up recomposition of the obtained results. In particular, when workflows are defined by well-nested composite blocks (i.e., composition of AND, SEQ, and XOR blocks), the exact e2e response time distribution can be evaluated by recursive numerical analysis.

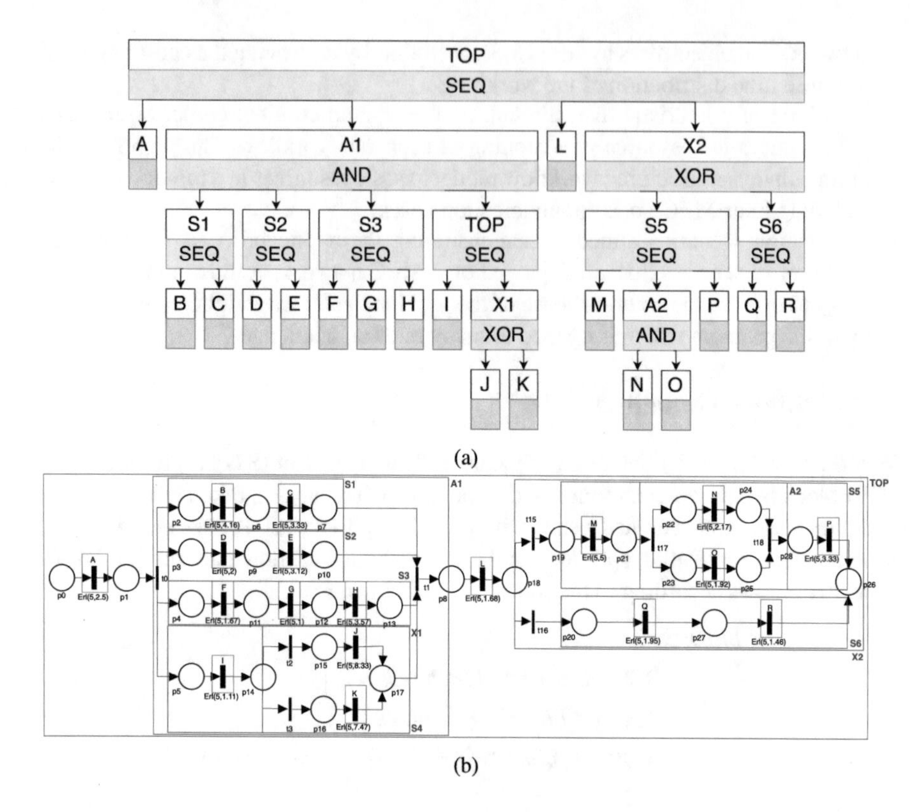


**Fig. 1.** (a) Structure tree and (b) STPN modeling a synthetic workflow.

# 3 A Communication Costs Aware Scaling Approach

## 3.1 A Performance Model

We propose a coordinated approach to determine the optimal resource provisioning of a workflow of services, that can be subject to communication costs incoming from distinct placements of the services on an edge network. The approach determines how to distribute an arbitrary amount of resources $R^{\text{in}}$ to the elementary activities of a workflow, i.e., the leaves of the workflow structure tree, with the intent of minimizing the e2e response time distribution of the workflow. Resource allocations are determined by performing a top-down approach that splits the input resources to the child nodes, exploiting both topological information and the response time characterization of each node. In particular, the approach leverages on the assumption that the job size of an activity is invariant with respect to the provisioned resources, i.e. given two different resource allocations $R_i, R_i'$ resulting in the expected response times $T_i$ and $T_i'$, then $T_i R_i = T_i' R_i' = X_i$, for each node $n_i$ of the workflow. This enables to express the expected response time of a node as a function of the child node resources, which is employed as the objective function of an optimization problem. To deal with communication costs, we include a probability distribution representing the time spent to reach services that are hosted on different MEC nodes. In particular, $D_i$ is introduced to represent the expected value of such communication time, for each node $n_i$. Consequently, the performance model of a node $n_i$ is defined as: $T_i = X_i/R_i + D_i$, where $X_i/R_i$ is the computation time of the activity. Communication times can be specified both on simple and composite activities, and are propagated bottom-up on the workflow nodes. In particular:

- For a sequence of $k$ activities, $D_{\mathrm{SEQ}(n_0,n_1,\ldots,n_k)} = \sum_{i=0}^{k} D_i$;
- For a fork/join of $k$, $D_{\mathrm{AND}(n_0,n_1,\ldots,n_k)} = \max(D_0, D_1, \ldots, D_k)$;
- For a random choice between $k$ activities, $D_{\mathrm{XOR}(\langle n_0,p_0\rangle,\langle n_1,p_1\rangle,\ldots,\langle n_k,p_k\rangle)} = \sum_{i=0}^{k} p_i D_i$;

Note that, since it is not possible to determine the exact expected communication time of a fork/join activity, we select the maximum of its children. In so doing, we design the approach to deal with safety-critical contexts, where worst case scenario is typically considered.

To summarize, the approach proceeds as follows:

- Initially, an arbitrary amount of resources $R_0^{\mathrm{in}}$ is assigned to the root node $n_0$.
- For each non-leaf node $n_k$, the amount of input resources $R_k^{\mathrm{in}}$ is split by assigning an amount $R_j^\star$ to each child node $n_j$, i.e., $\sum_{n_j \in C_k} R_j^\star = R_k^{\mathrm{in}}$; the assignment exploits a performance models where the job size of an activity is invariant with respect to any resource variation and communication costs changes the way resources are distributed.

By induction, the sum of the amounts of resources allocated to the leaf nodes is equal to the amount of resources of the root node, i.e., $\sum_{n_k | C_k = \emptyset} R_k^\star = R_0^{\mathrm{in}}$.

### 3.2 Resource Allocation Decisions

We characterize different rules to determine the resource provisioning of a node, that are based on the workflow pattern represented by the node.

**Elementary Activities.** Let $n_k$ be an elementary activity, when a new resource allocation $R^\star$ is evaluated by the approach, then its response time changes as

$$T_k^\star = \frac{R^\star}{R_k} T_i \tag{1}$$

which leads to a transformation of the node distribution parameters.

**Sequential Activities.** Let $n_k = SEQ(n_i, n_j)$ be the sequence of $n_i$ and $n_j$. Then:

$$T_k = T_i + T_j = \frac{X_i}{R_i} + D_i + \frac{X_j}{R_k^{\mathrm{in}} - R_i} + D_j \tag{2}$$

has a minimum with the following allocation

$$R_i^\star = \frac{\sqrt{X_i}}{\sqrt{X_i} + \sqrt{X_j}} R_k^{\mathrm{in}}. \tag{3}$$

The result is obtained by imposing $\frac{dT_k}{dR_i} = 0$, and it can be extended by induction to the sequence of $K > 2$ activities, $SEQ(n_1, \ldots, n_K)$:

$$R_i^\star = \frac{\sqrt{X_i}}{\sum_{i=1}^K \sqrt{X_i}} R_k^{\text{in}}. \tag{4}$$

**Concurrent Activities.** Let $n_k = \text{AND}(n_i, n_j)$ be a fork/join between $n_i$ and $n_j$. Then:

$$T_k = \max(T_i, T_j) = \max\left(\frac{X_i}{R_i} + D_i, \frac{X_j}{R_k^{\text{in}} - R_i} + D_j\right) \tag{5}$$

Since response time is not defined as an explicit function of $R_i$, the minimum can not be evaluated exploiting the Fermat theorem. Hence, we provide a heuristics evaluation of $R_i^\star$ which imposes equality between response times of $n_i$ and $n_j$:

$$\frac{X_i}{R_i} + D_i = \frac{X_j}{(R_k^{\text{in}} - R_i)} + D_j. \tag{6}$$

This leads to the allocation:

$$R_i^\star = \begin{cases} \frac{X_i}{X_i + X_j} R_k^{\text{in}} & D_i = D_j \\ \frac{\Delta D R_k^{\text{in}} - X_i - X_j \pm \sqrt{4 \Delta D R_k^{\text{in}} X_i + (X_i + X_j - \Delta D R_k^{\text{in}})^2}}{2 \Delta D} & D_i \neq D_j \end{cases} \tag{7}$$

where $\Delta D = D_i - D_j$. In particular, when $D_i \neq D_j$, it is chosen the solution for which $R_i^\star > 0$ and $R_i^\star < R_k^{\text{in}}$. The solution is extended to $m$ activities by rearranging the topology of the fork/join into a 2-children fork/join pattern, i.e., $\text{AND}(n_1, n_2, \ldots n_N) = \text{AND}(n_1, \text{AND}(n_2, \ldots, n_N))$.

**Alternative Activities.** Let $n_k = \text{XOR}(\langle n_i, p_i \rangle, \langle n_j, p_j \rangle)$ be a random alternative choice $n_i$ and $n_j$, with probabilities $p_i$ and $p_j = 1 - p_i$. Then:

$$T_k = p_i T_i + p_j T_j = p_i \left(\frac{X_i}{R_i} + D_i\right) + p_j \left(\frac{X_j}{R_k^{\text{in}} - R_i} + D_i\right) \tag{8}$$

has the minimum:

$$T_k^\star = \frac{\left(\sqrt{p_i X_i} + \sqrt{p_j X_j}\right)^2}{R_k^{\text{in}}} \tag{9}$$

for the following resource allocation:

$$R_i^\star = \frac{\sqrt{p_i X_i}}{\sqrt{p_i X_i} + \sqrt{p_j X_j}} R_k^{\text{in}} \tag{10}$$

which is in turn obtained by exploiting the Fermat theorem. As the solution shows, the optimal allocation of a XOR node is a generalization of the optimal allocation for a SEQ node. Hence, resource allocation for $n$ activities, can be derived as done for sequential nodes. Note that resources $R_k^{\text{in}}$ are split among activities $i$ and $j$, independently from how they occur or not. This typically occurs in service-oriented applications, where each service has a reserved amount of resources. In contrast, for Function as a Service (FaaS) solutions, resources are allocated on-demand only when a service is executed: in this case, resource costs are accrued only for the selected service, and the expected cost is $p_i R_i + p_j R_j$ instead of $R_i + R_j$, resulting in a different optimal allocation.

## 4 Application Scenarios

We illustrate a preliminary methodology to explore the space design of a MEC network hosting services of a workflow. We consider a synthetic and well-nested workflow (see Figs. 1a and 1b), combining 19 web services, each distributed as a 5-phases Erlang distribution with rate randomly selected in $[0, 10]$, for a maximum concurrency degree equals to 4. It is assumed that the considered response time arises by provisioning 1 resource to each service. We consider a QoS requirement obtained as 5-phase Erlang distribution whose rate is chosen to fit an expected value of 12.96 ms, which is the half of the workflow expected response time. Then, we consider two scenarios. In the first, each service of the workflow is deployed on a single MEC node; in the second, services B, C, M, N, P, Q are placed in a different MEC node, thus introducing some communication costs in the QoS requirement fit. In particular, communication times are characterized as uniform distributions, where support bounds are randomly selected in $[0, 5]$. In both scenarios, we assume that each node has an residual availability of 20 resources. Figure 2a illustrates the impact of limited MEC node resource availability to the problem of resource scaling. By applying the proposed technique, the QoS requirement could be met without significant effort (green line), by provisioning a total amount of 22.91 resources. However, this quantity overcomes the considered availability, requiring to adjust the computed provisioning, but ending up worsening the e2e response time distribution (blue line), which results not to fulfill the QoS requirement. Figure 2b illustrates how communication costs enable to mitigate the impact of limited resource availability. The presence of transmission costs implies a higher resource demand for the entire workflow to fulfill the QoS requirement, which is met with a total amount of 27.01 resources (green line). However, transmission costs affect the ways resources are distributed among the services and allocated to the considered MEC nodes. In particular, the total amount of resources allocated on the node that hosts services that causes transmission costs is 8.46, while in the other node 18.64 resources are allocated. Despite the higher resource costs, transmission costs mitigate MEC node saturation, which may be a desirable implication in the prospect of solving other problems such as service offloading or dynamic service placement. Table 1 reports the resource allocations evaluated for the considered scenarios. Column 2 provides the provisioning of resources when no constraints are given on the resource availability of the nodes. Column 3 reports the resource provisioning when all services are deployed on the single MEC node *A*. Finally, column 5 reports the resource allocation in the case services B, C, M, N, P, Q are deployed on the MEC node *B*. The last row of the table reports the cumulative resources provisioning of each considered scenario.

The proposed scenarios allow us to highlight the inherent complexity of the problem of provisioning resources to services hosted on MEC nodes, by illustrating a methodology through which exploring the design space of a workflow of services that is deployed at the edge of the network.

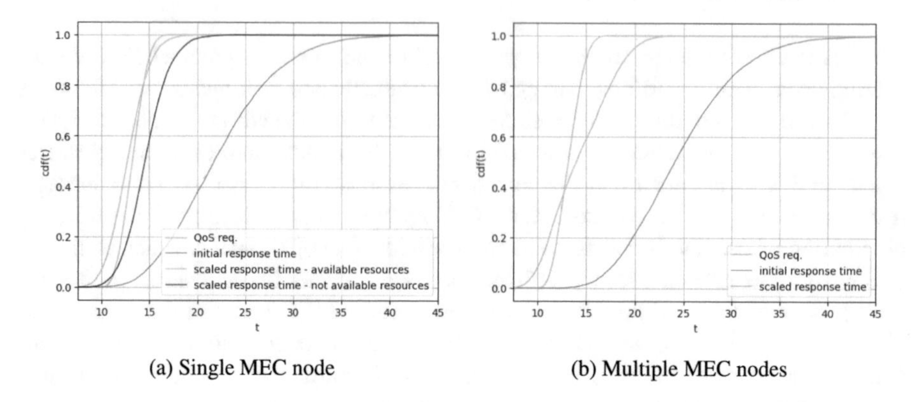


(a) Single MEC node

(b) Multiple MEC nodes

**Fig. 2.** Application of the proposed approach to different MEC network scenarios.

**Table 1.** Resources provisioning obtained by applying the proposed approach without constraint on the resource availability (column 2), when services are deployed on a single MEC node (column 3) and when services are deployed on different MEC nodes (column 5).

| SERVICE | OPTIMAL PROVISIONING | RESOURCES | NODE | RESOURCES | NODE |
|---|---|---|---|---|---|
| A | 1.87 | 1.66 | A | 2.03 | A |
| D | 0.80 | 0.69 | | 1.00 | |
| E | 1.96 | 1.68 | | 0.80 | |
| F | 0.56 | 0.48 | | 1.97 | |
| G | 2.53 | 2.17 | | 2.54 | |
| H | 1.34 | 1.15 | | 1.34 | |
| I | 1.52 | 1.31 | | 1.53 | |
| J | 0.35 | 0.30 | | 0.35 | |
| K | 0.45 | 0.39 | | 0.46 | |
| L | 2.29 | 2.03 | | 2.49 | |
| Q | 1.49 | 1.32 | | 1.91 | |
| R | 1.88 | 1.52 | | 2.21 | |
| B | 0.63 | 0.54 | | 1.36 | B |
| C | 1.00 | 0.85 | | 1.52 | |
| M | 0.93 | 0.82 | | 1.19 | |
| N | 1.08 | 0.95 | | 1.38 | |
| O | 1.22 | 1.08 | | 1.56 | |
| P | 1.14 | 1.01 | | 1.46 | |
| TOTAL | 22.91 | 20.00 | | 27.01 | |

## 5 Conclusions

In this paper, we propose a coordinated approach to evaluate the resource provisioning of a workflow of web services that are deployed at the edge of a network. Tha approach is experimented on a workflow that exhibits a well-nested topology with a non-negligible degree of complexity, with a maximum concurrency degree equals to 4. Evaluation of resource provisioning is performed in a top-down fashion, exploiting a hierarchical formalism, termed structure tree. To identify resource allocations for elementary web services, the approach solves an optimization problem which aims at minimizing the response time of structure tree nodes. In so doing, we leverage on an assumption of invariance of the job size of each node. We also characterize transmission costs to deal with services deployed on different MEC nodes.

Finally, the experimented workflow allows us to illustrate a methodology to explore the design space of a workflow that is deployed on an MEC network, illustrating how the approach can be exploited to analyze different scenarios. The results prove the applicability of the method, and highlight the inherent complexity of evaluating the optimal resource provisioning for a workflow deployed on an edge network.

## References

1. Carnevali, L., Paolieri, M., Picano, B., Reali, R., Scommegna, L., Vicario, E.: A quantitative approach to coordinated scaling of resources in complex cloud computing workflows. In: Iacono, M., Scarpa, M., Barbierato, E., Serrano, S., Cerotti, D., Longo, F. (eds.) EPEW ASMTA 2023 2023. LNCS, vol. 14231, pp. 309–324. Springer, Cham (2023). https://doi.org/10.1007/978-3-031-43185-2_21
2. Carnevali, L., Paolieri, M., Reali, R., Vicario, E.: Compositional safe approximation of response time probability density function of complex workflows. ACM Trans. Model. Comput. Simul. **33**, 1–26 (2023)
3. Espadas, J., Molina, A., Jiménez, G., Molina, M., Ramírez, R., Concha, D.: A tenant-based resource allocation model for scaling software-as-a-service applications over cloud computing infrastructures. Future Generation Computer Systems **29**(1), 273–286 (2013). https://doi.org/10.1016/j.future.2011.10.013, https://www.sciencedirect.com/science/article/pii/S0167739X1100210X. including Special section: AIRCC-NetCoM 2009 and Special section: Clouds and Service-Oriented Architectures
4. Fantacci, R., Picano, B.: Edge-based virtual reality over 6G terahertz channels. IEEE Netw. **35**(5), 28–33 (2021). https://doi.org/10.1109/MNET.101.2100023
5. Jiang, J., Lu, J., Zhang, G., Long, G.: Optimal cloud resource auto-scaling for web applications. In: 2013 13th IEEE/ACM International Symposium on Cluster, Cloud, and Grid Computing. pp. 58–65 (2013). https://doi.org/10.1109/CCGrid.2013.73
6. Lee, D.Y., Jeong, S.Y., Ko, K.C., Yoo, J.H., Hong, J.W.K.: Deep q-network-based auto scaling for service in a multi-access edge computing environment. Int. J. Netw. Manage **31**(6), e2176 (2021)
7. Russell, N., Ter Hofstede, A.H., Van Der Aalst, W.M., Mulyar, N.: Workflow control-flow patterns: a revised view. BPM Center Report BPM-06-22, BPMcenter. org **2006** (2006)
8. Subramanya, T., Harutyunyan, D., Riggio, R.: Machine learning-driven service function chain placement and scaling in MEC-enabled 5G networks. Comput. Netw. **166**, 106980 (2020)

9. Subramanya, T., Riggio, R.: Centralized and federated learning for predictive VNF autoscaling in multi-domain 5G networks and beyond. IEEE Trans. Netw. Serv. Manage. **18**(1), 63–78 (2021). https://doi.org/10.1109/TNSM.2021.3050955
10. Vicario, E., Sassoli, L., Carnevali, L.: Using stochastic state classes in quantitative evaluation of dense-time reactive systems. IEEE Trans. Softw. Eng. **35**(5), 703–719 (2009)
11. Yuan, Q., Ji, X., Tang, H., You, W.: Toward latency-optimal placement and autoscaling of monitoring functions in MEC. IEEE Access **8**, 41649–41658 (2020). https://doi.org/10.1109/ACCESS.2020.2976858

# Decentralized Updates of IoT and Edge Devices

Francesco Bruschi(✉), Marco Zanghieri, Michele Terziani, and Donatella Sciuto

Politecnico di Milano, Piazza L. Da Vinci, 31, 20133 Milano, MI, Italy
francesco.bruschi@polimi.it

**Abstract.** This paper explores software updates for Internet of Things (IoT) and Edge devices, focusing on the concept of device ownership. Our analysis of existing solutions reveals that, although asymmetric cryptography addresses the basic security and authenticity challenges of updates, it lacks flexibility in supporting a wide range of conditions or protocols, such as those needed for owner delegation and revocation of update privileges. As a potential solution, we propose the use of smart contracts on blockchain technology to articulate control policies, utilizing the blockchain's decentralization, security, and Turing-complete expressiveness for managing firmware updates. Additionally, we investigate how IoT devices with limited resources can interact with the blockchain, initially through a gateway model deployed on an ESP32 architecture, and subsequently via a direct connection by proposing a light consensus node for more capable architectures like the ARM Cortex-A72. Our findings suggest that smart contracts offer a viable and innovative method for firmware update control, presenting new opportunities for secure, efficient, and adaptable device management in IoT and Edge computing environments.

**Keywords:** smart contracts · blockchain · software update · OTA · device ownership

## 1 Introduction

In the age of digital transformation, the importance of the Internet of Things (IoT) and edge computing cannot be overstated. The increasing pervasiveness of computers, embedded throughout our environment, marks a significant evolution in how technology integrates into our daily lives. This transformation underscores the critical role of software updates in ensuring these devices operate efficiently and securely.

The behavior of programmable devices is fundamentally determined by their software or firmware. This distinction between firmware and software, while nuanced, highlights the layers of complexity in device programming. Control over this software is crucial for several reasons. Firstly, it enables the flexibility to reconfigure or repurpose devices, allowing them to adapt to new functions or


L. Barolli (Ed.): AINA 2024, LNDECT 203, pp. 161–170, 2024.
https://doi.org/10.1007/978-3-031-57931-8_16

behaviors. Secondly, it provides the means to correct or fix problems, ensuring devices operate as intended.

However, the significance of software control extends beyond these practical considerations. The ability to control a device's software increasingly defines its ownership. In a world where digital devices not only determine functionality but also regulate access to physical spaces, this concept of ownership becomes profoundly meaningful. Consider an IoT-enabled lock that controls access to a room, programmable to unlock under specific conditions. The control over this lock equates to a significant form of ownership over the room's space, as defining the lock's behavior closely relates to possessing the room itself. This is even more true if the devices can be reprogrammed without physical access, as is the case with the increasingly widespread over-the-air updates (OTA).

In traditional scenarios, ownership transfer of a lock might involve handing over a physical key. For IoT devices, the notion of changing ownership raises questions about data control and the implications of software management. Ownership in the context of IoT involves not just the physical device but also the data it generates and the control over its functionalities, including software updates.

Ownership, in essence, can be seen as a form of control. Yet, this control, particularly regarding software updates, introduces a nuanced debate. For instance, while an owner may have the ultimate say in device updates, they might delegate this responsibility under specific conditions. This delegation opens discussions on the prerequisites for secure and efficient update processes, the potential for unauthorized access, and the mechanisms to prevent attacks.

The conventional reliance on signatures for authorizing updates, while secure, offers limited flexibility. Advanced scenarios might require more complex authorization processes, such as allowing one party to propose updates subject to another's review within a specified timeframe. Implementing such processes challenges traditional security models, suggesting a move towards more dynamic and programmable solutions.

This evolution points towards the utilization of blockchain technology and smart contracts. Blockchains offer a decentralized and verifiable platform for managing transactions and data, overcoming the limitations of traditional computing infrastructure. By encoding ownership and control mechanisms within smart contracts, IoT devices can operate according to transparent and immutable rules. This approach ensures that device behavior aligns with predetermined regulations, secured not by the integrity of a single server or entity but by the distributed and resilient nature of the blockchain network.

In this paper, we explore the possibility of defining and controlling the rules for software updates of Edge and IoT devices through the use of smart contracts executed on blockchain platforms. This objective introduces several challenges, the foremost being the need to enable secure and decentralized access for devices to the state of a smart contract. Consequently, we propose and verify multiple options for connecting devices with varying computational and connectivity limitations to the consensus mechanism of a blockchain.

## 2 IoT Updates State of the Art

In traditional firmware update mechanisms, particularly those reliant on wired connections, initiating the update process requires manual intervention by the device owner. This method is commonly used for devices with limited network connectivity. A significant drawback of wired firmware updates is their time-consuming nature, especially when compared to automatic firmware update mechanisms. The most notable disadvantage is the necessity for the device owner to physically retrieve and update the device, requiring the selection of an appropriate maintenance window to minimise downtime impacts. Furthermore, the manual nature of this process increases the potential for human error during the update. In contrast, automatic update mechanisms, such as Over-The-Air (OTA) updates, are becoming increasingly popular due to their efficiency and reduced error likelihood. OTA updates offer substantial cost advantages, enabling centralised management of firmware across a broad range of devices through a cohesive, unified interface, significantly simplifying the update process. Currently, the OTA system is emerging as the optimal solution for addressing IoT firmware update and management challenges [4]. On the other hand, increasing adoption of OTA impacts the definition of device ownership and how it is excercised, since physical access is not a condition for updating its firmware or software, and makes the problem of defining access rules more pressing.

### 2.1 Roles and Structure of over the Air Updates

In the domain of updates, especially in the Internet of Things (IoT) context, three primary actors orchestrate the process: 1) The IoT device, such as sensors, actuators or computer devices which support OTA updates, is the update recipient. 2) The gateway acts as an intermediary, particularly in scenarios where devices are not directly internet-connected. This role is crucial in complex networks that serves as a connection point between multiple IoT devices and cloud where direct device management is impractical. 3) The cloud server, housing the firmware, serves as a repository and management point for updates. It stores various firmware versions and controls their distribution.

The firmware update method can be categorised into PUSH and PULL methods. The key difference lies in who initiates the firmware update process. The PUSH method can be seen as a server-initiated mode in which the update server pushes the updates to the client when new patches are available. Then, the same process of downloading, verifying, and installing the new image begins. In contrast, the PULL method is a client-initiated mode in which the client queries the update server for new updates on a regular basis, and if updates are available, the client downloads,verifies, and installs the new firmware.

In [13], an application of blockchain technology to update firmware of IoT devices was proposed. In this scheme, an hybrid PUSH-PULL update mechanism is adopted. In this scheme, the IoT device is assumed to execute a PUSH update mechanism by default. In case that any IoT device in the network could not obtain the firmware update due to unavailability of connection or other issues, the IoT device can still obtain the firmware update through a PULL update mechanism. Solution as such allows to obtain secure and performative update processes in IoT environments.

## 2.2 Security Aspects of IoT Updates

The main security goal in Over-The-Air updates is to establish a trust chain from the the subject legitimately tasked with oversight to the device, considering the capabilities of each participant in the process [5]. Currently, many available solutions and frameworks generally rely on a single centralised server that acts as a dispatcher of firmware updates on behalf of the device's manufacturer. However, this model is vulnerable and becomes a single Point-Of-Failure if the server is compromised. Additionally, it necessitates secure communication to avoid tampering and eavesdropping between the manufacturer and the devices, requiring encryption on every piece of data exchanged during the update process. Such a setup demands that the device itself has some processing capacity to perform encryption/decryption of sensitive data along with a secure internet-connected protocol to communicate with a server.

**2.2.1 Digital Signatures and Their Limits** Ensuring secure communication, with guarantees on the identity of the counterpart and integrity of the data, is today effectively addressed by cryptographic technologies. In particular, public key cryptography provides established tools for creating channels resistant to eavesdropping, through mechanisms such as key exchange and symmetric cryptography (e.g., TLS), and to impersonation, via digital signatures. Increasingly, even computationally limited devices are equipped with components or subsystems capable of handling cryptographic primitives, enabling the production and verification of digital signatures and the creation and use of secure cryptographic channels.

However, these tools exhibit limited expressiveness. For instance, a device can verify that a certain update originates from an entity controlling a specific public key, enabling the implementation of update control policies such as "subject A is authorized to update". This facilitates straightforward policies where an update is deemed legitimate if signed by entity A.

Yet, envisioning a slightly more complex policy, such as "subject A owns the update and authorizes subject B to push updates, but retains the right to revoke B's authorization at any time," becomes challenging when relying solely on cryptographic keys. Similarly, multi-stage control protocols, such as those requiring an update proposed by A to be available for B's review for a week, after which it can be pushed in the absence of objections, highlight the limitations in the expressiveness of cryptographic signatures for implementing nuanced access policies.

These examples illustrate that while cryptographic signatures can serve as a powerful and decentralized tool for device control, they also have inherent limitations in the complexity and flexibility of the access policies they can enforce.

## 2.3 Blockchain for Software Updates

The role, contribution, and function of blockchain technology encompass a broad spectrum of descriptions, from a decentralized ledger system for assets like Bitcoin [citare whitepaper] to a distributed registry where the sequence of records is determined through a permissionless consensus mechanism. For our discussion, it's beneficial to consider blockchain as a distributed state machine with several defining characteristics.

At any moment, the blockchain exists in a specific state, referred to as $S$. It operates based on a next state function, whereby applying this function to the current state and an input results in a new state. This process is essential for the progression and adaptability of the blockchain. Any user can submit a transaction $t$, which, when applied to $S$, transitions the blockchain to a new state, $S' = f(S, t)$. A consensus mechanism ensures the emergence of a unique, stable, and universally accepted ordering of user-submitted transactions. This mechanism resolves the inherent challenge of achieving a unified sequence of events within a decentralized network. The expressiveness of the state $S$ and of the next state function varies, accommodating different implementations and use cases. The consensus mechanism's properties allow for the verification of the correct sequence of transactions by anyone. In complex blockchain systems like Ethereum [3], the state $S$ can contain Turing-complete executable code, enabling users to submit transactions that execute arbitrarily complex logic. This feature allows for the creation of decentralized applications where the state's evolution is transparent and verifiable.

In particular, an entity can submit a transaction to the network that contains code, which is then uploaded to the blockchain's state, and from that moment on represents a process with which it is possible to interact, through other transactions. Such a process is called a *smart contract*. The evolution of a smart contract's state is subject to the same guarantees as the overall state of the blockchain. In this way, smart contracts can implement coordination logics between actors in a transparent and tamper-proof manner.

In recent years, the integration of blockchain technologies into the IoT ecosystem has been studied to improve the overall security. In [11], the authors face the main challenges of current IoT frameworks including centralised architectures, constrained devices, and privacy concerns due to the constant transmission of personal data across the network and propose a new architecture based on Ethereum. The main advantage claimed, beyond the elimination of a central authority over the network in favour of a decentralised system, is the local processing of data. This solution ensures that no personal data is transmitted outside the home network, thus enhancing privacy and providing a more secure communication protocol compared to commonly used methods.

Other studies have focused on the firmware update aspect of IoT devices as it represents a critical component. Blockchain-based update frameworks proposed in these studies aim to mitigate, and possibly eliminate, the major security concerns previously mentioned. Within these frameworks, blockchain technologies are mainly employed to securely share and authenticate the version and the integrity of the distributed firmware binary during the update process [8]. Additionally, other solutions advise for the use of smart contracts to manage the entire process, leveraging on their immutability and transparency. In [12], every phase of a firmware update is initiated and validated through smart contracts deployed on the Ethereum Network, thereby making the process accessible and visible to anyone. The main achievement of this solution is that it ensures the verification of the update during the entire process and offers a method to record each firmware version released by the manufacturer in a smart contract, presenting a general view of all the previous updates made which can be useful for data integrity checks.

## 3 Proposed Approach

To enable arbitrarily articulated upgrade policies to be implemented in a secure manner, we propose a smart contract-based upgrade control system. The system is structured as follows:

- A smart contract is deployed on blockchain, offering a public function that allows to identify which firmware/software version is updated. The version is identified via an untamperable software fingerprint, such as as a collision resistent hash function. The contract address is used as a reference to be incorporated into the device. The contract code describes the logic that defines the update/ownership policy (for instance, a delegation policy like the one presented in section ??. The different entities that participate in the update proposal process can access the smart contract and perform their role. When conditions are met, a new version of the software gets authorized.
- In the simplest configuration, the update manager process/component queries the status of the update contract. The contract address plays a similar role to public keys in the classic case of signature checking.
- If the device detects the availability of a new software version, it searches for it on a distribution network, checks the hash, and proceeds with the update.

The protocol is deliberately general, for example in the use of the distribution network. One of the problems in the application of this system concerns the device's access to the blockchain.

Traditionally, interacting with blockchain networks requires running a full node on one's machine, to directly follow the evolution of the consensus process. With this setup users can have their own copy of the ledger for instant querying and provides direct access to the blockchain network, enabling participation in the block-by-block validation and verification process. However, running a full node requires a significant amount of memory, storage, and CPU, meaning that it is not feasible for everyone to run their own node [6].

Another approach is to have the device query the blockchain through a gateway, i.e. a third-party synchronized node which therefore knows the state of the smart contract. This has the advantage of requiring limited resources from the device (little more than an http client), but requires trusting the gateway.

To address this limitation new protocols were studied and defined, leading to the introduction of so called *light clients*, that are able to verify the evolution of the consensus without having to witness all transactions history. In particular, light clients are able to verify the validity of a compressed *commitment* of the current state of the network, such as a Merkle root. Subsequently, they can query a full node about the current value of a specific portion of the state (for example, the result of the function call that returns the version of the latest software available). The full node returns the answer, with a proof that validates the result against the state's commitment. In this way, it is not necessary to trust the correctness of the node responding to the query.

his capability opens the possibility to deploy such clients on constrained environments, including and not limited to browser, mobile apps and even Edge and IoT devices.

Currently, there are several light clients for Ethereum, written in a variety of programming languages [10].

### 3.1 Using IPFS for Decentralized Distribution

Typically, in the phase of data retrieval of the OTA update process, the IoT device relies on a centralised cloud server architecture for retrieving these updates, which can present vulnerabilities and concerns regarding data privacy and security. To enhance the decentralisation of the OTA update processor we propose to use IPFS as a cloud server for storing updates. IPFS is a peer-to-peer (P2P) file distribution protocol that was designed by the Protocol Labs company and Juan Benet in 2014. It is built on distributed peer-to-peer network content addressing, assigning unique cryptographic hashes to files based on their content, enabling efficient retrieval and verification. It decentralises firmware update storage, reducing reliance on centralised cloud servers. Overall, IPFS provides a robust, scalable, and fault-tolerant mechanism, aligning with edge computing networks' decentralised nature [2]. To address the challenges using this technology, is the exponential increase of available files. Because each file is identified with an unique hash, even small changes of files would result in a completely different hash [9].

## 4 Experiments: Deployment on MCUs and Edge Devices

TThe purpose of these experiments is to assess the feasibility of deploying light clients on microcontrollers (MCUs) and edge devices to enable direct blockchain access for devices with limited capabilities. This approach allows these devices to independently verify transactions for over-the-air (OTA) updates without needing to trust third parties. However, challenges have been encountered, especially in integrating light clients with existing OTA update mechanisms and overcoming the inherent limitations of MCUs. The main challenge was integrating a light client for transaction verification with the OTA firmware update process. Initially, we faced compatibility issues among libraries due to differences in programming languages, package dependencies, and limited board support. To address this, we explored cross-compilation techniques to enable seamless integration of light client functionality with OTA updates. We investigated frameworks like Espruino, DeviceScript, and Mongoose OS, known for their cross-compilable micro-runtime libraries. However, these frameworks often fell short in executing light clients or supporting OTA updates, highlighting the inconsistency in MCU library capabilities and compatibility across various boards from different manufacturers. This diversity requires significant customization and testing to achieve the necessary compatibility and functionality. To overcome compatibility issues, we explored compiling the light client into WebAssembly (WASM) before deployment. WASM is designed to be portable, compact and efficient, making it suitable for small IoT devices. However, the size of the WASM executable turned out to be in the order of megabytes, exceeding the few hundred kilobytes typically available in the device's ROM. Therefore, choosing programming languages that produce the smallest possible binary files is crucial, as optimizing the WASM file size without compromising functionality is difficult. This led us to conduct two distinct experiments.

For the first experiment we chose an Espressif ESP32 MCU device with Xtensa architecture which represents one of the most common and used devices in the IoT environment. The device has Wi-Fi connectivity and a 2 MB Flash memory with double partitions for OTA updates. ESP32-based boards are highly compatible with a variety of programming languages, which gave us the chance of trying different light clients. We first considered cpp-ethereum [1], entirely written in C++. This library allows us to create a small RPC client which is able to interact with the Ethereum Network and any deployed smart contract via a trusted RPC end-points. The light client handles the security aspects of this experiment and it can be integrated with OTA libraries written in the same language as well as with the official Espressif libraries that manage the interaction with the device's hardware. However cpp-ethereum does not include a consensus cliente, and thus has to trust the end-point. To overcome this security problem, we considered adding the consensus layer of light clients.

### 4.1 Consensus Client on Edge Device

We then focused on deploying a light client with both the execution and consensus layer in our device. Currently there are several clients available for this purpose and for our experiment we used the Helios Light Client [7]. Helios is a portable client that is able to convert an untrusted centralised RPC endpoint into a safe local RPC for its user. It syncs in seconds, requires no storage and is lightweight enough to run on small devices. The entire library is written in Rust, a programming language that is commonly used in the microcontroller ecosystem and can run natively on some ESP32 devices, thanks to cargo-espflash tool, or can be compiled into a WASM file and run on a microenvironment. However, some dependencies of Helios could not be compiled into an executable with espflash and the WASM compilation resulted in an executable file that was in the order of a few megabytes, therefore too large for the ESP32 and similar devices with the same characteristic.

This led us to develop the experiment in a device with more resources. We chose the Raspberry Pi4 to develop a consensus client that is able to interact with deployed smart contracts while guaranteeing the trustworthiness of every transaction sent. Given the high security of this client, we employed it in the creation of an update verification system. The executable size was 14,5 MB, and it was able to sync and to read the state of the update manager contract in about 10 s. During the process, it scored a peak heap memory consumption of 1,25 MB.

Metrics collected during execution suggest that it is entirely possible to use light clients capable of following and verifying Ethereum consensus, making our approach decentralized and not dependent on third parties beyond the device and smart contract.

## 5 Conclusions

In this paper, we have explored the critical issue of software updates for IoT and Edge devices, emphasizing the concept of device ownership. Our examination of the current state of the art revealed that, although fundamental security and authenticity challenges of updates are addressed through asymmetric cryptography, this approach falls short in terms of the variety of conditions or protocols it can support. For example, it struggles to implement straightforward protocols where an owner delegates an operator to propose updates and later revokes this delegation.

As a potential solution, we have proposed the use of smart contracts running on blockchain technology to express control policies. This approach leverages the blockchain's decentralization, security, and Turing-complete expressiveness, offering a novel avenue for managing firmware updates.

Further, we delved into the challenge of enabling IoT devices with limited resources to communicate with the blockchain. Initially, we considered access through a gateway and successfully deployed it on a constrained architecture, such as Espressif's ESP32. We then explored a configuration that eliminates the need to trust the gateway by implementing a light consensus node directly on the device. Our findings indicate that while highly constrained architectures, like those based on the ESP32, currently cannot host light nodes, more capable ones based on the ARM Cortex-A72 can.

In conclusion, our research demonstrates that employing smart contracts for firmware update control is a viable and promising solution. This approach not only addresses the limitations of current methods but also opens up new possibilities for secure, efficient, and flexible device management in the IoT and Edge computing landscapes.

## References

1. a16z. Helios: A fast, secure, and portable light client for ethereum (2024). https://github.com/a16z/helios. Accessed 04 Feb 2024
2. Abhilash, P.K., Chidananda, K., Sandeep, K., Molaka, N.R., Awasthi, Y.K., Rajabhishek, S.: Secure and sustainable decentralized cloud using IPFS. In: E3S Web of Conferences, vol. 430, pp. 01097. EDP Sciences (2023)
3. Buterin, V., et al.: Ethereum white paper. GitHub Repository **1**, 22–23 (2013)
4. de Sousa, M.J.B., Gonzalez, L.F.G., Ferdinando, E.M., Borin, J.F.: Over-the-air firmware update for IoT devices on the wild. Internet of Things **19**, 100578 (2022)
5. El Jaouhari, S.: Toward a secure firmware OTA updates for constrained IoT devices. In: 2022 IEEE International Smart Cities Conference (ISC2), pp. 1–6. IEEE (2022)
6. Ethereum Foundation. Light clients (2024). https://ethereum.org/developers/docs/nodes-and-clients/light-clients. Accessed 05 Feb 2024
7. ethereumproject. cpp-ethereum: Ethereum c++ client (2024). https://github.com/ethereumproject/cpp-ethereum. Accessed 04 Feb 2024
8. Mtetwa, N.S., Sibeko, N., Tarwireyi, P., and Abu-Mahfouz, A.M.: OTA firmware updates for lorawan using blockchain. In: 2020 2nd International Multidisciplinary Information Technology and Engineering Conference (IMITEC), pp. 1–8. IEEE (2020)
9. Nanopoulos, S.A.: Código Network: a Decentralized Firmware Update Framework for IoT Devices. PhD thesis, Thesis Master of Science School of Informatics, University of Edinburgh (2018)
10. Paavolainen, S., Carr, C.: Security properties of light clients on the ethereum blockchain. IEEE Access **8**, 124339–124358 (2020)
11. Raj, A., Maji, K., Shetty, S.D.: Ethereum for Internet of Things security. Multimedia Tools Appl. **80**, 18901–18915 (2021)
12. Yohan, A., Lo, N.W.: An over-the-blockchain firmware update framework for IoT devices. In: 2018 IEEE Conference on Dependable and Secure Computing (DSC), pp. 1–8. IEEE (2018)
13. Yohan, A., Lo, N.-W.: FOTB: a secure blockchain-based firmware update framework for IoT environment. Int. J. Inf. Secur. **19**(3), 257–278 (2020)

# AI-SPRINT: Design and Runtime Framework for Accelerating the Development of AI Applications in the Computing Continuum

Francesco Lattari, Matteo Matteucci, and Danilo Ardagna(✉)

Dipartimento di Elettronica, Informazione e Bioingegneria (DEIB), Politecnico di Milano, 20133 Milan, Italy
{francesco.lattari,matteo.matteucci,danilo.ardagna}@polimi.it

**Abstract.** Artificial Intelligence (AI) and edge computing have recently emerged as major trends in the ICT industry. Enterprise applications increasingly make intensive use of AI technologies and are often based on multiple components running across a computing continuum. However, the heterogeneity of the technologies and software development solutions in use are evolving quickly and are still a challenge for researchers and practitioners. Indeed, lack of solutions tailored for AI applications is observed in the areas of applications placement and design space exploration with performance guarantees, both under-developed. The aim of the AI-SPRINT "Artificial Intelligence in Secure PRIvacy-preserving computing coNTinuum" project is to develop a framework composed of design and runtime management tools to seamlessly design, partition and operate Artificial Intelligence (AI) applications among the current plethora of cloud-based solutions and AI-based sensor devices (i.e., devices with intelligence and data processing capabilities), providing resource efficiency, performance, data privacy, and security guarantees. AI-SPRINT is intended to accelerate the development of AI applications, whose components are spread across the edge-cloud computing continuum, while allowing trading-off application performance and AI models accuracy. This is accomplished by the thorough suite of design tools provided by the AI-SPRINT framework, which exposes a set of programming abstractions with the goal of hiding as much as possible the computing continuum complexity, while further providing a simple interface to define desired constraints upon which the application design is guided.

## 1 Introduction

The convergence of Artificial Intelligence (AI) and edge computing stands as a transformative force within the Information and Communication Technology (ICT) industry, significantly influencing the landscape of enterprise applications. The AI-SPRINT[1] "Artificial Intelligence in Secure PRIvacy-preserving computing coNTinuum") project emerges as a proactive response to the critical need for tailored solutions for AI applications distributed across the edge-cloud computing continuum. The project envisions

[1] This work is funded by EU Grant no. 101016577. The AI-SPRINT website is accessible at https://ai-sprint-project.eu.


L. Barolli (Ed.): AINA 2024, LNDECT 203, pp. 171–181, 2024.
https://doi.org/10.1007/978-3-031-57931-8_17

accelerating the development of AI applications while allowing a nuanced trade-off between application performance and the accuracy of AI models.

To this end, AI-SPRINT allows enriching AI applications by quality of service (QoS) annotations to express performance, accuracy, and security constraints that could be evaluated to drive the application deployment. The suite of annotations enables setting requirements in terms of execution times, for single or multiple application components, data-flow rate, deep neural networks partitioning, application with alternative implementations, and secured executions.

The AI-SPRINT design workflow involve key tools for the automatic generation of Docker images for the containerized execution of application components (TOSCARIZER[2]), for the automatic profiling of application performance (OSCAR-P[3] [1]), and for the automated design space exploration to provide the optimal production deployments (SPACE4AI-D[4] [2]). Design tools, further enriched with tools for neural architecture search (POPNAS [3]), federated learning, and privacy preserving models execution, make AI-SPRINT an advanced framework that fills known technology gaps, by providing solutions to efficiently develop and deploy AI enterprise applications across the computing continuum.

This article highlights the key design tools in the AI-SPRINT framework, focusing on the programming abstractions driving AI application design. It also gives an overview of the main tools and steps in the AI-SPRINT design workflow, starting from using AI-SPRINT annotations, followed by the generation of the Docker images, performance profiling, and optimal components placement.

## 2 Related Work

Advanced solutions developed for the design of AI applications encompass a spectrum of methodologies tailored to address the challenges posed by the complexity of Deep Learning (DL) models. These encompass diverse methodologies ranging from network architecture search and federated learning to DNN partitioning, AI application performance modeling, and component placement within computing continua. In the following sections, we provide an overview of the related methodologies in two pivotal fields of the DNN partitioning, application performance modeling, and optimal component placement, key aspects of the developed framework.

### 2.1 Deep Neural Networks Partitioning

Recent research addresses the resource-intensive nature of Deep Neural Networks (DNNs) for inference. [4] discusses challenges in inference serving systems, including model selection, heterogeneous hardware integration, load variability, and startup latency. It introduces managed and model-less inference systems for resource optimization and user simplicity. Partitioning DNNs and distributing computations across edge-cloud devices have emerged as a common strategy to tackle their resource demands.

[2] Source code available at https://github.com/grycap/im.

[3] Source code available at https://github.com/ai-sprint-eu-project/OSCAR-P.

[4] Source code available at https://github.com/ai-sprint-eu-project/space4ai-d.

The work in [5] proposes DDNN, a distributed deep neural network architecture that maps trained DNNs onto diverse physical devices spanning local, edge, and cloud environments. Authors in [6] introduce a lightweight scheduler to dynamically partition DNN computations between mobile devices and data centers. Their approach leverages performance prediction models to optimize partition points based on latency or energy consumption targets. Techniques like early exiting and pruning [7, 8] further enhance efficiency. Advancements include LcDNN [9] for energy-efficient mobile web inference, JointDNN [10] for optimizing both inference and training phase partitioning, and hybrid parallelism [11] for optimal resource utilization across edge, server, and cloud. Lastly, [12] presents CNN placement on IoT devices, minimizing latency and accommodating constraints through strategic layer assignment and early exits.

### 2.2 AI Applications Performance Modeling

The rising prominence of Deep Learning (DL) across various domains, from image and voice recognition to text processing, and the prediction of DL application performance has gained significant interest in the last years. While both training and inference of Neural Networks (NNs) typically occur on GPUs, the literature offers limited studies on how hardware configurations impact performance. Given the complexity of GPU hardware, recent trends favor black box approaches based on Machine Learning (ML) over analytical models. These approaches derive performance models from data, enabling predictions without prior knowledge of the target system's internals. However, ML models necessitate initial profiling campaigns to gather training data. The work in [13] offers a comparative analysis of analytical and ML-based models. A high-level approach is adopted in [14], predicting forward pass execution times based on matrix multiplication computations for CNN deployment on mobile devices. Similarly, in [15] linear regression models are derived for estimating CNN training times based on layer computational complexity. In microservices and Function-as-a-Service (FaaS) systems, ML-based performance modeling is gaining traction. The works in [16] and [17] develop temporal and steady-state performance models for FaaS platforms, aiming to optimize QoS and utilization while reducing operating costs.

### 2.3 AI Applications Components Placement

The recent surge of interest in computing continua has led to significant research attention on component placement and design exploration. A recent survey [18] offers a classification of literature proposals based on various factors such as the layers involved and the relevant metrics (e.g., latency, energy consumption, profit-cost analysis). Several studies focus on Mobile Edge Clouds (MECs), particularly emphasizing network resource optimization. In MECs, user mobility plays a crucial role. Authors in [19] explore fog node placement around a central cloud in a square planar layout to maximize data rate and minimize transmission delay. The work in [20] offers both offline and online algorithms for component placement, while [21] focuses on edge-only placement, proposing algorithms to map application components to physical devices optimally. The work in [22] optimizes performance and cost of serverless applications by considering budget limitations and Quality of Service (QoS). Finally, the authors in [23] target cost optimization of serverless applications in AWS Lambda while maintaining latency thresholds.

# 3 AI-SPRINT Design Workflow

This section outlines how the design tools within the AI-SPRINT Studio[5] interact, forming the framework that provides essential functionalities to the AI-SPRINT users. These tools facilitate the development, profiling, deployment, and execution of applications across the computing continuum.

## 3.1 Quality of Service Annotations

Quality of Service annotations in AI-SPRINT empower developers to specify quality constraints for deployment. These annotations serve to abstract resource management while granting developers control over resource allocation in the cloud continuum. Implemented as Python decorators, they enrich user code, simplifying the definition of quality constraints. Developers annotate application components and, depending on the annotation, provide additional configuration files.

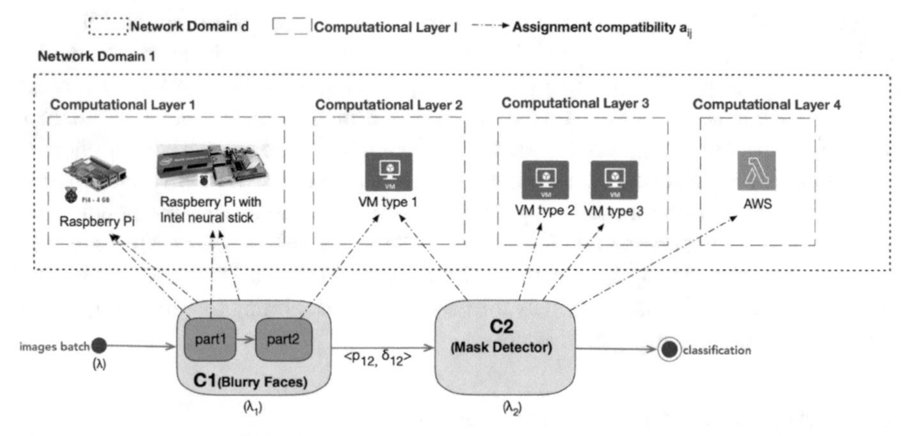


**Fig. 1.** Example of AI-SPRINT application with two components: the "Blurry Faces" and the "Mask Detector". The two components are Python applications that run sequentially. When a new video is uploaded to the input bucket, the first component anonymizes detected faces and saves them to another bucket. This action then triggers the second component, which detects masks in the anonymized frames. Each component can use various resources across different computing layers.

Indeed, an AI-SPRINT application[6] is a set of Python components whose execution is orchestrated by OSCAR[7], an open-source platform that supports the FaaS computing

[5] Source code is available at https://github.com/ai-sprint-eu-project/AI-SPRINT-STUDIO.
[6] Examples applications are provided at https://gitlab.polimi.it/ai-sprint/ai-sprint-examples.
[7] OSCAR is available at http://github.com/grycap/oscar.

model for file-processing applications, and SCAR[8], a framework to transparently execute containers out of Docker images in AWS Lambda. The execution of a component is triggered by a file upload. The output is also stored in the output bucket, which as consequence, triggers the execution of the next component in the application workflow, and so on. An example of an application workflow is provided in Fig. 1. It is possible to observe how a component and its partitions (see *Partitionable Model* annotation later) can have different candidate execution layers. Finding the optimal assignment is the goal of a specific AI-SPRINT tool called SPACE4AI-D (see Sect. 3.4).

In the following we present some of the annotations provided by AI-SPRINT to annotate the application components and to enable specific functionalities to drive the design and deployment of the application.

**Execution Time.** The execution time annotation allows defining time constraints, i.e., upper bounds for the execution time of application components, which, besides guiding the design of the deployment, will be monitored at runtime by the AI-SPRINT Monitoring Subsystem (AMS[9]) to detect any violation. Users apply the *exec_time* annotation to specify maximum execution times. The annotation wraps components, collecting runtime information sent to the AMS for storage in InfluxDB.

Users define both local and global constraints through the annotation. Local constraints set maximum execution times for single components, while global constraints set times for multiple consecutive components. Global constraints, annotated on the last component, can specify preceding components involved. Local and global constraints can be combined, allowing components to be subject to both types. During the design pipeline, AI-SPRINT parses annotated code to generate configuration files summarizing time constraints for each component and computational layer. These files inform AMS to define monitoring constraints and rules for runtime violation triggers.

**Partitionable Model.** AI-SPRINT provides the users with the possibility of partitioning DNNs to guarantee memory and QoS constraints. Indeed, a neural network may be split, depending on the needs, and consecutive layers may be placed on different resources in the computing continuum.

The SPACE4AI-D-partitioner is the tool responsible for finding the optimal partitions of a considered neural model, i.e., the optimal partitioning point (the network layer) at which the network is split. AI-SPRINT enables users to specify that a DNN-based component can be partitioned by using the *partitionable_model* annotation. AI-SPRINT requires the partitionable networks to be provided in the Open Neural Network Exchange (ONNX) format, which is an intermediate representation providing a powerful sharing mechanism among frameworks. Partitioning is obtained by generating new ONNX files, one for each partition, representing the computation of the single network segments. Furthermore, AI-SPRINT provides an automatic mechanism to create new consecutive components from the obtained partitions, by generating the new Python scripts, one for each partition, where the ONNX file corresponding to a specific network segment is executed. All this procedure is masked to the developer, thus reducing the complexity from the implementation perspective, that simply requires annotating the component

[8] SCAR is available at http://github.com/grycap/scar.

[9] Source code available at https://github.com/ai-sprint-eu-project/monitoring-subsystem.

function. As standard components, the partitions may be placed on different computing infrastructures (see Fig. 1).

**Early-Exits Model.** AI-SPRINT provides support for the deployment of DNNs with early exits. Some architectures may be developed by including the so-called early-exits points, which are additional output layers that are trained to predict based on intermediate features without the need to compute the whole network. Indeed, these kinds of models are provided to stop the execution earlier in the network under some conditions, thus reducing the computational burden when the evaluation of all the network layers is not necessary.

Through the use of a specific annotation, named *early_exits_model*, the user is able to explicitly declare a component whose computation is based on a model with early-exits nodes. Similarly to a partitionable component, the users must provide the ONNX file of the model, which is specified as one of the arguments of the provided annotation. In addition, the name of the nodes of the computational graph corresponding to the early exits must be provided by the users to allow AI-SPRINT to correctly retrieve the corresponding output. Furthermore, the user must provide the function implementing the evaluation of the condition used at runtime to automatically decide whether the computation should stop or not at a specific early exit.

As in the case of a partitionable model, AI-SPRINT is able to automatically split the original ONNX model into segments. A new component, together with the corresponding Python code, is automatically generated for each obtained segment, which will be placed on a specific device based on the candidate resources. During the execution each segment evaluate the early-exit condition on the output and, in the case it is verified, the segment corresponding to the considered early exit executes the post-processing part of the original component and stores the output to the final output bucket, otherwise it continues the execution by triggering the next segment.

**Device Constraints.** AI-SPRINT components based on the execution of DL models may have strict constraints on the needed resources. AI-SPRINT allows the users to specify a set of minimal resources that the device on which the component will be deployed must have. This is done by using the *device_constraints* annotation, whose arguments are the *ram*, the minimum memory in GB required to run the annotated component, and the *vram*, minimum video memory in GB required to run the annotated component. A $vram > 0$ implicitly highlights the need for a Graphics Processing Unit (GPU).

**Model Performance.** AI-SPRINT enables the development of AI applications that can have multiple alternative implementations with degraded performance (e.g., from the AI model accuracy perspective), which can switched at runtime in order to compensate for the high computational costs of the original application. Nevertheless, measuring the application-level performance for each alternative combination of the components is not an easy task, and the difficulty increases even more if we consider more than two components. For this reason, AI-SPRINT considers, for this particular feature, specific kinds of applications for classification tasks involving only two components and for which computing a combined application-level metric is feasible.

The *model_performance* annotation allows the user to provide different implementations for each component. Users provide confusion matrices for both original and degraded components, enabling automatic computation of application performance across all possible workflows. Results are summarized in a YAML file, listing deployments in descending order of performance. This file aids decision-making in both design and runtime, helping select the optimal deployment for production based on classification performance.

**Detect Metric Drift.** The AI-SPRINT framework enables the detection of data drift at runtime through the identification of a drift of a particular monitored metric, i.e., a relevant change of the monitored compared to its historical behavior. The drift-detection tool allows the user to be aware of the drift at runtime, enabling the re-training of the model thanks to the automatic collection of data after the drift.

The *detect_metric_drift* annotation indicates the need to detect drift on specific metrics. It is applied to the first component of the application workflow, as one drift detection algorithm monitors all user-defined metrics, monitored by the AMS and whose time series are stored in InfluxDB. Users specify metrics and statistical tests (e.g., Kolmogorov-Smirnov, Fisher's Exact Test) along with thresholds and detection intervals. During the design pipeline, the parsing of the annotation triggers the automatic generation of the Drift Detector tool. The tool is automatically added to the user application by AI-SPRINT as a special component, which is deployed to the computational layer assigned to the last component in the application workflow. At runtime, based on the result of the detection, the user can decide to collect new data after the drift to re-train the metric model in order to adapt to the new data distribution.

### 3.2 Parsing the Annotations

AI-SPRINT automatically parses the code provided by the users to search for QoS annotations. Then, based on the defined parameters and additional configuration files filled by the programmer and sysop starting from the application template, i.e., the description of available resources (*candidate_resources.yaml*), the candidate assignments of the components to the resources (*candidate_deployments.yaml*), application wokflow (DAG), and the list of libraries required for running the application components (*requirements.txt* and *requirements.sys*), AI-SPRINT triggers specific design functionalities.

The first part of AI-SPRINT design pipeline terminates with the generation of all the partitioned components, in the case a partitionable or early-exits components exist, the generation of the execution time constraints for the AMS, if local or global constraints have been defined, the generation of all the possible alternative deployments with degraded performance, if components with alternative implementation are provided, and the generation of the Drift Detector component, if the drift detection is required by the user. The scheme summarizing the design pipeline until this stage is provided Fig. 2.

At the end of this stage, the TOSCARIZER, included in the AI-SPRINT Studio, is executed to generate the images for all the components, the corresponding partitions, and the Drift Detector, considering the candidate underlying platforms where they can be executed (AMD64 and ARM64). It also pushes the images to the Docker registry specified by the user and, finally, generates an output file with the full image names. The images will be then used by OSCAR to execute the containerized components at runtime.

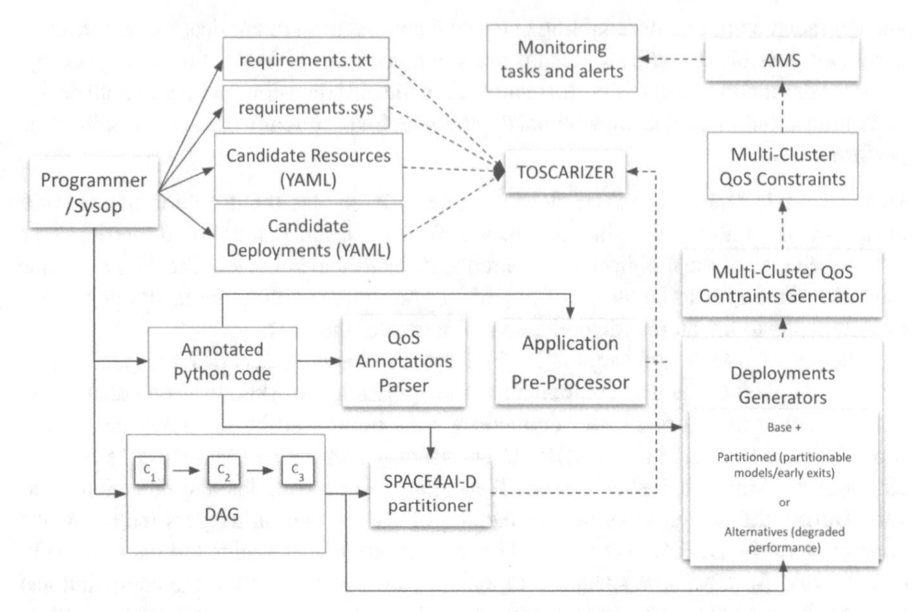


**Fig. 2.** Scheme of the AI-SPRINT design workflow. The programmer and sysop provide the annotated code and required configuration files describing the application and the available resources. Then the annotations are automatically parsed by AI-SPRINT to generate components partitions, alternative deployments, monitoring constraints, and configuration files needed by other AI-SPRINT tools.

### 3.3 OSCAR-P and Performance Models

The aim of OSCAR-P is to simplify and fully automate the testing of specific OSCAR/SCAR application workflows on different hardware configurations and collect data to train machine learning performance models. OSCAR-P takes as input the configuration files relative to the candidate resources for each component, as defined by the application architect, and the output files produced after the parsing of the annotated code, including the Docker images generated by the TOSCARIZER. The tests are performed by varying the resources used, by either changing the number of active nodes or swapping one resource with another. This step is achieved either through the Infrastructure Manager (IM[10]) (for cloud virtual clusters) or directly through Kubernetes (for local physical clusters). Once all the required resources are in place and correctly configured, the testing campaign is controlled by a single YAML configuration file containing the list of services, the description of the clusters and their worker nodes and information on how the application needs to be tested.

The datasets collected by OSCAR-P are then used by the aMLLibrary[11] to train a performance model for every service/resource pair. Performance models provide to the application manager and the AI-SPRINT runtime components means to choose an

[10] Source code is available at https://github.com/grycap/im.

[11] Source code is available at https://gitlab.polimi.it/ai-sprint/a-mllibrary.

appropriate configuration to avoid applications performance violations, under or overestimation of the utilization of the continuum resources, and to predict the execution time of AI components on a target configuration.

### 3.4 SPACE4AI-D and Deployment

SPACE4AI-D is the AI-SPRINT tool which tackles the joint component placement and resource selection problem in the computing continuum at design time, dealing with different AI application requirements. SPACE4AI-D receives as inputs the resource description, performance models, performance constraints, application DAG and specific configuration YAML files, and it determines the minimum-cost solution for the resource selection and component placement problem while guaranteeing performance constraints (namely, requirements on the maximum admissible response times of single components or sequences of components) using, in the current release, several heuristic algorithms (i.e., random greedy, local search, tabu search, etc.). The output of this tool is the optimal production deployment, i.e., the description of the selected resources, the component placement, and the optimal number of nodes/Virtual Machines (VMs) which should be considered for the application execution. This information is used by the TOSCARIZER to generate the TOSCA files for each component and partition to enable the deployment of all the required components (Kubernetes, OSCAR, etc.). The generated TOSCA files are then submitted and the virtual infrastructure provisioning module, i.e., IM, creates the needed infrastructure.

## 4 AI-SPRINT Use Cases

A thorough evaluation of all the tools within the AI-SPRINT framework was conducted as an integral part of the development. In particular, three domain-specific use cases have been provided to assess the functionality, compatibility, and efficiency of each tool in its respective use case environment. Other than evaluating the single tools, this process consisted also in evaluating how they collectively enhanced the workflow, meeting the unique demands of different scenarios presented in the following.

**Personalized Healthcare**[12]**.** This use case involves leveraging AI models for health monitoring through wearable devices connected to mobile phones, with a specific focus on personalized stroke risk assessment and prevention. By integrating wearable and mobile devices, the initiative aims to provide AI-powered insights into stroke patient care while prioritizing privacy and security through federated learning.

**Maintenance and Inspection**[13]**.** In this use case, AI models are utilized to identify windmill blade damage based on vision and thermal images collected by drones. To optimize bandwidth usage and inspection time, relevant images are selected at the edge, reducing operator effort and enabling more real-time feedback. By providing image quality feedback to operators and implementing AI-powered image analysis for autonomous mission control, the initiative aims to improve the overall inspection process.

[12] Details at https://www.ai-sprint-project.eu/use-cases/personalised-healthcare.

[13] Details at https://www.ai-sprint-project.eu/use-cases/maintenance-inspection.

**Farming 4.0**[14]**.** Demonstrating the application of AI in agriculture, this use case focuses on enhancing the efficiency and sustainability of pesticide spraying in vineyards. A Smart Farming Device (SFD) installed on a tractor enables smart spraying and data collection, adjusting pesticide doses for each plant based on computer vision methods. The tractor edge device communicates with a farm computer, allowing for monitoring, tuning of the spraying process, and data transfer. Cloud-based AI models predict yield and detect diseases using data collected during spraying, contributing to improved crop management practices.

## 5 Achievements and Conclusions

The AI-SPRINT framework facilitates easy development and deployment of AI applications for developers with limited ML expertise. It reduces skill shortage and learning curves, shortens time to market, and lowers costs. Through open-source tools, models, and methodologies, it supports application design and deployment on computing continuum. AI-SPRINT tools provide high-level programming abstractions for defining requirements, annotations for execution time constraints and DNN partitioning, and a design pipeline guiding users from creation to deployment. The AI-SPRINT framework and its tools were successfully tested across the project. The Personalized Healthcare use case showcased AI-SPRINT's effectiveness in managing patient data with GDPR compliance. Maintenance & Inspection demonstrated the framework's ability to transition from batch-based to real-time processing. Farming 4.0 highlighted AI-SPRINT's role in building digital services integrating edge and cloud processing. QoS annotations were essential across all use cases, optimizing execution time and performance monitoring. TOSCARIZER streamlined deployment processes, while OSCAR-P provided effective profiling and performance prediction. SPACE4AI-D automated design space exploration, optimizing execution costs and ensuring response time reliability. Insights from these experiences provide valuable guidance for future AI developments, solidifying its impact on the evolving landscape of artificial intelligence.

## References

1. Galimberti, E., et al.: OSCAR-P and amllibrary: performance profiling and prediction of computing continua applications. In: Companion of the 2023 ACM/SPEC International Conference on Performance Engineering (ICPE 2023 Companion) (2023)
2. Sedghani, H., Filippini, F., Ardagna, D.: A random greedy based design time tool for AI applications component placement and resource selection in computing continua. In: IEEE Edge 2021 Proceedings (2021 IEEE International Conference On Edge Computing), pp. 32–40. Guangzhou, China (online) (2021)
3. Falanti, A., Lomurno, E., Samele, S., Ardagna, D., Matteucci, M.: POPNASv2: an efficient multi-objective neural architecture search technique. In: 2022 International Joint Conference on Neural Networks (IJCNN), pp. 1–8 (2022)
4. Yadwadkar, N., Romero, F., Li, Q., Kozyrakis, C.: A case for managed and model-less inference serving. In: HotOS 2019 (2019)

[14] Details at https://www.ai-sprint-project.eu/use-cases/farming-40.

5. Teerapittayanon, S., McDanel, B., Kung, H.: Distributed deep neural networks over the cloud, the edge and end devices. In: IEEE 37th ICDCS (2017)
6. Kang, Y., Hauswald, J., Gao, C., Rovinski, A., Mudge, T., et al.: Neurosurgeon: collaborative intelligence between the cloud and mobile edge. In: ACM ASPLOS 2017 (2017)
7. Li, E., Zeng, L., Zhou, Z., Chen, X.: Edge AI: on-demand accelerating deep neural network inference via edge computing. IEEE Trans. Wirel. Commun.Wirel. Commun. **19**(1), 447–457 (2019)
8. Shiand, W., Hou, Y., Zhou, S., Niu, Z., et al.: Improving device-edge cooperative inference of deep learning via 2-step pruning. In: IEEE INFOCOM 2019 Proceedings (2019)
9. Huang, Y., Qiao, X., Ren, P., Liu, L., et al.: A lightweight collaborative deep neural network for the mobile web in edge cloud. IEEE Trans. Mob. Comput. (2020)
10. Eshratifar, A.E., Abrishami, M.S., Pedram, M.: JointDNN: an efficient training and inference engine for intelligent mobile cloud computing services. IEEE Trans. Mob. Comput.Comput. **20**(2), 565–576 (2019)
11. Liu, D., Chen, X., Zhou, Z., Ling, Q.: HierTrain: fast hierarchical edge AI learning with hybrid parallelism in mobile-edge-cloud computing. IEEE Open J. Commun. Soc. **1**, 634–645 (2020)
12. Disabato, S., Roveri, M., Alippi, C.: Distributed deep convolutional neural networks for the internet-of-things. IEEE Trans. Comput.Comput. **14**(8), 1–14 (2015)
13. Madougou, S., Varbanescu, A., de Laat, C., et al.: The landscape of GPGPU performance modeling tools. J. Parallel Comput. **56**, 18–33 (2016)
14. Lu, Z., Rallapalli, S., Chan, K., La Porta, T.: Modeling the resource requirements of convolutional neural networks on mobile devices. In: Proc. Conf. Multimedia (2017)
15. Gianniti, E., Zhang, L., Ardagna, D.: Performance prediction of GPU-based deep learning applications. In: 30th Int'l Symp. Computer Architecture and High Performance Computing (SBAC-PAD 2018) (2018)
16. Mahmoudi, N., Khazaei, H.: Temporal performance modelling of serverless computing platforms. In: Sixth International Workshop on Serverless Computing (WoSC 2020) (2020)
17. Mahmoudi, N., Khazaei, H.: Performance modeling of serverless computing platforms. In: IEEE Transactions on Network and Service Management (2020)
18. Bellendorf, J., Mann, Z.Á.: Classification of optimization problems in fog computing. Future Gener. Comput. Syst. **107**, 158–176 (2020)
19. Balevi, E., Gitlin, R.D.: Optimizing the number of fog nodes for cloud-fog-thing networks. IEEE Access **6**, 11173–11183 (2018)
20. Bahreini, T., Grosu, D.: Efficient placement of multi-component applications in edge computing systems. In: The Second ACM/IEEE Symposium (SEC 2017) (2017)
21. Wang, S., Zafer, M., Leung, K.K.: Online placement of multi-component applications in edge computing environments. IEEE Access **5**, 2514–2533 (2017)
22. Lin, C., Khazaei, H.: Modeling and optimization of performance and cost of serverless applications. IEEE Trans. Parallel Distrib. Syst.Distrib. Syst. **32**(3), 615–632 (2021)
23. Elgamal, T., Sandur, A., Nahrstedt, K., Agha, G.: Costless: optimizing cost of serverless computing through function fusion and placement. In: IEEE/ACM Symposium on Edge Computing (SEC) (2018)

# Workflow Scheduling in the Cloud-Edge Continuum

Luca Zanussi[1(✉)], Daniele Tessera[2], Luisa Massari[1], and Maria Carla Calzarossa[1]

[1] Department of Electrical, Computer and Biomedical Engineering, University of Pavia, Pavia, Italy
luca.zanussi01@universitadipavia.it, {luisa.massari,mcc}@unipv.it
[2] Department of Mathematics and Physics, Catholic University of Sacred Heart, Brescia, Italy
daniele.tessera@unicatt.it

**Abstract.** Scheduling in the cloud-edge continuum is a challenging problem. In fact, scheduling has to cope with the peculiarities of these complex ecosystems and satisfy at the same time the desired service levels. In this paper, we investigate the benefits of the cloud-edge continuum for deploying workflows with different characteristics, e.g., computation or communication-intensive. In detail, we formulate a multi-objective optimization problem solved using a Genetic Algorithm. This problem is aimed at identifying the scheduling plans that minimize two conflicting objectives, namely, the expected workflow execution time and monetary cost associated with the cloud and edge resources to be provisioned. Our experiments have shown that the plans that exploit both cloud and edge resources represent a good tradeoff between the two objectives. In addition, the workflow characteristics strongly influence these plans. Similarly, the uncertainties that might affect the infrastructure performance are responsible of significant changes in the corresponding Pareto fronts.

## 1 Introduction

Edge computing brings processing capabilities closer to the sources of data and reduces network delays and bandwidth usage. On the contrary, cloud computing offers potentially unlimited processing and storage capabilities at the expense of increased network delays. Therefore, to fully exploit these complex ecosystems, a seamless integration of these technologies is necessary [11].

In this paper we investigate the benefits of the cloud-edge continuum for deploying distributed applications characterized by different requirements in terms of computation and communication. In particular, we focus on workflows and on their scheduling. For this purpose, we formulate a multi-objective optimization problem aimed at identifying the "best" scheduling plans that minimize two conflicting objectives, that is, expected workflow execution time and monetary cost. The results of our experiments clearly demonstrate that the combined usage of cloud and edge resources represents a good compromise between the two objectives even under performance variability of the infrastructure.

The rest of the paper is organized as follows. Section 2 briefly discusses the state of the art in the area of cloud-edge scheduling. Section 3 presents the proposed schedul-


L. Barolli (Ed.): AINA 2024, LNDECT 203, pp. 182–190, 2024.
https://doi.org/10.1007/978-3-031-57931-8_18

ing framework, while Sect. 4 explains the setup of the experiments and discusses their results. Finally, Sect. 5 summarizes the paper and outlines future research directions.

## 2 Related Work

Scheduling policies for cloud environments have been extensively studied for more than a decade. Many diverse approaches have been presented in the literature. These approaches mainly differ in terms of optimization models, number and types of objectives as well as type of workloads being scheduled (see, e.g., [1,4,14,16] for detailed surveys).

In the context of cloud-edge continuum, scheduling policies have been researched to a more limited extent (see, e.g., [2,3,8,12,15,17,18]. Most papers focus on independent tasks, whereas only few consider more complex workloads consisting of workflows with tasks characterized by precedence constraints. As discussed in [13], genetic-based optimizations are often adopted. For example, Ali et al. [3] propose a task scheduling model based on an extension of the NSGA-II algorithm. Makespan and overall cost are considered as objectives to be minimized. Ijaz et al. [15] focus on workflow scheduling by formulating a multi-objective optimization problem that considers both makespan minimization and reduction in energy consumption. In particular, a weighted bi-objective cost function is introduced for selecting the processing node that minimizes task completion time and energy consumption based on a user-defined weighting factor. A further reduction in energy consumption is obtained by applying deadline constrained frequency scaling.

In this paper we study the benefits of the cloud-edge continuum for workflow scheduling by solving a multi-objective optimization problem that takes into account the performance variability that might affect these complex ecosystems.

## 3 Scheduling Framework

The proposed scheduling framework is based on the formulation of an optimization problem aimed at minimizing two conflicting objectives, namely, the expected values of the workflow execution time and overall monetary cost under uncertain conditions. For solving this problem, we rely on a Genetic Algorithm applied in combination with a Monte Carlo simulation of all possible solutions. As a result, we obtain multiple "optimal" solutions, i.e., scheduling plans, distributed on the so-called Pareto front. In what follows we introduce the workload and architectural models and we discuss the details of the problem and its solution.

### 3.1 Workload and Architectural Models

The workload considered in this study is represented by a workflow $W$ consisting of $n$ tasks. This workflow is modeled by a Directed Acyclic Graph (DAG) whose nodes $T_i\,(i=1,\dots,n)$ correspond to the tasks and whose edges represent the control and data dependencies between the various tasks (see, e.g., Fig. 1).

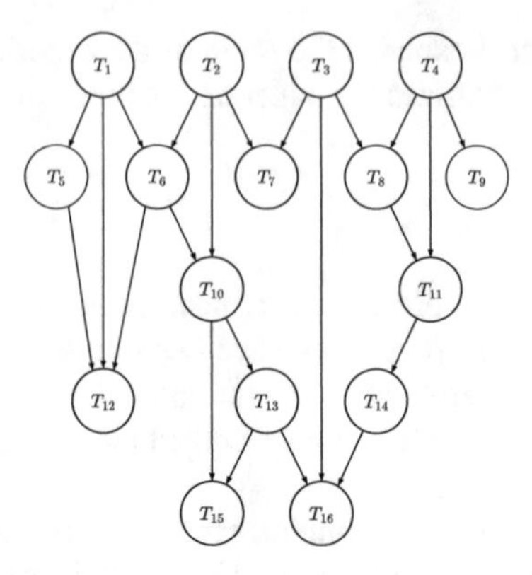


**Fig. 1.** Example of a workflow with 16 nodes.

Each task is described by demands corresponding to the requirements in terms of computation, i.e., processing, communication, i.e., volume of data exchanged with other tasks, input, i.e., data volume transferred from external sources, and output, i.e., data volume transferred to storage devices.

The architectural model of the cloud-edge continuum is represented by multiple instances of physical and logical resources organized according to a layered structure. Some resources are located at the edge/fog layers, whereas some others are inside cloud data centers. Independently of its location, each resource is characterized by processing capacity, transfer rates towards other resources and transfer rates from external sources and towards storage devices.

A resource is also described by its monetary cost as well as the costs for the various types of data transfers.

For each task, we then obtain its computation, communication, input and output times which depend on the resources being allocated. From these values we compute the overall execution time of each task as well as the cost for its deployment. The times and costs of individual tasks together their dependencies are then used to derive the workflow execution time $T_W$ and its overall cost $C_W$.

As discussed in [6,10], the demands of individual tasks and the characteristics of the infrastructure are often affected by performance variability and uncertainty, thus we will model them as random variables. Hence, both $T_W$ and $C_W$ are random variables.

Another important aspect considered in the workload and architectural models deals with the characterization of the workflows in terms of computation and communication. In detail, we define the Communication to Computation Ratio as

$$\mathrm{CCR_W} = \frac{\sum_i^n \overline{t_i^{\mathrm{comm}}}}{\sum_i^n \overline{t_i^{\mathrm{comp}}}}$$

where $\overline{t_i^{\mathrm{comm}}}$ and $\overline{t_i^{\mathrm{comp}}}$ denote the average communication and computation times of task $T_i$ computed by considering all possible communication links and resources available in the infrastructure, respectively. A value of $\mathrm{CCR_W}$ greater than 1 denotes a communication-intensive workflow, while a value less than 1 denotes a computation-intensive workflow.

## 3.2 Scheduling Problem

The scheduling problem is formulated as a multi-objective optimization problem as follows:

$$\text{minimize} \quad (\mathbb{E}[T_W], \mathbb{E}[C_W])$$

where $\mathbb{E}[T_W]$ and $\mathbb{E}[C_W]$ denote the expected values of the workflow execution time and overall monetary cost. As already mentioned, both $T_W$ and $C_W$ are random variables since they are derived from the random variables describing the characteristics of the tasks and of the resources.

The main components of the solution framework are summarized in Fig. 2. As can be seen, the solution of the optimization problem is based on the combined application of a meta-heuristic able to cope with multi-objective problems, i.e., the Genetic Algorithm (GA), and the Monte Carlo simulation, that provides the probabilistic evaluation of the objectives. As a result, "optimal" scheduling plans distributed on a Pareto front are obtained.

In detail, starting from the provisioned resources and the scheduling plans generated by the Genetic Algorithm, the Monte Carlo simulation provides their evaluation by properly combining samples of the random variables that describe the tasks and the infrastructure. Hence, multiple realizations of the workflow execution time and cost are obtained and used to derive the empirical distributions of the execution time $T_W$ and of the cost $C_W$.

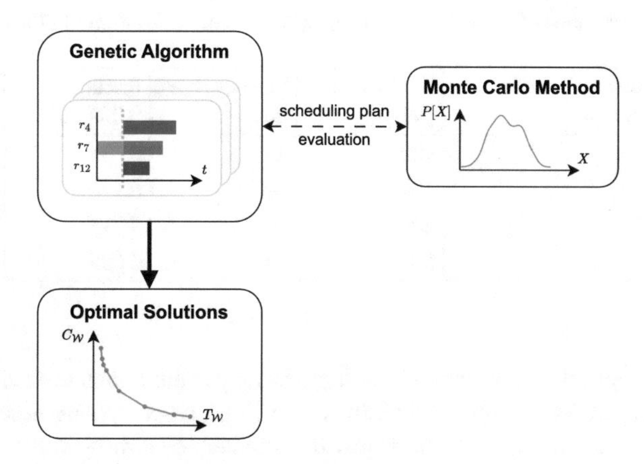


**Fig. 2.** Main components of the proposed solution framework.

## 4 Experimental Results

To assess the benefits of deploying workflows in the cloud-edge continuum, we perform several experiments varying the characteristics of the workflow and the uncertainty affecting the infrastructure performance. In what follows, we describe the workflow and infrastructure used in the experiments as well as the choices made for solving the optimization problem. We also present the results of the experiments and discuss the main findings.

### 4.1 Workload and Infrastructure Characteristics

For the experiments, we consider the IoT data processing workflow proposed in [17] and displayed in Fig. 1. Each task performs some computation followed by communication with one or more tasks according to the precedence constraints depicted in the figure. On average, each task requires 2.24 Million Instructions and exchanges 5.14 GB of data with its neighboring tasks.

We also outline that the input data of the four entry tasks is generated by external sources associated with IoT devices, while the five exit tasks transfer their output data to storage devices associated with cloud infrastructure. In our experiments, we set the output data to 5 GB, while we vary the input data.

The infrastructure used for the experiments consists of multiple instances of different resource types grouped in two layers, namely, four types of resources at the fog/edge layer and eight types at the cloud layer, whose characteristics are summarized in Tables 1 and 2.

**Table 1.** Characteristics of edge devices of the infrastructure considered in the experiments.

| | Processing Capacity [MIPS $\times 10^3$] | Bandwidth [Mbps] | Latency [ms] | Pricing [USD/hr] | Instance Count |
|---|---|---|---|---|---|
| *Fog/Edge* | 10 | 400 | 20 | 0.035 | 5 |
| | 8 | 100 | 100 | 0.020 | 10 |
| | 5 | 200 | 50 | 0.015 | 20 |
| | 2 | 50 | 200 | 0.005 | 20 |

As can be seen, the resources at the fog/edge layer are characterized by a limited processing capacity and lower bandwidth, while cloud resources are much more powerful and well connected in terms of bandwidth and latency (whose value is set to 2ms), although at higher costs. Note that the tables present the cost per hour of the resources. Nevertheless, in our experiments we use a per-minute billing, as typically adopted by cloud providers.

The solution of the multi-objective scheduling problem is based on the NSGA-II algorithm, an elitist non-dominated sorting Genetic Algorithm [9], properly customized as described in [5,7]. Concerning the GA operators, we choose single point crossover,

**Table 2.** Characteristics of cloud devices of the infrastructure considered in the experiments.

| | Processing Capacity [MIPS $\times 10^3$] | Bandwidth [Mbps] | Pricing [USD/hr] |
|---|---|---|---|
| *Cloud A* | 400 | 20,000 | 4.25 |
| | 100 | 4,000 | 1.00 |
| | 50 | 4,000 | 0.45 |
| | 25 | 3,000 | 0.20 |
| *Cloud B* | 240 | 24,000 | 2.00 |
| | 120 | 12,000 | 0.90 |
| | 60 | 6,000 | 0.40 |
| | 30 | 3,000 | 0.18 |

random mutation and a binary tournament selection method. In addition, we set the initial population to 100 individuals, i.e., scheduling plans, and the probabilities associated with crossover and mutation to 0.9 and 0.02, respectively.

To model the variability affecting the performance of the infrastructure, we build a uniform distribution using the nominal performance as its maximum value and computing its minimum value by subtracting from the nominal performance its value multiplied by a penalty factor corresponding to the desired variability. For example, a 10% variability applied to a processing capacity of 100 MIPS leads to a uniform distribution in the range $[90, 100]$ MIPS, whereas for a 90% variability we obtain the range $[10, 100]$ MIPS. In our experiments, we assume both processing capacity and network bandwidth affected by variability.

### 4.2 Results

The first set of experiments considers a balanced workflow in terms of communication and computation, that is, with $\mathrm{CCR}_W = 1$, whose entry tasks receive in total 15 GB of input data from external sources. Each of the four experiments exploits different resources, namely, the resources of Cloud A, Cloud B, fog/edge layer and of the entire cloud-edge continuum. The main objective of these experiments is to investigate the impact of these infrastructures on the "optimal" scheduling plans identified by the Genetic Algorithm and the benefits of the cloud-edge continuum. Figure 3 displays the corresponding Pareto fronts. We can observe in Fig. 3(a) that the two cloud-only Pareto fronts and the fog/edge front differ significantly. As expected, the monetary cost of the scheduling plans that exploit Cloud A resources are generally higher than their Cloud B counterparts. Nevertheless, the corresponding execution times only slightly benefit of these expensive resources. On the contrary, for the fog/edge resources, the costs are rather limited, but the execution times are much longer. There is also an evident gap between the cloud fronts and the fog/edge front which suggests the lack of solutions. As shown in Fig. 4(b) the cloud-edge continuum is able to fill this gap. In fact, the GA identifies "hybrid" solutions, that is, scheduling plans that exploit the cloud-edge continuum. These plans represent a good compromise between time and cost.

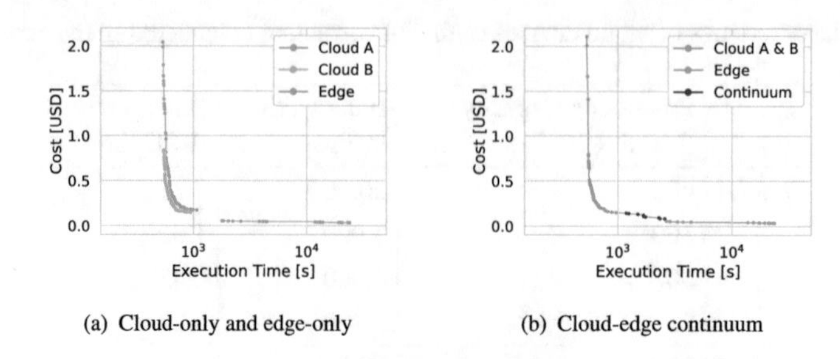


(a) Cloud-only and edge-only

(b) Cloud-edge continuum

**Fig. 3.** Pareto fronts obtained for the balanced workflow, i.e., with $\text{CCR}_W = 1$, as a function of the resources being provisioned, i.e., belonging to Cloud A, Cloud B or to the fog/edge layer (a) and to the cloud-edge continuum (b). Log scale is used on the $x$-axis.

Another set of experiments aims at analyzing the Pareto fronts obtained in the cloud-edge continuum as a function of the workflow characteristics, that is, balanced, computation or communication-intensive, as well as of the input data received from external

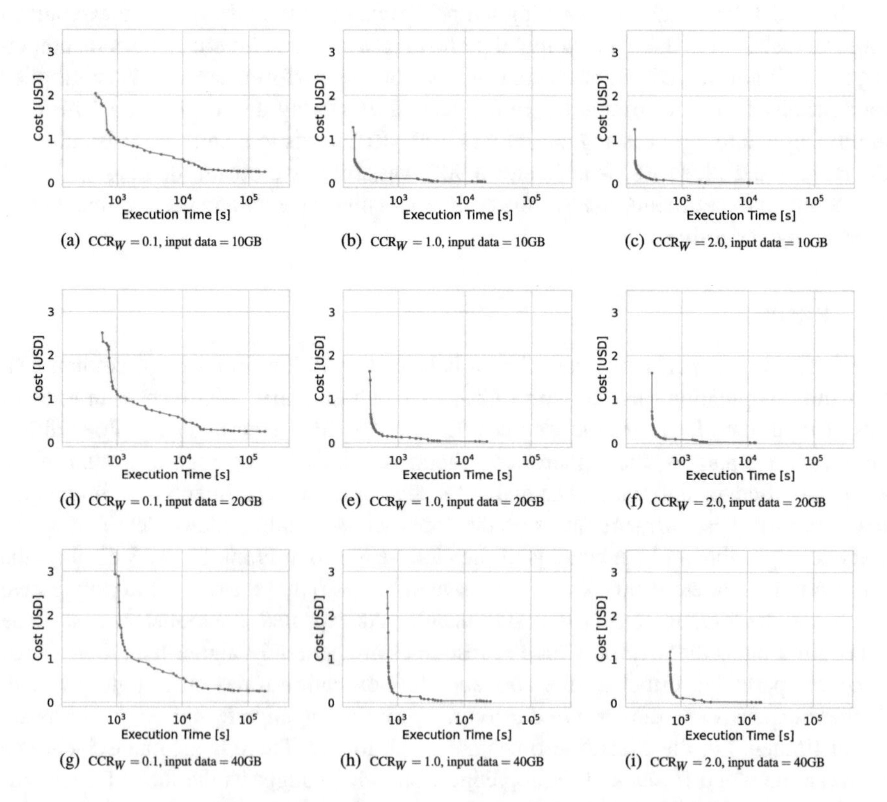


(a) $\text{CCR}_W = 0.1$, input data = 10GB

(b) $\text{CCR}_W = 1.0$, input data = 10GB

(c) $\text{CCR}_W = 2.0$, input data = 10GB

(d) $\text{CCR}_W = 0.1$, input data = 20GB

(e) $\text{CCR}_W = 1.0$, input data = 20GB

(f) $\text{CCR}_W = 2.0$, input data = 20GB

(g) $\text{CCR}_W = 0.1$, input data = 40GB

(h) $\text{CCR}_W = 1.0$, input data = 40GB

(i) $\text{CCR}_W = 2.0$, input data = 40GB

**Fig. 4.** Pareto fronts as a function of the value of $\text{CCR}_W$ describing the workflow and the input data size. Log scale is used on the $x$-axis.

sources. Figure 4 summarizes the behaviors of the corresponding Pareto fronts. We can easily notice that, for a given input data size, the fronts change with the characteristics of the workflow because of the different resource requirements. For example, the scheduling plans of computation-intensive workflows are generally more expensive and characterized by a longer execution time. Concerning the effects of the data size, the diagrams suggest that, as the size increases, the cost tends to increase.

Finally, we analyze the effects of the variability affecting the infrastructure, namely, processing capacity and network bandwidth. Figure 5 displays the Pareto fronts obtained for the computation-intensive workflow, i.e., with $\mathrm{CCR}_W = 0.1$, as a function of the variability. As expected, the variability of the processing capacity strongly affects the Pareto fronts, whereas the variability of the network bandwidth has more limited effects.

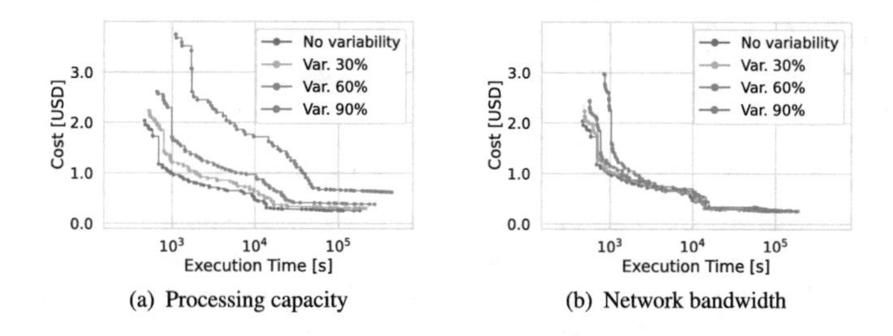


**Fig. 5.** Pareto fronts obtained for the computation-intensive workflow as a function of the variability introduced in the processing capacity (a) and in the network bandwidth (b) of the infrastructure. Log scale is used on the $x$-axis.

## 5 Conclusion

Workflow scheduling in the cloud-edge continuum is a very challenging problem especially whenever multiple requirements have to be satisfied simultaneously. In this paper, we identified scheduling plans that minimize the workflow execution time and monetary cost by formulating a multi-objective optimization problem solved through a combined application of the Genetic Algorithm and the Monte Carlo simulation. Our experiments have demonstrated that workflows greatly benefit of the cloud-edge continuum especially in presence of large inputs from external data sources, such as IoT devices.

As future research directions, we plan to further study the scheduling problem in the cloud-edge continuum by introducing different types of variability and considering various types of constraints, e.g., related to data privacy and confidentiality.

**Acknowledgments.** This work was partly supported by the Italian Ministry of University and Research (MUR) under the PRIN 2022 grant "Methodologies for the Parallelization, Performance Evaluation and Scheduling of Applications for the Cloud-Edge Continuum" (Master CUP: B53D23013090006, CUP: J53D23007110008, CUP: F53D23004300006) and by the European Union - Next Generation EU.

## References


1. Adhikari, M., Amgoth, T., Srirama, S.N.: A survey on scheduling strategies for workflows in cloud environment and emerging trends. ACM Comput. Surv. **52**(4) (2019)
2. Agarwal, G., Gupta, S., Ahuja, R., Rai, A.: Multiprocessor task scheduling using multi-objective hybrid genetic algorithm in fog-cloud computing. Knowl.-Based Syst. **272**, 110563 (2023)
3. Ali, I., Sallam, K., Moustafa, N., Chakraborty, R., Ryan, M., Choo, K.K.R.: An automated task scheduling model using Non-dominated Sorting Genetic Algorithm II for fog-cloud systems. IEEE Trans. Cloud Comput. **10**(4), 2294–2308 (2022)
4. Arunarani, A., Manjula, D., Sugumaran, V.: Task scheduling techniques in cloud computing: a literature survey. Futur. Gener. Comput. Syst. **91**, 407–415 (2019)
5. Calzarossa, M.C., Della Vedova, M.L., Massari, L., Nebbione, G., Tessera, D.: Multi-objective optimization of deadline and budget-aware workflow scheduling in uncertain clouds. IEEE Access **9**, 89891–89905 (2021)
6. Calzarossa, M.C., Della Vedova, M.L., Tessera, D.: A methodological framework for cloud resource provisioning and scheduling of data parallel applications under uncertainty. Futur. Gener. Comput. Syst. **93**, 212–223 (2019)
7. Calzarossa, M.C., Massari, L., Nebbione, G., Della Vedova, M.L., Tessera, D.: Tuning genetic algorithms for resource provisioning and scheduling in uncertain cloud environments: challenges and findings. In: Proceedings of the 27th Euromicro International Conference on Parallel, Distributed and Network-Based Processing (PDP), pp. 174–180 (2019)
8. De Maio, V., Kimovski, D.: Multi-objective scheduling of extreme data scientific workflows in Fog. Futur. Gener. Comput. Syst. **106**, 171–184 (2020)
9. Deb, K., Pratap, A., Agarwal, S., Meyarivan, T.: A fast and elitist multiobjective genetic algorithm: NSGA-II. IEEE Trans. Evol. Comput. **6**(2), 182–197 (2002)
10. Della Vedova, M.L., Tessera, D., Calzarossa, M.C.: Probabilistic provisioning and scheduling in uncertain cloud environments. In: Proceedings of the 2016 IEEE Symposium on Computers and Communication - (ISCC), pp. 797–803 (2016)
11. Esposito, A., et al.: Methodologies for the parallelization, performance evaluation and scheduling of applications for the cloud-edge continuum. In: Barolli, L. (ed.) AINA 2024. LNDECT, vol. 203, pp. XX–YY. Springer, Cham (2024). https://doi.org/10.1007/978-3-031-57931-8_25
12. Goudarzi, M., Wu, H., Palaniswami, M., Buyya, R.: An application placement technique for concurrent IoT applications in edge and fog computing environments. IEEE Trans. Mob. Comput. **20**(4), 1298–1311 (2021)
13. Guerrero, C., Lera, I., Juiz, C.: Genetic-based optimization in fog computing: current trends and research opportunities. Swarm Evol. Comput. **72**, 101094 (2022)
14. Hosseinzadeh, M., Ghafour, M.Y., Hama, H.K., Vo, B., Khoshnevis, A.: Multi-objective task and workflow scheduling approaches in cloud computing: a comprehensive review. J. Grid Comput. **18**, 327–356 (2020)
15. Ijaz, S., Munir, E., Ahmad, S., Rafique, M., Rana, O.: Energy-makespan optimization of workflow scheduling in fog-cloud computing. Computing **103**(9), 2033–2059 (2021)
16. Masdari, M., ValiKardan, S., Shahi, Z., Azar, S.: Towards workflow scheduling in cloud computing: a comprehensive analysis. J. Netw. Comput. Appl. **66**, 64–82 (2016)
17. Stavrinides, G.L., Karatza, H.D.: A hybrid approach to scheduling real-time IoT workflows in fog and cloud environments. Multimed. Tools Appl. **78**, 24639–24655 (2019)
18. Sun, Y., Lin, F., Xu, H.: Multi-objective optimization of resource scheduling in fog computing using an improved NSGA-II. Wirel. Pers. Commun. **102**, 1369–1385 (2018)

# On the Performance of STAC-FastAPI and PgSTAC Using a Cloud-Native Deployment

Alexandru Munteanu[1,2](✉), Silviu Panica[2,3], and Gabriel Iuhasz[1,2,3]

[1] Department of Computer Science, West University of Timişoara, Timişoara, Romania
[2] Innovation Qube SRL, Timişoara, Romania
alexandru.munteanu@e-uvt.ro, silviu@innoqube.ro
[3] Research Institute e-Austria, Timişoara, Romania

**Abstract.** Storing, processing and delivering geospatial data in large quantities can present a significant challenge due to the sheer amount of available Earth observation satellites, extensively producing copious amounts of data on a daily basis. Given the popularity of cloud-based solutions, it was inevitable that approaches relying on delivering data from public and private clouds have been adopted to facilitate the dissemination of Earth observation data as well. The purpose of this paper is to investigate the performance of recent, cloud native implementations of the data cube concept utilizing technologies developed in the Spatio-Temporal Asset Catalogs (STAC) ecosystem. Specifically, within this paper, we evaluate the performance of STAC-FastAPI and PgSTAC in terms of their scalability and data delivery capabilities by employing the Locust.io framework for testing substantial, concomitant user loads.

**Keywords:** STAC · PgSTAC · STAC-FastAPI · performance · locust · cloud native

## 1 Introduction

Geospatial data provides valuable insights into a variety of areas [5,11,20], including urban planning, environmental monitoring, precision agriculture, and disaster management. In an urban context, it facilitates decision-making by informing the spatial allocation of infrastructure and urban development [4]. Environmental monitoring benefits from geospatial data by providing a comprehensive view of changes in land cover/land use [31], aiding in ecosystem health assessment and conservation efforts. Precision agriculture relies on geospatial data for optimizing crop management, irrigation strategies, and resource allocation, contributing to increased agricultural productivity and sustainability [26]. Additionally, geospatial data is critical in disaster management for rapid and effective response efforts by delineating affected areas, planning evacuation routes, and facilitating resource distribution [14]. Essentially, geospatial data


L. Barolli (Ed.): AINA 2024, LNDECT 203, pp. 191–200, 2024.
https://doi.org/10.1007/978-3-031-57931-8_19

plays various roles, acting as a guide for informed decision-making across multiple industries and ultimately resulting in more resilient, sustainable, and effective practices.

Efficient storage, indexing, and retrieval of big geospatial data constitute a significant challenge, particularly with the evergrowing constellations of satellites dedicated to Earth observation. The continuous generation of large amounts of data coming from initiatives such as the European Space Agency (ESA) Copernicus program that by itself is capable of generating roughly 16TB of data on a daily basis[1] is indicative of a need to develop optimised solutions to manage, process, store and deliver this kind of data for exploiting the valuable underlying information. Adding to this necessity is the adoption of Machine Learning techniques by the geospatial community, specifically Deep Learning for solving Computer Vision tasks, requiring big datasets in order to train and evaluate. Multiple curated examples have been so far developed [23,24,29,32] to name but a few. The location of these datasets spans across object-storage, FTP servers, or zip archives available via HTTP. Retrieval and further processing is left up to the end user. Data filtering or partial downloads are often not offered.

Multiple concepts for harnessing cloud resources in order to solve the aforementioned tasks have been disseminated within the geospatial community [6,10,17]. Among these approaches, the *data cube* concept has gained popularity [2,8,9,12,13,16]. Data cube deployments have been effectively utilised in the context of various use cases such as land use/land cover mapping [3], change detection [15] and surface water mapping [19], showcasing their potential and usefulness. However, as of recently, data cube approaches based on the Spatio-Temporal Asset Catalog (STAC) [27] specification have been garnering attention, being largely adopted within the community [1,7,21,25,28].

As of January 2023[2], ESA has announced the ceasement of operations for the Data Hub Software (DHuS) [30], utilised up to this point as the main method of distributing Copernicus programme data. Access to the data is now available through the Copernicus Data Space Ecosystem[3], offering a STAC compliant API and through Data and Information Access Service (DIAS) cloud-based platforms.

STAC-FastAPI offers an HTTP API compliant with the STAC specification. It has been implemented based on FastAPI, and as of the time of writing, this paper can be utilized with three backends relying on either Postgres (PgSTAC), SQLAlchemy[4], or ElasticSearch[5] in order to store metadata for facilitating rapid indexing and querying functionalities. Due to the rich ecosystem built around the STAC specification, there are numerous consumers available for various technologies: pystac-client, stackstac, odc-stac, Rstac, STAC.jl, xpystac and GDAL.

This paper sets out to assess the performance and scalability of cloud native deployments of STAC-FastAPI and PgSTAC in terms of geospatial data delivery.

---

[1] https://defence-industry-space.ec.europa.eu/eu-space-policy/copernicus_en.
[2] https://scihub.copernicus.eu/news/News01146u.
[3] https://dataspace.copernicus.eu.
[4] https://www.sqlalchemy.org.
[5] https://www.elastic.co/guide/en/elasticsearch/reference/current/index.html.

## 2 Related Work

Modern approaches taken towards solving the problems that arise when dealing with big Earth observation data, such as the Australian Geoscience Data Cube (AGDC) [16] describe efficient storage, processing, indexing, querying and retrieval mechanisms. The AGDC served as a foundational effort, and a proof of concept for shaping up the Open Data Cube (ODC) [22] initiative, steered by the Committee on Earth Observation Satellites (CEOS), as described in [12].

A brief comparison of download time ratio for retrieval of Earth observation data from both object-store and HTTP inventories between the United States Geological Survey (USGS)[6], Google Earth Engine (GEE) [10] and Amazon Web Services (AWS)[7] is included in [8]. In addition, it is highlighted that the data distributed by different providers is not stored in the same format, sometimes consisting of a compressed archive, which delays the acquisition of partial data.

### 2.1 Spatio-Temporal Asset Catalogs (STAC)

The popularity of the STAC specification can be attributed to the numerous benefits it provides. STAC offers a standardised way for cataloging and describing spatio-temporal assets, which is essential in order to manage large-scale Earth observation data. This standardisation facilitates the interoperability and integration of data from various sources. The STAC specification also facilitates data discovery by enabling users to search, browse, and access geospatial data easily. The adoption of a common metadata format ensures that metadata, such as spatial extent, temporal coverage, and other relevant properties, are readily available and easily accessible. STAC implementations can also be paired with frameworks and libraries, such as Dask[8], for scaling processing resources on-demand, enabling efficient analysis of large-scale and high-resolution datasets.

A scalable remote sensing data processing platform intended to be used in conjunction with high-performance computing (HPC) is described by [1]. The solution utilises the STAC specification for efficient data filtering and retrieval. Furthermore, the datasets provided within Microsoft's Planetary Computer [18] are exposed through a STAC API implementation deployed in Azure Kubernetes.

## 3 Elastic STAC Deployment and Performance

In this section we propose a deployment of two STAC based solutions, namely PgSTAC and STAC-FastAPI, using cloud-native technologies in order to support dynamic loads with fewer resources committed with respect to the existent monolith approaches. Our approach leverages Kubernetes[9], an orchestration system used to deploy, scale, and manage containerized applications. Kubernetes

[6] https://earthexplorer.usgs.gov.
[7] https://aws.amazon.com/earth.
[8] Dask - library for parallel computing - https://www.dask.org.
[9] Kubernetes - Open-source system for automating deployment, scaling, and management of containerized applications - https://kubernetes.io.

clusters are normally deployed on top of cloud resources that use dynamic allocation policies to optimise resource usage. Finally, we describe the load testing process employing Locust[10] to benchmark the proposed solution's performance.

### 3.1 Cloud-Native STAC-FastAPI and PgSTAC Testbed

In Fig. 1, the proposed deployment architecture is depicted with a typical cloud-native platform to support the STAC based components. In our use case, we have deployed three main components:

- PgSTAC[11] - a PostgreSQL based STAC backend, using a RDBMS to store and retrieve STAC items;
- STAC-FastAPI[12] - frontend API which enforces STAC standard specifications, used to store and consume STAC items, hosted on PgSTAC;
- Object Storage - a distributed object store solution to host the ESA Copernicus Sentinel-2 products that exposes an S3 interface. In our use case we used MinIO Object Storage[13]

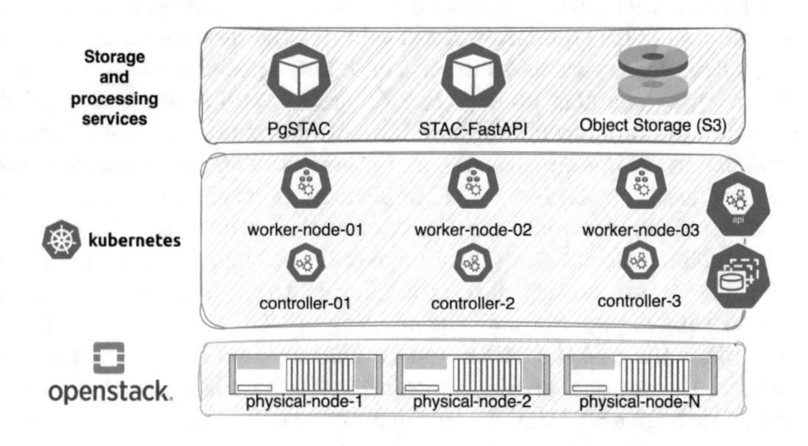


**Fig. 1.** Cloud-native STAC deployment using Kubernetes using cloud resources

All of the described components are deployed and managed dynamically using Kubernetes. For PgSTAC and STAC-FastAPI implementations, we created a set of Kubernetes templates to deploy and distribute the workload using the available resources. As for the object storage component, we used the official deployment templates provided by MinIO. Kubernetes testbed is deployed on

[10] Locust - A modern load testing framework - https://locust.io.
[11] PgSTAC - https://github.com/stac-utils/pgstac.
[12] STAC-FastAPI - https://github.com/stac-utils/STAC-FastAPI.
[13] MinIO - High Performance, Kubernetes Native Object Storage - https://min.io.

cloud infrastructure, in our case using OpenStack technology, and the committed resources have the following hardware specifications: (a) 3 Kubernetes controllers with 4 cores @2 Ghz AMD EPYC Rome CPU and 8 GB RAM memory and 32 GB SSD storage and (b) 3 Kubernetes worker nodes with 8 cores @2 Ghz AMD EPYC Rome CPU, 32 GB RAM memory and 128 GB SSD storage.

In terms of resources, each of the three components was configured with the following resource consumption thresholds: (i) PgSTAC: 6 cores and 12 GB RAM memory, 128 GB SSD storage, deployed using one instance; (ii) STAC-FastAPI: 8 cores and 16 GB RAM memory, 4 GB SSD storage, deployed using one instance; (iii) Object Storage: 3 storage nodes, each with 2TB of SSD storage, used to create a pool of 3TB cloud-native object storage to store the satellite images; (iv) communication network: dual 25 Gbps aggregated connections to support both processing and storage communication channels.

## 3.2 Locust Load Testing

In order to assess the performance of our proposed cloud-native deployment, we have employed the Locust load testing framework for simulating thousands of simultaneous users. In Locust, each simulated user picks tasks (usually in the form of HTTP requests) from a predefined set and performs them. Several tasks were defined for Earth observation data querying, filtering, and retrieval based on the available indexed product metadata.

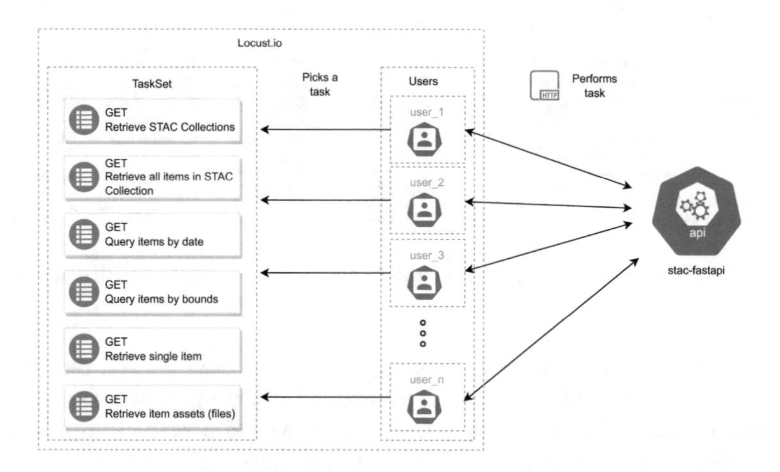


**Fig. 2.** Locust load testing.

Figure 2 illustrates the Locust users randomly picking tasks and performing them in the form of HTTP requests directed to the STAC-FastAPI. For the tests, Locust was configured to run in parallel using 1 master and 30 workers. Multiple configurations, where the number of users was gradually increased, were tested in order to saturate the available resources allocated for the deployment.

For the experiments, we have used 1.789 ESA Copernicus Sentinel-2 archives, amounting to a total 2.3TB of data, uploaded in the object storage component. These datasets were processed to extract the corresponding metadata. The collected metadata was further transformed into STAC items and later uploaded into the PgSTAC component, consisting of a total of 21.468 inserted STAC items.

### 3.3 Results

In our experiments, we simulated a heavy load with 3000, 5000, and 7000 simultaneous users. The users randomly perform 7 distinct operations, stressing the three components of our deployment. To ensure consistent comparison results, we performed ten consecutive iterations for each of the configurations. The duration of each iteration was 1 h. The operations performed by Locust users are likely to occur in a real-life scenario, being specific to a regular user or application behavior that utilises a STAC-compliant service for consuming ESA Copernicus Sentinel-2 data. The scope of these tests is to identify potential performance issues that could be addressed using specific tuning configurations.

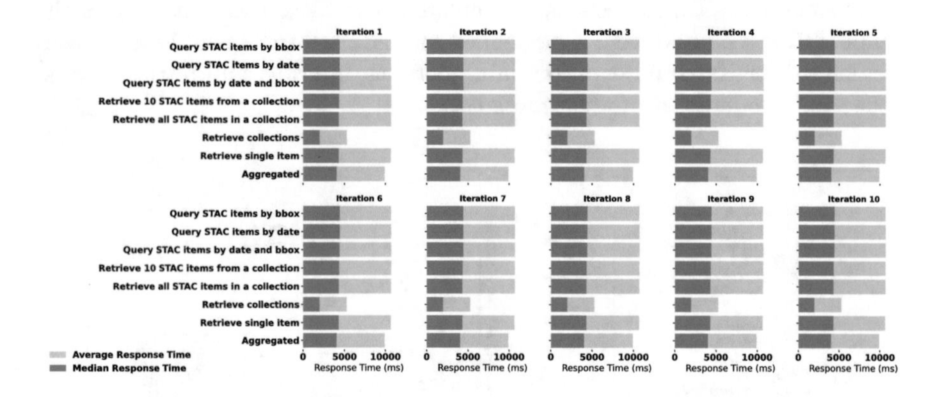


**Fig. 3.** Median and average response time of the requests - 3000 simultaneous users.

Figure 3 shows the median and average response time for the requests performed by the users. The response times are presented separately by operation.

Based on the collected metrics, we have computed a rate of approximately 200 requests per second issued by 3000 simultaneous users to be the limit that the deployment can handle while offering a reasonable response time given the allocated resources. In Fig. 4, the resource consumption graph shows that the PgSTAC component is CPU-intensive, with memory usage being far lower compared to STAC-FastAPI, which uses both CPU and RAM memory to sustain the load. These results are consistent with higher usage patterns, as illustrated in Fig. 5, showing the average consumption for each configuration (10 iterations).

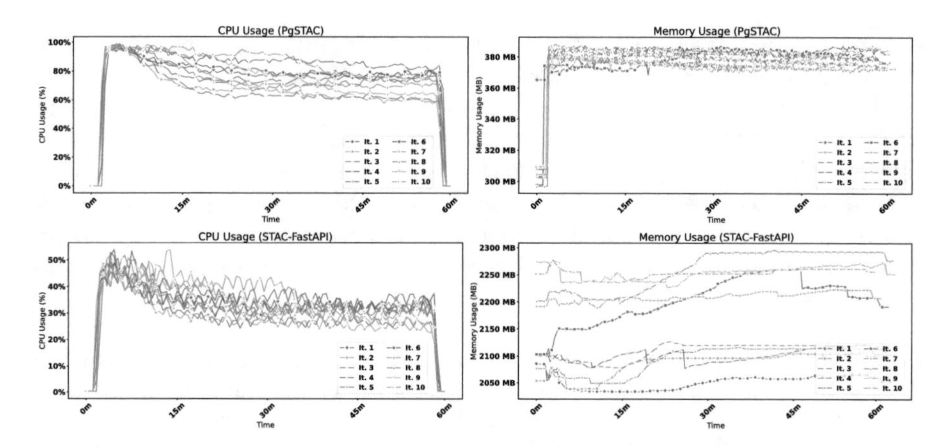


**Fig. 4.** CPU and Memory usage for both the PgSTAC and STAC-FastAPI components for each iteration during load testing with 3000 simultaneous users.

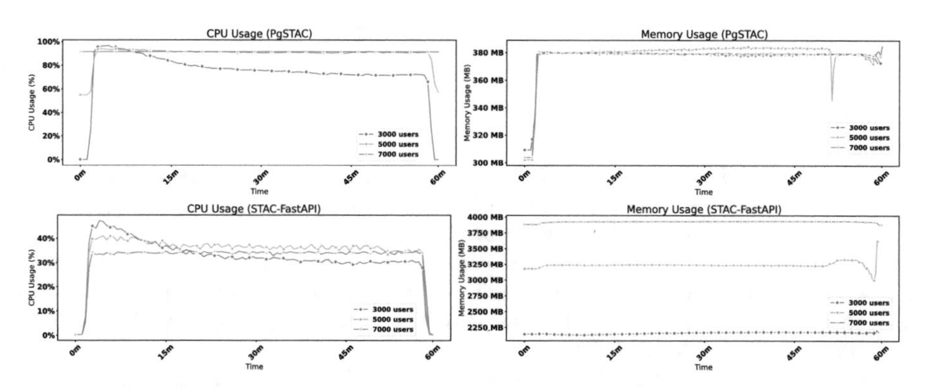


**Fig. 5.** CPU and Memory usage of PgSTAC and STAC-FastAPI components during load testing. Each line represents an average of the 10 iterations ran for a configuration.

## 4 Conclusions

In this paper, we have presented incipient work on a cloud native deployment for distributing geospatial data by employing STAC-based solutions using state of the art orchestration and deployment technologies. By employing the Locust load testing framework, we explored the performance of this deployment, simulating heavy user loads carrying out requests as they would occur on such platforms.

The results show that all three components of the proposed deployment are capable of achieving a good performance while serving a significantly high user load. Monitoring resource consumption per component has provided valuable insights that should be taken into consideration for both further development of scaling strategies and resource allocation. The results of the tests carried out in this paper indicate that further investigation of cloud native approaches for deploying STAC based applications, as well as their performance with respect to delivering large quantities of geospatial data should be taken into consideration.

For future work, we intend to extend the testing methodology in order to include Kubernetes horizontal scalability for adjusting the resource usage dynamically. Another point of interest is to analyse the cloud usage costs in various scenarios involving different deployment models, monolith versus cloud native. Additionally, we intend to compare the performance results by further employing different existing STAC backends based on SQLAlchemy or Elastic-Stack (ELK).

**Acknowledgement.** The work presented in this paper was partially supported by a grant of the EU's Horizon 2020 Research and Innovation programme under grant agreement No. 101003517, acronym HARMONIA, The Romania Competitiveness Operational Programme SMIS 120725 - SCAMP-ML (Advanced computational statistics for planning and tracking production environments) as well as a grant of EU's Horizon 2020 Research and Innovation programme under grant agreement No. 101017168, acronym SERRANO.

## References

1. Astsatryan, H., Lalayan, A., Giuliani, G.: Scalable data processing platform for earth observation data repositories. Scalable Comput. Pract. Exp. **24**(1), 35–44 (2023). https://doi.org/10.12694/scpe.v24i1.2041, number: 1
2. Baumann, P., et al.: Fostering cross-disciplinary earth science through datacube analytics. In: Mathieu, P.-P., Aubrecht, C. (eds.) Earth Observation Open Science and Innovation. ISRS, vol. 15, pp. 91–119. Springer, Cham (2018). https://doi.org/10.1007/978-3-319-65633-5_5
3. Chaves, M.E.D., Soares, A.R., Sanches, I.D., Fronza, J.G.: CBERS data cubes for land use and land cover mapping in the Brazilian Cerrado agricultural belt. Int. J. Remote Sens. **42**(21), 8398–8432 (2021). https://doi.org/10.1080/01431161.2021.1978584
4. Chrysoulakis, N., et al.: A conceptual list of indicators for urban planning and management based on earth observation. ISPRS Int. J. Geo-Inf. **3**(3), 980–1002 (2014)
5. Denis, G., et al.: Towards disruptions in earth observation? new earth observation systems and markets evolution: Possible scenarios and impacts. Acta Astronaut. **137**, 415–433 (2017). https://doi.org/10.1016/j.actaastro.2017.04.034
6. Evangelidis, K., Ntouros, K., Makridis, S., Papatheodorou, C.: Geospatial services in the cloud. Comput. Geosci. **63**, 116–122 (2014). https://doi.org/10.1016/j.cageo.2013.10.007
7. Ferreira, K.R., et al.: Earth observation data cubes for Brazil: requirements, methodology and products. Remote Sens. **12**(24), 4033 (2020). https://doi.org/10.3390/rs12244033
8. Giuliani, G., et al.: Building an earth observations data cube: lessons learned from the swiss data cube (SDC) on generating analysis ready data (ARD). Big Earth Data **1**(1), 100–117 (2017). https://doi.org/10.1080/20964471.2017.1398903

9. Giuliani, G., Chatenoux, B., Piller, T., Moser, F., Lacroix, P.: Data cube on demand (DCoD): generating an earth observation data cube anywhere in the world. Int. J. Appl. Earth Obs. Geoinf. **87**, 102035 (2020). https://doi.org/10.1016/j.jag.2019.102035
10. Gorelick, N., Hancher, M., Dixon, M., Ilyushchenko, S., Thau, D., Moore, R.: Google earth engine: planetary-scale geospatial analysis for everyone. Remote Sens. Environ. **202**, 18–27 (2017). https://doi.org/10.1016/j.rse.2017.06.031
11. Kansakar, P., Hossain, F.: A review of applications of satellite earth observation data for global societal benefit and stewardship of planet earth. Space Policy **36**, 46–54 (2016). https://doi.org/10.1016/j.spacepol.2016.05.005
12. Killough, B.: Overview of the open data cube initiative. In: IGARSS 2018 - 2018 IEEE International Geoscience and Remote Sensing Symposium, pp. 8629–8632 (2018). ISSN: 2153-7003
13. Kopp, S., Becker, P., Doshi, A., Wright, D.J., Zhang, K., Xu, H.: Achieving the full vision of earth observation data cubes. Data **4**(3), 94 (2019). https://doi.org/10.3390/data4030094
14. Le Cozannet, G., Kervyn, M., Russo, S., Ifejika Speranza, C., Ferrier, P., Foumelis, M., Lopez, T., Modaressi, H.: Space-based earth observations for disaster risk management. Surv. Geophys. **41**, 1209–1235 (2020)
15. Lewis, A., et al.: Rapid, high-resolution detection of environmental change over continental scales from satellite data - the earth observation data cube. Int. J. Digit. Earth **9**(1), 106–111 (2016). https://doi.org/10.1080/17538947.2015.1111952
16. Lewis, A., et al.: The Australian geoscience data cube - foundations and lessons learned. Remote Sens. Environ. **202**, 276–292 (2017). https://doi.org/10.1016/j.rse.2017.03.015
17. Lunga, D., Gerrand, J., Yang, L., Layton, C., Stewart, R.: Apache spark accelerated deep learning inference for large scale satellite image analytics. IEEE J. Sel. Top. Appl. Earth Obs. Remote Sens. **13**, 271–283 (2020). https://doi.org/10.1109/JSTARS.2019.2959707
18. Microsoft, O.S., McFarland, M., Emanuele, R., Morris, D., Augspurger, T.: microsoft/planetarycomputer, October 2022. https://doi.org/10.5281/zenodo.7261897
19. Mueller, N., et al.: Water observations from space: mapping surface water from 25 years of Landsat imagery across Australia. Remote Sens. Environ. **174**, 341–352 (2016)
20. Pandey, P.C., Koutsias, N., Petropoulos, G.P., Srivastava, P.K., Ben Dor, E.: Land use/land cover in view of earth observation: data sources, input dimensions, and classifiers-a review of the state of the art. Geocarto Int. **36**(9), 957–988 (2021). https://doi.org/10.1080/10106049.2019.1629647
21. Picoli, M.C.A., et al.: CBERS data cube: a powerful technology for mapping and monitoring Brazilian biomes. ISPRS Ann. Photogramm. Remote Sens. Spat. Inf. Sci. **V-3-2020**, 533–539 (2020). https://doi.org/10.5194/isprs-annals-V-3-2020-533-2020
22. Ross, J., Killough, B., Dhu, T., Paget, M.: Open data cube and the committee on earth observation satellites data cube initiative. IAC (2017)
23. Rottensteiner, F., et al.: The ISPRS benchmark on urban object classification and 3D building reconstruction. ISPRS Ann. Photogramm. Remote Sens. Spat. Inf. Sci. I-3 **1**(1), 293–298 (2012)
24. Schmitt, M., Hughes, L.H., Qiu, C., Zhu, X.X.: SEN12ms-a curated dataset of georeferenced multi-spectral sentinel-1/2 imagery for deep learning and data fusion. arXiv preprint arXiv:1906.07789 (2019)

25. Schramm, M., et al.: The openEO API-harmonising the use of earth observation cloud services using virtual data cube functionalities. Remote Sens. **13**(6), 1125 (2021). https://doi.org/10.3390/rs13061125
26. Shafi, U., Mumtaz, R., García-Nieto, J., Hassan, S.A., Zaidi, S.A.R., Iqbal, N.: Precision agriculture techniques and practices: from considerations to applications. Sensors **19**(17), 3796 (2019). https://doi.org/10.3390/s19173796
27. STAC Contributors: SpatioTemporal Asset Catalog (STAC) specification (2021). https://stacspec.org
28. Sudmanns, M., Augustin, H., van der Meer, L., Baraldi, A., Tiede, D.: The Austrian semantic EO data cube infrastructure. Remote Sens. **13**(23), 4807 (2021). https://doi.org/10.3390/rs13234807
29. Sumbul, G., Charfuelan, M., Demir, B., Markl, V.: Bigearthnet: a large-scale benchmark archive for remote sensing image understanding. In: IGARSS 2019 - 2019 IEEE International Geoscience and Remote Sensing Symposium, pp. 5901–5904 (2019). ISSN: 2153-7003
30. Tona, C., Bua, R.: Open source data hub system: free and open framework to enable cooperation to disseminate earth observation data and geo-spatial information. In: Geophysical Research Abstracts, vol. 20 (2018)
31. Vali, A., Comai, S., Matteucci, M.: Deep learning for land use and land cover classification based on hyperspectral and multispectral earth observation data: a review. Remote Sens. **12**(15), 2495 (2020). https://doi.org/10.3390/rs12152495
32. Van Etten, A., Hogan, D., Martinez-Manso, J., Shermeyer, J., Weir, N., Lewis, R.: The multi-temporal urban development SpaceNet dataset. https://doi.org/10.48550/arXiv.2102.04420, http://arxiv.org/abs/2102.04420

# The Efficiency of Building Maintenance Using Digital Twins: A Literature Review

Ionica-Larisa Puiu[1(✉)] and Teodor-Florin Fortiş[1,2]

[1] West University of Timişoara, 300223 Timişoara, Romania
ionica.puiu@e-uvt.ro

[2] Research Institute e-Austria Timişoara, 300223 Timişoara, Romania

**Abstract.** As buildings severely impact global energy consumption and greenhouse gas emissions, effective maintenance becomes more important for their performance and sustainability. On the other hand, digital twins, as a promising technology, enhance building operation and maintenance by providing near real-time data and insights. To better understand the efficiency of building maintenance using digital twins, we analyzed twelve recent papers, highlighting benefits (such as energy efficiency, reduced costs), challenges (data quality, interoperability), and opportunities (integration with AI, IoT) in residential, commercial, and industrial buildings.

**Keywords:** Digital twin · BIM · Building maintenance · Energy efficiency · Sustainability

## 1 Introduction

According to recent reports of the International Energy Agency (IEA), under the umbrella of the United Nations Environment Programme, a large share of the global energy use and greenhouse gas emissions is generated by building operation [1]. The 2022 report emphasize that an all-time high in $CO_2$ emissions have been reached, with an important increase of around 2% compared to the previous peak, of 2019, up to a value of $10GtCO_2$.

The energy efficiency of buildings depends largely on the operation and maintenance (O&M) phase, which includes the activities and processes for monitoring and optimizing the building systems. The O&M phase can last for decades and can constitute up to 80% of the total lifecycle cost of a building. Therefore, effective and efficient O&M of buildings is essential to reduce the energy consumption, operational cost and environmental impact of buildings.

Numerous challenges can be identified, such as: (a) the lack of reliable data, (b) the complexity and diversity of building systems and components which can lead to poor performance, (c) increased energy waste, (d) lower occupant comfort and shorter lifespan of the building systems and equipment , and others. To address these challenges, an extensive implementation of Building Information Modeling (BIM) technologies, integrating recent IoT applications, offers a series


L. Barolli (Ed.): AINA 2024, LNDECT 203, pp. 201–211, 2024.
https://doi.org/10.1007/978-3-031-57931-8_20

of new perspectives and capabilities aimed at improving decisions in the different stages of the architectural life cycle.

Recently, building digital twins have emerged as a promising technology that could improve building O&M and provide the capability of real-time connectivity to online sensors.

There are several definitions for the digital twins, as explained in [2]. In one of the simplest characterizations, "a digital twin (DT) is a computerized model of a physical device or system that represents all functional features and links with the working elements". [3] By using DTs, data can be integrated and automated among different systems and predictive analysis and simulation of the physical entity can be conducted.

The target of our work is to offer a comprehensive and critical evaluation of recent academic work on the use of digital twins in the context of building maintenance research. In order to reach this goal, our research is based on analysing the relevant literature on how the digital twins contribute to improving the efficiency of building maintenance, analysis which is following a predefined protocol based on the PRISMA guidelines and standards.

The remainder of this paper is structured as follows:

- Section 2 offers some background information about systematic review methodologies;
- Section 3 explains the methods used in our approach for the literature review;
- A presentation and comparison of the main findings and information extracted is included in Sect. 4, data which is synthesized from the considered papers. These findings are complemented by discussions on the results of the synthesis and highlights the main themes, patterns, and inconsistencies identified in the literature;
- The conclusions of our research are included in Sect. 5, together with some recommended steps for further investigations.

## 2 Background Information

In this section, we discuss several concepts and tools that are related to Digital Twins and systematic reviews. Digital Twins, being complex entities, require precise definitions and organization for effective research. Systematic reviews are methods of synthesizing information from multiple sources that require thorough evaluation for quality and reliability. Systematic reviews, on the other hand, are methods of synthesizing evidence from multiple sources, that require thorough assessment for quality and reliability. We will present three tools that can offer support for these tasks: PRISMA, AMSTAR 2, and ROBIS.

PRISMA, which is a comprehensive set of guidelines, serves as a robust framework for authors engaged in the construction of systematic reviews and meta-analyses. It includes a 27-item checklist, an abstract checklist, and a flow diagram, which are important tools in order to ensure the quality and transparency of these research methods. By emphasizing key elements, such as eligibility criteria, search strategy, selection of studies, data extraction, and conclusions, PRISMA enhances the efficacy of systematic review processes. [4]

The AMSTAR 2 is a robust tool for the methodological quality evaluation for systematic reviews, which originally involved both randomized and non-randomized studies of healthcare intervention [5]. This tool includes 16 items that covers various aspects of systematic reviews, such as the protocol, strategy, selection, extraction, synthesis, and reporting of evidence. Furthermore, it integrates a rating system to assess the confidence levels in the findings, which can range from high to critically low confidence. [5–7]

Another tool, which was specifically designed to assess the risk of bias within systematic reviews, is ROBIS. It uses 16 questions to evaluate different aspects of the review process, including the criteria, selection, extraction, synthesis, and reporting. Similarly with AMSTAR 2, ROBIS employs a rating system to gauge the confidence in the results, which is based on the performance in both critical and non-critical domains. The ratings span from high to critically low confidence levels. These tools aid in determining the trustworthiness of systematic review findings. [8]

## 3 Methods

Our literature review methodology, which is aligned with the PRISMA standards, investigates the enhancement in building maintenance efficiency through the implementation of digital twins across various contexts and scenarios. The methods that were considered for the review are as follows: a) formulation of a clear research question; b) development of a protocol; c) search for relevant studies in multiple databases; d) screening of the studies for eligibility; e) assessment of the quality of studies; f) extraction and synthesis of data; g) interpretation and presentation of results.

### 3.1 The Search Strategy

A comprehensive literature review was executed on relevant databases including Scopus, ScienceDirect, and Google Scholar. The search strategy was based on a set of selected keywords: "digital twin", "building maintenance", "energy efficiency", "Building Information Modelling", and "sustainability". The review targeted English English peer-reviewed articles and conference papers that were published in an interval of five years, and yielded a total of 207 relevant entries. The search operation was initially conducted on December 2, 2023.

The selection of an interval of recent five years is relevant for the evolving Digital Twin Simulation (DTS) technologies in the context of building maintenance, thereby ensuring the use of updated insights. The choice of English, due to its widespread acceptance, ensures the accessibility of the research. The search terms used for the literature search were derived from the research question and the keywords of the topic. The search terms were also refined and modified by using the Boolean operators – AND, OR, and NOT – to increase the precision and recall of the search outcomes. The search protocol implemented for Scopus included the following steps, linked with the process described in Sect. 3.3:

- An initial search for `KEY ( "digital twin*" )`, returning 12,754 entries;
- Limit the year range of results to 2018–2022 (8,051 documents):
  `PUBYEAR > 2017 AND PUBYEAR < 2023`;
- Consider only articles or conference papers (7,404 documents):
  `( LIMIT-TO ( DOCTYPE, "ar" ) OR LIMIT-TO ( DOCTYPE, "cp" ) )`
- Filter based on the most significant keywords (207 entries):
  ```
  ( LIMIT-TO ( EXACTKEYWORD, "Building Information Modelling" )
  OR LIMIT-TO ( EXACTKEYWORD, "BIM" )
  OR LIMIT-TO ( EXACTKEYWORD, "Sustainability" )
  OR LIMIT-TO ( EXACTKEYWORD, "Building maintenance" ).
  ```

### 3.2 Keyword Co-occurrence

An investigation of keyword relevancy was undertaken to assess the alignment between the achieved outcomes and the established search guidelines. Such an investigation also aimed to identify clusters of relevant terms that are significant to the subject domain. Keyword co-occurence analysis is used to map the current state of research and highlight research directions. Additionally, it offers support in constructing of a knowledge map before the implementation of a comprehensive systematic review. [9, 10]

We considered a network map for recurring keywords, each with a minimum of five occurrences, aided by VOSviewer[1], and illustrated in Fig. 1. The size of the nodes and words are indicative for their weight. The closeness of nodes indicates the strength of their relationships, with shorter distances denoting stronger connections. Lines between nodes represent pairs of co-occurring keyword, with thicker lines indicating a higher probability of co-occurrence. Nodes sharing the same color are grouped into clusters.

The highest frequency keywords from Fig. 1 are "building information modelling" (BIM) and "digital twin". These keywords are linked with five distinct clusters of related terms:

1. "Sustainability", "manufacture", and "life cycle" are associated with "digital twin". This suggests that the adoption of these technologies may improve building maintenance in scenarios related with these keywords.
2. "Smart city", "artificial intelligence", "machine learning", and "energy utilization" are linked with BIM. These connections refer to processes aimed at improving the energy efficiency of a building.
3. "Construction industry", "Internet of Things", "built environment", and "project management" are linked with both BIM and digital twins.
4. "Maintenance", "facility management", "digital transformation", and "information management" offer the final cluster of terms.

These findings highlight relevant research areas within the broader context of digital twin investigations, highlighting robust connections among these terms.

[1] https://www.vosviewer.com.

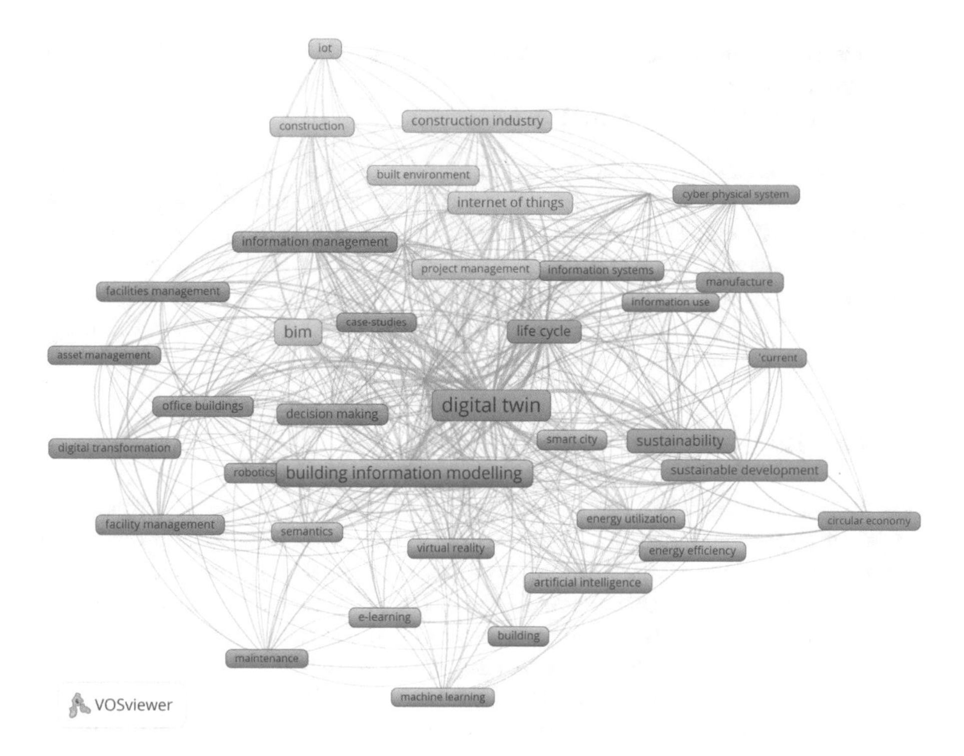


**Fig. 1.** The keyword co-occurrence network.

## 3.3 Screening, Selection and Data Extraction

The initial screening process involved the examination of titles and abstracts, based on which we were able to exclude less relevant papers. For the remaining papers a full-text assessment for quality and eligibility was performed. Methodology, validity, and quality were evaluated using quality checklist and tools such as AMSTAR 2, ROBIS, and others. Articles that met a predetermined quality threshold were retained, while the others were excluded. The screening and selection criteria and process are as follows:

**Relevance** Screening of titles and abstracts of the retrieved papers (*stage 1*);
**Language** Excluding papers not written in English (*stage 2*);
**Type** Excluding papers not published in peer-reviewed journals or conference proceedings (*stage 3*);
**Time-frame** Excluding papers that are not published in the specified time-frame (i.e., 2018 — 2023, *stage 4*);
**Topic** Excluding papers unrelated to the utilization of digital twins for building maintenance (*stage 5*);
**Objective** Excluding papers that do not evaluate the effectiveness of digital twins for building maintenance (*stage 6*);

**Quality** Excluding papers that score below the cut-off point on the quality assessment checklist or tool.
Using the AMSTAR 2 or ROBIS (*stage 7*).

Data was extracted, synthesized, and organized from the various papers considered. Our analysis included information such as the details of the papers, use cases of digital twin, their advantages and disadvantages, opportunities, tools, findings, limitations, and recommendations. A method of thematic synthesis was employed, which involved the coding, organization, and description of information to identify primary patterns and inconsistencies. Our consolidated findings are presented and compared in Sect. 4.

## 4 Results

This section presents and compares the data extracted and synthesized from the set of papers included in the literature review. Additionally, it provides an overview of the characteristics and features of the selected papers, including details such as the publication source, type, aim, context, and scenario.

The research methods and approaches applied for the set of papers include review, case study, and experiment. Among these, review papers constitute 50%, followed by experiment and case study papers, each with 25%.

Case study papers focused on the use and assessed of digital twins (DTs) for building maintenance within real-world contexts, experiment papers involved testing and measurement of DTs for building maintenance, while review papers usually synthesized existing literature on DTs for building maintenance, identifying the main trends, gaps and opportunities.

Regarding the aims and objectives of these papers, they primarily considered to: explore DTs for building maintenance (around 66.66%); develop systems or tools for DTs in building maintenance (25%); propose frameworks or models for the use of DTs in building maintenance (8.33%). propose a framework or a model for the use of DTs for building maintenance (8.33%).

### 4.1 Benefits of Digital Twins for Building Maintenance

Digital Twins can help improving building maintenance practices through the use of various technologies [11–13], such as:

**Component Identification.** By linking physical and digital components using Building Information Modeling and RFID tags, we can enable efficient maintenance checks and real-time data access;
**BIM and Building Management System (BMS).** Connecting BIM with the BMS bridges physical and digital objects, streamlining maintenance operations;
**Augmented Reality (AR).** AR overlays digital information onto the physical world, enhancing perception and aiding in problem visualization and troubleshooting.

In recent research, as mentioned by Zhao et al. [14], the benefits of integrating BIM to streamline facilities management (FM) processes are emphasized. Specifically, BIM can reduce the time and costs associated with updating FM databases during the operations and maintenance (O&M) phase. Also, it can facilitate data integration and enhance decision-making efficiency across stakeholders.

In the context of Building Information Models (BIM), the study from [11] specify three areas of improvements for building maintenance. Such improvements are linked with monitoring and optimizing building performance using DTs.

*Monitoring improvements* – researchers have reported important improvements in building monitoring and equipment management through the adoption of DTs. For instance, in [15], Peng et al. developed a real-time platform that visualizes equipment status and room occupancy within a hospital A centralized database also helps in filtering information from smart sensors and select relevant data based on user roles. Project actors can collaboratively use the digital twins to monitor building performance effectively [16]. Furthermore, centralization enables data-to-equipment links, enabling complex analyses. These benefits of monitoring also help optimizing energy consumption and costs during inspection and maintenance planning phases [11].

*Improvements for Inspection* – digital twin algorithms can provide automatic assessment for monitored systems using data from smart sensors. Researchers have proposed various methods that use smart sensors and image capture devices to evaluate building damage [11]. For example, Shim et al. introduced a visual assessment method for bridges, as mentioned in [11], by using a code-based system to inventory the damages. Additionally, algorithms have been proposed to detect deviations and recalibrate systems accordingly [11,16].

In the case of predictive maintenance, DTs offer a proactive approach to prevent faults [17]. Researchers have highlighted the advantages of BIM-based DT approaches, in the context of planning. Prediction algorithms, such as machine learning and neural networks, enable visualization of equipment degradation and lifespan, as mentioned by Coupry et al., in [11]. For example, Tahmasebinia et al. employed these algorithms to estimate the impact of dead load creep and shrinkage on the Sydney Opera House. By exploiting these predictive capabilities, maintenance planning becomes proactive, shifting from reactive responses to preventive strategies. [18]

In the context of energy optimization in building design, the paper [19] identifies several studies related with DTs and their impact on energy efficiency. *Tariq et al.* developed a DT model for a solar chimney, using artificial intelligence techniques to maximize energy efficiency and minimize emissions. Their study correlated results with various issues. In their work, *Zhao et al.* created a building energy model, assessing the feasibility of retrofitting based on nearly zero-energy building concepts. Their scan-to-BIM-based DT demonstrated a 14.1% energy reduction potential. *Massafra et al.* focused on the integration of Heritage Building Information Modeling (HBIM) and Building Performance Simulation (BPS) tools. to improve energy efficiency in an Italian context. Such an integration aims to enhance the energy efficiency in the given context.

Investigating energy savings and occupant comfort, *Wang et al.* employed a DT approach in intelligent buildings to evaluate environmental satisfaction. They also highlighted the use of Data Fusion Algorithms within Wireless Sensor Networks. Their discussions centered around strategies for energy-efficient DTs, where BIM plays a role.

In the case of *Zaballos et al.*, by proposing a concept for smart campuses, the authors integrated BIM tools with the Internet of Things. Their goal was to monitor comfort and emotion detection systems, providing insights into comfort levels. The Smart Readiness Indicator for university buildings was presented by *Martínez et al.* This indicator evaluates the capability of a structure to optimize its functionality according to the demands of its inhabitants, while enhancing energy efficiency. It also allows for energy flexibility based on parameters such as $CO_2$ levels, temperature and humidity.

### 4.2 Challenges of Digital Twins for Building Maintenance

Digital twin technology allows organizations to create virtual replicas of physical assets. These digital twins serve as dynamic models, enabling real-time monitoring, optimization, and predictive capabilities. Thus, they play an important role in building energy management, leading to cost savings and sustainability [18].

Digital twins present challenges that organizations must address, such as: (a) *upfront investment*: organizations need to invest in technology, infrastructure, and expertise to create and maintain accurate digital twins; (b) *organizational integration*: integrating digital twins across different teams or departments can be a difficult task. Breaking down silos and fostering collaboration are essential for successful implementation; (c) *stakeholder familiarity*: many stakeholders may lack familiarity with digital twin concepts and technologies. Training and education are important to exploit their potential; (d) *static to dynamic data transformation*: traditional building data often exists in static and scattered formats. Transforming this data into dynamic, real-time digital twins requires significant effort and data synchronization; (e) *system integration complexity*: connecting digital twins with existing systems, such as facility management software, is a challenging task. Ensuring smooth data flow and compatibility is a major challenge.

Aditionally, when using a BIM-based digital twins in conjunction with XR devices, further challenges arise: data integration, sensor selection, and smooth interaction between the user, DT, and application framework. Researchers emphasize the need of strong data integration systems and smart decision-making processes [11].

### 4.3 Opportunities of Digital Twins for Building Maintenance

According to [19], urban digital twins (DTs) are more than virtual models; they let cities simulate policies and projects, previewing their impacts before decisions. The future of smart cities depends on urban DTs, but there's still much to do. To build these DTs, large datasets analyzed by algorithms and computer models are essential, and Cloud Computing, IoT, and sensors become key enabling technologies.

Researchers are exploring algorithms to assess a city's environmental impact and suggest green solutions. Machine learning (ML) models are commonly used, with supervised ML algorithms like regression, classification, and probabilistic methods. Similarly, [20] mentions that digital twin applications, which were mainly used in manufacturing and industry, are now being used in the context of smart cities. These DTs are essential for urban infrastructure. Like in manufacturing, smart city DTs integrate 3D models, simulations, and analyses. They also use bidirectional data from sensors across IoT devices in city buildings and machinery.

Digital-twin based methods and approaches are being investigated in [21], to optimize and automate energy management in a residential district. The study uses a three-dimensional data model includign technologies like IoT, AI, and ML. The case study is based on a residential area in Rome with 16 buildings and 216 units, buildings that are using 70% self-renewable energy. The research uses integrated dynamic analysis algorithms to evaluate scenarios for energy efficiency. The goal is to achieve a virtuous energy system for the complex with optimal comfort and climate. The research aims to develop and implement economical information technology (IT) infrastructure that utilizes dependable data through the utilization of edge computing techniques. This ensures connectivity and real-time information. According to [11], data analysis techniques, including machine learning, can be used to create energy prediction models. These algorithms can forecast energy consumption, find anomalies, and optimize maintenance.

### 4.4 Discussions

Most papers mentioned in our study focus on macro-scale urban digital twin (DT) development, allowing cities to simulate policies and infrastructure impacts before real-world decisions. As expected, the future smart cities depend on large datasets, cloud computing, the IoT and the ground sensors. Researchers aim to develop algorithms that assess cities' environmental impact and propose green solutions, which are widely used algorithms such as Machine Learning (ML), especially supervised algorithms like Artificial Neural Networks (ANN); BIM is a common DT foundation, with Autodesk Revit frequently chosen; IoT devices, mainly sensors in smart buildings, track real-time data in case studies, often in residential and university buildings.

Furthermore, we have observed that a BIM-based DT solution provides an improvement for the maintenance operations. With the new technologies such as XR devices, the maintenance inspections were enhanced by displaying linked data by inspected equipment. Additionally, a BIM-based digital twin enables access to the entire system architecture, offering valuable data on upstream and downstream equipment. This information aids in identifying cause-and-effect relationships, preventing potential ripple effects in case of failure, and pinpointing potential origins of issues.

## 5 Conclusions

In this research we review the role and potential of digital twins in improving the building maintenance. Buildings use a lot of energy and emit greenhouse gases, so efficient maintenance is important for sustainability. DTs are a new technology that could optimize building performance by providing real-time data and recommendations.

The paper identifies and examines 12 relevant papers published in the last five years and reveals key aspects of digital twins in building maintenance. It shows several benefits, such as increased energy efficiency and lower operation costs. It also shows some challenges, such as data quality, security and lack of standards and interoperability; and identifies promising opportunities for the future.

Integrating DTs with advanced technologies such as artificial intelligence and the IoT can enhance their performance. Applying digital twins to various building types, such as residential and commercial, can show their scalability and versatility.

The paper provides implications and recommendations for theory, practice and policy in building maintenance using digital twins. These insights can guide future research, implementation and help achieve the full potential of digital twins in making building maintenance more efficient, sustainable and occupant-friendly.

**Acknowledgements.** The Romania Competitiveness Operational Programme partially supported this paper under project number SMIS 120725 - SCAMP-ML (Advanced computational statistics for planning and tracking production environments). The research conducted in this paper was partially supported by the UVT 1000 Develop Fund of the West University of Timişoara.

## References

1. United Nations Environment Programme. 2022 global status report for buildings and construction: Towards a zero-emission, efficient and resilient buildings and construction sector (2022)
2. Fuller, A., Fan, Z., Day, C., Barlow, C.: Digital twin: enabling technologies, challenges and open research. IEEE Access **8**, 108952–108971 (2020)
3. Chen, Y.: Integrated and intelligent manufacturing: perspectives and enablers. Engineering **3**(5), 588–595 (2017)
4. Moher, D.: Preferred reporting items for systematic reviews and meta-analyses: the Prisma statement. Ann. Intern. Med. **151**(4), 264 (2009)

5. Shea, B.J.: AMSTAR 2: a critical appraisal tool for systematic reviews that include randomised or non-randomised studies of healthcare interventions, or both. BMJ **358**, j4008 (2017)
6. Li, L., et al.: AMSTAR 2 appraisal of systematic reviews and meta-analyses in the field of heart failure from high-impact journals. System. Rev. **11**(1), 147 (2022)
7. Reeves, B.C., Shea, B.J.: AMSTAR 2: a critical appraisal tool for systematic reviews that include randomised and/or non-randomised studies of healthcare interventions (2017)
8. Whiting, P., et al.: ROBIS: a new tool to assess risk of bias in systematic reviews was developed. J. Clin. Epidemiol. **69**, 225–234 (2016)
9. Catone, M.C., Diana, P., Giordano, G.: Keywords co-occurrence analysis to map new topics and recent trends in social research methods. In: Barolli, L., Amato, F., Moscato, F., Enokido, T., Takizawa, M. (eds.) AINA 2020. AISC, vol. 1151, pp. 1078–1088. Springer, Cham (2020). https://doi.org/10.1007/978-3-030-44041-1_93
10. Radhakrishnan, S., Erbis, S., Isaacs, J.A., Kamarthi, S.: Novel keyword co-occurrence network-based methods to foster systematic reviews of scientific literature. PLoS ONE **12**(3), e0172778 (2017)
11. Coupry, C., Noblecourt, S., Richard, P., Baudry, D., Bigaud, D.: BIM-based digital twin and XR devices to improve maintenance procedures in smart buildings: a literature review. Appl. Sci. **11**(15), 6810 (2021)
12. Hosamo, H.H., Imran, A., Cardenas-Cartagena, J., Svennevig, P.R., Svidt, K., Nielsen, H.K.: A review of the digital twin technology in the AEC-FM industry. Adv. Civil Eng. **2022**, 1–17 (2022)
13. Kaewunruen, Sakdirat, Ningfang, Xu.: Digital twin for sustainability evaluation of railway station buildings. Front. Built Env. **4**, 77 (2018)
14. Zhao, J., Feng, H., Chen, Q., de Soto, B.G.: Developing a conceptual framework for the application of digital twin technologies to revamp building operation and maintenance processes. J. Build. Eng. **49**, 104028 (2022)
15. Peng, Y., Zhang, M., Fangqiang, Yu., Jinglin, X., Gao, S.: Digital twin hospital buildings: an exemplary case study through continuous lifecycle integration. Adv. Civil Eng. **2020**, 1–13 (2020)
16. Drobnyi, V., Zhiqi, H., Fathy, Y., Brilakis, I.: Construction and maintenance of building geometric digital twins: state of the art review. Sensors **23**(9), 4382 (2023)
17. Errandonea, I., Beltrán, S., Arrizabalaga, S.: Digital twin for maintenance: a literature review. Comput. Ind. **123**, 103316 (2020)
18. Tahmasebinia, F., Lin, L., Shuo, W., Kang, Y., Sepasgozar, S.: Exploring the benefits and limitations of digital twin technology in building energy. Appl. Sci. **13**(15), 8814 (2023)
19. Bortolini, R., Rodrigues, R., Alavi, H., Vecchia, L.F.D., Forcada, N.: Digital twins' applications for building energy efficiency: a review. Energies **15**(19), 7002 (2022)
20. Mylonas, G., Kalogeras, A., Kalogeras, G., Anagnostopoulos, C., Alexakos, C., Munoz, L.: Digital twins from smart manufacturing to smart cities: a survey. IEEE Access **9**, 143222–143249 (2021)
21. Agostinelli, S., Cumo, F., Guidi, G., Tomazzoli, C.: Cyber-physical systems improving building energy management: digital twin and artificial intelligence. Energies **14**(8), 2338 (2021)

# Structuring the Continuum

Marco Danelutto(✉), Patrizio Dazzi, and Massimo Torquati

Department of Computer Science, University of Pisa, Pisa, Italy
{marco.danelutto,patrizio.dazzi,massimo.torquati}@unipi.it
https://www.di.unipi.it

**Abstract.** This paper addresses challenges arising from inherent heterogeneity in the rapidly evolving landscape of computing infrastructures, spanning diverse environments across the compute continuum. As a result of the current transition from centralized architectures to contemporary distributed and edge computing models, the demand for innovative programming paradigms becomes increasingly apparent. These paradigms must efficiently harness available resources while seamlessly accommodating their heterogeneity, thus relieving programmers of management burdens. Our work encompasses various resource types within the computational environment to maximize efficiency, responsiveness, and availability. At its core, our contribution introduces a structured programming approach that facilitates the integration and exploitation of dynamically enlisted resources. A dynamic allocation policy and a communication abstraction layer orchestrate resources and workloads adaptively. To address the heterogeneity of the execution environment, we adopt a fat binary that encapsulates diverse executable formats in a single deployment package. Empirical evaluation validates the efficacy of our programming pattern in meeting real-world application needs within the compute continuum. Our approach demonstrates significant promise in navigating the complexities posed by the diverse and dynamic nature of modern computing environments.

**Keywords:** Compute continuum · Parallel computing · Parallel design patterns · Communication libraries

## 1 Introduction

In today's dynamic and complex computing landscape, where resources span diverse environments across the compute continuum, this paper tackles challenges intrinsic to continuum heterogeneity. From the evolution of centralized architectures to the contemporary era of distributed and edge computing, a pressing need emerges for innovative programming paradigms. These paradigms must efficiently harness available resources while enabling seamless utilization, alleviating programmers from the management burdens associated with heterogeneity.

Operating within a computational environment that offers a rich array of resource types, our study aims to uncover opportunities embedded in this diversity. Driven by imperatives such as efficiency, responsiveness to user demands, and availability/characteristics of many modern applications (e.g., next-generation applications,


L. Barolli (Ed.): AINA 2024, LNDECT 203, pp. 212–223, 2024.
https://doi.org/10.1007/978-3-031-57931-8_21

streaming services, etc.), our work emphasizes the strategic utilization of computational resources.

At the core of our contribution is a structured programming approach that allows developers to seamlessly integrate and exploit a dynamically enlisted set of resources. Through dynamic allocation policies for application workloads and a communication abstraction layer that decouples application logic from information exchange media, our approach adeptly orchestrates resources and workloads with efficiency and adaptability. Managing the complexities arising from the heterogeneity of the target execution environment, including different executable formats, is achieved by employing a fat binary. This approach incorporates various executable formats, each supporting a different architecture, within a single deployment package. To assess our approach, we conducted an empirical evaluation of the developed programming pattern, evaluating its ability to support the needs of real-world applications and demonstrating its efficacy in the compute continuum environment.

The remainder of this paper is organized as follows: Sect. 2 delves into the scientific underpinnings of compute continuum architectural approaches, addressing challenges in programming continuum resources. It discusses a few existing programming models, information exchange, application representation, and resource management. The section also identifies relevant scientific contributions and surveys in the field. Section 3 introduces the structured programming pattern adopted in response to the identified challenges. It outlines the dynamic resource management approach implemented and details the communication abstraction layer adopted. Section 4 provides a detailed description of the implementation of the proposed approach. It presents experimental evaluation results and engages in a thorough discussion of performance metrics. The final section (Sect. 5) summarizes the objectives, methods, and results.

## 2 Background and Literature Review

In this section, we provide background information and a literature review on the main topics related to our paper: the compute continuum and its essential architectural design, the benefits deriving from a structured programming approach in such a context, and the technological solutions that enable exploitation of continuum resources (including fat binary, and communication abstractions).

The compute continuum is a paradigm that describes the evolution of computing infrastructures from centralized to distributed and edge architectures, where resources span diverse environments across the network. The compute continuum promises to offer a broad spectrum of deployment opportunities for workloads that can leverage the strengths of cloud (scalable infrastructure, high reliability) and edge (energy efficient, low latencies) computing. However, the compute continuum also poses significant challenges for the development and management of applications, such as the heterogeneity and dynamism of resources, the variability and uncertainty of network conditions, complexities deriving from the trade-off between performance and cost, and security and privacy issues. Several research efforts have been devoted to addressing the challenges of the compute continuum, from different perspectives and levels of abstraction [16]. Some of them focus on the architectural and infrastructural aspects, such as the design and deployment of cloud-edge platforms [9], the resource provisioning and allocation strategies [4, 12, 17], the workload scheduling [15] and offloading mechanisms [3, 6, 14], and the performance and energy optimization techniques [8]. In our contribution, we mainly focus on the ability to exploit the resources available in the continuum dynamically by employing an auto-scheduling-based allocation methodology based on a well-consolidated policy suitable for highly heterogeneous and dynamic environments like the compute continuum. We will shed light on the process we adopted in Sect. 3.

Another relevant challenge, highly considered in the scientific literature related to the compute continuum, focuses on the complexities that hinder the programmability of computing continuum environments. The diversity of resources presents a multifaceted challenge for developers. These resources exhibit variations in architecture, format, capacity, performance, and availability. Consequently, developers must craft applications capable of adapting to diverse resource types and configurations. To address this challenge, it is essential to employ suitable tools and frameworks that abstract and manage this inherent heterogeneity [2]. Furthermore, the dynamic nature of resources and workloads adds an extra layer of complexity. This dynamism stems from changes over time and space, influenced by factors such as mobility, volatility, and unpredictability within the network [10]. This phenomenon not only impacts resource management algorithms but also significantly influences how applications are conceptualized and written. Developers must, therefore, design and implement applications with the flexibility to navigate the variability and uncertainty inherent in network conditions.

In this paper, we propose a structured parallel programming approach for the compute continuum that, while easing the task of developing parallel applications, aims at simplifying the integration and exploitation of heterogeneous and dynamic resources, while ensuring efficient and adaptive orchestration of resources and workloads. The key advantages coming from the exploitation of a structured parallel programming approach include enhanced code readability and maintainability, and a more straightforward management of complex parallel execution, relieving the programmer from the burden of managing the parallel computation by hand. This approach has been at the core of many frameworks for parallel programming proposed by the research community over the years. Notable examples include SkePU [5], Muesli [11] and FastFlow [1]. By relying on a set of predefined patterns (templates) that characterize parallel programming (such as farm, map, divide-and-conquer, etc.), structured parallel programming simplifies the programming task. This well-established approach offers a familiar and proven foundation for expressing parallelism, allowing developers to leverage existing knowledge and best practices. The use of recognized templates provides a structured and intuitive way to design parallel applications, promoting reusability and expediting the development process.

However, such a programming approach *per se* is not sufficient to cope with the complexity and diversity of the compute continuum, where resources and workloads can vary over time and space, and where different architectures and formats may coexist. Therefore, we embodied two additional features in our proposed approach: a dynamic allocation policy for application workload and a fat binary that embeds different executable formats for different architectures.

The fat binary is a concept that we adopted to deal with the heterogeneity of the target execution environment in the compute continuum. It is a single deployment package that contains different executable formats, each supporting a different architecture, such as x86, ARM, or PowerPC. The fat binary enables the deployment and execution of the same application on different nodes of the compute continuum without requiring recompiling or repackaging the application for each node. The fat binary also allows us to dynamically select the most suitable executable format for each node, depending on the availability and performance of the resources.

To ensure the communication between the computational entities interacting within the compute continuum, which are involved in the computations we are supporting, we leveraged the MCTL[1] library [7], that serves as an abstraction layer for communications. The MTCL layer is a fundamental component in dealing with the dynamism and variability of the network conditions in the compute continuum. The communication abstraction layer decouples the application logic from the information exchange media and provides a uniform and transparent interface for data transmission and reception. MTCL supports different communication protocols, such as TCP, MPI, MQTT [13], and others, and allows to dynamically switch between them, depending on the network quality and the application requirements. The communication abstraction layer can also be used in conjunction with other software and formats for data representation, such as JSON [18], XML, Protobuf[2] Cereal[3] and others, and allows to automatically serialize and deserialize the data, depending on the application needs.

## 3 Proposed Approach: Structured Programming Solutions for Dynamic Provisioning and Execution

In this section, we present the structured programming approach that we adopted to address the problem of developing and executing applications in the compute continuum. We describe the primary components and features of our approach and illustrate them with a simple example application.

The adopted structured parallel programming model plays a central role in our approach, offering key features that enhance the development and execution of applications within the compute continuum. This model allows to raise the level of abstraction for application developers that can avoid directly dealing with the burden of managing the dynamicity and heterogeneity characterizing the continuum. Furthermore, structured models allow the embedding of cost models to drive workload decomposition and assignment, contributing to the overall efficiency and adaptability of our proposed solution.

Figure 1 depicts the overall approach we propose. The computational entities depicted in the figure represent some of the different architectures that can be found in the continuum that we are targeting in our proposed approach. A wide range of different architectures encompassing small devices, laptops, and large servers populate the

---

[1] MTCL library home: https://github.com/ParaGroup/MTCL.

[2] Protobuf library home https://github.com/protocolbuffers/protobuf.

[3] https://uscilab.github.io/cereal/index.html.

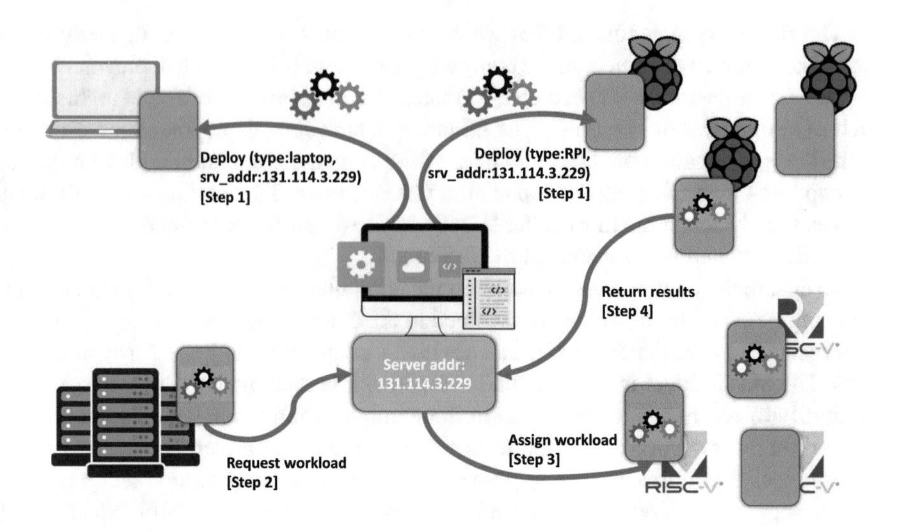


**Fig. 1.** Our proposed approach

continuum ecosystem. In the center of the picture, there is a desktop server that drives the computation.

In our proposed approach, we deliberately omit the auto-discovery phase, opting for a direct transition to the deployment stage. This means that the initial step of the process is immediately engaged in deploying application executors on available resources. However, it's worth acknowledging the conceptual consideration of a preliminary step dedicated to resource discovery and recruitment.

Auto-discovery mechanisms could entail various strategies, such as network scanning, service announcements, or the utilization of directory services commonly employed in cloud environments. Techniques like DNS-based Service Discovery (DNS-SD), or cloud platforms' built-in discovery mechanisms, allow dynamic identification and recruitment of available resources within the compute continuum.

For example, a preliminary stage could involve automated network scans using protocols like Bonjour or UPnP, similar to strategies in local network environments. Alternatively, in cloud environments, leveraging a directory service such as AWS Cloud Directory or Azure Active Directory could facilitate systematic resource discovery. Although our current implementation omits this step, recognizing its conceptual relevance opens avenues for future adaptations and enhancements to the approach, incorporating systematic resource discovery mechanisms.

The process is organized around four different Steps:

1. **Application deployment** [*server* $\rightarrow$ *resources*]**:** The driver of the computation deploys on (a subset of) the available resources (aka Processing Elements, PEs) application executors (AEs). The number and nature of executors depend on the type of PE exploited. The server also communicates to the AE the endpoint to connect to in order to retrieve the workload to process.

2. **Request workload** [*server* $\leftarrow$ *resources*]**:** Once an AE is running, it connects to the driver endpoint to signal its availability to receive workload to process.
3. **Workload assignment** [*server* $\rightarrow$ *resources*]**:** The driver of the computation, depending on the amount of workload to process and the actual availability of resources, performs workload assignments. As stated before, the policy we adopted in this work for workload assignment is quite simple and based on an auto-scheduling behavior.
4. **Result delivery** [*server* $\leftarrow$ *resources*]**:** The last step is related to the delivery of results, that PEs perform after completing the result processing phase.

# 4 Implementation and Results

We implemented a *proof-of-concept* prototype of a manager supporting the implementation of data or task parallel computations with independent (sub)tasks written in C++ on top of the communication library MTCL implementing the logical design in Fig. 1.

## 4.1 Implementation Details

Remote workers behave as servers. They are started by the computation manager and cycle receiving tasks to compute up to the reception of a special *EOS* (End-of-Stream) task. Upon reception of a regular task, they compute the task and send back the result(s). If an EOS is received, they simply terminate. Machines that are planned to be used as remote workers are assumed to run some Linux-based distribution and to be reachable via `ssh` protocol.

The manager starts a number of remote workers, as dictated by user parameters that include the names of the machines to be used along with the number of distinct workers to be used (per machine). The manager also takes two objects as parameters, one providing the methods to get input stream or task partitions to be computed (with methods `bool hasNext()` and `Task getNext()`, this is the *source* object) and the second one providing a `void process(Result)` method taking care of processing the task/partition results (this is the *drain* object).

Targeting of heterogeneous remote "worker" nodes is achieved by using a kind of "fat binary" approach. Indeed, we assume to have the code implementing the function(s) to be computed at the worker sites and we recompile a dedicated worker binary for each one of the remote architectures we plan to use. If needed, serialization of tasks and results is implemented using an open-source serialization library such as Cereal[4]. Specialized worker binaries are automatically managed in the `init` phase of the manager activities, again exploiting `ssh` and the associated `scp` tools.

[4] https://uscilab.github.io/cereal/index.html.

Auto scheduling policy is implemented in the proof-of-concept prototype by implementing one thread per remote worker in the manager. The thread peeks available tasks from the *source* object, delivers a message to the associate remote worker, waits for the results, and invokes the *drain* object. In case the source has no more tasks to be computed, the thread terminates the remote worker by sending the special *EOS* message.

Despite the very basic implementation of the auto-scheduling policy, notable results have been achieved, as detailed in Sec. 4.2. We are currently implementing a second version of the auto-scheduler capable of better masking the times involved in tasks and results communications with standard double-buffering techniques.

## 4.2 Experimental Evaluation

We conducted a series of experiments utilizing various worker machines, employing a total of 5 different hardware systems, each detailed in Table 1.

**Table 1.** Different types of hardware used

| name | CPU | core no. | location |
|---|---|---|---|
| `riscv` | StarFive JH7110 | 4 | office |
| `c2` | ARM Cortex A59 | 4 | office |
| `nuc` | INTEL i7-1165G7 | 4 | office |
| `titanic` | ADM EPYC 7551 | 32 | wan |
| `pianosa` | INTEL E5-2650 | 10 | lan |

The `nuc` has served as the manager, overseeing the operations, while the remaining machines have been dedicated to hosting different numbers of workers. A straightforward floating-point number crunching benchmark was employed to gauge diverse computing capabilities among remote workers. The computations within this benchmark closely align with those encountered in conventional convolution codes characteristic of the AI domain.

In Fig. 2, the outcomes of running 1K tasks across all utilized machines are presented. The blue bars illustrate the ratio between the tasks actually computed and those that would have been assigned under a uniform distribution. Smaller machines undertake a proportionally lower number of tasks-approximately half or a quarter of what would have been assigned with uniform distribution-driven workload allocation. Moreover, it is evident that more powerful machines handle a greater quantity of tasks. Nevertheless, the latency from the machine hosting the computation manager also plays a significant role. Observably, machines farther away in terms of network round-trip time (RTT) process fewer tasks than expected, considering their processing power. Notably,

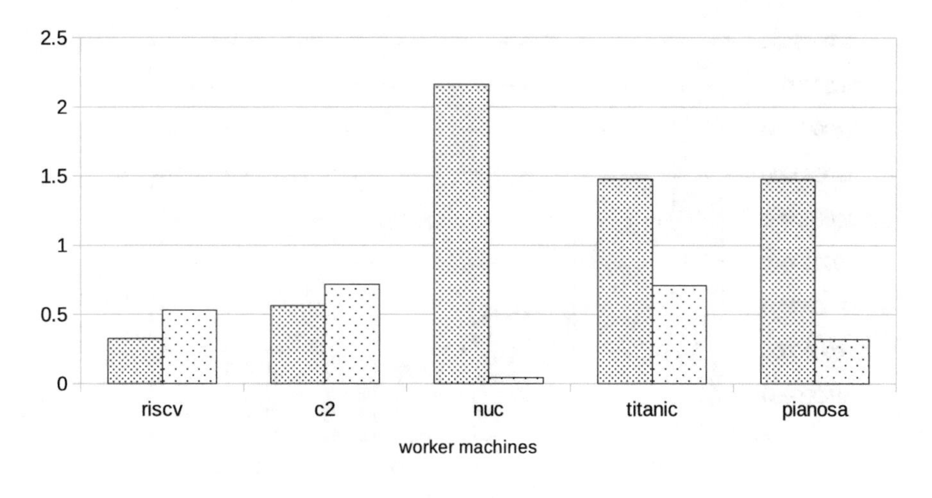


**Fig. 2.** Distribution of tasks w.r.t. network distance and machine computing power

the desktop machine (`nuc`), equipped with a processor featuring recent and modern technology, surpasses expectations by computing more than twice the tasks that would have been assigned under a uniform distribution.

Figure 3 shows the actual impact of the auto-scheduling policies implemented in the manager. The time spent to compute the tasks is *de facto* the one spent on the slowest worker. Auto-scheduling achieves a speedup greater than 3x w.r.t. fixed, uniform task scheduling. Figure 4 shows the scalability achieved while using more and more workers. In this case,"small" and "big" refer to the riscv+arm and AMD+Intel machines, respectively. All the machines run the amount of workers in the x-axis labels. Therefore the first point is relative to 4 workers and the last one is relative to 24 workers in total. Finally, Fig. 5 outlines the impact of communication to computation ratio. The bars refer to computations whose tasks and results have the same size in the three cases, but the actual computation performed at the worker is 1x, 10x, and 100x times, respectively. With smaller computations, the time required to schedule remote tasks dominates, and therefore, the number of tasks computed (on the y-axis per worker is more or less the same. When the computation time of a single task increases, less powerful devices start to compute fewer and fewer tasks than the more powerful ones.

Overall, the results outlined in this section demonstrate two things:

- The manager actually implements realistic policies that allow the exploitation of available resources depending on their actual computational power and the reachability of the devices on the network.

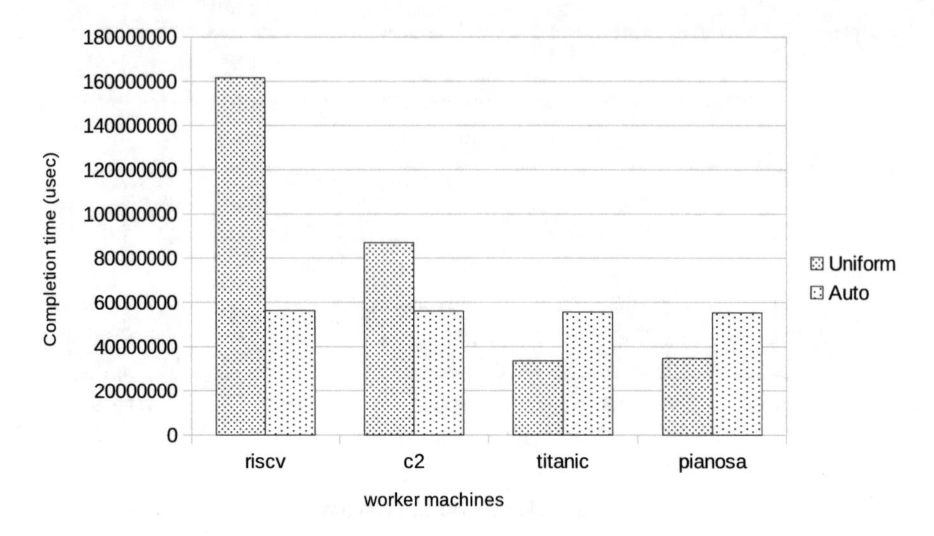


**Fig. 3.** Fixed vs. auto scheduling

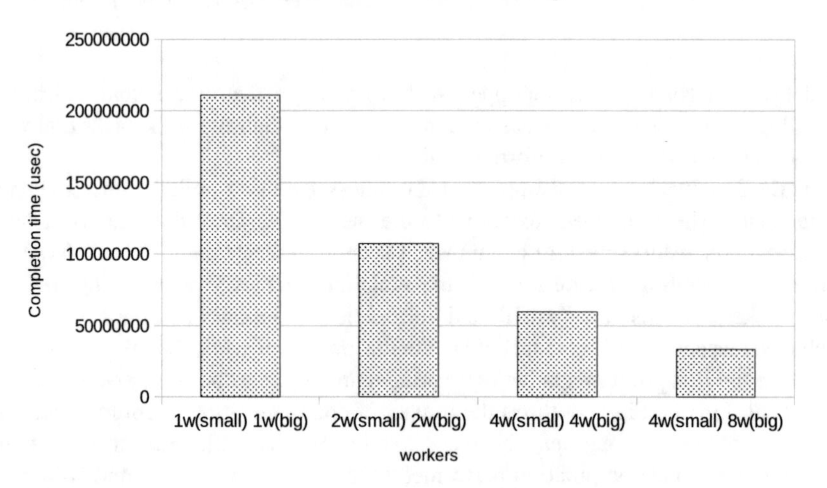


**Fig. 4.** Scalability

- The manager encapsulates all the know-how, mechanisms, and policies needed to implement the parallel computation pattern. The application programmer is eventually only required to provide the *source* and *drain* objects and the `Result f(Task)` function to be computed at the worker side, along with a list of available machines. By instantiating the manager object using these parameters, the application programmer directly gets a working application with the performances discussed above.

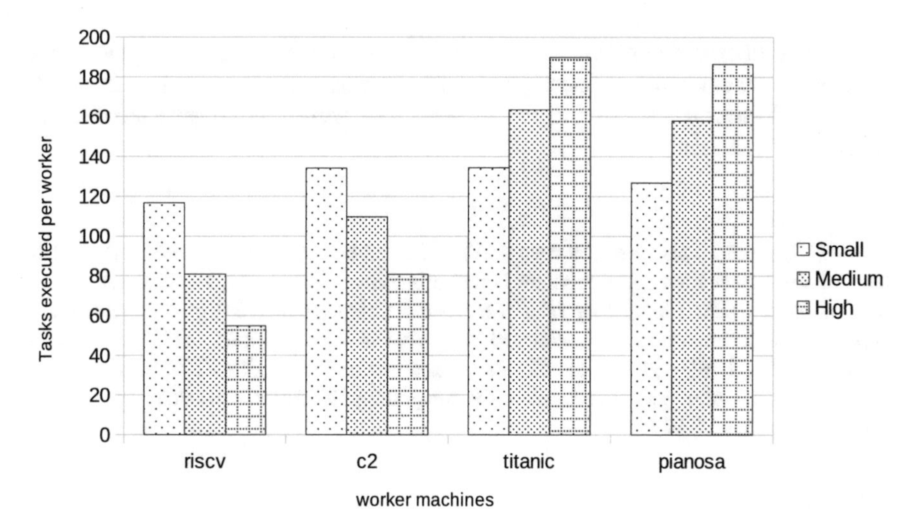


**Fig. 5.** Impact of computation/communication grain

## 5 Conclusions and Future Work

This paper introduces a structured programming approach to address the challenges of developing and executing applications within the compute continuum. The programming model proposed provides a set of key features that enhance efficiency and adaptability. Overall, the approach is based on four steps that orchestrate the deployment, workload assignment, and result delivery seamlessly across diverse architectures within the continuum. While our current implementation excludes an auto-discovery phase, recognizing the conceptual importance of resource discovery and recruitment lays the foundation for future adaptations. The empirical evaluation of our programming pattern validates its efficacy in meeting real-world application needs, showcasing promise within the dynamic and heterogeneous nature of modern computing environments.

Building upon our current work, several possibilities for future research and development emerge: *Enhanced Resource Discovery Mechanisms*: Investigate and integrate advanced auto-discovery mechanisms, such as those based on directory services, cloud-specific approaches, or peer-to-peer protocols. These mechanisms could dynamically identify and recruit available resources within the compute continuum, possibly improving the efficiency of the initial deployment stage. *Optimization of Workload Assignment Policies*: Explore and implement more sophisticated workload assignment policies to address specific application and resource characteristics. Investigate adaptive strategies that dynamically adjust to workload and resource availability, optimizing the overall performance of applications. *Scalability and Performance portability*: Conduct extensive scalability and performance testing of our programming pattern in larger and more diverse compute continuum environments. Evaluate the approach's effectiveness in scenarios with increased numbers of computational entities and varying workloads.

*Security and Privacy Considerations*: Investigate and address security and privacy considerations inherent in the compute continuum. Develop strategies to ensure the secure exchange of information and protect sensitive data across the diverse architectures within the continuum.

**Acknowledgments.** This work has been partially funded by the Spoke 1 "FutureHPC & BigData" of the Italian Research Center on High-Performance Computing, Big Data and Quantum Computing (ICSC) funded by MUR Missione 4 Componente 2 Investimento 1.4: Potenziamento strutture di ricerca e creazione di "campioni nazionali di R&S (M4C2-19)" - Next Generation EU (NGEU), and by the NOUS (A catalyst for EuropeaN ClOUd Services in the era of data spaces, high-performance and edge computing) HORIZON-CL4-2023-DATA-01-02 project, G.A. n. 101135927

## References

1. Aldinucci, M., Danelutto, M., Kilpatrick, P., Torquati, M.: Fastflow: high-level and efficient streaming on multicore. In: Programming Multi-core and Many-core Computing Systems, pp. 261–280. Wiley Online Library (2017). https://doi.org/10.1002/9781119332015.ch13
2. Beckman, P., et al.: Harnessing the Computing Continuum for Programming Our World, chap. 7, pp. 215–230. John Wiley & Sons, Ltd. (2020). https://doi.org/10.1002/9781119551713.ch7
3. Carlini, E., Coppola, M., Dazzi, P., Mordacchini, M., Passarella, A.: Self-optimising decentralised service placement in heterogeneous cloud federation. In: 2016 IEEE 10th International Conference on Self-Adaptive and Self-Organizing Systems (SASO), pp. 110–119 (2016). https://doi.org/10.1109/SASO.2016.17
4. Cohen, I., Chiasserini, C.F., Giaccone, P., Scalosub, G.: Dynamic service provisioning in the edge-cloud continuum with bounded resources. In: IEEE/ACM Transactions on Networking, pp. 1–16 (2023). https://doi.org/10.1109/TNET.2023.3271674
5. Enmyren, J., Kessler, C.W.: Skepu: a multi-backend skeleton programming library for multi-gpu systems. In: Proceedings of the Fourth International Workshop on High-level Parallel Programming and Applications, pp. 5–14 (2010)
6. Ferrucci, L., Mordacchini, M., Dazzi, P.: Decentralized replica management in latency-bound edge environments for resource usage minimization. IEEE Access (2024)
7. Finocchio, F., Tonci, N., Torquati, M.: MTCL: a multi-transport communication library. In: 1st International Workshop on Scalable Compute Continuum (WSCC 2023), Co-located with Euro-Par 2023, Cyprus. LNCS vol. 14351. Springer (2023). ISBN 978-3-031-48802-3
8. Hou, S., Li, H., Yang, C., Wang, L.: A new privacy-preserving framework based on edge-fog-cloud continuum for load forecasting. In: 2020 IEEE Wireless Communications and Networking Conference (WCNC), pp. 1–8 (2020). https://doi.org/10.1109/WCNC45663.2020.9120680
9. Jansen, M., Al-Dulaimy, A., Papadopoulos, A.V., Trivedi, A., Iosup, A.: The spec-rg reference architecture for the compute continuum. In: 2023 IEEE/ACM 23rd International Symposium on Cluster, Cloud and Internet Computing (CCGrid), pp. 469–484 (2023). https://doi.org/10.1109/CCGrid57682.2023.00051
10. Jansen, M., Wagner, L., Trivedi, A., Iosup, A.: Continuum: automate infrastructure deployment and benchmarking in the compute continuum. In: Companion of the 2023 ACM/SPEC International Conference on Performance Engineering, ICPE 2023 Companion, pp. 181–188. ACM, New York (2023). https://doi.org/10.1145/3578245.3584936

11. Kuchen, H.: Parallel programming with algorithmic skeletons. In: Bergener, K., Räckers, M., Stein, A. (eds.) The Art of Structuring: Bridging the Gap Between Information Systems Research and Practice, pp. 527–536. Springer, Cham (2019). https://doi.org/10.1007/978-3-030-06234-7_49
12. Massa, J., Forti, S., Dazzi, P., Brogi, A.: Declarative and linear programming approaches to service placement, reconciled. In: 2023 IEEE 16th International Conference on Cloud Computing (CLOUD), pp. 1–10 (2023). https://doi.org/10.1109/CLOUD60044.2023.00033
13. Mishra, B., Kertesz, A.: The use of MQTT in M2M and IoT systems: a survey. IEEE Access **8**, 201071–201086 (2020)
14. Mordacchini, M., Carlini, E., Dazzi, P.: A mathematical model for latency constrained self-organizing application placement in the edge. In: FRAME 2022, pp. 29–32. Association for Computing Machinery, New York (2022). https://doi.org/10.1145/3526059.3533620
15. Nastic, S., et al.: Polaris scheduler: edge sensitive and SLO aware workload scheduling in cloud-edge-IoT clusters. In: 2021 IEEE 14th International Conference on Cloud Computing (CLOUD), pp. 206–216. IEEE (2021)
16. Taleb, T., et al.: Toward supporting XR services: architecture and enablers. IEEE Internet Things J. **10**(4), 3567–3586 (2022)
17. Theodoropoulos, T., et al.: Cloud-based XR services: a survey on relevant challenges and enabling technologies. J. Netw. Netw. Appl. **2**(1), 1–22 (2022)
18. Viotti, J.C., Kinderkhedia, M.: A survey of JSON-compatible binary serialization specifications. arXiv preprint arXiv:2201.02089 (2022)

# Empowering Sustainable Mobility: Exploring MaaS as a Big Data Application in Transportation Planning

Antonella Falanga[1], Ilaria Henke[2](✉), and Armando Cartenì[3]

[1] Department of Engineering, University of Campania "Luigi Vanvitelli", Aversa, Italy
antonella.falanga@unicampania.it

[2] Department of Civil, Construction and Environmental Engineering, University of Naples "Federico II", Naples, Italy
ilaria.henke@unina.it

[3] Department of Architecture and Industrial Design, University of Campania "Luigi Vanvitelli", Aversa, Italy
armando.carteni@unicampania.it

**Abstract.** In today's data-rich landscape, the term "big data" encompasses diverse information sources like mobile devices, sensors, and online transactions. Its utilization is crucial, offering substantial advantages across fields, enhancing comprehension, decision-making, efficiency, and fostering innovation. The 3Vs framework – velocity, volume and value – spotlights the rapid, vast, diverse, and valuable nature of these datasets, transforming industries globally, unlocking possibilities and driving innovation in various sectors. Within transportation, big data optimizes vehicle routing, traffic management, logistics, and sustainability. It supports public administration, smart city development, and urban planning by enabling precise decision-making through access to extensive data reservoirs, amplified by cloud computing and artificial intelligence. The study centers on the Naples' Mobility as a Service (MaaS) project, named "MaaS4Naples", highlighting big data's crucial role within this sustainable mobility solution. This project proposes ad ad-hoc App integrating transport modes and ticketing in a unique useful personalized service.

## 1 Introduction

In recent years, copious volumes of data, known as big data, has revolutionized diverse domains, garnering attention from scientific circles, governmental bodies, and industries. This surge of information denotes exceedingly extensive and intricate datasets demanding specialized methodologies for effective gathering, administration, processing, and analysis (e.g., [1]) and it has transformed various sectors like emergency response, enhancing real-time crisis/event management.

The concept of "big data", coined in 2001 by Doug Laney, a pioneer in technology and data analytics, has evolved with the formulation of the "3Vs" model – "volume", concerning the enormity of data, whether organized, disorganized, or partially structured, gathered or produced; "velocity", relating to the pace at which data is created, processed, and scrutinized, encompassing real-time data or swiftly flowing data streams;


L. Barolli (Ed.): AINA 2024, LNDECT 203, pp. 224–233, 2024.
https://doi.org/10.1007/978-3-031-57931-8_22

"variety", embodying the assortment of data types, covering organized data (like traditional databases), unorganized data (such as text, images, videos), and partially structured data (like XML or JSON files), (e.g., [2]). Laney's model, encompassing vast and intricate datasets, provides a foundational understanding of data complexities. Additional attributes, including "veracity" and "value", emphasize data quality and diverse benefits (e.g., [3–5]). In particular, veracity ensures data excellence, including precision, dependability, and trustworthiness. Value encompasses the diverse benefits derived from handling and analyzing big data, spanning financial, social, research, and educational values. Moreover, the integration of big data with cloud computing has become instrumental in managing and processing these vast datasets (e.g., [6, 7]). Cloud computing, a paradigm that delivers computing services, including storage, processing power, and analytics over the internet, has streamlined data management and democratized access to advanced computing resources. This shift is a result of the accessibility of web-based remote computing sites. Instead of installing software individually on each computer, a straightforward application installation on the local device is sufficient, transmitting processing data to a singular cloud-based computer (e.g., [8–11]). The dynamic synergy between big data and cloud computing propels transformative impacts across diverse sectors, such as business, finance, healthcare, education, media, entertainment, and transportation, that is the main focus of this research.

Big data transforms business by refining marketing, elevating customer experiences, and optimizing decision-making through insightful analysis of extensive datasets (e.g., [12–14]). Big data in finance is essential for risk analysis, fraud detection, and personalized services, enhancing security, enabling tailored solutions, and informing data-driven decisions in areas like investment and product development (e.g., [15, 16]). Big data revolutionizes healthcare, enabling personalized treatments, real-time monitoring, and early disease detection. Neural network models support diagnostics and pandemic forecasting (e.g., [17–20]). Furthermore, big data advances personalized medicine, tailoring treatments for improved patient outcomes (e.g., [21, 22]). In education, it is applied for analyzing students' behavior and performance for personalized learning experiences. Institutions use it to understand individual strengths, customize materials, and enhance teaching based on academic data (e.g., [23–25]). Big data revolutionizes media and entertainment, powering personalized content recommendations for improved user experiences. Platforms analyze user preferences, deliver tailored content, and shape delivery strategies to enhance satisfaction. (e.g., [26, 27]). In transportation and logistics, big data optimizes delivery routes, cuts costs, and boosts performance. Real-time traffic management uses data for instant insights, aiding route adjustments and traffic flow. Data-driven insights enhance logistics efficiency and supply chains, reducing travel times through suggested alternative routes (e.g., [28–30]). Integrating data improves infrastructure planning and resource allocation for overall transportation system enhancement (e.g., [31–33]). Additionally, it plays a key role in mitigating pollutant emissions, contributing to environmental sustainability by optimizing transportation operations (e.g., [34–36]).

Starting from these considerations, this study aims to review the main applications of big data in transportation sector and discuss the Mobility as a Service (MaaS) for Naples (Italy) project named "MaaS4Naples", highlighting big data's crucial role within this sustainable mobility solution. The MaaS4Naples project is an experimental project worth over 3 million euros and aims to improve mobility for its citizens, enhancing accessibility to the public and shared transport system, thereby improving mobility for all population segments, including the disadvantaged.

The paper is organized as follows: Sect. 2 describes several big data applications in transportation; Sect. 3 discusses Italian Naples' MaaS project. Finally, the conclusions are reported in Sect. 4.

## 2 Big Data Applications Within Transport Sector

In the expansive realm of transportation, the integration of big data applications has become a transformative force, offering multifaceted solutions across planning, design, and management domains. These innovations yield substantial advantages, spanning from the optimization of travel times and the reduction of road incidents to addressing environmental concerns. Their impact extends across pivotal areas within the transportation sector, including real-time traffic monitoring, route optimization for commercial vehicles, road safety, autonomous and connected vehicles, as well as travel demand estimation, mobility habits, and transportation planning.

Big data analytics in transportation drives innovation, efficiency, and solutions for key challenges. Utilizing large-scale datasets, stakeholders make informed decisions, enhancing safety, efficiency, and resource allocation across diverse transport modes. Data-driven analytics play a pivotal role in real-time traffic monitoring, route optimization, road safety, and integrating autonomous vehicles, refining travel demand estimation, mobility habits, and overall transportation planning. This approach offers advantages in optimizing routes for commercial vehicles using real-time insights from GPS, traffic cameras, and IoT devices. This enhances operational efficiency, reduces costs, and ensures timely deliveries (e.g., [37, 38]). By utilizing big data and machine learning, the study of [39] provides valuable insights for the state's Traffic Management and Public Works Department, allowing informed decision-making, efficient fund allocation, and long-term transport policy formulation. Real-time traffic monitoring benefits from big data, improving sustainability by providing timely information to users. [2] enhanced travel time predictions using IoT sources; leveraging big data, the study of [40] reveals a positive non-linear causality between traffic congestion and road accidents in Latin American cities; reducing traffic delay by 10% could cut road accidents by 3.4%. Big data contributes significantly to road safety by analyzing diverse data sources such as traffic patterns and weather conditions. It predicts accident-prone areas, understands factors affecting safety, and develops proactive measures (e.g., [41]). Autonomous and connected vehicles rely on big data for safe navigation, using V2V and V2I connections. [42] explores VANETs for road safety and self-driving systems, emphasizing efficient big data transmission. It discusses methods, including machine learning, to enhance VANET performance and address communication issues, enriching vehicles' understanding of traffic for predictive models. This proficiency guarantees not only enhanced safety during operations but also an overall boost in effectiveness (e.g., [43–45]). Furthermore, big data enhances connected/autonomous vehicle safety by comprehending their surroundings and overcoming traditional analytics limits (e.g., [46]). Transport demand estimation and mobility habits benefit from big data, providing insights into travel patterns and preferences. [37] investigated big data's role in Intelligent Transportation Systems (ITS), enhancing safety, efficiency, and profitability. Predictive models enable accurate forecasting of future mobility needs, empowering transportation agencies to proactively optimize services and infrastructure [47, 48].

# 3 The MaaS4Naples Project

The term "Smart Mobility" implies the use of technology to create more efficient, flexible, and sustainable transportation options and it refers to a set of new services related to transportation made possible by Information and Communications Technology (ICT). The definition is broad enough to encompass diverse services, which can be grouped into four classes: info mobility, providing information related to mobility; sharing mobility, involving sharing vehicles, such as car or bike sharing, as well as ride-sharing ("on-demand"), which is the transportation of third parties by private vehicle owners, with or without profit motives (a more evolved form of taxi services), path sharing, like carpooling, typically without profit motives, also falls under this category; smart pricing, including advanced road or area pricing mechanisms; Mobility as a Service (MaaS), encompassing integrated services that offer end-to-end mobility solutions, allowing users to plan, book, and pay for various modes of transportation through a single digital platform.

The focus is on leveraging ICT to enhance information availability, sharing resources, implementing intelligent pricing strategies, and providing comprehensive mobility services through platforms like Mobility as a Service (MaaS). The MaaS paradigm represents a revolutionary approach to mobility, involving a seamless integration of diverse transportation services through a singular digital platform, often an app. The aim is to provide comprehensive information and simplify the management of travel across various modes of transportation. With MaaS, it becomes feasible to plan customized journeys, optimizing travel times and costs in real-time while maximizing sustainability. It is analogous to carrying a versatile toolkit in your pocket – complete with a car, bicycle, scooter, bus, and subway – allowing users to craft personalized travel experiences in real-time. In the realm of MaaS, accessibility to transportation services is predominantly digital, facilitated by user-friendly apps. These applications not only offer tailored travel recommendations but also serve as hubs for planning, reserving, and making payments, all while providing crucial route information. The overarching goals of MaaS extend beyond convenience. They encompass a broader vision to cultivate more sustainable mobility practices, diminish the reliance on private cars, encourage diverse and multimodal commuting, and unify disparate transport services – ranging from public transit and ride-sharing to bike-sharing, scooter-sharing, taxis, car rentals, and ride-hailing – under a comprehensive mobility umbrella. The purpose is to perfectly integrate these services into a cohesive, user-centric mobility experience. The keyword of MaaS is integration, which can occur at four levels: (*i*) information integration through information, whether static or dynamic (e.g., Google Maps); multimodal journey planner; price information; direct connection to the MSP (Mobility Service Provider) platform; (*ii*) booking and payment integration through platforms enabling booking and payment for a single integrated multimodal journey (e.g., Moovit, MyCicero); single trips: find, book, and pay; deep integration of PAYG (Pay-As-You-Go) pricing; (*iii*) service offer integration through integration of transport services into predefined packages available through subscription with a single centralized billing (e.g., Whim, Uber, UbiGo); PAYG pricing; mobility packages and subscriptions; (*iv*) social goals integration through promotion of societal goals through specific mobility policies such as dynamic pricing, demand control, congestion charging; financial incentives; $CO_2$ challenges.

In Mobility as a Service (MaaS), Big Data and Cloud Computing play crucial roles in enhancing efficiency and user experience. Big Data is utilized to analyze vast amounts of mobility-related information, including user preferences, travel patterns, and environmental data. This analysis provides valuable insights for tailoring offerings, optimizing routes, and anticipating user needs. Cloud Computing serves as a key infrastructure for managing and storing these massive datasets in a scalable manner, accessible from any location. Moreover, it enables swift data processing, facilitating the delivery of real-time services such as trip planning, reservations, and digital payments. Essentially, the integration of Big Data and Cloud Computing in MaaS aims to create a more efficient, personalized, and accessible mobility ecosystem (Fig. 1). This means that the extensive data about user behaviors, travel habits, and other relevant factors are processed and stored in the cloud. This centralized and scalable storage allows MaaS providers to analyze patterns, make informed decisions, and offer customized services to users. Additionally, the cloud-based infrastructure enables the seamless provision of services in real time, ensuring a smoother and more responsive MaaS experience for users.

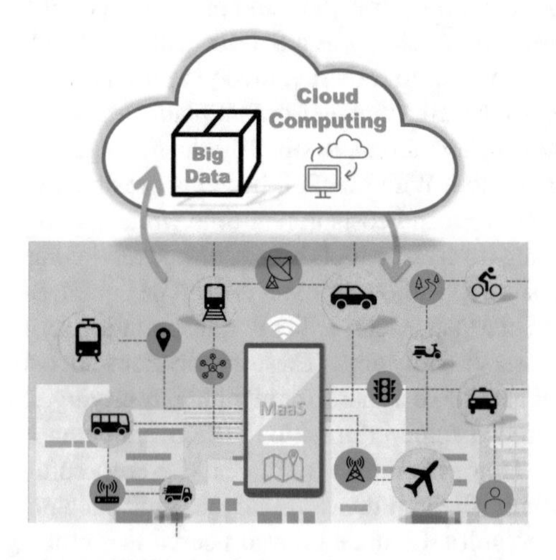


**Fig. 1.** The integration of Big Data and Cloud Computing within the MaaS paradigm

The project "Mobility as a Service for Italy" is a strategy by the Italian Government [49] and part of the broader strategy "Digital Italy 2026", incorporated in the National Recovery and Resilience Plan (NRRP), that allocates a portion of the investments (totaling 40 million euros plus an additional 16.9 million allocated from the Complementary Fund), aimed at transforming the mobility sector through the implementation of the Mobility as a Services concept. The project entails experimenting with MaaS in different pilot cities, creating an open platform for mobility data, and enhancing digital public transport. The government acts as a regulator and enabler, funding an open platform named Data Sharing and Service Repository Facilities (DS&SRF). The project unfolds in three main phases: firstly, experimenting with MaaS in territories involves introducing digital platforms, new business models, and data sharing to assess environmental and socio-economic impacts. Secondly, a national open platform for mobility data will be established, facilitating interaction among industry players and enabling various services. Lastly, the digital dimension of public transportation will be enhanced in selected territories, promoting MaaS dissemination. The overarching goal is to transform mobility habits, offering an accessible, modern, and sustainable service for citizens, businesses, and the environment. The multi-territorial approach encourages innovation and sustainability in transportation.

The experimentation began in July 2023 and will conclude in September 2025. Initially, 3 pilot cities were selected: Naples, Rome, and Milan. With the Complementary Fund, another 3 cities were chosen to experiment with MaaS services: Bari, Florence, and Turin. The approval of the rankings was published in July 2022. On May 31, 2023, the public notice to select 7 territories (broad areas) to participate in the MaaS experimentation concluded. The metropolitan city of Naples is the first "pilot" territory taking a forefront role in developing a digital intermodal mobility system. The MaaS4Naples project is an experimental project worth project is an experimental project worth over 3 million euros and aims to improve mobility for its citizens, enhancing accessibility to the public and shared transport system, thereby improving mobility for all population segments, including the disadvantaged. The MaaS4Naples project was developed in collaboration between the Municipality of Naples and various partners including the Campania Region, Ente Autonomo Volturno EAV, Unicocampania Consortium, GE.S.A.C. S.p.A., ANM S.p.A., BE CHARGE srl, Reby Italia srl, M.C. Consulting srl, LIME Technology srl, Helbiz Italy srl, GESCO Cooperative Sociali Consortium, University of Campania "Luigi Vanvitelli", University of Naples "Federico II" and University "Suor Orsola Benincasa". Three specific Apps were implemented for the scope: MooneyGo (myCicero), Unico myCicero, and MaaS4UNI. Users input their origin and destination, choose their preferred trip, and purchase ticket(s) for the journey. Naples had advantages over other pilot cities thanks to the Unicocampania Consortium (Project's Trip Integrator) that has already provided integrated and discounted fares for TPL movements, utilizing multiple modes of transportation, and also because the metropolitan stations (turnstiles) were already equipped with infrastructure for dematerialized tickets.

The experimentation started on July 14, 2023, with available and downloadable apps and the beginning of the ex-ante investigation. 25,000 residents of the Municipality of Naples were interviewed (via phone interviews). The Municipality of Naples has provided incentives for the entire experimentation period to encourage virtuous mobility behaviors by increasing the use of local public transport (metro, funiculars, buses), with resulting benefits on external impacts generated by movements (e.g., reduction of emissions, traffic congestion). Incentives will be provided in the form of cashback, up to a maximum of 35 euros per month, which will cancel out at the end of the month for the unused amount in the reference month. This amount can be used on the app to which the experimental subject has previously registered and is valid for only journeys combining at least 2 different services (public transport, electric scooter, electric bike, car-sharing, electric charging, parking), of which at least 1 must be local public transport (bus, metro, funiculars).

In the initial months of experimentation, a modest usage of MaaS was observed. This occurred for two main reasons: *1*) lack of time. Users, even if potentially interested in using MaaS, had not felt the need to take a MaaS trip; *2*) low awareness of the service. Despite the extensive communication campaign launched by the municipality, more than half of the surveyed users stated that they did not use MaaS because they were not aware that the service had started. Furthermore, it is interesting to note that almost the entirety of the interviewed users who used MaaS stated that the service is useful and innovative. The Maas4Naples project is currently in the operational phase, with the service already available and users actively experiencing it. The experimentation phase will conclude in September 2025.

## 4 Conclusion

The fusion of big data with Mobility as a Service (MaaS) is a catalyst for advancing sustainable mobility initiatives. Big data's comprehensive analysis of mobility patterns, user preferences, and environmental factors empowers MaaS to tailor services, optimize routes, and promote eco-friendly transportation choices. This synergy extends beyond customization, as big data aids in predicting mobility demands, dynamically adapting MaaS offerings to encourage the adoption of low-emission modes. Additionally, it identifies traffic bottlenecks, offering insights for infrastructure improvement and resource optimization.

In essence, the integration of big data into MaaS not only enhances efficiency and user-centric experiences but also champions initiatives for sustainable urban mobility, paving the way for a greener and more environmentally conscious transportation landscape.

## References


1. Kaisler, S., Armour, F., Espinosa, J.A., Money, W.: Big data: issues and challenges moving forward. In: 2013 46th Hawaii International Conference on System Sciences, pp. 995–1004. IEEE (2013)
2. Kitchin, R., McArdle, G.: What makes Big Data, Big Data? Exploring the ontological characteristics of 26 datasets. Big Data Soc. **3**(1), 2053951716631130 (2016)
3. Anuradha, J.: A brief introduction on Big Data 5Vs characteristics and Hadoop technology. Procedia Comput. Sci. **2015**, 319–324 (2015)
4. Debattista, J., Lange, C., Scerri, S., Auer, S.: Linked'Big'Data: towards a manifold increase in big data value and veracity. In: 2015 IEEE/ACM 2nd International Symposium on Big Data Computing (BDC), pp. 92–98. IEEE (2015)
5. Liu, X., et al.: Enhancing veracity of IoT generated big data in decision making. In: 2018 IEEE International Conference on Pervasive Computing and Communications Workshops (PerCom Workshops), pp. 149–154. IEEE (2018)
6. Liu, H.: Big data drives cloud adoption in enterprise. IEEE Internet Comput. **17**(4), 68–71 (2013)

7. O'Driscoll, A., Daugelaite, J., Sleator, R.D.: 'Big data', Hadoop and cloud computing in genomics. J. Biomed. Inform. **46**(5), 774–781 (2013)
8. Hashem, I.A.T., Yaqoob, I., Anuar, N.B., Mokhtar, S., Gani, A., Khan, S.U.: The rise of "big data" on cloud computing: review and open research issues. Inf. Syst. **47**, 98–115 (2015)
9. Talia, D.: Clouds for scalable big data analytics. Computer **46**(5), 98–101 (2013)
10. Jadeja, Y., Modi, K.: Cloud computing-concepts, architecture and challenges. In: 2012 International Conference on Computing, Electronics and Electrical Technologies (ICCEET), pp. 877–880. IEEE, Kumaracoil, India (2012)
11. Mathur, P., Nishchal, N.: Cloud computing: new challenge to the entire computer industry. In: 2010 First International Conference On Parallel, Distributed and Grid Computing (PDGC 2010), pp. 223–228. IEEE (2010)
12. Niu, Y., Ying, L., Yang, J., Bao, M., Sivaparthipan, C.B.: Organizational business intelligence and decision making using big data analytics. Inf. Process. Manage. **58**(6), 102725 (2021)
13. Sayyad, S., Mohammed, A., Shaga, V., Kumar, A., Vengatesan, K.: Digital marketing framework strategies through big data. In: Pandian, A.P., Senjyu, T., Islam, S.M.S., Wang, H. (eds.) Proceeding of the International Conference on Computer Networks, Big Data and IoT (ICCBI - 2018). LNDECT, vol. 31, pp. 1065–1073. Springer, Cham (2020). https://doi.org/10.1007/978-3-030-24643-3_127
14. Hu, Y.: Marketing and business analysis in the era of big data. Am. J. Ind. Bus. Manage. **8**(07), 1747 (2018)
15. Mahalakshmi, V., Kulkarni, N., Kumar, K.P., Kumar, K.S., Sree, D.N., Durga, S.: The role of implementing artificial intelligence and machine learning technologies in the financial services industry for creating competitive intelligence. Mater. Today Proc. **56**, 2252–2255 (2022)
16. Indriasari, E., Gaol, F.L., Matsuo, T.: Digital banking transformation: application of artificial intelligence and big data analytics for leveraging customer experience in the Indonesia banking sector. In: 2019 8th International Congress on Advanced Applied Informatics (IIAI-AAI), pp. 863–868. IEEE (2019)
17. Krumholz, H.M.: Big data and new knowledge in medicine: the thinking, training, and tools needed for a learning health system. Health Aff. **33**(7), 1163–1170 (2018)
18. Hampel, H., et al.: A precision medicine initiative for Alzheimer's disease: the road ahead to biomarker-guided integrative disease modeling. Climacteric **20**(2), 107–118 (2017)
19. Ahmed, I., Ahmad, M., Jeon, G., Piccialli, F.: A framework for pandemic prediction using big data analytics. Big Data Res. **25**, 100190 (2021)
20. Barrett, M.A., Humblet, O., Hiatt, R.A., Adler, N.E.: Big data and disease prevention: from quantified self to quantified communities. Big Data **1**(3), 168–175 (2013)
21. Hassan, M., et al.: Innovations in genomics and big data analytics for personalized medicine and health care: a review. Int. J. Mol. Sci. **23**(9), 4645 (2022)
22. Inomata, T., et al.: Using medical big data to develop personalized medicine for dry eye disease. Cornea **39**, S39–S46 (2020)
23. Huang, A.Y., Lu, O.H., Huang, J.C., Yin, C.J., Yang, S.J.: Predicting students' academic performance by using educational big data and learning analytics: evaluation of classification methods and learning logs. Interact. Learn. Environ. **28**(2), 206–230 (2020)
24. Gaftandzhieva, S., Doneva, R., Petrov, S., Totkov, G.: Mobile learning analytics application: using students' big data to improve student success. Int. J. Inf. Technol. Secur. **10**(3), 53–64 (2018)
25. Baig, A.R., Jabeen, H.: Big data analytics for behavior monitoring of students. Procedia Comput. Sci. **82**, 43–48 (2016)
26. Lippell, H.: Big data in the media and entertainment sectors. New Horizons for a Data-Driven Economy: A Roadmap for Usage and Exploitation of Big Data in Europe 245–259 (2016)

27. Ghani, N.A., Hamid, S., Hashem, I.A.T., Ahmed, E.: Social media big data analytics: a survey. Comput. Hum. Behav. **101**, 417–428 (2019)
28. Cartenì, A.: Updating demand vectors using traffic counts on congested networks: a real case application. WIT Trans. Built Environ. **96**, 211–221 (2007). https://doi.org/10.2495/UT070211
29. Chen, Y.T., Sun, E.W., Chang, M.F., Lin, Y.B.: Pragmatic real-time logistics management with traffic IoT infrastructure: big data predictive analytics of freight travel time for Logistics 4.0. Int. J. Prod. Econ. **238**, 108157 (2021)
30. Cantelmo, G., Viti, F.: A big data demand estimation model for urban congested networks. Transp. Telecommun. J. **21**(4), 245–254 (2020)
31. Cartenì, A., Henke, I., Di Francesco, L.: A sustainable evaluation processes for investments in the transport sector: a combined multi-criteria and cost–benefit analysis for a new highway in Italy. Sustainability **12**(23), 1–27 (2020). https://doi.org/10.3390/su12239854
32. Cartenì, A., Henke, I., Molitierno, C., Di Francesco, L.: Strong sustainability in public transport policies: an e-mobility bus fleet application in Sorrento Peninsula (Italy). Sustainability **12**, 7033 (2020). https://doi.org/10.3390/su12177033
33. Cascetta, E., Carteni, A., Henke, I.: Acceptance and equity in advanced path-related road pricing schemes. In: 5th IEEE International Conference on Models and Technologies for Intelligent Transportation Systems, MT-ITS 2017 - Proceedings, art. no. 8005722, pp. 492–496 (2017). https://doi.org/10.1109/MTITS.2017.8005722
34. Bibri, S.E., Krogstie, J.: Environmentally data-driven smart sustainable cities: applied innovative solutions for energy efficiency, pollution reduction, and urban metabolism. Energy Inform. **3**, 1–59 (2020)
35. Cartenì, A., Henke, I., Errico, A., Bartolomeo, M.I.D.: A big data and cloud computing model architecture for a multi-class travel demand estimation through traffic measures: a real case application in Italy. Int. J. Comput. Sci. Eng. **26**(5), 482–493 (2023)
36. Chung, S.H.: Applications of smart technologies in logistics and transport: a review. Transp. Res. E Logist. Transp. Rev. **153**, 102455 (2021)
37. Zhu, M., Liu, X.Y., Qiu, M., Shen, R., Shu, W., Wu, M.Y.: Traffic big data based path planning strategy in public vehicle systems. In: 2016 IEEE/ACM 24th International Symposium on Quality of Service (IWQoS), pp. 1–2. IEEE (2016)
38. Aleyadeh, S., Oteafy, S.M., Hassanein, H.S.: Scalable transportation monitoring using the smartphone road monitoring (SRoM) system. In: Proceedings of the 5th ACM Symposium on Development and Analysis of Intelligent Vehicular Networks and Applications, pp. 43–50 (2015)
39. Ramesh, R., et al.: Real-time vehicular traffic analysis using big data processing and IoT based devices for future policy predictions in smart transportation. In: 2019 International Conference on Communication and Electronics Systems (ICCES), pp. 1482–1488. IEEE (2019)
40. Sánchez González, S., Bedoya-Maya, F., Calatayud, A.: Understanding the effect of traffic congestion on accidents using big data. Sustainability **13**(13), 7500 (2021)
41. Picone, M., Errichiello, A., Cartenì, A.: How often are ADAS used? Results of a car drivers' survey. WSEAS Trans. Syst. **22**, 566–577 (2023). https://doi.org/10.37394/23202.2023.22.57
42. Cheng, N., et al.: Big data driven vehicular networks. IEEE Netw. **32**(6), 160–167 (2018)
43. Picone, M., Cartenì, A.: Users' propensity to use self-driving systems of SAE automation level 1 and 2 cars: results of an Italian survey. WSEAS Trans. Environ. Dev. **19**, 479–488 (2023). https://doi.org/10.37394/232015.2023.19.46
44. Cascetta, E., Cartenì, A., Di Francesco, L.: Do autonomous vehicles drive like humans? A turing approach and an application to SAE automation level 2 cars. Transp. Res. Part C Emerg. Technol. **134**, 103499 (2022). https://doi.org/10.1016/j.trc.2021.103499

45. Cartenì, A.: The acceptability value of autonomous vehicles: a quantitative analysis of the willingness to pay for shared autonomous vehicles (SAVs) mobility services. Transp. Res. Interdiscip. Persp. **8**, 100224 (2020). https://doi.org/10.1016/j.trip.2020.100224
46. Lian, Y., Zhang, G., Lee, J., Huang, H.: Review on big data applications in safety research of intelligent transportation systems and connected/automated vehicles. Accid. Anal. Prev. **146**, 105711 (2021)
47. Anda, C., Erath, A., Fourie, P.J.: Transport modelling in the age of big data. Int. J. Urban Sci. **21**(sup1), 19–42 (2017)
48. Zhao, Y., Zhang, H., An, L., Liu, Q.: Improving the approaches of traffic demand forecasting in the big data era. Cities **82**, 19–26 (2018)
49. Mobility as a Service for Italy. https://innovazione.gov.it/progetti/mobility-as-a-service-for-italy/. Accessed Dec 2023

# Mobile Phones in the Cloud-Edge Continuum: Understanding the TEE's Role

Alessandro De Crecchio[1(✉)], Giovanni Maria Cristiano[2], and Alfredo Petruolo[2]

[1] Scuola IMT Alti Studi Lucca, Piazza S. Ponziano, Lucca, Italy
alessandro.decrecchio@imtlucca.it

[2] Centro Direzionale Isola E2, University of Naples "Parthenope", Naples, Italy
{giovannimaria.cristiano001,alfredo.petruolo001}@studenti.uniparthenope.it

**Abstract.** Recent years have seen the cloud computing landscape shift from a centralized model, primarily based on large data centers, to a more distributed and heterogeneous architecture that incorporates edge and IoT units. This transition has given rise to the *Cloud-Edge Continuum* or *Computing Continuum*, effectively narrowing the gap between expansive data centers and end-user devices. This distributed computation requires the establishment of a chain of trust spanning from the cloud to the edge. A key enabling approach for building such trustworthy distributed infrastructure is exploiting Trusted Execution Environments.

It is common to equate edge devices with IoT devices, nevertheless, a big fraction of such category is represented by mobile phones which have unique characteristics in the TEE usage strongly influenced by commercial logic. This research aims to analyze TEE-supported features available for users and developers of legacy mobile phones, also identifying their constraints and limitations.

## 1 Rationale and Motivation

In recent years, cloud computing has shifted from a centralized model centred on large data centers to more distributed architectures. This evolution is blurring the traditional boundaries between distributed data centres and end-user devices, creating the challenge of ensuring the seamless and secure flow of both information and computation between cloud infrastructures and edge devices [1]. While guaranteeing security for information flows (data in transit) is a common practice, preserving information while in use is an emerging challenge which requires the establishment of a chain of trust among the different nodes within this evolving infrastructure. To address this challenge, a possible solution involves the Trusted Execution Environment (TEE). TEE acts as a secure enclave within computing systems, creating isolated and protected execution environments for handling sensitive operations. TEEs establish a secure space where cryptographic operations and other sensitive functions can take place with a high level of trust [2]. The objective of this paper is to focus on the edge part of the chain, where the most widely used technologies are ARM TrustZone [3] and more recently ARM Confidential Compute Architecture (CCA) [4]. Moreover, a significant portion of edge devices, commonly termed as IoT devices, is dominated by mobile phones, and the


L. Barolli (Ed.): AINA 2024, LNDECT 203, pp. 234–243, 2024.
https://doi.org/10.1007/978-3-031-57931-8_23

nuanced differences in TEEs utilization become evident. In contrast to other devices equipped with TEE, mobile phones showcase distinct features in their utilization of TEE technology, shaped significantly by commercial considerations.

This work aims to demonstrate to developers, who need to interface with these security solutions, the current possibilities offered by the mobile environment. At the same time, it aims to shed light on the limitations that exist in this area. The motivation behind this research is to provide an overview, illustrating how, from a technological point of view, it is possible to establish a seamless chain of trust between all nodes within an infrastructure that adheres to the paradigm of cloud-edge continuum.

The rest of the paper is structured as follows. Section 2 presents a comprehensive overview of the current TEE usage paradigm in the realm of IoT Devices. Section 3 delves into the emerging paradigm of the Cloud Edge Continuum, while in Sect. 4, we outline the current limitations and potential steps for facilitating the prototyping of TEE in mobile phones. Lastly, Sect. 5 summarizes the conclusions drawn from this study.

## 2 Related Work

This section examines surveys on edge device security, and hardware-assisted technologies for confidential computing. Furthermore, initiatives regarding the improvement of the TEE in mobile phones are discussed, where its limitations were addressed and ways to extend its capabilities in the context of edge security were explored. Starting from the analysis of edge device security, in their work Zeyu et al. [5] have examined the influence of the Internet of Everything on edge computing, which presents significant challenges, including security ones. Privacy protection is regarded as a primary concern, with edge nodes being the most vulnerable to potential attacks on sensitive information. Continuing the analysis of system security from cloud to edge, reference is made to Coppolino et al.'s work [6]. With the increasing processing of sensitive data on devices beyond user control, particularly in the Internet of Things, security risks arise in both Cloud and Edge Computing deployments. This paper conducts a comprehensive survey of hardware-assisted technological solutions from vendors like Intel, AMD, and ARM for both embedded edge devices and hosting machines such as cloud servers. The different approaches are classified based on the type of attacks prevented and the enforced techniques. An analysis of their mechanisms, issues, and market adoption is provided, supporting researchers approaching this field of systems security. Moving to the analysis of the TEE's usage for developing secure mobile applications, to the best of our knowledge, only Three studies are worth mentioning. The first is TruZ-Droid [7] in which the authors integrate TrustZone TEE with Android OS, eliminating the need for specialized coding. The work has been demonstrated on development board, and it shows how to enhance security for user input and data transmission without vendor Software Development Kit (SDK). The second is TruzCall [8] in which the authors demonstrate how to secure mobile VoIP, enabling end-to-end calls from compromised devices. Leveraging Android OS and TrustZone TEE, they provide a practical solution for VoIP communication risks. The last is SANCTUARY [9] which overcomes vendor limitations on ARM TrustZone, proposing an innovative security architecture. Allowing unconstrained TEE use and leveraging Address-Space Controller, unlocks advanced mobile services potential, addressing deployment policy restrictions.

While these three studies provide valuable insights into harnessing TEE for secure mobile application development, it is essential to acknowledge that limitations in Trusted Application (TA) deployment policies still exist. In the following sections of this paper, we will elaborate on these limitations and explore potential solutions required to fully unleash the capabilities of TEE in context of the Cloud Edge Continuum.

## 3 Cloud Edge Continuum

The Computing Continuum is characterized by its intrinsic heterogeneity and decentralization, featuring multiple layers, associated technologies, and distributed data and entities spanning nodes, processes, and objects [10,11]. Building upon this foundation, the Cloud Continuum emerges as an extension of traditional cloud computing, including diverse entities like Edge, Fog, and IoT [12]. These entities contribute analysis, processing, storage, and data generation capabilities to the continuum. Two perspectives define this continuum: one views it as the distribution of resources across various network elements, including IoT, Fog, Edge, and data centers; the other perceives it as an extension of processing power to different nodes. To elucidate further, it encompasses the proficient allocation of resources and services across a diverse network of elements, concomitantly extending processing capabilities across discrete nodes.

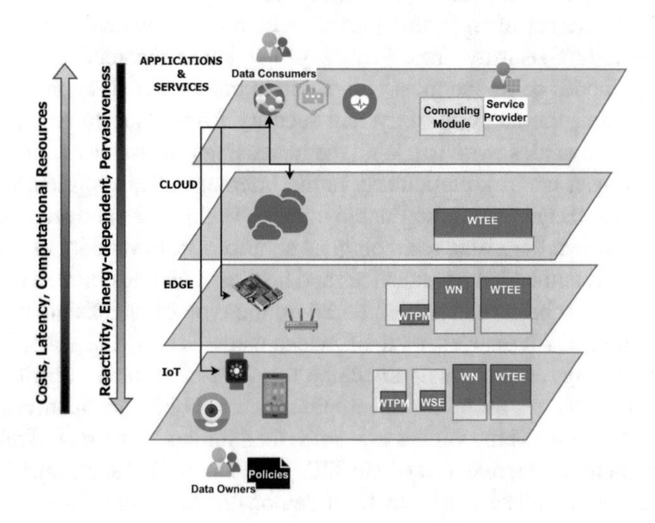


**Fig. 1.** Computing Continuum architecture

The depicted architecture in Fig. 1 necessitates an enhancement in security to address the increased attack surface. This requires implementing measures to ensure the confidentiality, integrity, and availability of data. Devices in this scenario exhibit diverse technological configurations. Specifically, they are categorized as follows:

WTEE (devices with a commercial off-the-shelf Trusted Execution Environment integrated into the CPU, such as Intel SGX and TDX, e.g. using Java [13], AMD SEV, ARM TrustZone, and CCA); WTPM (Devices equipped with a Trusted Platform Module external to the CPU); WSE (Devices equipped with a Secure Element); and WN (Devices with no external security measure). This categorization commonly applies to smartphones, tablets, and hardware crypto-wallets. In this work, the focus is on mobile phones falling under the first category, WTEE, owing to the prevalent usage of ARM processors featuring TrustZone or CCA. As computation seamlessly transitions across various nodes within the Computing Continuum, the establishment of a chain of trust, where each node in the infrastructure represents a link in the chain, becomes a foundational element in safeguarding data. Each link in this chain must adhere to stringent security standards, ensuring a reliable and secure computational flow. An illustrative use case within the Computing Continuum centers on the seamless continuation of computation across various nodes. Rather than fragmenting tasks, this scenario envisions a unified computational process that smoothly shifts from one node to the next. The Computing Continuum effectively manages this process, ensuring a strong and uninterrupted computational flow. Such an approach proves invaluable for tasks demanding substantial processing power, showcasing the Computing Continuum's capacity to optimize resource utilization and minimize latency throughout the continuous computational progression [14].

In this paradigm, where different devices are involved, there is a need for consistency in the technologies that are used to ensure adoption without any difficulties in implementation. To address the challenges posed by the cloud-edge continuum paradigm architecture, Menetrey et al. [15] proposed an interoperable solution, emphasising the need for seamless operation across different hardware and software environments. They highlighted the critical requirements of optimal performance and

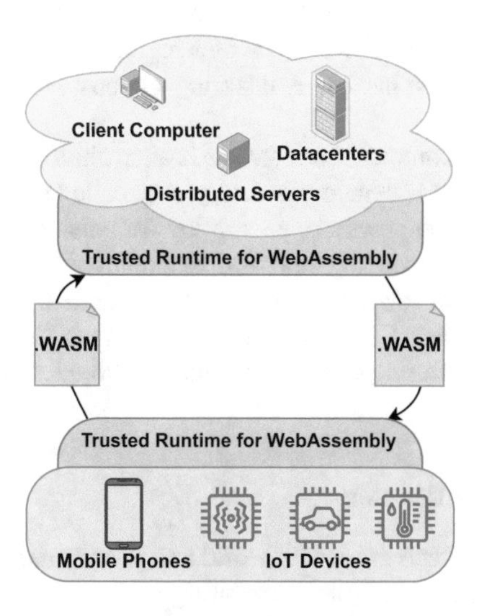


**Fig. 2.** WebAssembly interoperability across devices

high-level security for off-site code and data processing. In particular, the authors introduced the use of WebAssembly on modern virtual machines, incorporating a core set of supporting services and libraries to improve interoperability. WebAssembly is a binary instruction format that serves as a portable compilation target for programming languages, allowing developers to run their code with near-native performance in a browser environment.

WebAssembly's compatibility with privacy-preserving technologies like SGX and TrustZone is of technical significance. As shown in Fig. 2 Trusted WebAssembly runtimes like TWINE [16], to integrate a Wasm runtime into Intel SGX enclaves, and WATZ [17], a trusted runtime designed for Arm TrustZone, facilitate the creation of such chain of trust. This chain includes secure compilation, transmission, and execution of WebAssembly code within these trusted environments.

## 4 TEE in Mobile Phones

Software security mechanisms often fail to cope with advanced attacks [18]. In these cases, creating robust and secure solutions requires the incorporation of secure hardware components, and over the years, several have been developed (e.g. MTM [19], simTPM [20]). However, for various reasons such as limited resources, energy consumption or lack of anti-tampering certification, these solutions have never spread widely. Hence, GlobalPlatform [21], a non-profit industry association, defined specifications for secure chip technologies known as Trusted Execution Environment. In the realm of mobile phones, two key enabling technologies emerge i.e. ARM TrustZone and ARM Confidential Compute Architecture.

ARM's TrustZone technology is a security feature embedded in system-on-chip (SoC) architectures. It operates by partitioning the system into two distinct areas: the Secure world and the Normal world. The Secure world is designed for sensitive operations and data, while the Normal world handles regular, non-sensitive operations. This partition is enforced through hardware, ensuring a robust isolation between these two environments.

ARM's Confidential Compute Architecture is an evolution of TrustZone, introduced in ARMv9-A in 2021. It adds dynamic isolated space called realms, which are managed at runtime and offer per-page memory encryption for enhanced security. CCA extends the TrustZone model by introducing new worlds: a realm world for dynamically created realms and a root world for low-level firmware management.

Key differences between CCA and TrustZone include CCA's dynamic realm creation and management, offering more flexibility and broader application development possibilities.

### 4.1 Secure OS Standardization

Given the separation between the secure world and the normal world in TrustZone, the presence of a secure operating system executing on top of the secure world becomes necessary for running secure applications. Initially, GlobalPlatform directed its attention towards standardizing TEE. Subsequently, it introduced a specification for the

Secure OS, which includes the internal API and TEE applications, allowing for developing Trusted Applications within a TEE in a standardized manner.

Several Secure OS solutions have been developed on top of TrustZone for Android, each with its own characteristics and considerations. Some of these OS implementations are listed in Table 1.

**Table 1.** Overview of Secure OS Solutions on TrustZone for Android.

| Secure OS | Compliance | Suitability | License |
|---|---|---|---|
| Open-TEE | GP compliant | Research only | Open source |
| OP-TEE | GP compliant | Research/Prototype/IoT | Open source |
| Qualcomm's QSEE | GP compliant | Widely adopted in Mobile | Commercial |
| Samsung Teegris | GP compliant | Samsung/Exynos specific devices | Commercial |
| Trusty | Not GP compliant | Fully supported by Google | Open source |

The first distinction that can be made between implementations is their licensing. OpenTee [22], created for the development of secure applications with the Android operating system, and OPTEE [23], which works mainly under the Linux system, are both open source developed to be a complete TEE on top of TrustZone. Lastly, developed by Google, Trusty TEE [24], which is Open Source and available for testing and developing custom applications, can be installed on devices with necessary modification. However, OpenTee is still active but its support is limited to Android version 5.0-5.1.1 only, OPTEE had Android support for a limited number of boards with which to essentially make prototypes, and Trusty is not GP compliant making the prototyping in a standardized manner difficult.

Different matters for the implementations of Qualcomm and Samsung which are widely spread in the mobile phone market but their use is regulated by license. This restricts access to a select group of developers, limiting the ability to employ the TEE for custom applications.

### 4.2 TEE Features and Limitations

As discussed above, even if the TEE technology is ready, there's a huge gap in developing secure applications due to agreement problems or support limitations. In Android TAs are embedded in Operating Systems and they need to be loaded at boot, hence the only way to get them loaded for development purposes is to have specific agreements. This need for agreements is a limitation for the end developer [25] who can not design and implement robust security measures for the global market. Beyond these limitations, and more generally with the use of TEE, the mobile device acquires functions given by that technology. These include, for example, isolated execution and secure storage.

Regarding isolated execution, a clear is Widevine DRM [26]. This offers key functions and APIs, including Content Encryption and Decryption, Authentication and

Device Capability, License Management etc. An example of secure storage available in the Android framework is Android Keystore [27]. This framework provides developers with cryptographic tools to protect user data by transferring cryptographic primitives from the Android OS software libraries to secure hardware. In addition, Keystore allows the specification of application-related restrictions on key usage, further enhancing security measures. Additionally, Android offers the Key Attestation feature, which enables the verification of hardware-backed key pairs through an attestation mechanism. Essentially, it confirms that the cryptographic keys used by an app are securely stored in the device's hardware-backed Keystore.

Given the strong commercial footprint that the mobile sector has, the potential of the TEE has been limited by deployment policies enforced by device vendors both for application developers and end-users. This restriction impacts the possible level of data in-use protection, as it prevents applications from fully leveraging the security benefits of TEEs for processing sensitive data in a secure, isolated environment. As mentioned earlier, when mobile devices are an important part of the edge devices in the Computing Continuum paradigm, these limitations affect the required chain of trust. While features for securing data in transit and at rest are sufficient, limitations arise when there is a need to protect data during computation itself, given the inability to run custom code in the secure world. This limitation arises when considering data processing applications that "continue" across different nodes. It becomes clear that mobile phones cannot be used effectively for such purposes, despite the stability and consolidation of the technology.

### 4.3 Prototyping with AOSP and TrustZone

Acknowledging the issues that have been presented in this work, there is still the possibility to explore the concrete opportunities that the TEE can offer. This subsection investigates the practical application of TEEs in mobile devices, focusing on prototyping using the Android Open Source Project (AOSP) with TrustZone using two Secure OS solutions: OP-TEE and Trusty.

For OP-TEE, the initial step involves building the secure OS on a supported board, i.e. HiKey620 - LeMaker, HiKey620 - CircuitCo, and HiKey960, by downloading the relevant manifests as outlined in the documentation. After the environment setup, developers can proceed to craft custom Trusted Applications using the standardized GlobalPlatform APIs. Nonetheless, as stated by Amacher and Schiavoni in their work [28], two significant limitations arise in OPTEE. Firstly, the REE kernel lacks several fundamental features for security reasons, resulting in the absence of basic syscalls like fopen and msgget. Consequently, it becomes crucial to minimize syscall dependencies during the development of Trusted Applications. Secondly, there are existing constraints related to memory allocation and addressing within TrustZone. Together these limitations may pose challenges to the deployment of more complex TAs, impacting the overall capability within this isolated environment.

Focusing on Trusty, there are two methods to explore its capabilities and potentially prototype secure applications. The first approach involves building and installing Trusty directly onto a device through firmware flashing. Alternatively, one can build Trusty on the Quick EMUlator (QEMU) [29], utilizing the Android image along with an official script supplied by Google. This flexibility allows developers to choose the most suitable testing environment for their needs. However, Trusty's non-compliance with GlobalPlatform results in specific APIs for creating Trusted Applications. These APIs allow the development of two categories of applications or services: Trusted Applications or services, which operate on the TEE processor and Normal or untrusted applications, which run on the main processor and utilize the services offered by trusted applications.

## 5 Summary and Conclusions

This paper investigates the role of Trusted Execution Environments within the mobile Cloud-Edge Continuum, emphasizing the necessity for a secure chain of trust amid the evolution of distributed architectures. It evaluates the standardization of TEEs and the development of Secure Operating Systems for Android, such as OpenTee, OPTEE, and Google's Trusty TEE.

While highlighting the advantages of TEEs, this work tries to provide an overview to developers, presenting some ways of prototyping with this technology, regarding their role in the mobile world, addressing the problem of the limitations of the commercial and open source ones available. The paper points out the limitations due to vendor-specific deployment policies and licensing restrictions, particularly with Qualcomm and Samsung's implementations, which limit access to a narrow group of developers and complicate the prototyping process with standards compliance issues. Constraints that hinder their broader application in mobile devices are discussed. It advocates an open approach to TEE deployment, aiming to empower developers, bolster data security, and address the evolving demands of distributed computing. The identified challenges and limitations offer insights for future developments and suggest a path toward more standardized, accessible, and secure solutions.

## References


1. Alamouti, S.M., Arjomandi, F., Burger, M.: Hybrid edge cloud: a pragmatic approach for decentralized cloud computing. IEEE Commun. Magaz. **60**(9), 16–29 (2022)
2. Sabt, M., Achemlal, M., Bouabdallah, A.: Trusted execution environment: what it is, and what it is not. In: 2015 IEEE Trustcom/BigDataSE/Ispa, vol. 1, pp. 57–64. IEEE (2015)
3. Pinto, S., Santos, N.: Demystifying arm trustzone: a comprehensive survey. ACM Comput. Surv. **51**(6), 1–36 (2019)
4. Li, X., et al.: Design and verification of the arm confidential compute architecture. In: 16th USENIX Symposium on Operating Systems Design and Implementation (OSDI 22), Carlsbad, pp. 465–484. USENIX Association (2022)
5. Huang, Z., Xia, G., Wang, Z., Yuan, S.: Survey on edge computing security. In: 2020 International Conference on Big Data, Artificial Intelligence and Internet of Things Engineering (ICBAIE), pp. 96–105 (2020)
6. Coppolino, L., D'Antonio, S., Mazzeo, G., Romano, L.: A comprehensive survey of hardware-assisted security: from the edge to the cloud. Internet of Things **6**, 100055 (2019)
7. Ying, K., Ahlawat, A., Alsharifi, B., Jiang, Y., Thavai, P., Du, W.: Truz-droid: integrating trustzone with mobile operating system. In: Proceedings of the 16th Annual International Conference on Mobile Systems, Applications, and Services (MobiSys 2018), pp. 14–27. Association for Computing Machinery, New York (2018)

8. Ahlawat, A., Du, W.: Truzcall: secure voip calling on android using arm trustzone. In: 2020 Sixth International Conference on Mobile And Secure Services (MobiSecServ), pp. 1–12 (2020)
9. Brasser, F., Gens, D., Jauernig, P., Sadeghi, A.-R., Stapf, E.: Sanctuary: arming trustzone with user-space enclaves. In: Proceedings 2019 Network and Distributed System Security Symposium (2019)
10. Khalyeyev, D., Bureš, T., Hnětynka, P.: Towards characterization of edge-cloud continuum. In: Batista, T., Bureš, T., Raibulet, C., Muccini, H. (eds.) Software Architecture. ECSA 2022 Tracks and Workshops, pp. 215–230. Springer, Cham (2023). https://doi.org/10.1007/978-3-031-36889-9_16
11. Gkonis, P., Giannopoulos, A., Trakadas, P., Masip-Bruin, X., D'Andria, F.: A survey on IoT-edge-cloud continuum systems: status, challenges, use cases, and open issues. Future Internet **15**(12), 383 (2023)
12. Bittencourt, L., et al.: The internet of things, fog and cloud continuum: integration and challenges. Internet of Things **3**, 134–155 (2018)
13. Coppolino, L., D'Antonio, S., Mazzeo, G., Romano, L.: A comparative analysis of emerging approaches for securing java software with intel SGX. Future Gen. Comput. Syst. **97**, 620–633 (2019). Cited by: 16
14. Beckman, P., et al.: Harnessing the computing continuum for programming our world. In: Fog Computing: Theory and Practice, pp. 215–230 (2020)
15. Ménétrey, J., Pasin, M., Felber, P., Schiavoni, V.: Webassembly as a common layer for the cloud-edge continuum. In: Proceedings of the 2nd Workshop on Flexible Resource and Application Management on the Edge (FRAME 2022), pp. 3–8. Association for Computing Machinery, New York (2022)
16. Ménétrey, J., Pasin, M., Felber, P., Schiavoni, V.: Twine: an embedded trusted runtime for webassembly. In: 2021 IEEE 37th International Conference on Data Engineering (ICDE), pp. 205–216 (2021)
17. Ménétrey, J., Pasin, M., Felber, P., Schiavoni, V.: Watz: a trusted webassembly runtime environment with remote attestation for trustzone. In: 2022 IEEE 42nd International Conference on Distributed Computing Systems (ICDCS), pp. 1177–1189 (2022)
18. Ahvanooey, M.T., Li, Q., Rabbani, M., Rajput, A.R.: A survey on smartphones security: software vulnerabilities, malware, and attacks. arXiv preprint arXiv:2001.09406 (2020)
19. Kim, M., Kim, Y., Ju, H., Park, Y.: Design and implementation of mobile trusted module for trusted mobile computing. In: 2010 Digest of Technical Papers International Conference on Consumer Electronics (ICCE), pp. 255–256 (2010)
20. Chakraborty, D., Hanzlik, L., Bugiel, S.: simTPM: User-centric TPM for mobile devices. In: 28th USENIX Security Symposium (USENIX Security 19), pp. 533–550. USENIX Association, Santa Clara (2019)
21. Global Platform. https://globalplatform.org/
22. McGillion, B., Dettenborn, T., Nyman, T., Asokan, N.: Open-tee–an open virtual trusted execution environment. In: 2015 IEEE Trustcom/BigDataSE/ISPA, vol. 1, pp. 400–407 (2015)
23. OPTEE. https://optee.readthedocs.io/en/latest/index.html
24. Google Trusty TEE. https://source.android.com/docs/security/features/trusty
25. Ekberg, J.-E., Kostiainen, K., Asokan, N.: Trusted execution environments on mobile devices. In: Proceedings of the 2013 ACM SIGSAC Conference on Computer and Communications Security (CCS 2013), pp. 1497–1498. Association for Computing Machinery, New York (2013)
26. Patat, G., Sabt, M., Fouque, P.-A.: Exploring widevine for fun and profit. In: 2022 IEEE Security and Privacy Workshops (SPW), pp. 277–288 (2022)

27. Cooijmans, T., de Ruiter, J., Poll, E.: Analysis of secure key storage solutions on android. In: Proceedings of the 4th ACM Workshop on Security and Privacy in Smartphones and Mobile Devices (SPSM 2014), pp. 11–20. Association for Computing Machinery, New York (2014)
28. Amacher, J., Schiavoni, V.: On the performance of ARM TrustZone: (practical experience report). In: Pereira, J., Ricci, L. (eds.) Distributed Applications and Interoperable Systems: 19th IFIP WG 6.1 International Conference (DAIS 2019), Held as Part of the 14th International Federated Conference on Distributed Computing Techniques (DisCoTec 2019), pp. 133–151. Springer, Cham (2019). https://doi.org/10.1007/978-3-030-22496-7_9
29. Bellard, F.: Qemu, a fast and portable dynamic translator. In: Proceedings of the Annual Conference on USENIX Annual Technical Conference (ATEC 2005), p. 41. USENIX Association, USA (2005)

# Cloud-Native Software Development Life Cycle: A Case Study with Italian Ministry of Justice

Dario Branco[1(✉)], Salvatore D'Angelo[1], Beniamino di Martino[2,3,4], Antonio Esposito[1], Vincenzo de Lisi[5], and Gianluca Paravati[5]

[1] Università Degli Studi Della Campania Luigi Vanvitelli, Department of Engineering, Aversa, Italy
{dario.branco,salvatore.dangelo,antonio.esposito}@unicampania.it
[2] Università della Campania Luigi Vanvitelli, Department of Engineering, Aversa, Italy
beniamino.dimartino@unina.it
[3] Department of Computer Science and Information Engineering, Asia University, Taichung, Taiwan
[4] Department of Computer Science, University of Vienna, Vienna, Austria
[5] Ministero della Giustizia - Dipartimento per la transizione digitale, l'analisi statistica e le politiche di coesione - Direzione generale per i sistemi informativi automatizzati, Rome, Italy
{vincenzo.delisi,gianluca.paravati}@giustizia.it

**Abstract.** In recent years, the Cloud Computing paradigm has witnessed rapid and extensive adoption across diverse industries and sectors. This transition to Cloud-based solutions has fundamentally transformed the landscape of software development and deployment. Consequently, prioritizing Cloud compatibility in the software development lifecycle has become imperative for developers and organizations. This paper aims to offer a comprehensive and detailed overview of the software development lifecycle tailored for Cloud-based software. We will delve into the entire process, spanning from conception to deployment, elucidating key stages, methodologies, and best practices. Throughout the manuscript, we will underscore the significance of employing the most effective tools and techniques available to streamline development processes and maximize the benefits of Cloud Computing. Through the utilization of Cloud-native technologies, such as containerization and orchestration frameworks, developers can augment scalability, reliability, and performance in their Cloud-based applications. Additionally, we will present a case study exemplifying a software development lifecycle for Cloud-based software. This case study is based on the logical infrastructure devised by the Italian Ministry of Justice to manage the introduction of new functionalities in their services, following a DevOps cycle.


L. Barolli (Ed.): AINA 2024, LNDECT 203, pp. 244–253, 2024.
https://doi.org/10.1007/978-3-031-57931-8_24

## 1 Towards a Cloud-Native Development

The realm of software development is undergoing a significant transformation with the increasing adoption of Cloud-native approaches. This shift represents a departure from traditional development paradigms, emphasizing scalability, agility, and resilience in the Cloud environment. As organizations strive to leverage the full potential of Cloud Computing, the journey towards Cloud-native development becomes a strategic imperative [1–3]. Cloud-native development often embraces **microservices architectures**, enabling applications to be broken down into modular, independently deployable components[4]. This fosters flexibility, scalability, and easier maintenance. **Containers** play a pivotal role in Cloud-native development, offering a consistent and portable environment. Kubernetes, as an orchestration tool, facilitates the efficient deployment, scaling, and management of containerized applications. Cloud-native development emphasizes collaboration and integration between development and operations teams. **DevOps** practices, such as Continuous Integration and Continuous Delivery (CI/CD), become instrumental in achieving rapid and reliable software delivery. Managing infrastructure through code streamlines resource provisioning and configuration, promoting consistency and reproducibility. **Infrastructure as Code** (IaC) is a foundational practice in Cloud-native development, enhancing efficiency and reducing manual errors. Cloud-native applications can scale dynamically based on demand, optimizing resource utilization and cost-effectiveness. The modular nature of Cloud-native development allows for quicker development cycles and faster deployment, fostering a culture of continuous innovation, helped also by Artificial Intelligence[5]. Fault tolerance and automated recovery mechanisms enhance the resilience of Cloud-native applications, ensuring uninterrupted service even in the face of failures. Cloud-native development enables organizations to pay for resources on a usage basis, leading to cost savings compared to traditional on-premises infrastructure.

However, it is not always simple to adopt new technologies and development strategies, especially in contexts where specific practices have been applied for a long time, and change is more often seen as a problem, rather than an opportunity.

Embracing Cloud-native development often requires a cultural shift within organizations, emphasizing collaboration, innovation, and a mindset of continuous improvement. While Cloud providers implement robust security measures, developers must remain vigilant to address potential vulnerabilities in their applications and configurations. Adopting Cloud-native technologies involves a learning curve for development teams, necessitating training and skill development.

This article explores the key considerations, benefits, and challenges associated with this paradigm shift and delves into the tools and methodologies shaping the landscape of cloud-native development, by also presenting a simplified case study of their application.

## 2 Source Code Management

The first thing that enables cloud-native development is the ability to share source code and collaborate with it in a distributed way because the cloud itself is a distributed environment. To do this, we need to manage effectively and efficiently the source code of our application. Source code management (SCM) is the process of storing, managing, and tracking changes to the source code, and it is a fundamental part of the software development process. The execution of the process is made possible by the use of a version control system (VCS), which is a tool that helps the developers in their tasks, like tracking changes, managing versions, and coordinating the work of multiple developers.

There are different approaches to version control, but the most common are: **centralized version control systems** (CVCS) and **distributed version control systems** (DVCS). The first one is based on a client-server architecture with a server that stores all the source code and the developers can work on a single file or a subfolder, there are several advantages in terms of performance and security but a big disadvantage is the single point of failure of the server unless there is a backup strategy. The second one, instead is based on a peer-to-peer where the source code is replicated on each developer's machine with a bit loss in terms of performance and security.

The DVCS became the de facto standard for source code management, in particular, GIT is the most used VCS, open-source, and free.

One of the most important characteristics of a VCS is the ability to manage different versions of the source code, and this is done through the use of **branches**. A branch is a parallel version of the source code, and it is used to isolate the development work without affecting other branches. The main branch is called **master** or **main** and it is the one that contains the official release history. The other branches are used to develop new features or to fix bugs, and they are called **feature** branches. When the development of a feature is completed, the feature branch is merged into the master branch. Figure 1 shows an example of a branch.

To enhance the collaboration between developers, a collection of guidelines and best practices are defined, and they are called **workflows**. This work [6], presents a systematic mapping study of the most used workflows, but we will focus on the **Gitflow workflow** [7] because it is the most used in the industry and it is one that we will use in our case study.

The Gitflow workflow is a branching model for Git, which is a DVCS, that defines a strict branching model designed around the project release. It is composed of two main branches, the **master** branch, which contains the official release history, and the **develop** branch, which contains the latest delivered development changes for the next release. The **feature** branches are used to develop new features for the upcoming or a distant future release, the **release** branches are used to prepare a new production release, and the **hotfix** branches are used to quickly patch production releases. The Fig. 2 shows the Gitflow workflow.

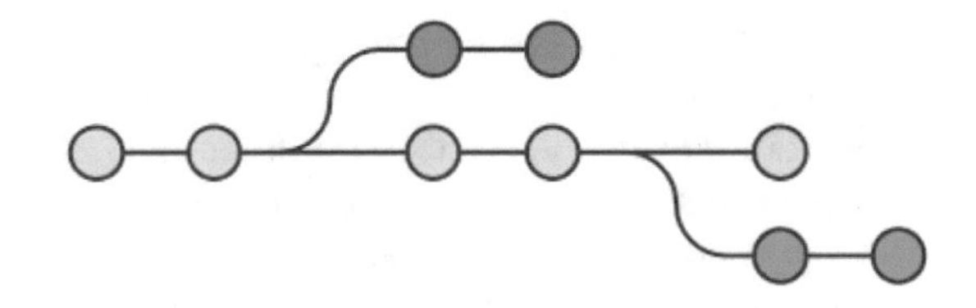

**Fig. 1.** Example of a branch

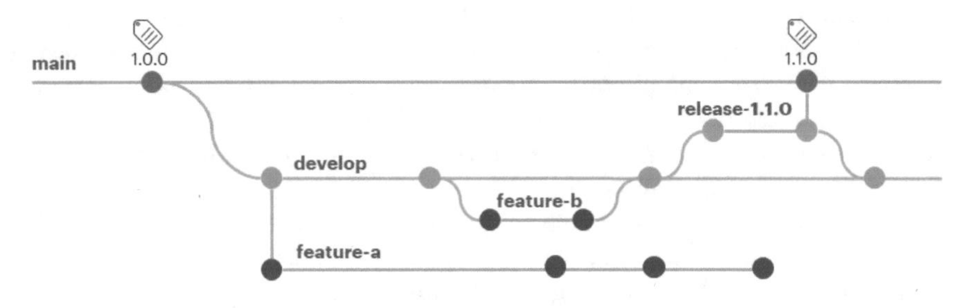


**Fig. 2.** Gitflow workflow[7]

## 3 Continuous Everything

Continuous operations, within the context of software development, encompasses a series of interconnected phases, notably Continuous Integration (CI), Continuous Delivery (CD), and Continuous Deployment (CD) [8]. Each one is a set of steps that are executed continuously, and the goal is to ensure that the software is always in a state that is ready to be deployed (Fig. 3).

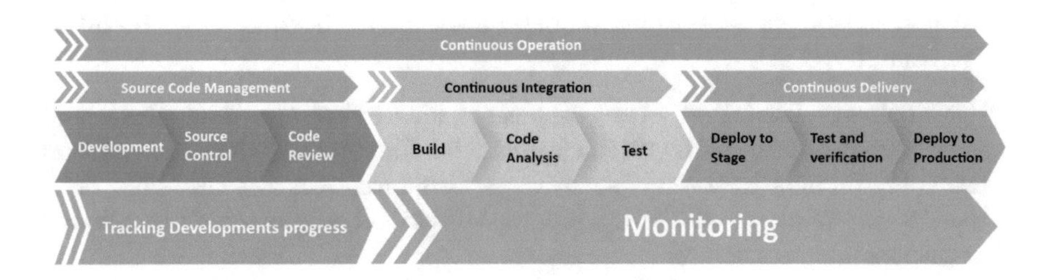


**Fig. 3.** CI/CD Pipeline in DGSIA

Beginning with **Continuous Integration**, this development practice involves frequently merging and integrating the latest work to facilitate a shorter release cycle. The software must be versioned and support the collaboration, which is done with Source Code Management (SCM), presented in the previous section. In this phase the source code is validated with some qualitative metrics, is tested, and is built to produce the artifacts [9] that are needed for the delivery or deployment phase. In this phase can be also applied some security measures like code review, code quality, etc.

Moving to **Continuous Delivery**, the focus shifts to the build and testing of software artifacts; these products are tested in a production-like environment and undergo various tests to ensure that quality standards are met. Once the tests have passed, the artifacts are deployed and are ready for release. The final release is done manually and after that end users can access the produced artifact.

The final phase **Continuous Deployment**, involves automation of the delivery process. As in Continuous Delivery, the software undergoes the same testing and quality standard assurance procedures for meeting metrics. This practice, however, differs from the previous one in that it requires no manual intervention to deploy the artifact to the production environment. The automation introduced at this stage streamlines the process and aims to improve efficiency by reducing the risk of human error.

The concept of Continuous Operations, in general, in addition to the previous phases, incorporates additional practices that are in the development practice called **Continuous Monitoring**. Continuous Monitoring involves tasks that perform continuous surveillance of software performance and behavior in the production environment. This practice ensures early identification of any problems and implements corrective mechanisms within the timeframe established by quality standards. Continuous monitoring, increasingly used in pipelines, has thus become an indispensable element of software health and reliability.

## 4 Requirements Traceability

End-to-end traceability is an essential process in software development, allowing the connection of all steps from the initial requirement to the testing and validation phase, ensuring that every part of the system has been adequately tested and meets the established requirements. Here's how to perform end-to-end traceability:

- Requirements Definition: start by clearly identifying and documenting all project requirements. Requirements can be functional (what the software must do) or non-functional (such as performance, security, usability, etc.). Each requirement should be unique, well-described, and assigned to a source or responsible person.
- Requirements Assignment: Link each requirement to a specific software feature or component that must fulfill it. This step helps establish a clear mapping between requirements and implementation.
- Software Development: During software development, programmers translate requirements into code. Ensure that each developed feature or component aligns with the assigned requirements.
- Testing and Validation: create a set of tests based on identified requirements. These tests should cover all expected features and conditions. Execute tests to verify that the software meets the established requirements. Document test results thoroughly.

- Traceability between Requirements and Tests: Link each test to the requirements it intends to verify. This can be done using a traceability system or a requirements management application.
- Change Management: Throughout the project, requirements may be modified or added. It's crucial to track these changes and update documentation and tests accordingly.
- Final Testing: Before software release, conduct comprehensive testing to verify that all features have been tested, and the software meets all requirements. Document the results of the final testing.
- Complete Documentation: Maintain comprehensive and organized documentation of all requirements, their sources, changes made, and test results.
- Automation: Test automation can significantly simplify the end-to-end traceability process, enabling efficient and repeatable test execution.
- Requirements Management Tools: use dedicated software tools for managing and tracing requirements.

End-to-end traceability is crucial to ensure software quality and compliance with established requirements, facilitating future maintenance and change management processes.

Traceability between requirements and tests is essential in the software development life cycle. This process systematically establishes and documents relationships between initial requirements and subsequently executes tests to verify them. Here are some common techniques for implementing traceability:

### 4.1 Traceability Matrix

The traceability matrix is a table that connects requirements to test cases. Each requirement is represented in a row, while test cases are listed in columns. You can indicate which test case verifies each requirement. The intersections between requirements and test cases are then used to highlight which requirements are covered by which tests.

Let's assume we have an online booking system for a travel agency with defined requirements. Here's how a simplified traceability matrix might look (Fig. 4):

In this example, we listed requirements along the horizontal axis (RQ-01, RQ-02, etc.) and test cases along the vertical axis (Test 1, Test 2, Test 3).

The "X" markers in the cells indicate which test cases cover each requirement. For instance, requirement RQ-01 is covered by all three tests (Test 1, Test 2, Test 3), while RQ-03 is covered only by Test 2.

### 4.2 Requirements Tagging

Requirements tagging is a technique that involves assigning unique identifiers or labels (tags) to requirements to facilitate traceability and management. These tags provide an efficient way to identify and link requirements to different parts

| | A | B | C | D |
|---|---|---|---|---|
| 1 | Requisiti | Test 1 | Test 2 | Test 3 |
| 2 | RQ-01: Prenota | X | X | X |
| 3 | RQ-02: Annulla | X | | X |
| 4 | RQ-03: Modifica | | X | |
| 5 | RQ-04: Visualizza | X | X | X |
| 6 | RQ-05: Pagamento | X | X | X |

**Fig. 4.** Traceability Matrix

of the software development process, including test cases, project documents, changes, and more.

Assume you have a list of requirements for a task management application. Each requirement is labeled with a unique identifier in square brackets:

- **[RQ-01]** The application must allow users to create new tasks.
- **[RQ-02]** Users must be able to assign a due date to tasks.

In this example, each requirement has been assigned a unique tag, like "[RQ-01]," "[RQ-02]," and so on. These tags can be used in various contexts:

***Linking to Test Cases.*** Suppose we want to link these requirements to corresponding test cases. Test cases could be labeled similarly:

- **[TC-01]** Test for creating a new task
- **[TC-02]** Test for assigning a due date

Now we can establish traceability between requirements and test cases using the tags. For example, requirement [RQ-01] would be linked to test case [TC-01], as the test verifies the creation of a new task.

***Change Management.*** If changes to requirements occur in the future, tags can help easily identify which requirements are affected by the changes and which parts of the system need to be updated accordingly.

Requirements tagging provides a flexible and organized way to trace and manage requirements in a project. It's important to establish a clear and consistent naming convention for tags and use them diligently throughout the documentation and the software development process.

## 5 Case Study: Application of Methodology to Software Development Lifecycle in the Italian Ministry of Justice

In this case study we will present the application of the methodology to the software development lifecycle in the Italian Ministry of Justice, in particular, in the General Directorate for Automated Information Systems (DGSIA). The DGSIA is the central body of the Ministry of Justice responsible for the management of the information systems and the digital transformation of the Ministry.

During the last years, the DGSIA started a process of digital transformation, and the first step was the migration of the information systems to a private cloud infrastructure. The migration was done with the help of a third-party company, and the DGSIA has the intention to develop the new information systems in a cloud-native way. To do this, the DGSIA has to change the way of developing software, and the first step is to define a new software development lifecycle.

Figure 5 reports the main structure of the logical architecture devised by DGSIA for the management of new functionalities. In particular, Evolutive Maintenance activities (MEV) and Corrective Maintenance activities (MAC) are taken into consideration.

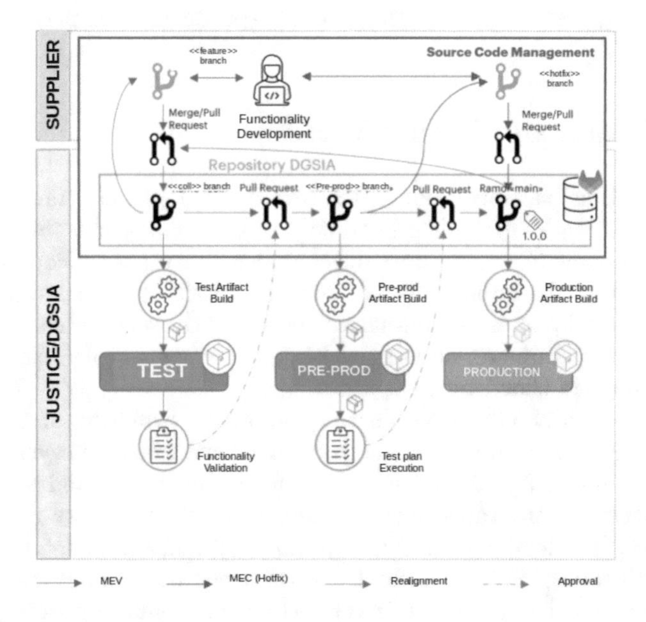


**Fig. 5.** Logical Structure for the development of new functionalities

The Information Systems (SI) owned by the DGSIA are developed and maintained by a wide range of suppliers. A structured working methodology that regulates the phases of development and source code transfer within the Ministry's Infrastructure, taking into account the specific contexts, is needed.

The DGSIA ecosystem is divided into two main domains:

- **Supplier Domain**: represents the set of environments used by suppliers during the internal development phase, i.e., before the code is transferred to justice repositories. The supplier is free to decide which and how many environments to set up within its infrastructure;
- **Justice Domain**: represents the set of environments and branches owned by the Ministry of Justice, regardless of the type of installation (cloud or on-premises).

The branching strategy is directly related to the number of development environments. In particular, the following environments have been defined for the Justice domain:

- **Testing**: represents the integration point between supplier systems and the DGSIA. repository and is the environment where the initial tests are performed;
- **Pre-production**: an intermediate multifunctional environment between testing and production. Not always present in legacy projects;
- **Production**: the environment exposed to the end user. Deployment occurs after passing tests in the Testing and Pre-production (if present) environments.

## 6 Conclusion and Future Work

Cloud Computing has revolutionized software development, transforming the landscape of software development and deployment. This paper provides a comprehensive overview of the software development lifecycle for Cloud-based software, highlighting the importance of Cloud compatibility and leveraging effective tools and techniques. Cloud-native technologies, such as containerization and orchestration frameworks, are crucial for enhancing scalability, reliability, and performance in Cloud-based applications.

Future research should focus on the continuous evolution of Cloud-native technologies and their impact on the software development lifecycle, exploring the scalability and adaptability of methodologies in various industry contexts and organizational structures. Additionally, addressing security protocols and vulnerabilities is crucial as Cloud Computing continues to shape the future of software engineering. Future research should focus on refining and extending methodologies, incorporating the latest technologies, and addressing emerging challenges to ensure excellence in Cloud-based software development practices.

## References


1. Di Martino, B., Esposito, A.: An overview of reference architectures for cloud continuum interoperability. IEEE Cloud Continuum **2**(3), 15–19 (2024)
2. Di Martino, B., Cretella, G., Esposito, A.: Advances in applications portability and services interoperability among multiple clouds. IEEE Cloud Comput. **2**(2), 22–28 (2015)
3. Di Martino, B., Venticinque, S., Esposito, A., D'Angelo, S.: A methodology based on computational patterns for offloading of big data applications on cloud-edge platforms. Future Internet **12**(2), 28 (2020)
4. Di Martino, B., Esposito, A., D'Angelo, S., Maisto, S.A., Nacchia, S.: A compiler for agnostic programming and deployment of big data analytics on multiple platforms. IEEE Trans. Parallel Distrib. Syst. **30**(9), 1920–1931 (2019)
5. Martino, B.D., Esposito, A., Damiani, E.: Towards AI-powered multiple cloud management. IEEE Internet Comput. **23**(1), 64–71 (2019)
6. Cortés Ríos, J.C., Embury, S.M., Eraslan, S.: A unifying framework for the systematic analysis of git workflows. Inform. Softw. Technol. **145**, 106811 (2022)
7. A successful git branching model. https://nvie.com/posts/a-successful-git-branching-model/
8. Shahin, M., Ali Babar, M., Zhu, L.: Continuous integration, delivery and deployment: a systematic review on approaches, tools, challenges and practices. IEEE Access **5**, 3909–3943 (2017)
9. Di Martino, B., Esposito, A.: A rule-based procedure for automatic recognition of design patterns in UML diagrams. Softw. Pract. Experience **46**(7), 983–1007 (2016)

# Methodologies for the Parallelization, Performance Evaluation and Scheduling of Applications for the Cloud-Edge Continuum

Antonio Esposito[1(✉)], Rocco Aversa[1], Enrico Barbierato[2], Maria Carla Calzarossa[3], Beniamino Di Martino[1], Luisa Massari[3], Ivan Giuseppe Mongiardo[3], Daniele Tessera[2], Salvatore Venticinque[1], Luca Zanussi[3], and Rasha Zieni[3]

[1] Università Degli Studi Della Campania "Luigi Vanvitelli", Aversa (CE), Italy
{antonio.esposito,rocco.aversa,beniamino.dimartino, salvatore.venticinque}@unicampania.it

[2] Università Cattolica del Sacro Cuore, Milan, Italy
{enrico.barbierato,daniele.tessera}@unicatt.it

[3] Università di Pavia, Pavia, Italy
{mcc,luisa.massari}@unipv.it, {ivangiuseppe.mongiardo01,luca.zanussi01, rasha.zieni01}@universitadipavia.it

**Abstract.** This paper outlines a project, started in October 2023, and entitled Methodologies for the Parallelization, Performance Evaluation and Scheduling of Applications for the Cloud-Edge Continuum (MPE-SACC), aimed at developing a holistic approach to parallelize sequential code across the Cloud-Edge Continuum. MPESACC entails the formulation of a Directive-based programming model, integration of performance analysis, bottleneck identification, scheduling, and optimization techniques for the execution of parallelized code. The proposed model, tailored for Cloud-Edge computing, involves the design of a Compiler software capable of decomposing serial codes, annotated with directives, into distributed components for remote execution. Parallelization techniques conveyed through Parallel Patterns and communication templates in the form of Code Skeletons, simplify transformations and adhere to best practices in code parallelization and distribution. The performance analysis models consider crucial aspects of the Cloud-Edge continuum, such as bandwidth, processing capacity, and energy constraints, providing essential feedback for directive and pattern selection. Scheduling considerations address resource availability and computational needs, optimizing time, energy, storage, and bandwidth constraints. Robust mathematical models support scheduling optimization to prevent task failure or idleness due to temporary resource shortages. The proposed Directives guide scheduling algorithms by providing execution constraints, ensuring informed decisions for optimal task allocation on the Cloud-Edge continuum. Throughout the project, an end-to-end example, involving procedural code, performance analysis, and optimal scheduling, will be


L. Barolli (Ed.): AINA 2024, LNDECT 203, pp. 254–263, 2024.
https://doi.org/10.1007/978-3-031-57931-8_25

developed to demonstrate the feasibility, applicability, and potential of the proposed approach.

## 1 Introduction

The overall objective of the MPESACC project is to provide an encompassing methodology that, starting from sequential, imperative code, provides an efficient decomposition of the code, aiming at modelling its execution over a Cloud-Edge continuum.

In recent years the attention of the IT community has moved from Cloud to Edge, mostly thanks to the improvement of smart devices and of the interconnection networks. An outstanding example is represented by 5G networks that have been spreading throughout the world. Edge also answers to the need for data locality deriving from privacy issues, that make it infeasible to transfer data from local devices to a centralised server for elaboration. The European community has rapidly moved towards the adoption of Cloud-Edge continuum-related technologies, pushing researchers to investigate this field through Research and Innovation Actions, such as the Horizon Europe RIA call "Programming tools for decentralised intelligence and swarms", which is strongly centred on Edge, Swarms and 5G technologies. In particular, it seems clear that tools, programming models and languages for the development of edge-oriented environments are quite requested, as no real open, reliable and robust solution has been yet proposed. The research motivation underpinning this Project goes exactly in the direction of researching and defining an efficient and reliable methodology that, through the definition of a programming model to support users in annotating their sequential code, would not only help them in decomposing such code in parallelizable software components, but would also provide valuable information on the performances of the obtained parallelization, and on the optimal scheduling plan for the execution of the components over Edge resources. At the end of the research activities, the expected results will be represented by:

- A programming model based on standard procedural languages, augmented with Pragmas and Directives for the parallelization of code and the expression of execution constraints;
- The design of a Code Compiler, able to interpret the Pragmas and Directives and to exploit well-known parallelization solutions, namely Parallel Patterns;
- A methodology for the evaluation of the workflow of parallelized components, able to analyse the performances of the parallel program and detect execution bottlenecks;
- A methodology for the identification of the optimal scheduling of software execution, given a pool of available resources managed by the user.

There are three main aspects that the proposed methodology addresses, which have been deemed vital for the general purpose of the proposal, and that are strictly interrelated. These aspects are summarised by the Conceptual Architecture provided in Fig. 1.

The Modelling Module is where a Directive-based procedural programming model is exposed to an external user, to annotate sequential imperative code and start a Pattern and Skeleton guided code decomposition. More specific methodological aspects are exposed in the following sections.

The Performance Analysis Module takes the code components that have been proposed by the Modelling Module, described in Sect. 3, and analyses them through analytical approaches and techniques. The general objective of the Performance Analysis Module, discussed in Sect. 4, is to evaluate the decomposition proposed by the Modelling module and to provide immediate feedback that can be used to guide the user in choosing different annotations for the code.

The Scheduling Module, described in detail in Sect. 5, has the fundamental role to guide the Performance Analysis, by providing scheduling of the code execution over the Cloud-Edge continuum, through an optimization of the use of the available resources. The results of the Scheduling Optimization algorithms will be necessary to provide feedback to the Users and to help them choose the best execution plan in consideration of the available resources.

Interactions between the users and the modules composing the Conceptual architecture are described by the Business Process Model Notation (BPMN) diagram reported in Fig. 2. All the interactions start when the users, bringing their sequential procedural code, Cloud-Edge resources and specific constraints on them, employ the Programming Paradigm exposed by the Modelling Module. The original code, annotated with the provided Pragmas and Directives, expressing both parallelism and constraints, is passed to the Compiler that creates an application workflow. Such a workflow is passed to the Performance module that evaluates it and reports the results as feedback to the users. The users can either deem the performances unsatisfying and then modify the annotations made to the code to get better results or accept the proposed solution. The accepted workflow is then passed to the Scheduling module, which analyses it and identifies the Optimal scheduling, taking into consideration the workflow produced by the compiler, the constraints imposed by the users, and the available resources. The Optimal Scheduling is proposed to the users that can use it to parallelize their original application over the Cloud-Edge continuum.

The remainder of this paper is organised as follows: Sect. 2 reports the State of the Art that is at the base of the presented methodology; Sect. 3 describes the Modelling Module that is used by the proposed logical architecture; Sect. 4 focuses on the Performance Analysis Module, describing its connections with other components of the logical architecture; Sect. 5 provides insight on the Scheduling Module, in charge of sorting the deployment and execution of tasks in the Cloud Continuum, Sect. 6 describes possible applications of the presented methodology and its foreseen impacts; finally, Sect. 7 closes the paper with concluding remarks.

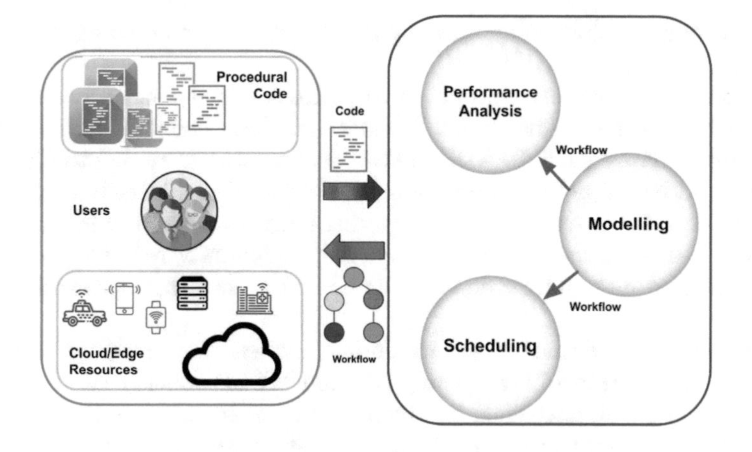


**Fig. 1.** Conceptual Architecture of the proposal

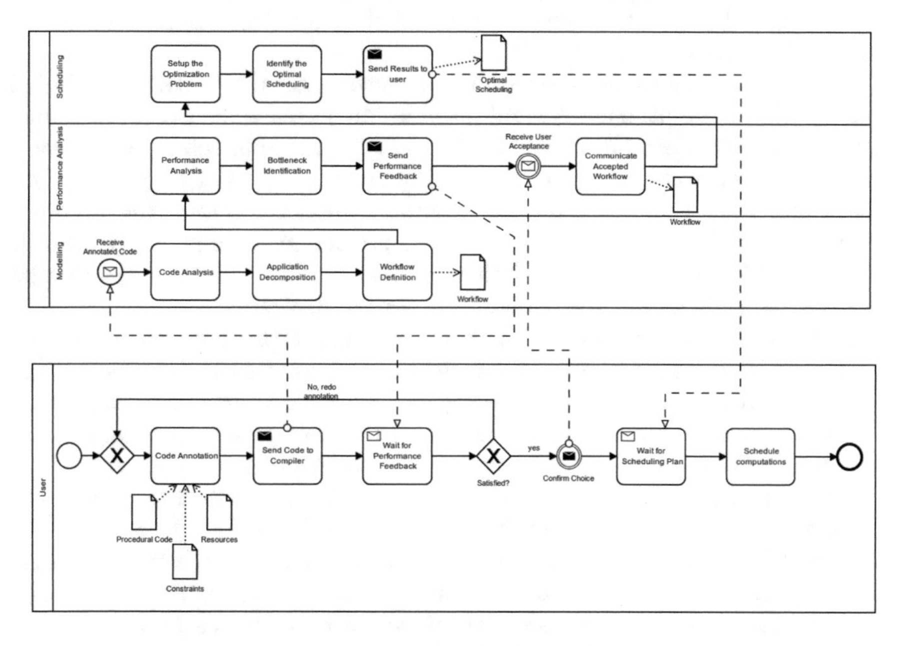


**Fig. 2.** Description of the Interactions between Users and the Modules

## 2 State of the Art

The ubiquity of smart devices and the Internet of Things (IoT) necessitates intricate development and management tools and infrastructures. The imperative to maintain data proximity to its source, driven by privacy and security concerns, has expedited the shift from a Cloud-centric programming paradigm to the Cloud-Edge continuum. The intricacies of contemporary Edge frameworks cannot be overlooked, given the non-linear and intricate nature of programming and managing multiple computational devices dynamically entering or leaving the Edge network.

Various tools have been proposed to address this challenge, employing the concepts of actors and/or agents. Notable solutions include JADE, JaCaMo, Jadex [16], SARL, PADE/SPADE, SAGE, among others, aiming to fill the void of an Edge-oriented programming language or paradigm. Despite the introduction of several Agent-Oriented Programming Languages (AOPs) in recent years, their acceptance in the software engineering community remains limited, and current application practices do not fully exploit their advantages and technical potential.

Additionally, alternative programming models for decentralized smart devices, such as Protocol Over Thing [17] and MOLE (Microservice Orchestration LanguagE) [18], have emerged. Inspired by nature, some approaches model the behaviour of software components against swarms or colonies of insects, known as swarm intelligence, defined by Beni and Watt as the property of a system where the collective behaviour of unsophisticated agents produces global functional schemes.

Frameworks implementing swarm intelligence target UAV, MAV, and Robot devices, with some utilizing general-purpose languages while others introduce custom languages (e.g., Buzz [15] and Meld [3]). Existing solutions, however, face major drawbacks, often requiring programmers to learn proprietary languages and modify or rewrite their original code. In contrast, the proposed Pragmas/Directives-based approach aims to provide an open language built on existing well-known languages, eliminating the need for programmers to alter their original code and instead annotate it for parallelization.

While the Cloud-Edge paradigm offers numerous advantages, challenges persist. Determining which computations should be transferred to the Cloud versus those handled at the Edge becomes crucial. Task scheduling and data distribution are central concerns to ensure overall software performance aligns with Quality of Service standards and energy requirements remain manageable.

The intricate problem of resource provisioning and scheduling for scientific workflows in cloud environments has been extensively addressed in the literature from various perspectives [11]. Single and multi-objective optimization problems have been tackled using meta-heuristics or purpose-specific algorithms [2,12,13,19]. Furthermore, the impact of performance uncertainty has been explored, with robust scheduling approaches proposed, such as in [9], where workflow task processing times are assumed to be unknown but bounded within a given uncertainty interval.

Efforts to define efficient scheduling techniques, accounting for performance metrics, are ongoing. Particularly, exploring the possibility of expressing computational constraints and organizing scheduling based on code analysis supported by ad-hoc Pragmas and Directives is an area of active research.

## 3 The Modelling Module

The proposed approach aims to facilitate the development and distribution of code on Cloud-Edge environments by providing users with tools to annotate and instrument their original code. This involves using a Directive/Pragma-based approach to model computation, indicate resource requirements, and express constraints. The Directives target Python and GO languages, with a focus on Cloud-Edge Continuum.

The proposed Directives cover information on the computational model, deployment model, and available resources. They enable users to express constraints, which are utilized by Performance and scheduling modules for parallel code execution and optimal scheduling. The chosen languages, Python and GO, support different mechanisms such as Decorators and function management.

A Compiler is essential to interpret and decompose annotated code, transforming sequential code into parallel code suitable for Cloud-Edge deployment. The Compiler relies on Directives to guide decomposition, and it is emphasized that models are necessary for effective transformation. Computational and deployment primitives are mapped to specific Patterns, such as Map-reduce and Producer-Consumer, providing guidance for code transformation.

The use of Skeletons, connected to Directives and Patterns, aids in rapid code decomposition, reducing implementation dependency on communications and auxiliary functionalities. The design of the Compiler considers reference Patterns, including those from literature on Cloud and Edge environments.

Resource-related Directives play a crucial role in orchestrating components and scheduling computational tasks. The Modelling module aims to expose a programming model to users, facilitating code distribution over Cloud-Edge continuum environments. The research outcomes include the definition of Pragmas/Directives for languages like Python and Go Lang, offering a programming model for Cloud-Edge development.

The research efforts also result in the design of a Compiler using Directives, Skeletons, and Parallel Patterns to decompose computation and generate software artifacts compatible with Cloud-Edge deployment. The anticipated benefits for users include simplified coding, accelerated development, reduced maintenance costs, and improved program reliability [6–8,10].

## 4 The Performance Analysis Module

The Performance Analysis module will address the initial coarse grain evaluation of the performance achieved by the parallelization strategies translating the structure of a sequential application into a workflow of components to be deployed on Cloud-Edge continuum devices.

Due to the large heterogeneity of the edge computing devices, the performance will be measured in terms of communication bandwidth, processing capacity, energy consumption, and reliability. Furthermore, it will be possible to denote classes of devices sharing similar performance.

The information gathered from independent evaluations will be used to review the different deployed directives (pragmas) of the parallelization strategies, providing feedback to the programmers. The metrics used to evaluate the performance will be modelled by random variables distributed according to probability distributions, whose composition can be analysed by Monte Carlo techniques. The performance will be studied to identify possible bottlenecks concerning the individual resource provisioning and scheduling plan.

## 5 The Scheduling Module

An important phase that takes place before the actual deployment of an application in cloud/edge environments refers to the decisions about the resources to be provisioned and the mapping between resources and application components [4,5].

These decisions are very challenging because they have to take into account several aspects related to the application to be deployed (e.g., characteristics of the individual components, precedence constraints), the scheduling of cloud-edge resources that can be provisioned (e.g., resource types, performance, cost) and the QoS metrics to be optimised (e.g., execution time). In addition, some of these aspects might be affected by uncertainties [20].

In this Project, these issues will be addressed as an optimization problem aimed at minimising the metrics of interest and satisfying the constraints set by the user on these metrics. In detail, a multi-objective constrained optimization problem will be formulated. By solving this problem multiple solutions laying on the so-called Pareto front will be obtained. In particular, these solutions represent a set of "optimal" scheduling plans satisfying the execution ordering of the application components and the objectives and constraints of the problem. Heuristics will be developed for guiding users in the choice of the most suitable plan according to their preferences and needs [14]. In fact, the scheduling plans will be identified offline, that is, prior to the execution of the application.

## 6 Applications and Impact

The widespread adoption of the Cloud-Edge Continuum paradigm is evident across diverse domains, particularly in scenarios where proximity between computations and data sources is imperative for enhancing efficiency, privacy, and reducing energy consumption. A pivotal advancement in the methodologies applied to Cloud-Edge software development, ranging from application design to optimal resource scheduling, promises significant impacts on the adoption and exploitation of existing Cloud-Edge environments. The absence of standardized, open, and easily applicable Programming Models and Tools has been a hindrance, and the introduction of a Directive-based methodology coupled with Performance checking and Scheduling Optimization techniques holds the potential for substantial scientific and technological advancements. This approach is poised to attract interest from programming communities eager to experiment and validate the proposed methodology. Interested groups can be represented by:

- Artificial Intelligence experts who want to develop their AI algorithms for execution in a decentralised Edge environment, and who will only need to use standard imperative and non-natively distributed programming languages to define their applications, by also describing their available resources and computation constraints.
- Multi-access Edge Computing (MEC)-based systems developers would benefit from a directive-based programming model that would greatly impact the way that applications and network components are implemented over cellular networks, in particular the recent 5G networks and emerging 6G networks.
- Infrastructure providers and planners could apply the methodology to evaluate different deployment options and resource allocations. While possible target groups of users can be identified, we deem it important to also find possible application domains in which our methodology can be applied.

## 7 Conclusions

In conclusion, the identified impact areas highlight the versatility and significance of the proposed Cloud-Edge Continuum methodology. To further extend its applicability, three distinct domains have been identified for focused exploration: Microgrids, Energy and Health.

The primary objective of smart grids is to coordinate the needs of generators, grid operators, end-users, and electricity market stakeholders, optimizing the entire system for cost-effectiveness while minimizing environmental impacts and maximizing reliability, resilience, and stability. In the specific context of this application domain, our team leverages expertise gained from the CoSSMic project (Collaborating Smart Solar-powered Micro-grids, FP7-SMARTCITIES-2013). Addressing the challenge of aligning production and consumption for each user, the project aimed to maximize self-consumption within the neighborhood [1]. This system necessitates information sharing for exchanging excess power production and storage capacity, following policies defined by each cluster member. The distributed nature of this application, coupled with Cloud-Edge continuum devices, can greatly benefit from the design and deployment support provided by our project's methodology.

In the realm of Smart Healthcare, a crucial domain, Cloud-Edge continuum applications play a pivotal role in assisting clinicians by providing intelligent healthcare analytics from real-time data collected through smart sensors, wearables, and the Internet of Medical Things. Processing the substantial data generated by these devices poses new challenges in computing capability, bandwidth, latency, and reliability, especially in rural clinics lacking high-speed WANs. Regulatory constraints in the healthcare domain, along with privacy considerations, may restrict the transmission of patient data to public clouds. Consequently, innovative computing approaches, such as federating machine learning and Cloud-Edge continuum applications extracting anonymous features from medical records, have been proposed to address these issues. It is essential to aid programmers in translating standard cloud-centric coding paradigms into novel approaches for exploiting Cloud-Edge continuum environments. This ensures that the performance requirements of clinicians, caregivers, and physicians are met seamlessly, irrespective of location and time, without interruptions.

The proposed methodology serves as a bridge to translate standard cloud-centric coding paradigms into innovative frameworks, ensuring performance requirements for clinicians, caregivers, and physicians are met seamlessly, regardless of location and time constraints. Overall, the Cloud-Edge Continuum methodology emerges as a transformative force with broad applications and tangible impacts across diverse sectors, paving the way for enhanced efficiency and privacy in decentralized computing environments.

**Acknowledgments.** This work was partly supported by the Italian Ministry of University and Research (MUR) under the PRIN 2022 grant "Methodologies for the Parallelization, Performance Evaluation and Scheduling of Applications for the Cloud-Edge Continuum" (Master CUP: B53D23013090006, CUP: J53D23007110008, CUP: F53D23004300006) and by the European Union - Next Generation EU.

## References

1. Amato, A., et al.: Software agents for collaborating smart solar-powered micro-grids. In: Caporarello, L., Di Martino, B., Martinez, M. (eds.) Smart Organizations and Smart Artifacts. LNISO, vol. 7, pp. 125–133. Springer, Cham (2014). https://doi.org/10.1007/978-3-319-07040-7_14
2. Anwar, N., Deng, H.: Elastic scheduling of scientific workflows under deadline constraints in cloud computing environments. Future Internet **10**(1), 5 (2018)
3. Ashley-Rollman, M.P., Goldstein, S.C., Lee, P., Mowry, T.C., Pillai, P.: Meld: a declarative approach to programming ensembles. In: 2007 IEEE/RSJ International Conference on Intelligent Robots and Systems, pp. 2794–2800. IEEE (2007)
4. Calzarossa, M.C., Della Vedova, M.L., Massari, L., Nebbione, G., Tessera, D.: Multi-objective optimization of deadline and budget-aware workflow scheduling in uncertain clouds. IEEE Access **9**, 89891–89905 (2021)
5. Calzarossa, M.C., Della Vedova, M.L., Tessera, D.: A methodological framework for cloud resource provisioning and scheduling of data parallel applications under uncertainty. Futur. Gener. Comput. Syst. **93**, 212–223 (2019)
6. Casadei, R., Pianini, D., Viroli, M., Natali, A.: Self-organising coordination regions: a pattern for edge computing. In: Riis Nielson, H., Tuosto, E. (eds.) COORDINATION 2019. LNCS, vol. 11533, pp. 182–199. Springer, Cham (2019). https://doi.org/10.1007/978-3-030-22397-7_11
7. Di Martino, B., Esposito, A.: Applying patterns to support deployment in cloud-edge environments: a case study. In: Barolli, L., Woungang, I., Enokido, T. (eds.) AINA 2021. LNNS, vol. 227, pp. 139–148. Springer, Cham (2021). https://doi.org/10.1007/978-3-030-75078-7_15
8. Erl, T., Cope, R., Naserpour, A.: Cloud Computing Design Patterns. Prentice Hall Press (2015)
9. Fard, H.M., Ristov, S., Prodan, R.: Handling the uncertainty in resource performance for executing workflow applications in clouds. In: Proceedings of the 9th International Conference on Utility and Cloud Computing, pp. 89–98 (2016)

10. Fehling, C., Leymann, F., Retter, R., Schupeck, W., Arbitter, P.: Cloud Computing Patterns. Springer, Vienna (2014). https://doi.org/10.1007/978-3-7091-1568-8
11. Hilman, M.H., Rodriguez, M.A., Buyya, R.: Multiple workflows scheduling in multi-tenant distributed systems: A taxonomy and future directions. ACM Comput. Surv. (CSUR) **53**(1), 1–39 (2020)
12. Liu, L., Zhang, M., Buyya, R., Fan, Q.: Deadline-constrained coevolutionary genetic algorithm for scientific workflow scheduling in cloud computing. Concurrency Comput. Pract. Experience **29**(5), e3942 (2017)
13. Meena, J., Kumar, M., Vardhan, M.: Cost effective genetic algorithm for workflow scheduling in cloud under deadline constraint. IEEE Access **4**, 5065–5082 (2016)
14. Alkhalaileh, M., Calheiros, R.N., Nguyen, Q.V., Javadi, B.: Performance analysis of mobile, edge and cloud computing platforms for distributed applications. In: Mukherjee, A., De, D., Ghosh, S.K., Buyya, R. (eds.) Mobile Edge Computing, pp. 21–45. Springer, Cham (2021). https://doi.org/10.1007/978-3-030-69893-5_2
15. Pinciroli, C., Beltrame, G.: Buzz: a programming language for robot swarms. IEEE Softw. **33**(4), 97–100 (2016)
16. Pokahr, A., Braubach, L., Jander, K.: Jadex: a generic programming model and one-stop-shop middleware for distributed systems. PIK-Praxis der Informationsverarbeitung und Kommunikation **36**(2), 149–150 (2013)
17. Smirnova, D., Chopra, A.K., Singh, M.P., et al.: Protocols over things: a decentralized programming model for the internet of things. Computer **53**(12), 60–68 (2020)
18. Song, Z., Tilevich, E.: A programming model for reliable and efficient edge-based execution under resource variability. In: 2019 IEEE International Conference on Edge Computing (EDGE), pp. 64–71. IEEE (2019)
19. Verma, A., Kaushal, S.: A hybrid multi-objective particle swarm optimization for scientific workflow scheduling. Parallel Comput. **62**, 1–19 (2017)
20. Zanussi, L., Tessera, D., Massari, L., Calzarossa, M.: Workflow scheduling in the cloud-edge continuum. In: Barolli, L., (ed.) Advanced Information Networking and Applications (AINA), pp. 182–190. Springer, Cham (2024)

# All for One, All at Once: A Pluggable and Referenceable Architecture for Monitoring Biophysical Parameters Across Intertwined Domains

Oscar Tamburis[1], Adriano Tramontano[1(✉)], Giulio Perillo[1(✉)], Arriel Benis[2], and Mario Magliulo[1]

[1] National Research Council of Italy, Institute of Biostructures and Bioimaging, Naples, Italy
{oscar.tamburis,adriano.tramontano,giulio.perillo,
mario.magliulo}@ibb.cnr.it

[2] Department of Digital Medical Technologies, Holon Institute of Technology, Holon, Israel
arrielb@hit.ac.il

**Abstract.** An architecture capable of monitoring biophysical parameters from both the human, veterinary, and environmental domains is described, which focuses on the intersection of Internet of Things (IoT) and Edge Computing. The main features of the architecture - pluggability, cross-referenceability, and measures accuracy and reliability - were developed to deal with the characteristics of the One Digital Health framework. The potential benefits of a pluggable and referenceable architecture have been emphasized for different kinds of stakeholders, as described in two different use-cases from the human/animal and the environmental domains. Potential benefits for the stakeholders from deploying the architecture compared to market and applied research solutions have been discussed.

**Keywords:** Edge Computing · Internet of Things · Use-cases · eHealth · One Digital Health

## 1 Introduction

In the last twenty years, the Internet of Things (IoT) has experienced a diffusion over many areas of society and a number of application fields, among which eHealth, Home Automation, Smart Cities and Industry 4.0 are the most emblematic [1, 2]. IoT applications are able to collect data in real time through sensors, to communicate with other devices and to implement services that aim to improve the quality of life. Edge computing (EC) is a distributed computing model in which data processing occurs the closer as possible to where the data is generated. A common ground in human, animal and environmental health projects is often the need for the institution stakeholder to be provided with a solution able to (i) allow cross-referenceability of different data sources, and (ii)

O. Tamburis and A. Tramontano—Share first co-authorship.


L. Barolli (Ed.): AINA 2024, LNDECT 203, pp. 264–276, 2024.
https://doi.org/10.1007/978-3-031-57931-8_26

be pluggable into any kind of Information System already on duty. Furthermore, some stakeholders may be forced to make use of certified devices which, far from being the goal of the research themselves, are instead able to produce data to support research objectives [3].

Herein, an architecture is introduced that allows collection, real-time transmission and processing of data using EC and IoT concepts. Such pervasive data monitoring infrastructure is supposed to be: (i) pluggable in any kind of Information System already in use in various possible fields of application, without requiring redesign or refitting of the existing solutions; and (ii) referenceable in that it allows to cross-reference the data gathered in automatic and unified way from different kind of sensors and contexts. Aim of the present work is to evaluate the applicability of the introduced cloud-edge architecture to different health-related scenarios, centered on a human/animal (health informatics/veterinary health informatics) or a green subject (environmental informatics). The scope is to test its capability to support the comprehensive monitoring of biophysical parameters in intertwined scenarios, according to the specifics of a One Digital Health (ODH) Intervention [4–6].

After the introduction and a brief analysis of the literature, the proposed architecture is described from a technical viewpoint, along with the two mentioned application use-cases. Some discussion and conclusions are then provided, especially for what concerns the role of the architecture within the specifics of the ODH framework.

## 2 Background

### 2.1 From Cloud to Edge

The most widespread architecture adopted for providing services through the Internet follows the Cloud Computing model, in which a localized cluster of Data Centers is tasked with receiving, processing, storing data and transmitting results. This cluster, called Data Center Network (DCN), is seen as a single, scalable and ever available service, and is strongly oriented to the client-server paradigm. Digital communications technology leaders have made evident how the number of connected devices is increasing, especially mobile and IoT devices, whose traffic is increasingly machine-to-machine, as devices communicate without human interaction [7]. Accordingly, the summation of the data created, captured or replicated is expected to grow from 33 Zettabytes in 2018 to 175 Zettabytes in 2025 [8]. This growth has raised two main concerns: (i) since each DCN is localized, time-critical applications are affected by degraded performance based on the client's geographical position; (ii) the amount of data transmitted worldwide is outgrowing the Internet infrastructure enhancement. To overcome these issues, the most straightforward idea is to move the service closer to the client, in order to require shorter transmission paths, decreasing both service delays and the network load, a concept later evolved in EC.

### 2.2 Edge and IoT, as a Whole

The IoT is the extension of the Internet to devices that communicate among themselves without human intervention, often organized as nets of sensors, actuators and wearables that produce and consume data in order to implement real world services. These devices need to transmit measurements occurring several times each second, therefore installing multiple sensors can add a severe bandwidth consuming load on the network. Also, the sensors usually need to be cheap and to have a long-lasting battery, thus they only support a Low-Power Wireless Personal Area Network (LoWPAN) [9]. Both those characteristics have usually led IoT infrastructures to include IoT Gateways, tasked to retrieve data from data producers, make decisions, actuate devices and send data to Cloud structures. This kind of infrastructure naturally overlaps with an EC application one, thus making it an enabler for sophisticated IoT applications like Smart Cities for environmental monitoring, Telemedicine, and Industry 4.0 [5, 9, 10].

With specific reference to the healthcare sector, data can be e.g., taken by means of IoT devices during an ongoing rehabilitation process far from the medical department, allowing for shorter hospitalization and cost reduction [11]; furthermore, non-invasive observations can be held and processed continuously in order to allow timely intervention or prevention [12], and the same data can be useful when consulted by medical staff when treatment is needed, even when they are not characterized by a medical grade accuracy [13]. Since that data is taken remotely, healthcare can be fairer in case of geographical limitations [11]. Yacchirema et al. [14] presented a system able to detect falls with 91.67% accuracy. Their architecture, embedding a smart IoT Gateway, can detect falls based on a decision tree updated and supplied by the Cloud Node. P-Ergonomics Platform, shows how EC enabled IoT health monitoring can take place in occupational wellbeing and can help mitigate work-related disease such as musculoskeletal disorders [15]. This platform uses an Infrastructure-as-a-Service (IaaS) EC platform such as well-known providers (e.g., Amazon, Google, Microsoft).

## 3 Proposed Architecture

While EC is all about bringing resources closer to the edge, implementations may differ greatly, having as common ground the presence of three main layers: *Edge Devices Layer, Cloud Layer* and *Edge Node Layer* (Fig. 1).

The proposed implementation of the architecture is shown in Fig. 2. It has been evaluated for the use in public projects where entities, covering the role of the customer, requested a pervasive data monitoring infrastructure without having a bleeding-edge Information System.

The *Edge Devices Layer* features devices connecting from the edge (or Endpoints), requesting a service. For the proposed architecture it is made up of sensors that can leverage low power PAN connection technologies in order to have a long-lasting battery life. Non-invasive devices could become set-and-forget.

For the deployed architecture it has been chosen to add a *Gateway* between the Edge Devices Layer and the Edge Node Layer. The choice was driven by the willingness of either, keeping most of the network load confined in the PAN and, allowing for quick responses for time-critical applications. The gateway is therefore responsible for managing a star-shaped network of sensors for, on the one side, performing data aggregation and, for the other side, routing both internally and through the Internet. In addition, gateways perform local analysis to react to field state changes while adapting its behavior

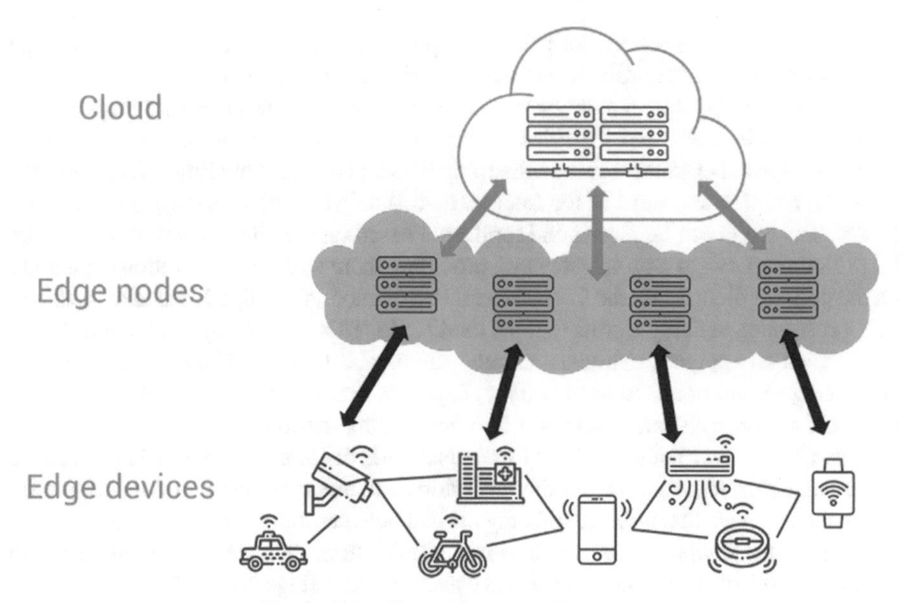


**Fig. 1.** Edge Computing general structure

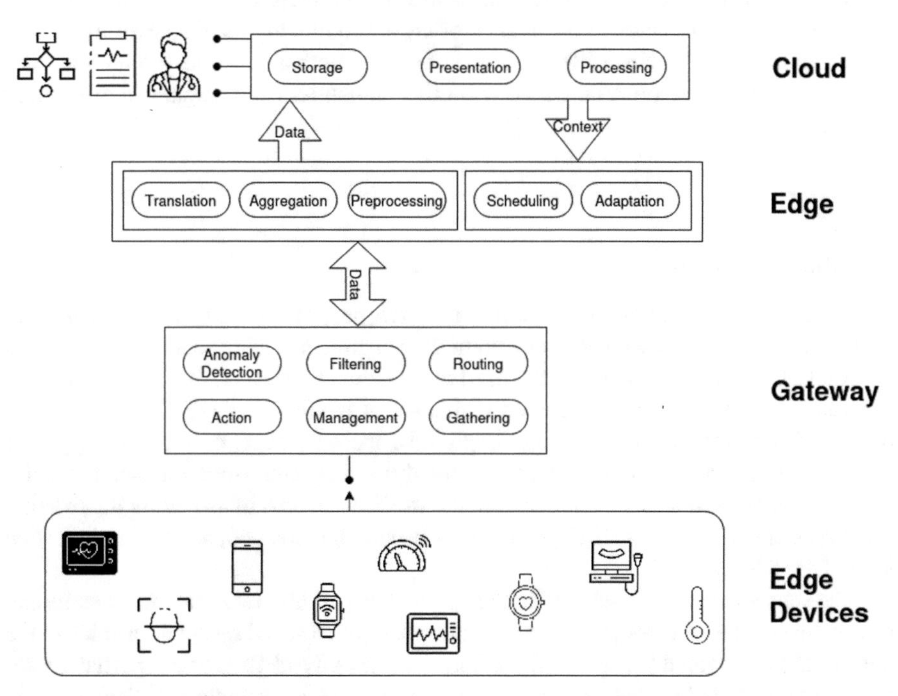


**Fig. 2.** General structure of the proposed architecture

depending on the context. Sensor values and policies requested by the customers shape such context so that a coordinator on the layer above is prescribed.

The *Edge Node Layer* features servers that are generically described as lying between the Cloud Node Layer and the Edge Devices Layer. Communication among layers is mainly composed of traffic between the Endpoints and one or more Edge Nodes. For the proposed architecture, most of the Internet load is between the Gateways and the Edge Nodes. The Node can be placed and replicated based on the client's requirements. The Edge Node supervises a set of gateways, providing them with a configuration depending on the policies dictated by the Cloud Node. Furthermore, the Edge Node takes a role in data-processing, sending results to the Cloud Node. This processing can be of different nature such as applying mathematical algorithms and taking decisions. However, the processing should not need to identify a subject and does not store sensitive data, thus leaks and data correlation attacks result in unprofitable efforts.

The *Cloud Layer* features a Data Center Network, or Cloud Node, and it has a central role in applying intensive processing and storing massive amounts of data. However, managed data have the property of being needed both centrally for data processing, and globally by the endpoints. In the proposed architecture, the customer is interested in receiving a set of data from both the Gateway and the Edge Node. The Cloud Node is installed directly into the customer's facilities in order to respect data protection regulations and policies. This Node can elaborate and store data being able to apply complex processing thanks to the chance of cross-referencing the data ever harvested. One main focus is allowing different kinds of agents, both algorithms and human staff, to consult and operate on data, which will differ in nature, from diagnostic imaging to time series.

## 4 Use Cases

### 4.1 Human Domain

A scenario has been figured out within a research project with a University Hospital, where elderly people have undergone a home care treatment in which data monitoring and Assisted Physical Activity have been used to implement a tele-rehabilitation prototype program. Patients have been provided with a set of medical IoT (IoMT) devices, and instructed to perform exercises and simply take measurements of some among their biophysical parameters, the way they were already used to; data is therefore automatically collected, processed and then included in the medical records managed by the existing Electronic Health Records (EHR) framework used in the hospital, and currently hosted in the Cloud Node.

The sensors currently included in the application fall into two categories: some can be considered "set-and-forget", such as a fitness tracker, a speed sensor embedded in the mini-cyclette provided for the Assisted Physical Activity (APA), or an inertial sensor that can be installed on the beddings in order to estimate heart rate and sleep quality; others need some interaction due to their nature, like a thermometer, a pulse oximeter, a blood pressure monitor and a scale. In order to deploy the latter category of sensors into the application, the instruments needed to look familiar and work like their non-smart counterpart: this means that patients would only be required to operate the device starting a measurement and any consequent setup or connection would be handled by the gateway. To act as the gateway, an appliance equipped with an ARM-based SoC and the Android operating system has been deployed. Many low-cost devices adhere to this description and some of them even support general purpose I/O ports, thus allowing the connection of custom inputs as screens or sensors. A framework running on such gateways features the appropriate interface to communicate with every BLE sensor to manage and retrieve data. The framework aggregates different kinds of data, both in the form of measurements and events e.g., devices moving out of range. The gateway behavioral policies are defined by the Edge Node, which provides it with a suitable configuration that can be updated during the operation time, whenever needed. Such a process makes it possible to (i) define new targets for data streams, (ii) change the scheduling of operations, and (iii) update security credentials, without requiring any user interaction. The gateway does not have enough computational resources for an effective data processing, so it can either transmit the data gathered from the devices to the Cloud Node (i.e., in case data do not need further processing), or it can send them to the Edge Node otherwise. However, gateways can make simple decisions based on non-medical interpretation of sensor readings, such as what to do on the occurrence of particular events, or when a value with a particular meaning has been gathered. For instance, the gateway can trigger a notification on the fitness tracker when a value exceeds a particular threshold. The installation of the gateway requires no more than a power cable and a LAN cable. In this regard, a cabled connection to the patient's home router has been chosen to let the gateway access the Internet without further configuration. A WiFi connection would have instead required little gateway configuration by means of a HID device. Each gateway at its first boot asks the Edge Node for a valid configuration to know the sensors it shall connect to. Authors refer to the "Gateway + sensors" set as "Kit". The role of an edge node can be fulfilled by a machine equipped with the computational power that can be commonly found in consumer electronics, and by open source software. What has been implemented in this case is a machine with two main aims: (i) communicating with the gateways via a message queue-based interface; and (ii) providing a REST service for the Cloud Node. The data received from the gateway needs non-medical and non-identifying processing, like unit conversion, data filtering, data aggregation and compression. The Edge Node receives from the Cloud Node via the REST interface new policies. Gateways will be updated with new configurations and schedules originated by the mentioned policies. As an example, some policies relate to the Kit composition and the sensors' sampling settings. The edge node is responsible for storing sensors' unique identifiers therefore setting up Kits and defining behaviors for the gateways. The Cloud Node exposes interfaces on two sides too: on the one side it receives data from both the Edge Node and the gateways; on the other side it should expose a set of interfaces that are appropriate to feed data to: medical staff, algorithms, and data visualization graphical interfaces. However, the Cloud Node has been left to the stakeholder to implement or commission, since as said it mainly hosts a EHR framework. What results is that medical staff have been provided with a unified user interface that aggregates (i) locally hosted data about patients, and (ii) data coming from the edge, gathered from the Kits patients are associated with. When a new patient needs to be included in the remote monitoring plan, the medical staff register their data on the platform and associate them with a Kit. The patient is instructed as to how to use the sensors of their Kit; no configuration or maintenance is needed besides basic battery management.

While the monitoring system operates normally and standard policies are set on the Cloud Node, the gateways receive measurements all-day-long from different Edge Devices as soon as they are made available from the equipment. Data collection happens simultaneously from all the devices operated by the patient so that, for instance, it is possible to retrieve weight from a scale while measuring temperature with a thermometer, thanks to the technicalities of the BLE protocol. Some of the Edge Devices have an embedded memory that allows multiple parameters to be stored for days, e.g. smartwatches and portable ECGs. When the devices are available in range, the whole memory content is retrieved by the gateway and processed to: (i) reduce redundancy, and (ii) update the Edge Node only on new data. As an example of a particular situation that has come across during the APA project, physicians decide to impose a policy at Cloud Node level for which each patient that had a severe cardiac episode must have their SpO2 and Heart Rate (HR) sampled while pedaling during their training sessions. Cloud Node provided, alongside the policy, a list of related Kit IDs to the Edge Node. The simultaneous sampling of both cycling rate and SpO2 and HR from the pulse oximeter was already possible but the Edge Node also imposed a subsequent policy: during bike sessions, all the devices capable of measuring SpO2 and HR - e.g. smartwatch - had to enable real-time sampling for the whole session as to have benchmarks. Policy group is then translated into a configuration sent to the interested gateways.

### 4.2 Green Environment Domain

A second scenario has been devised collaborating with the Municipal Administration of a metropolitan city in South Italy, for a research project aimed to establish a remote pervasive monitoring system for the urban greenery. The objective was to allow specialists, employed by the Municipal Administration, to remotely infer on the damages either unpredictable (e.g. broken branches caused by extreme weather conditions), too widespread to be monitored effectively (e.g. drying trees caused by high temperatures), or anthropic-related. The deployed system harvests, refines and feeds the cloud information system hosted in the eGovernment framework with biophysical parameters of the greenery including wind speed, temperature and humidity of both air and soil, acceleration and orientation of branches and trunks.

The application needed sensors to be installed non-intrusively on green subjects over a large area in open air, both at heights and near the roots, thus several requirements have been identified for the outdoor components. Since their particular placement on green subjects lowers the maintainability of the whole system, sensors needed to be durable, low-cost and require low maintenance. As for the latter, green computing strategies have been adopted to optimize longevity and power consumption of the sensors while still providing a suitable range for wireless communication. While a water and dust-resistant shell addresses the most immediate concern about exposure, an accurate choice of SoC with BLE support has been made aiming to maximize battery life, allowing for year-long maintenance cycles. The deployed sensors feature an additional signal amplification setting that can be remotely tuned to achieve a proper balance between power consumption and transmission range. A custom Gateway Station has been built using a dust and water-resistant shell that exposes the solar panel and an anemometer, in a compact package, that can be easily installed where suitable exposure to sunlight and wind is provided (e.g. lamp post). Cautions about power saving relate to gateways as well. The minimum hardware for the task has been identified in an ESP-32 microcontroller equipped with a GSM modem for internet connection, with long-term deployment being ensured by a 12Ah 12V battery and a solar panel that recharges it via a voltage controller module. In this Use Case, authors define as "Kit" each set featuring the gateway and exclusively the sensors it manages. The management tasks include: getting new configurations from the Edge Node, pushing sensor settings (e.g. signal attenuation and polling rate), gathering measurements from sensors and sending them to the Edge Node. In order to minimize the time in which the equipped GSM modem is turned on, the gateway buffers measurements in compressed units and turns the modem on only when needed. Additionally, the gateway acts as a local weather station, measuring parameters as a common context for the field. In case new sensors are declared on the Edge Node, they are dynamically added to the Kit of the geographically closest gateway, the configuration of which is timely updated. On the other hand, if a gateway is added to the greenery, a reassignment is performed to match each sensor to the closest gateway now that a new one is present in the field. The Edge Node is, again, made up of consumer electronics and open-source software. While, on one side, the Edge Node is tasked with operating on data and managing the gateways, on the other, it exposes a REST interface for communication with the Cloud Node. Data operation starts by decompressing the mentioned data unit. The measurements need then to be processed depending on the sensor's installation parameters that are stored in the Edge Node. This makes it possible to infer branches' and masts' inclinations time series. Finally, data is sent to the Cloud Node. Gateway management is a dynamic process that produces a configuration that minimizes power consumption and considers the policies received from the Cloud Node. Sensors declaration, administration and policies are received from the Cloud Node through the REST interface. The Cloud Node receives data from the Edge, while providing a sensible interface for the domain's experts to use. Authors did not implement the Cloud Node, as it lies in the Information System of the Municipal Administration along with its established framework which is subject to restrictive policies. However, a graphical user interface is available on the Cloud Node, which allows remote kit administration and locally saved data visualization.

The system, when the standard policies are in place on the Cloud Node, yields values of wind speed, acceleration, temperature and humidity with different frequencies, depending on the meaning of each of these parameters. For instance, branch acceleration measurements fluctuate faster than those of the trunks; temperature and humidity vary on a larger time scale. Based on this, sensors can be fine-tuned in order to improve battery life and reduce storage redundancy. However, policies can be specified to address specific scenarios improving precision botany intervention. In a first example, in order to measure the soil response to the water fed and actuate an efficient irrigation strategy, agronomy specialists have compiled an irrigation timesheet on the Cloud Node; the system reacts by increasing the humidity sampling rate. Another example pertains to anthropic risk during extreme windy conditions: when the wind speed exceeds a threshold, the trees' oscillations time series needs an increased definition, thus the polling rate is set to the maximum. In both examples, the policies are filled out on the Cloud Node through a graphical interface, then are sent to the Edge Node layer, from which configurations are calculated and delivered to the interested gateways, according to positioning within the greenery area. Gateways, then, dynamically orchestrate the sensors, even making choices and schedules based on the configuration received.

## 5 Discussion

The design of the proposed architecture was driven by the need to develop specific features to deal with the ODH paradigm [4], which brings in a syndemic scenario new questions, new solutions to be found. In this regard, in projects spanning at different extents human, animal, and environmental health, a frequently encountered requirement involves as said delivering solutions to institutional stakeholders, which make it possible to (i) seamlessly integrate into any existing Information System currently in use, and (ii) facilitate cross-referenceability across diverse data sources. Accuracy and reliability of measures is a further objective of utmost importance to be accounted for. Experiences moving in such directions can be either found for instance in [16, 17] for Veterinary medicine, and in [18] involving green subjects. In particular, the former shows how the human-based case study reported in this work can be extended to the animal domain as the logic underlying the two kinds of business processes are comparable [19]. Table 1 summarizes the main differences between the proposed architecture, the solutions currently existing on the market, and those developed in applied research.

**Table 1.** Features supported by different kinds of implementations.

| | Market Vendors | Applied Research | Proposed Architecture |
|---|---|---|---|
| Pluggability | N | Y | Y |
| Cross-referenceability | Y | Y | Y |
| Accuracy and reliability of measures | Y | N | Y |

Big vendors' solutions usually lack pluggability-related features for mere commercial reasons. There is little interest in developing solutions integrable with other systems, as the purpose is to sell a closed, single-package (although customizable within certain limits) to the final client. It is on the other hand highly likely that referenceability-related features are deployed, because in many cases it is noted that different devices from the same vendor show aggregated information via a single, shareable interface - e.g., websites/smartphones apps. Moreover, as big vendors aim to produce and commercialize devices that can be certified, accuracy and reliability of the measurements as well are at their best. Pluggable- and referenceable-related features can be detected in solutions developed in applied research-focused contexts: the devices embedded in such solutions are in fact mostly experimental in nature, and aimed at rapid prototyping, therefore they are highly customizable [13]. This makes it therefore possible to achieve high-level results in terms of both pluggability and referenceability. On the contrary, the major emphasis on the aspects of device integration simplicity and ease of platform creation may hinder a timely development of the aspects of measurement accuracy and reliability.

In our case, according to the principles of implementation of an ODH Intervention [5], the Edge Node Layer hosts those digital functionalities/digitalities that are operationalized to deliver specific activities (Specialities), by means of the IoT-based Edge Devices (Technologies). The sensors used in the *IoT* devices do not need significant throughput, but only require connectivity capable of being supported by backup batteries, and low-cost hardware. The study of this category of low-cost and low-energy devices has mainly focused on those communication protocols that enable the most effective strategies for this purpose, resulting in fertile ground for Bluetooth Low-Energy (BLE) [20]. Since the Bluetooth specifications prescribe hardware and connectivity protocols, but not the content of the transmitted packets, many sensors use proprietary data protocols and representations, requiring proprietary programs or smartphone applications to access said data: as previously stated, this was not favorable for the application. Finding different kinds of certified sensors capable of both producing *reliable and accurate measures* and to connect to a single gateway has been a lengthy task, and usually the manufacturer was requested to provide either the protocol documentation, or a library implementation of the communication interface. Many project stakeholders requested, similarly, data collection applications with low-cost non-invasive sensors, in which they would study, process and cross-reference the data using their Information System already on duty [21]. An IoT EC application can be connected to said Information System, not only letting IoT and EC complement each other, but resulting in easy adoption. Moreover, embodying characteristics of *pluggability* makes the adoption of the biophysical parameters monitoring solution, either within the same domain (yet in different application contexts), and in different but intertwined domains, easier and less traumatic for stakeholders within a real-world scenario.

In this sense, it is possible to agree that deploying an ODH Intervention by means of the proposed architecture, can (i) address One Health–related challenges; and (ii) ensure both economic and practical convenience for stakeholders. Likewise, the design and deployment of an ODH intervention also requires from the generated data, at each step, to be [22, 23]: (i) "Findable," as the digitalities involved contribute to the study and collection of all data pertinent to the interconnection between systems' needs; (ii) "Accessible" through standardized protocols, leveraging common substrates of data, information, and knowledge derived from digital biodiversity; (iii) "Interoperable," as a result of the deliberate establishment of an ecosystem capable of facilitating seamless and secure health data exchange and processing, addressing shared risks between animal and human populations; (iv) "Reusable," enabling a systematic, continuous, and intelligent integration of extensive, intelligent, and multidimensional data exchanges facilitated by the digitalities involved. At the semantic level, it is imperative that the nomenclature employed for (meta)data and the delineation of variables along with their attributes, aligns seamlessly with a controlled vocabulary [24]. The need to use data collected across scientific disciplines and related to different communities is in our case possible, as the proposed architecture has been designed with a focus on *referenceability*:

data collection from Edge devices is in fact accomplished by exploiting low-level communication protocols, thus leaving the authors full control of the (multi-terminology) thesaurus to be specifically designed [25]. In this way, data from different sources across the three One Health domains can essentially share common vocabulary and background knowledge - which matches with the requirements of Interoperability, actually critical to achieve data cross-referenceability.

Additionally, to all of the above, it is also possible to agree that deploying an ODH Intervention by means of the proposed architecture, can (i) achieve One Health–related important and strategic outcomes for clinical follow-up and practice, such as for technology improvements needed; and (ii) achieve FAIR uses of digital technologies by trying to be environmentally-respectful by design [26].

## 6 Conclusions

In the present paper the design of a Cloud Edge architecture embedding IoT devices was introduced, which consists of four layers of nodes in increasing distance from the data source. The main features of the architecture - pluggability, cross-referenceability, and measures accuracy and reliability - were developed to deal in the most appropriate way with the issues descending from the implementation of the One Digital Health framework. Two use cases were described to showcase the deployment of the architecture in human- and environment-centered scenarios, respectively. In the application of a ODH framework many opportunities can be found, as stakeholders may have non-bleeding edge Information Systems while subject to restrictive and complex privacy policies. However, having access to cross-referenceable data coming from different domains, collected by sensors that are both certified and already available on the market, makes stakeholders able to perform complex and precise data analysis having access to large amounts of data. Such amounts of data are of increasing importance if Artificial Intelligence-based approaches are also to be considered, and it certainly suggests a direction for future developments in One Health-focused scenarios of critical importance such as urban aquatic ecosystems, called to serve as vital connectors between people, animals, and plants, fostering biodiversity and sustainability in cities [27].

**Acknowledgments.** The research was partially funded by the European Union's Horizon Europe Research and Innovation Program (HORIZON-CL6- 2022-GOVERNANCE-01) under grant agreement No. 101086521 - OneAquaHealth (Protecting urban aquatic ecosystems to promote One Health).

## References

1. Lampropoulos, G., Siakas, K., Anastasiadis, T.: Internet of things in the context of industry 4.0: an overview. Int. J. Entrepre. Knowl. **7**, 4–19 (2019)
2. Umair, M., Cheema, M.A., Cheema, O., Li, H., Lu, H.: Impact of COVID-19 on IoT adoption in healthcare, smart homes, smart buildings, smart cities, transportation and industrial IoT. Sensors **21**, 3838 (2021)

3. Wilkes, M., et al.: One Health workers: innovations in early detection of human, animal, and plant disease outbreaks. J. Glob. Health Rep. **3**, e2019093 (2019)
4. Benis, A., Tamburis, O., Chronaki, C., Moen, A.: One digital health: a unified framework for future health ecosystems. J. Med. Internet Res. **23**, e22189–e22189 (2021). https://doi.org/10.2196/22189
5. Benis, A., Haghi, M., Deserno, T.M., Tamburis, O.: One digital health intervention for monitoring human and animal welfare in smart cities: viewpoint and use case. JMIR Med. Inform. **11**, e43871–e43871 (2023). https://doi.org/10.2196/43871
6. Tamburis, O., Magliulo, M., Tramontano, A., Perillo, G., Benis, A., Calabrò, S.: Precision grazing: when agriculture, livestock and technology unite. In: 2023 IEEE International Conference on Big Data (BigData), pp. 3510–3515. IEEE (2023)
7. Cisco: Annual Internet Report (201832023) (2020)
8. Reinsel, D., Gantz, J., Rydning, J.: The digitization of the world - from edge to core. International Data Corporation (2018)
9. Hayyolalam, V., Aloqaily, M., Ozkasap, O., Guizani, M.: Edge-assisted solutions for IoT-based connected healthcare systems: a literature review. IEEE Internet Things J. **9**, 9419–9443 (2022). https://doi.org/10.1109/JIOT.2021.3135200
10. Angelucci, A., Kuller, D., Aliverti, A.: A home telemedicine system for continuous respiratory monitoring. IEEE J. Biomed. Health Inf. **25**, 1247–1256 (2021). https://doi.org/10.1109/JBHI.2020.3012621
11. Singh, P.: Internet of Things based health monitoring system: opportunities and challenges. Int. J. Adv. Res. Comput. Sci. **9**, 224–228 (2018). https://doi.org/10.26483/ijarcs.v9i1.5308
12. abdulmalek, s., et al.: iot-based healthcare-monitoring system towards improving quality of life: a review. Healthcare (Basel) **10**, 1993 (2022). https://doi.org/10.3390/healthcare10101993
13. Tramontano, A., Tamburis, O., Cioce, S., Venticinque, S., Magliulo, M.: Heart rate estimation from ballistocardiogram signals processing via low-cost telemedicine architectures: a comparative performance evaluation. Front. Digital Health **5**, 1222898 (2023)
14. Yacchirema, D., De Puga, J.S., Palau, C., Esteve, M.: Fall detection system for elderly people using IoT and Big Data. Procedia Comput. Sci. **130**, 603–610 (2018). https://doi.org/10.1016/j.procs.2018.04.110
15. Vega-Barbas, M., Diaz-Olivares, J.A., Lu, K., Forsman, M., Seoane, F., Abtahi, F.: P-ergonomics platform: toward precise, pervasive, and personalized ergonomics using wearable sensors and edge computing. Sensors (Basel) **19**, 1225 (2019). https://doi.org/10.3390/s19051225
16. Ireifej, S.J., Krol, J.: Case studies of fifteen novel species successfully aided with the use of a veterinary teletriage service. Front. Veter. Sci. **10** (2023)
17. Tamburis, O., Masciari, E., Fatone, G.: Development of a decision tree model to improve case detection via information extraction from veterinary electronic medical records. In: Proceedings (2021). http://ceur-ws.org. ISSN. 1613-0073
18. Tramontano, A., Tamburis, O., Magliulo, M.: To the Green from the Bl(u)e: an innovative system for monitoring urban green areas. In: 2022 IEEE International Conference on Metrology for Extended Reality, Artificial Intelligence and Neural Engineering (MetroXRAINE), pp. 646–650. IEEE (2022)
19. Luzi, D., Pecoraro, F., Tamburis, O.: Appraising healthcare delivery provision: a framework to model business processes. In: Informatics for Health: Connected Citizen-Led Wellness and Population Health, pp. 511–515. IOS Press (2017)
20. Liu, C., Zhang, Y., Zhou, H.: A comprehensive study of bluetooth low energy. In: Journal of Physics: Conference Series, p. 012021. IOP Publishing (2021)

21. Koutalieris, G., et al.: Enhancing urban environmental sustainability through unified stakeholders needs co-creation process (AENEA). In: 2023 IEEE International Conference on Metrology for eXtended Reality, Artificial Intelligence and Neural Engineering (MetroXRAINE), pp. 899–904. IEEE, Milano (2023). https://doi.org/10.1109/MetroXRAINE58569.2023.10405595
22. Tamburis, O., Benis, A.: One digital health for more FAIRness. Methods Inf. Med. **61**, e116–e124 (2022). https://doi.org/10.1055/a-1938-0533
23. Wilkinson, M.D., et al.: The FAIR Guiding principles for scientific data management and stewardship. Sci. Data **3**, 1–9 (2016)
24. Motta, C., et al.: A framework for FAIR robotic datasets. Sci. Data **10**, 620 (2023)
25. Benis, A., et al.: Medical informatics and digital health multilingual ontology (MIMO): a tool to improve international collaborations. Int. J. Med. Informat. **167**, 104860 (2022)
26. Benis, A., Tamburis, O.: The need for green and responsible medical informatics and digital health: looking forward with one digital health. Yearb. Med. Inf. **32**, 007–009 (2023). https://doi.org/10.1055/s-0043-1768717
27. OneAquaHealth Project Website. https://www.oneaquahealth.eu/. Accessed 02 Feb 2024

# Enabling Precision Irrigation Through a Hierarchical Edge-to-Cloud System

Gabriele Penzotti(✉), Michele Amoretti, and Stefano Caselli

CIDEA, University of Parma, Parma, Italy
{gabriele.penzotti,michele.amoretti,stefano.caselli}@unipr.it

**Abstract.** Precision Agriculture (PA) leverages ICT innovations to optimize resource allocation, minimize environmental impact, and meet global food demands. PA faces significant challenges in aggregating and processing vast amount of sensor data and environmental inputs from diverse sources like sensors, satellites, weather stations, and drones. This paper describes a scalable Edge-to-Cloud (E2C) system designed for Precision Irrigation services and applications. The system integrates existing and new services, automating data flows to enable precision irrigation and support decision making. A detailed description of the service hierarchy and of service allocation along Cloud, Edge and intermediate Fog nodes is provided. E2C emerges as a key architectural solution to cope with the challenges of smart farming.

**Keywords:** Edge-to-Cloud · Smart Farming · Distributed System

## 1 Introduction

Water is a key resource for crop growth, and modern Precision Agriculture (PA) approaches [11] must rely on *Precision Irrigation (PI)* techniques to achieve sustainability. PI is based on the analysis of soil moisture and crop vigor sensed data, as well as on crop development models [2], to optimize the allocation of water to crops at plot, farm, and district level. Acquiring and processing extensive sensor data across large agricultural areas for PI faces logistical challenges to cover thousands of plots and farms, burdened by the complexity of providing appropriate crop management recommendations and real-time actions based on sophisticated biological models and large amounts of spatio-temporal data.

Indeed, data management in Smart Farming must address multiple technological issues (such as granularity, provenance, timeliness, accuracy, and security) stemming from the variety of information sources and proprietary technological silos. Indeed, an abundance of sources - e.g., sensors, satellite imagery, weather stations, and drones - provide multifaceted environmental and agricultural data. Integrating these disparate sources poses a significant challenge, yet it is mandatory to enable the full potential of PA.

Edge-to-Cloud (E2C) systems [3,6,9] are emerging as the key architectural solution for smart agriculture, as they encompass the diverse computational layers that connect the crop in the field with the decision support systems enabling


L. Barolli (Ed.): AINA 2024, LNDECT 203, pp. 277–286, 2024.
https://doi.org/10.1007/978-3-031-57931-8_27

operations at farm, district or regional level. This paper focuses on a scalable E2C system designed to support PI services and applications at the different levels. The system integrates existing and new services, and interconnects layers and components by means of automated data flows. These data flows replace human inspection and labour in the farms as well as manual entry of the crop-specific information needed to feed irrigation advisory services and other decision support systems.

## 2 Related Works

The *edge-to-cloud continuum* has emerged as a powerful paradigm to enable applications that are not suitable for all-in-cloud deployment. In this model, the computational, networking, storage and acceleration elements are known as *fog nodes*, positioned between the cloud layer and the IoT layer [4,8].

While most research in smart agriculture has focused on cloud-based and sensor applications, only a few research papers propose architectures in the edge-to-cloud continuum and Smart Agriculture. Alharbi and Aldossary [1] proposed an IoT-based Edge-Fog-Cloud architecture for Smart Agriculture, to solve real-time processing issues in terms of reducing CO2 emission, energy consumption and network traffic, compared to the traditional Cloud-based architecture. Other available architectures include the conceptual Things-Fog-Cloud based architecture that incorporates mechanisms for detecting and treating outliers [7], as well as the Latency-Adjustable Cloud/Fog Computing Architecture [10].

In a recent work [12], Kalyani et al. presented an edge-to-cloud architecture for Smart Agriculture based on a multi-agent system that incorporates the concept of Digital Twin, to address challenges like resource allocation, resource scheduling, and task scheduling. The main contribution of that work is the adoption of semantically interoperable JSON-LD based data, to foster service discovery and composition.

## 3 System Architecture

The proposed E2C system is an amalgamation of both existing, externally managed services and purposely designed services. Services are composed to enable large-scale soil-plant moisture assessment. The resulting system architecture has a hierarchical structure based on service characteristics, primarily resource capacity and functional complexity, defining four distinct layers.

*Cloud Layer.* In the proposed system, satellite data hubs are cloud nodes that act as reservoirs for vast amounts of data, enabling rapid dissemination crucial for irrigation monitoring service across wide geographic spans.

*Edge Layer.* These nodes operate in environments of limited or non-existent visibility. The computational capabilities of these nodes are minimal, accommodating tasks such as sensor data aggregation and localized operational processes within individual farms (as a farm management system). They serve as crucial endpoints, facilitating real-time interactions and immediate responses within the farming environment, typically without offering external services.

*Fog Layer.* Borrowing from the fog computing paradigm [4], these nodes inhabit the intermediate layer, boasting variable but moderate data processing and storage capabilities. Situated between the cloud and edge layers, they act as facilitators, bridging the gap between high-capacity cloud services and the constrained but immediate functionalities at the edge. Nodes that offer services to other nodes but do not have cloud capabilities are considered fog nodes.

*IoT Layer.* At the base of the system, IoT nodes are positioned across the agricultural landscape, enabling measurement and interaction with the environment. Typically, IoT nodes are sensors, acting as the primary data acquisition points, and actuators, the interactive components. IoT nodes should only interact with edge or fog nodes, possibly using short-medium range communication protocols to reduce latency, preserve privacy and reduce power consumption.

### 3.1 Services

Figure 1 overviews the designed services of the system. Raw farming data originate from edge nodes, whereas satellite data are sourced from cloud nodes. The primary stakeholders are *farmers* and *water reclamation consortia*, both exercising direct or indirect control over a range of private and public services pertinent to their interests. Other actors involved in the system are public data providers.

The main pertinent activities are the following. Farmers are tasked with identifying and applying precise water quantities to each plot within their operations, optimizing timings and methodologies. This necessitates collaboration with one or multiple consortia to secure access and permissions for utilizing irrigation infrastructure. This activity will be simply called *Precision Irrigation* (PI).

Water reclamation consortia shoulder the responsibility of planning and managing water resources throughout the agricultural season, especially when water is scarce. This mandate demands proactive planning of water allocation based on historical data and seasonal forecasts, and real-time adjustments to adapt to unforeseen and extreme scenarios, ensuring comprehensive and timely monitoring. This second activity will be identified as *Large Scale Irrigation Monitoring.*

Figure 1 highlights a suite of core services, illustrating an interaction scenario with one farmer and one consortium for sake of simplicity. While some services are already operational and publicly available, others must be improved to establish nodes capable of delivering effective services. Moreover, several services are purposely designed to implement a fully operational system.

**Agricultural Decision Support System.** One of the main services is the *Agricultural Decision Support System* (labeled DSS), providing information pertinent to agricultural production and crop monitoring. In this paper, emphasis is placed on the DSS ability to provide specialized irrigation advice tailored to specific plot conditions, leveraging both general and detailed agricultural information such as crop types and field-specific treatments. In Italy, IRRIFRAME [5] is an irrigation advisory service available on a national scale.

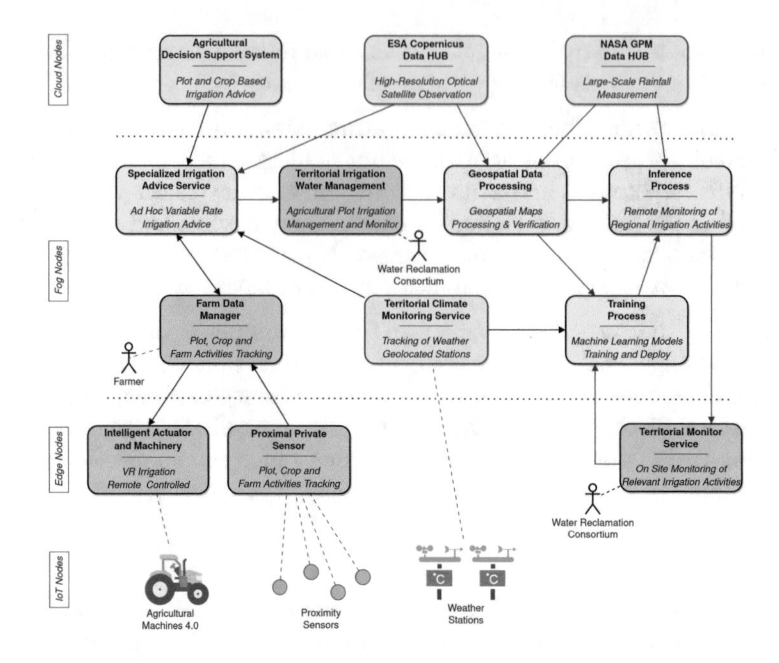


**Fig. 1.** Main services of the proposed E2C system for Precision Irrigation.

In soil-plant moisture monitoring applications, end users - local or regional entities like farmers and water reclamation boards - tend to withhold their data from the cloud. Consequently, these data are routed to and managed within fog nodes. Additionally, to ensure efficient performance, machine learning model training occurs within these fog nodes.

**Big Data Sources.** A range of services has been identified within the system, serving as data sources for the execution of various activities. Specifically, our focus has been on the selection of free, widely accessible services offering global coverage:

- **High-Resolution Optical Satellite Observation of the Earth**, indispensable for open field crops. Notably, the *ESA Copernicus Data HUB*[1] service stands out in this domain.
- **Large-Scale Rainfall Monitoring**, providing historical rainfall data. *NASA GPM Data HUB*[2] service stands out for its rapid and convenient data accessibility.
- **Territorial Climate Monitoring Service**, providing localized and detailed climate information, where available. In authors' experience, its access methods may be less immediate, making it more suited for non-real-time data assimilation.

[1] https://sentinel.esa.int/web/sentinel/missions/sentinel-2.

[2] https://gpm.nasa.gov/missions/GPM.

**Specialized Irrigation Advice Service.** To further refine irrigation advice, a dedicated *Specialized Irrigation Advice Service* becomes imperative, whether integrated within the DSS or as a standalone entity [2]. This specialized service aims to tailor irrigation recommendations based on detailed data points, optimizing water usage for varying agricultural contexts. This information may be derived from remote or proximal observations, enabling the transformation of broad irrigation advice into customized, site-specific recommendations.

**Farm Data Manager.** A Farm Data Manager serves as a centralized repository and orchestrator of agricultural data within the system. Its primary function involves aggregating, organizing, and storing data originating from farm-level sensors. This node should also have access and control of the advanced machinery capable of perform operational activities.

The preferred practice is to empower farmers with control over machine and sensor management services. This approach ensures that data collection and storage happen in close proximity to the entities generating it. By localizing these processes, we mitigate issues related to network congestion and reduce latency, facilitating faster and more efficient data processing. In fact, the *Farm Data Manager* should enable a complete access to historical and real-time farm-specific data, letting stakeholders to make informed decisions. data security, integrity, and accessibility for various stakeholders involved in agricultural operations should be mandatory.

**Services for Irrigation Monitoring.** At the heart of irrigation monitoring lies a series of interconnected services encapsulated under the umbrella term *Services for Irrigation Monitoring.* These services correlate remote sensing data with territorial agricultural information, employing advanced data analytics - such as machine learning (ML) - to track water utilization, distribution efficiency, and soil moisture levels spanning vast agricultural landscapes.

Within this framework, we have identified three main services:

- **Geospatial Data Processing**, with the aim of create a dataset using diverse source, essential for subsequent training procedures.
- **Training Service**, in charge of training a ML model able to classify plots based on spatial information and publicly accessible data.
- **Inference Process**, that utilizes the trained model to operationalize on-site monitoring. Through a service called **Territorial Monitor Service**, the proper consortium can conducts specific checks of unplanned irrigation activities, and provides feedback, enhancing subsequent training.

These tools support water reclamation consortia, enabling comprehensive surveillance and analysis of irrigation systems on a large scale. Real-time monitoring capabilities afford identification of inefficiencies or anomalies, enabling corrective actions to rectify issues as they arise. Conversely, a more comprehensive and in-depth analysis of season-end data empowers long-term planning for subsequent seasons, facilitating decision-making based on historical patterns and trends. This dual approach enables real-time interventions and informed long-term planning for effective irrigation management strategies.

# 4 Data Flow

This section analyzes the data flows within the two primary activities that define the irrigation management ecosystem.

## 4.1 Precision Irrigation

In Fig. 2, a Data Flow Diagram (DFD) is reported mainly concerning PI. With a view to achieving profitable PA practices, farmers should register their plot-specific data, recording daily crop management operations, including irrigation, within their farm information systems. Those data are essential to achieve a precise irrigation outcome, allowing to apply the correct amount of water (according to DSS models) at the correct time for each crop. Moreover, a Specialized Irrigation Advice Service (Fig. 1) must offer information leveraging the highest available precision. In fact, farms are increasingly adopting intelligent sensors and actuators, establishing a virtuous data flow. This commences with sampling specific land areas, followed by the customization of generic outputs - typically produced by DSS - based on the received site-specific information. Subsequently, this leads to the generation of machine-specific planning for intelligent actuators.

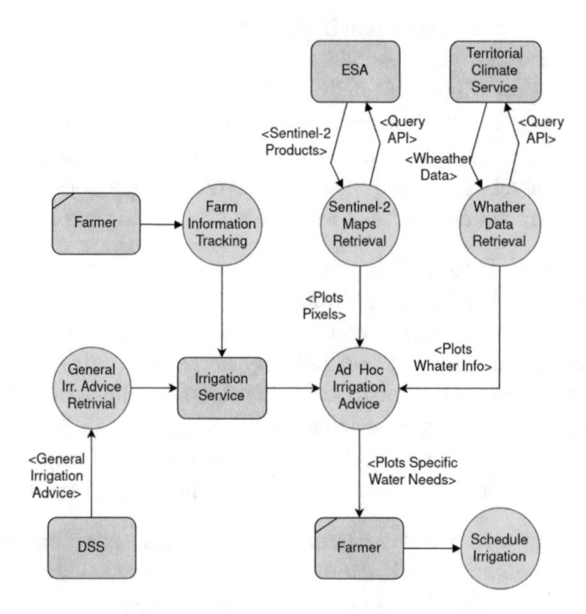


**Fig. 2.** Data flow diagram of the system for Precision Irrigation. Agents are highlighted in orange and main processes in blue.

These actuators are endowed with the capability to execute PI, enabling irrigation management based on localized data inputs.

In our conceptualization, the Irrigation Service entity (referenced in Fig. 2) aims to enable PI by integrating three primary data sources: geographical and agricultural plot data, remote space-varied observations, and territorial meteorological data. In the system design we emphasizes the data supplied by ESA Copernicus program, that are cost-effective (data are free to use), have large scale coverage, and high multispectral resolution. These features make them a valuable component in enabling PI implementation, despite some limitations (availability only for open fields, without cloud cover, etc.).

More specialized data source alternatives, such as proximal sensor networks or highly detailed drone observations, can be used for refining PI capabilities within the Irrigation Service. Such solutions require data provision by the farmers themselves, who retain ownership of these data. However, these fine grain approaches pose challenges in terms of uneven adoption and cost for small/medium size farms, making them less universally accessible compared to remote sensing.

The various entities can maintain a history of their data and provide it out when necessary, but in the DFD of Fig. 2 no external data stores have been highlighted. In fact, this activity can be carried out on-demand, i.e., given a certain irrigation advice it can be specialized on the basis of spatially varied observations of the plot and other data that can be collected with a single request, without the need for explicit data accumulation by the entity in charge of the Specialized Irrigation Advice Service.

### 4.2 Large Scale Irrigation Monitoring

In Fig. 3, a DFD concerning Large Scale Irrigation Monitoring is reported. This DFD does not involve the activities carried out by the DSS and Territorial Climate Monitoring Service for achieving a variable irrigation advice rate and therefore for actual irrigation activities. For a better comprehension, the DFD can be subdivided into three sub-activities.

The first one involves **plots and irrigation monitoring**. Each consortium should have knowledge of the cropped areas under its responsibility before the season starts. However, practical challenges often hinder this scenario (e.g., no communications regarding cultivated areas, meteorological unpredictability, extreme events). To face these challenges, a real-time monitoring service is required. This service should ideally operate independent of farm-specific information sources (e.g., a farm could have or not sensors in its plots), ensuring adaptability and responsiveness despite potential delays or inaccuracies in data arrival. Throughout the irrigation period, consortium and farmers must communicate about water usage, in order to monitor irrigation activities. Typically, this data collection relies on online forms filled out by individual farmers - a standard practice for both farm management and administrative purposes within the agricultural domain. Collaboration between farmers and water reclamation consortia is important in facilitating the acquisition of water from primary sources

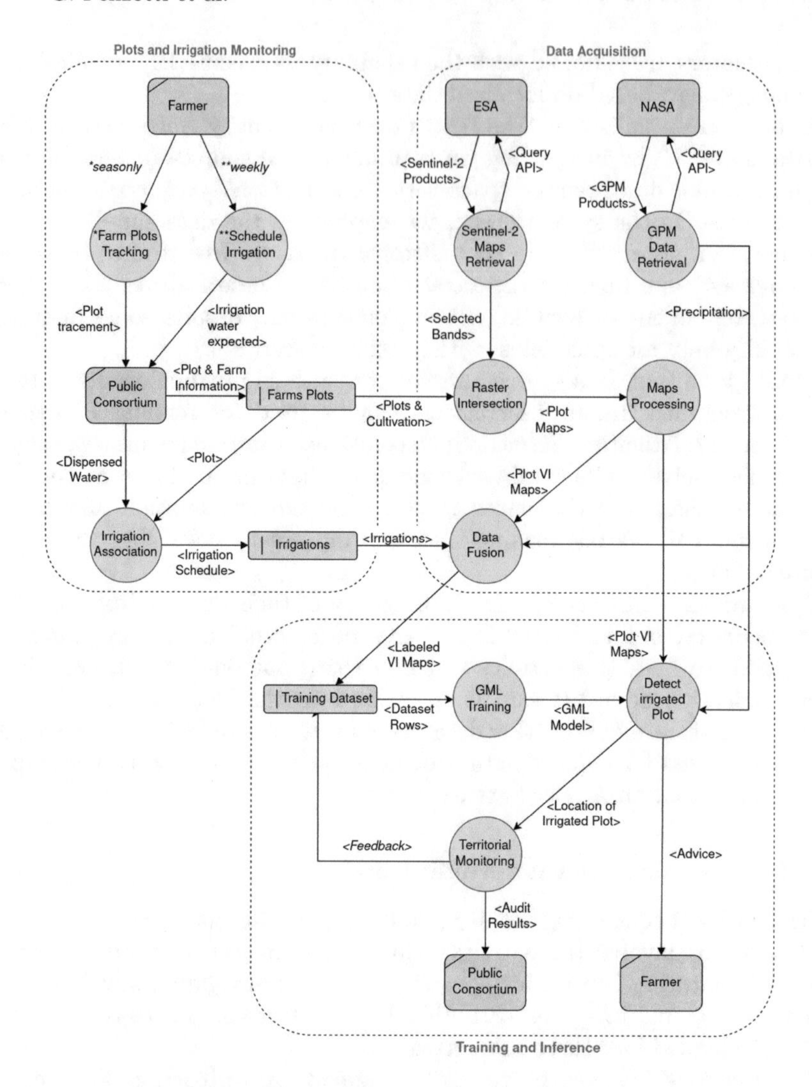


**Fig. 3.** Data flow diagram of the system for Large Scale Irrigation Monitoring. External agents are highlighted in orange, main processes in blue, data stores in red.

like irrigation canals, exemplified in the Emilia-Romagna region (Italy), where farmers submit specific claims for irrigation water with the relevant water reclamation consortium. In the proposed DFD, some data stores for data collection are highlighted. *Farm data* encapsulates essential information such as plot locations, ongoing crops and soil characteristics. Farms should employ personal Farm Data Manager. Usually, these nodes are kept private to ensure data security and integrity. *Irrigations* data store assumes a critical role for irrigation consortia, tracking the water supply planned for farms. Interactions and data exchanges between farms and consortia occur with varying frequencies, potentially spanning thousands of plots. These interactions necessitate correlation, associating irrigation (including date and water usage) with the information of the relevant plots (e.g., geographical position and crop type). Moreover, while irrigation data undergo weekly updates by irrigation consortia, plot data may be managed by other public entities, such as cadastral agencies, operating on seasonal or annual schedules for updates and maintenance.

The second sub-activity is **data acquisition**, which is executed within the system through the acquisition of the datasets maintained by consortia, alongside the measurements obtained by remote sensing (Sect. 3.1). The Sentinel-2 satellites capture multispectral images with resolution up to 10 m every 2–5 days. Accessing Sentinel-2 data involves querying the Copernicus hub API to acquire a historical and new products. The GPM mission provides global precipitation measurements with a 12-hour latency, accessible through platforms like the NASA GES DISC interface. The processing pipeline involves a series of tasks. Multispectral satellite maps intersect with farm plots, extracting pixels carrying information for each relevant band. These pixels undergo validation, ensuring absence of zero values, minimal cloud cover (below predefined thresholds), and no outliers. Subsequently, information such as vegetation indices (e.g., NDVI) is computed to construct datasets supporting periodic training of Geospatial Machine Learning models (GML). Furthermore, the association of plots with corresponding data, including irrigation and rainfall events, contributes to the enriched dataset aiding in comprehensive analysis and predictive modeling.

The last sub-activity is **training and inference**. The dataset created serves as the foundation for training GML models within a dedicated *Training Process*, which receives updates triggered by specific events: new irrigation instances, availability of fresh satellite observations, or receipt of new feedback. To ensure efficiency, methodologies emphasizing continuous model updates with minimal disruption, such as *Continuous Learning* approaches, can be employed. The goal is to produce a model enabling *Inference Process* capable of discerning the irrigation status of a field using satellite observations and rainfall data. Beyond detection, the model can provide finer insights, potentially including a quantitative estimation of the volume of water utilized. The primary objective of the Inference Process is to track irrigation activities, serving stakeholders - primarily farmers - to plan future operations by factoring in historical rainy events. Additionally, it aids public consortia in executing effective territorial monitoring and preempting improper water usage. To facilitate this process a *Territorial Monitor Service* is designed to gather direct feedback. This feedback mechanism operates either randomly or in response to alerts triggered by discrepancies between detected irrigation activity and the absence of corresponding requests or approvals.

## 5 Conclusion

We have described an E2C system for Precision Irrigation practices, enabling soil-plant moisture detection and classification. The overall system architecture is organized in a service hierarchy structured around diverse characteristics, outlining distinct layers along the E2C path. Data flow diagrams for both main activities, Precision Irrigation and Large Scale Irrigation Monitoring, have been presented and discussed. Precision Agriculture, with its inherent need of multiple data sources at different levels for smart and efficient operation, appears indeed the perfect yardstick where advanced E2C concepts must be leveraged upon.

**Acknowledgment.** Research carried out within Agritech Nat. Res. Center, funded by NextGenerationEU (PNRR, Mission 4, Component 2, Investment 1.4 - D.D. 1032 17/06/2022, Code CN00000022, CUP D93C22000420001).

## References

1. Alharbi, H.A., Aldossary, M.: Energy-efficient edge-fog-cloud architecture for IoT-based smart agriculture environment. IEEE Access **9**, 110480–110492 (2021)
2. Amoretti, M., Lodi Rizzini, D., Penzotti, G., Caselli, S.: A scalable distributed system for precision irrigation. In: Proceedings of the IEEE International Conference on Smart Computing (SMARTCOMP) (2020)
3. Balouek-Thomert, D., Renart, E.G., Zamani, A.R., Simonet, A., Parashar, M.: Towards a computing continuum: enabling edge-to-cloud integration for data-driven workflows. Int. J. High Perform. Comput. Appl. **33**(6), 1159–1174 (2019)
4. COM/EdgeCloud-SC: IEEE 1934-2018 - IEEE Standard for Adoption of OpenFog Reference Architecture for Fog Computing (2018)
5. Mannini, P., Genovesi, R., Letterio, T.: IRRINET: large scale DSS application for on-farm irrigation scheduling. Procedia. Environ. Sci. **19**, 823–829 (2013)
6. Milojicic, D.: The edge-to-cloud continuum. Computer **53**(11), 16–25 (2020)
7. Montoya-Munoz, A.I., Rendon, O.M.C.: An approach based on fog computing for providing reliability in IoT data collection: a case study in a Colombian coffee smart farm. Appl. Sci. **10**(24) (2020)
8. Penzotti, G., Caselli, S., Amoretti, M.: An N-tier fog architecture for smart farming. In: IEEE Symposium on Computers and Communications (ISCC) (2021)
9. Rosendo, D., Costan, A., Valduriez, P., Antoniu, G.: Distributed intelligence on the edge-to-cloud continuum: a systematic literature review. J. Parallel Distrib. Comput. **166**, 71–94 (2022)
10. Tsipis, A., Papamichail, A., Koufoudakis, G., Tsoumanis, G., Polykalas, S.E., Oikonomou, K.: Latency-adjustable cloud/fog computing architecture for time-sensitive environmental monitoring in olive groves. AgriEngineering **2** (2020)
11. Wolfert, S., Ge, L., Verdouw, C., Bogaardt, M.J.: Big data in smart farming - a review. Agric. Syst. **153**, 69–80 (2017)
12. Yogeswaranathan Kalyani, N.V.B., Collier, R.: Digital twin deployment for smart agriculture in cloud-fog-edge infrastructure. Int. J. Parallel Emerg. Distrib. Syst. **38**(6), 461–476 (2023)

# Performance Evaluation of Placement Policies for Cloud-Edge Applications

Ivan Mongiardo[1(✉)], Luisa Massari[1], Maria Calzarossa[1], Belen Bermejo[2], and Daniele Tessera[3]

[1] Department of Electrical, Computer and Biomedical Engineering, University of Pavia, Pavia, Italy
ivangiuseppe.mongiardo01@universitadipavia.it, {luisa.massari,mcc}@unipv.it

[2] Department of Computer Science, University of Balearic Islands, Palma, Spain
belen.bermejo@uib.es

[3] Department of Mathematics and Physics, Catholic University of Sacred Heart, Brescia, Italy
daniele.tessera@unicatt.it

**Abstract.** The applications commonly deployed in cloud-edge environments consist of multiple inter-dependent modules organized according to the Distributed Data Flow model. Decisions about the mapping between these modules and the available resources are quite difficult because of the resource-constrained nature of devices at the edge of the network and the timing requirements of the applications. In this paper we investigate the problem of application placement by proposing a lightweight heuristic that takes into account the volume of data exchanged between modules. In detail, among all possible devices, the proposed policy allocates a given module to the "best" device, that is, the device that ensures the minimum network delay. The policy has been evaluated in a simulated cloud-edge environment based on the iFogSim toolkit. We consider workloads consisting of applications with different demands in terms of processing and data exchanged between modules. The simulation results are promising and indicate that our policy offers competitive advantages when compared to other heuristics.

## 1 Introduction

The growing demand of smart pervasive applications capable of monitoring, filtering and analyzing in a timely manner the large volume of data flows produced by IoT devices opens new challenges. To ensure low and predictable delays to these applications, it is necessary to employ computation resources in the proximity of the data sources, that is, at the edge of the network [5]. Nevertheless, due the limited processing capacity of these resources, a seamless integration of cloud resources is required, thus a major driver towards the development of these innovative solutions is represented by the cloud-edge continuum [9]. These solutions strongly depend on the characteristics of the workloads being processed [6,7].


L. Barolli (Ed.): AINA 2024, LNDECT 203, pp. 287–296, 2024.
https://doi.org/10.1007/978-3-031-57931-8_28

In this paper, we explore the issues related to application placement in the cloud-edge continuum. In particular, we consider applications consisting of multiple inter-dependent lightweight modules organized according to the Distributed Data Flow (DDF) model and an infrastructure consisting of multiple resources organized in hierarchical layers. We propose a lightweight placement policy that takes into account the volume of data exchanged between modules and allocates a given module to the "best" resource, that is, the resource that ensures the minimum network delay. The experimental results have shown that our heuristic generally outperforms state-of-the-art heuristics, although its performance mainly depends on the characteristics of the applications being placed.

The rest of the paper is organized as follows. In Sect. 2 we discuss the state of the art in the area of application placement. In Sect. 3 we describe the proposed heuristic, while in Sect. 4 we present the experimental setup and results. In Sect. 5 we summarize the main outcomes and outline possible research directions.

## 2 Related Work

This section presents a brief review the state-of-the-art regarding placement heuristic in the cloud-edge continuum. During the last years, several reviews have been performed (e.g., [2,4,12,14,16,17]). These reviews covered various aspects, such as types of IoT application models and strategies for resource allocation, service placement problem, optimal application placement, architectural maintenance and AI-based heuristics.

In the context of placement heuristics, most papers focus on the minimization of various objectives, such as application execution time, resource usage, network latency, energy consumption. To achieve these objectives, different placement techniques have been proposed, mainly applied to fog/edge/cloud architecture. For example, in [18] authors proposed a module mapping algorithm for efficient resource usage in fog/cloud environments. The algorithm considers modules and devices sorted in ascending order with respect to their requirements and capacity, respectively. The mapping of modules to devices follows this order starting from the lightest module that is placed on the least powerful device. Similarly, the algorithm for module placement presented in [3] focuses on reducing the total network usage and the application execution time. According to this algorithm, the shortest modules in terms of processing requirements are placed on the available fog device located in the bottom-most layer.

A different placement approach was presented in [13] where a latency-aware module placement in distributed fog environments was proposed. This approach aims at satisfying service delivery deadlines for latency-sensitive and latency-tolerant applications. Modules can be placed either horizontally or vertically depending on the characteristics of the applications. Interdependent application modules characterized by a high data interaction rate were also considered in [8], where authors proposed a method that tries to place as many modules as possible in the same fog node with the aim of minimizing the network delay.

Other placement heuristics focus on aspects dealing with the impact of the communication requirements of the applications. This is the case of the

Communication-based policy proposed in [15] where the placement is based on the minimization of the data volume transferred by each module. More precisely, the modules with the minimum communication impact are placed on the highest layer of the cloud-fog computing infrastructure, thus reducing the network impact on the application delay.

A QoS-aware Greedy-edge placement scheme aimed at minimizing the end-to-end latency of real-time IoT applications in fog environments was proposed in [1]. The approach first assigns to each module a score obtained by combining its processing requirement with a priority that is set starting from the first module. Modules are then sorted in descending order according to their score. Similarly, the fog devices are sorted in ascending order according to their proximity to the IoT layer. Modules with the highest score are placed first by choosing the closest fog node that satisfies their requirements.

This policy, together with the Communication-based policy previously described, will be used as baselines in our experiments because they consider aspects similar to ours.

As a conclusion, to the best of our knowledge, it is important to devise a lightweight heuristic that considers complementary aspects dealing with the data volume exchanged between the application modules together with the processing capacity of various devices.

## 3 Placement Problem

The main goal of placement policies is to identify the most appropriate mapping between application modules and devices. These policies are usually driven by different objectives, e.g., minimizing communication delays, reducing the number of allocated devices, ensuring performance fairness across applications. In what follows we describe the application and architecture models considered in this study and the proposed placement policy.

### 3.1 Application and Architecture Models

We consider applications consisting of multiple inter-dependent modules organized according to the Distributed Data Flow model [10]. In this context, modules communicate by pushing and pulling data, thus triggering the data flows. More precisely, a module receives requests for processing the inputs generated by upstream modules and for producing outputs to be stored locally or to be sent to downstream modules. The only exceptions are represented by the modules at the start and at the end of the flow which only produce outputs or consume inputs, respectively.

Figure 1(a) shows an example of an application consisting of seven modules, including the so-called client module which receives the data being sensed and triggers actuators by sending the results of the processing.

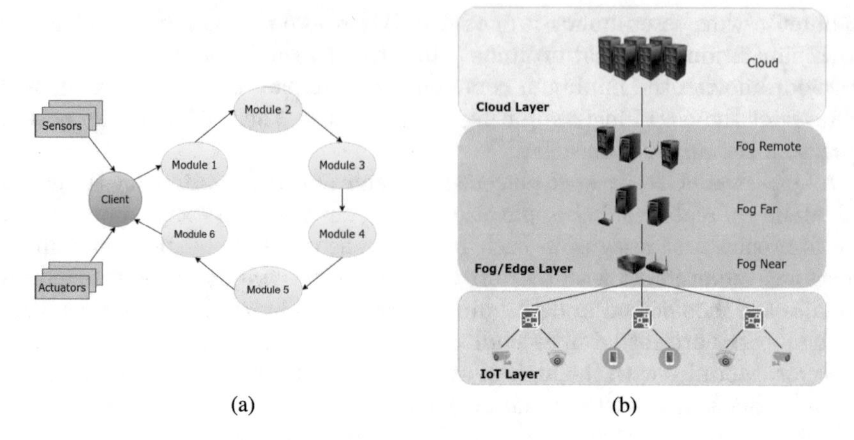


**Fig. 1.** Models of the applications (a) and of the cloud-edge continuum (b) considered in this study.

We model the cloud-edge continuum as a hierarchical architecture composed of three layers, namely, IoT, fog/edge and cloud layers. An example of such an architecture is sketched in Fig. 1(b). As can be seen, the fog/edge layer is characterized by a hierarchy of resources whose latency and processing capacity increase as the proximity from the IoT devices decreases.

### 3.2 Placement Policy

Our placement policy aims at minimizing the network delays over a set of applications by considering the network proximity of the upstream and downstream modules of a given application and placing one module at a time. This minimization problem is addressed by a lightweight heuristic. In fact, unlike optimization problems, heuristics are typically characterized by small computation requirements, thus they can provide timely placement decisions.

For our policy, we assume that placement starts from the first module of each application and proceeds one module at a time until the last module. We also assume that there are no network delays whenever a downstream module is placed in the same device of its upstream counterpart.

For each application the proposed policy focuses on pairs of modules (i.e., a module and its downstream counterpart) and searches for devices able to cope with the processing requirements of both. Under the assumption that module $i-1$ is placed in device $k$, the policy computes the network delays $D^i_{k,q}$ associated with the previously identified devices as:

$$D^i_{k,q} = L_{k,q} + \frac{V_{i-1,i}}{B_{k,q}}$$

where $L_{k,q}$ and $B_{k,q}$ denote the network latency and bandwidth between devices $k$ and $q$, while $V_{i-1,i}$ the volume of output data sent by module $i-1$ to module $i$.

The device with the minimum network delay is chosen as a possible candidate for placing modules $i$ and $i+1$.

According to our heuristic, module $i$ is placed on the identified device. This device might have insufficient processing capacity when module $i+1$ has actually to be placed, thus a re-evaluation of the placement decision has to be performed. In fact, placement decisions follow a rotation scheme across applications without implementing any resource reservation mechanism. For example, the first modules of all applications are placed before considering the second modules. As a consequence, these re-evaluations will increase the network delays because it will be necessary to place the module in a different device.

In summary, the placement decisions taken by the proposed heuristic are based on two complementary aspects dealing with the network delay that might be experienced by individual modules as well as the processing capacity to be provided in the cloud-edge continuum. These decisions ensure fairness across applications because of the rotation scheme.

## 4 Experimental Results

To evaluate the proposed policy, we performed several experiments in a simulated cloud-edge environment based on the iFogSim toolkit [11]. The simulated environment consists of devices organized according to the layers displayed in Fig. 1(b) whose characteristics are summarized in Table 1. Processing capacity refers to individual devices, whereas bandwidth and latency refer to adjacent devices. For example, the fog far device can exploit a bandwidth equal to 25 Mbps to communicate with the fog near device, whereas the bandwidth between the cloud and the fog remote device is equal to 10 Mbps. The gateway is dedicated to host the client modules only, thus it will not be considered in the placement decisions.

**Table 1.** Characteristics of the architecture considered in the experiments.

| | Processing Capacity [MIPS] | Bandwidth [Mbps] | Latency [ms] |
|---|---|---|---|
| Cloud | 400,000 | 10 | 70 |
| Fog Remote | 25,000 | 20 | 30 |
| Fog Far | 20,600 | 25 | 8 |
| Fog Near | 10,600 | 30 | 2 |
| Gateway | 1,000 | 5 | 1 |

To test the proposed policy, we initially focus on a toy application composed of six modules plus the client module whose characteristics have been chosen such as the policy needs to perform multiple re-evaluations. These characteristics are presented in Table 2. The output data volume refers to data exchanged by a

**Table 2.** Resource requirements of the modules of the toy application.

| | Processing Requirement [MIPS] | Output Data Volume [KBytes] |
|---|---|---|
| Client | 200 | 100 |
| Module 1 | 1,000 | 100 |
| Module 2 | 2,000 | 50 |
| Module 3 | 2,500 | 80 |
| Module 4 | 200 | 80 |
| Module 5 | 500 | 50 |
| Module 6 | 200 | 20 |

module with its downstream counterpart. Note that the client module is the downstream module of Module 6.

The performance of our policy as a function of the number of applications being placed is shown in Fig. 2. Performance refers to the network delay computed as an average over the applications being placed. The figure plots the network delays experienced by applications placed according to our policy and to two state-of-the-art policies, namely, Communication-based and Greedy-edge (see Sect. 2). For comparison purposes, we also include the results obtained with a random placement. As expected, the network delay increases with the number of applications, the only exception is the random placement whose average delay decreases because of the fewer available resources to choose from. In general, our policy outperforms the others even significantly, despite the re-evaluations that might be necessary because of the insufficient processing capacity on the device previously identified. For example, in these experiments the fraction of these re-evaluations reaches the 46% of the placement decisions.

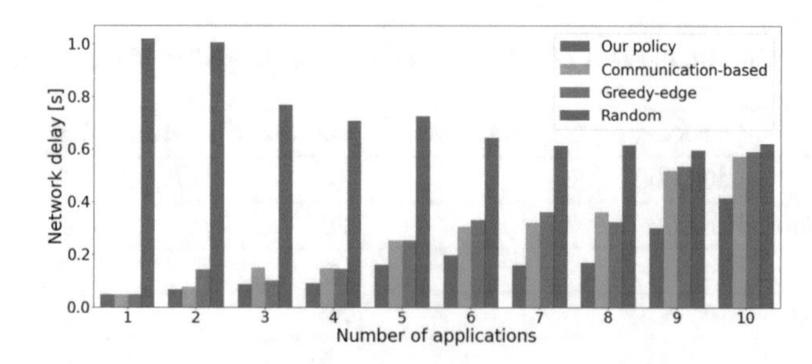


**Fig. 2.** Average network delays of the applications as a function of the placement policies and of the number of applications being placed.

To further assess the performance of our policy, we consider bigger applications, i.e., consisting of ten modules, whose processing requirements and output data volume are sampled from uniform distributions. Figure 3 displays the average values of the module characteristics together with the corresponding confidence intervals computed at the 95% confidence level.

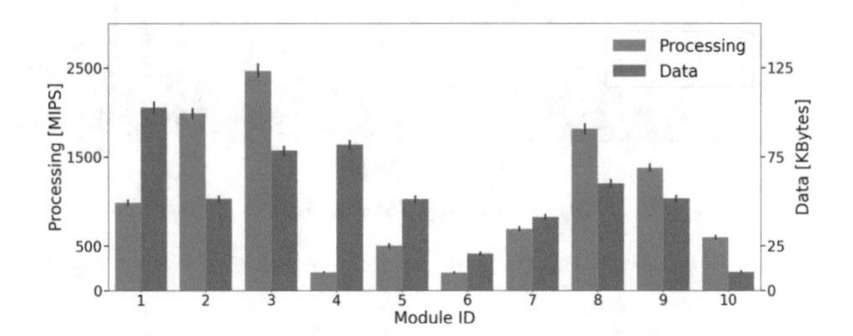


**Fig. 3.** Average processing requirement and output data volume of each module. Error bars represent the 95% confidence intervals.

The network delays as a function of the number of applications being placed for our policy and the Communication-based policy are shown in Fig. 4. Similarly to the previous experiments, these results suggest that our policy outperforms the other. The figure also shows the breakdown of the network delays as a function of cloud-edge continuum layers. Unlike our policy, the Communication-based policy tends to place modules across layers, thus it is affected by significant delays mainly between the fog/edge and cloud layers.

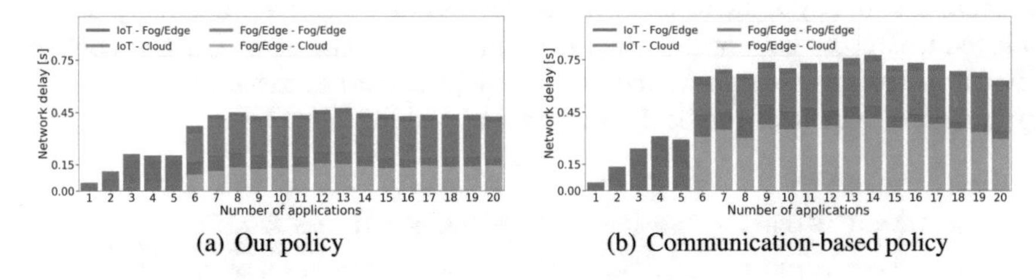


**Fig. 4.** Breakdown of the network delays as a function of the cloud-edge continuum layers and of the number of applications being placed.

Another set of experiments focuses on assessing the impact of the policies on the network delay when changing the communication patterns. We consider an application with ten modules exchanging, in total about 5.6 MBytes of data, ranging from 50 KBytes up to 1 MBytes per module. Figure 5 shows the results of these experiments where data volume decreases or increases with the modules, starting from the first one. The figure suggests that, unlike our policy, the start-of-the-art policies considered in this paper are affected by communication patterns, sometimes even considerably.

Finally, we analyze in Fig. 6 the mappings between the modules of ten identical applications and the cloud-edge devices obtained when applications are

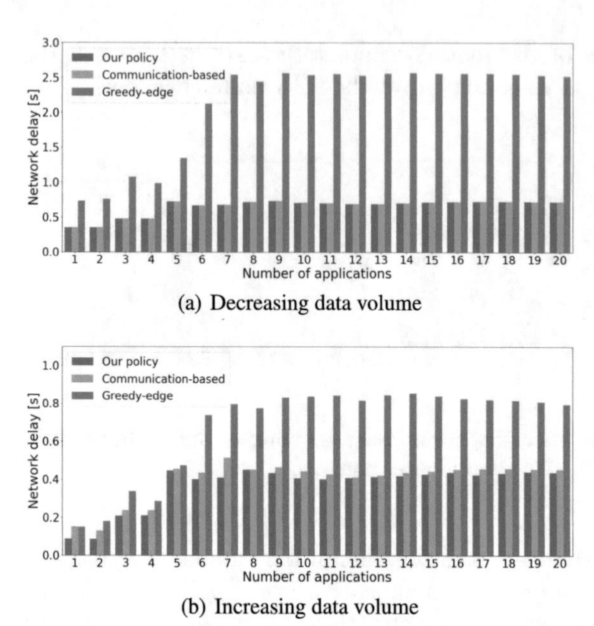


(a) Decreasing data volume

(b) Increasing data volume

**Fig. 5.** Network delays, as a function of the number of applications, with decreasing (a) and increasing (b) data volume patterns.

considered all together or one by one starting from their first module. As can be seen, our policy tries to exploit the devices of all layers for all applications independently of their order, thus ensuring fair placements. On the contrary, when applications are considered one by one, their order matters, thus only few applications benefit of the devices located at the fog/edge layer.

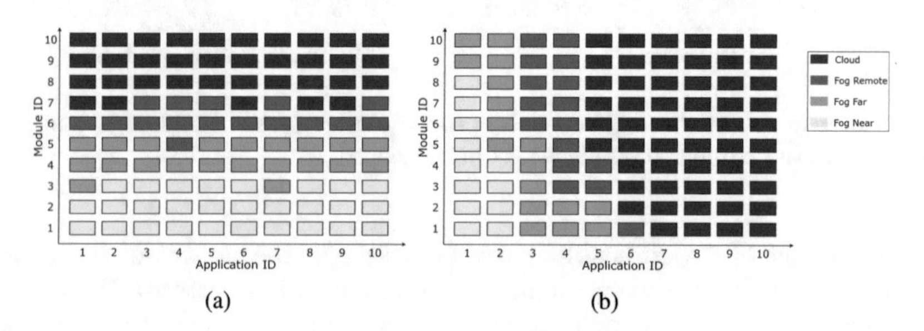


(a) (b)

**Fig. 6.** Mappings between modules and devices obtained with the proposed policy by considering applications all together (a) and one by one (b).

# 5 Conclusion

In this paper we proposed a lightweight placement policy that considers complementary aspects, that is, the network delay experienced by individual application modules and the processing capacity provided in the cloud-edge continuum. This heuristic generally outperforms state-of-the-art policies. In addition, our experiments have shown that, to minimize the network delays, placement decisions should adapt to the resource requirements of the applications, e.g., communication patterns.

As future research directions, we plan to further analyze the performance of the proposed policy using different mixes of heterogeneous applications. In addition, we will study more sophisticated policies that easily adapt to application requirements and to the cloud-edge continuum characteristics.

**Acknowledgments.** This work was partly supported by the Italian Ministry of University and Research (MUR) under the PRIN 2022 grant "Methodologies for the Parallelization, Performance Evaluation and Scheduling of Applications for the Cloud-Edge Continuum" (Master CUP: B53D23013090006, CUP: J53D23007110008, CUP: F53D23004300006) and by European Union - Next Generation EU.

# References

1. Abu-Amssimir, N., Al-Haj, A.: A QoS-aware resource management scheme over fog computing infrastructures in IoT systems. Multimed. Tools Appl. **82**(18), 28281–28300 (2023)
2. Apat, H., Nayak, R., Sahoo, B.: A comprehensive review on Internet of Things application placement in fog computing environment. Internet Things **23**, 100866 (2023)
3. Arora, U., Singh, N.: IoT application modules placement in heterogeneous fog-cloud infrastructure. Int. J. Inf. Technol. **13**(5), 1975–1982 (2021)
4. Bermejo, B., Juiz, C.: Improving cloud/edge sustainability through artificial intelligence: a systematic review. J. Parallel Distrib. Comput. **176**, 41–54 (2023)
5. Bonomi, F., Milito, R., Zhu, J., Addepalli, S.: Fog computing and its role in the Internet of Things. In: Proceedings of the First Workshop on Mobile Cloud Computing, MCC, pp. 13–16. Association for Computing Machinery (2012)
6. Calzarossa, M.C., Della Vedova, M.L., Massari, L., Petcu, D., Tabash, M.I.M., Tessera, D.: Workloads in the clouds. In: Fiondella, L., Puliafito, A. (eds.) Principles of Performance and Reliability Modeling and Evaluation. SSRE, pp. 525–550. Springer, Cham (2016). https://doi.org/10.1007/978-3-319-30599-8_20
7. Calzarossa, M.C., Massari, L., Tessera, D.: Workload characterization: a survey revisited. ACM Comput. Surv. **48**(3), 48:1–48:43 (2016)
8. Dadashi Gavaber, M., Rajabzadeh, A.: BADEP: bandwidth and delay efficient application placement in fog-based IoT systems. Trans. Emerg. Telecommun. Technol. **32**, e4136 (2021)
9. Esposito, A., et al.: Methodologies for the parallelization, performance evaluation and scheduling of applications for the cloud-edge continuum. In: Barolli, L. (ed.) AINA 2024. LNDECT, vol. 203, pp. 254–263. Springer, Cham (2024). https://doi.org/10.1007/978-3-031-57931-8_25

10. Giang, N., Blackstock, M., Lea, R., Leung, V.: Developing IoT applications in the fog: a distributed dataflow approach. In: Proceedings of the 5th International Conference on the Internet of Things, IOT, pp. 155–162 (2015)
11. Harshit, G., Dastjerdi, A., Ghosh, S., Buyya, R.: iFogSim: a toolkit for modeling and simulation of resource management techniques in the Internet of Things, edge and fog computing environments. Softw.: Pract. Exp. **47**, 1275–1296 (2017)
12. Islam, M.M., Ramezani, F., Lu, H.Y., Naderpour, M.: Optimal placement of applications in the fog environment: a systematic literature review. J. Parallel Distrib. Comput. **174**(C), 46–69 (2023)
13. Mahmud, R., Ramamohanarao, K., Buyya, R.: Latency-aware application module management for fog computing environments. ACM Trans. Internet Technol. **19**(1) (2018)
14. Mahmud, R., Ramamohanarao, K., Buyya, R.: Application management in fog computing environments: a taxonomy, review and future directions. ACM Comput. Surv. **53**(4) (2020)
15. Peixoto, M.L.M., Genez, T.A., Bittencourt, L.F.: Hierarchical scheduling mechanisms in multi-level fog computing. IEEE Trans. Serv. Comput. **15**(5), 2824–2837 (2021)
16. Salaht, F., Desprez, F., Lebre, A.: An overview of service placement problem in fog and edge computing. ACM Comput. Surv. **53**(3) (2020)
17. Smolka, S., Mann, Z.Á.: Evaluation of fog application placement algorithms: a survey. Computing **104**(6), 1397–1423 (2022)
18. Taneja, M., Davy, A.: Resource aware placement of IoT application modules in fog-cloud computing paradigm. In: Proceedings of the IFIP/IEEE Symposium on Integrated Network and Service Management, IM, pp. 1222–1228. IEEE (2017)

# Combining Federated and Ensemble Learning in Distributed and Cloud Environments: An Exploratory Study

Fiammetta Marulli(✉), Lelio Campanile, Stefano Marrone, and Laura Verde

Department of Maths and Physics, Università della Campania "L. Vanvitelli", Caserta, Italy
{fiammetta.marulli,lelio.campanile,stefano.marrone, laura.verde}@unicampania.it

**Abstract.** Conventional modern Machine Learning (ML) applications involve training models in the cloud and then transferring them back to the edge, especially in an Internet of Things (IoT) enabled environment. However, privacy-related limitations on data transfer from the edge to the cloud raise challenges: among various solutions, Federated Learning (FL) could satisfy privacy related concerns and accommodate power and energy issues of edge devices. This paper proposes a novel approach that combines FL and Ensemble Learning (EL) to improve both security and privacy challenges. The presented methodology introduces an extra layer, the Federation Layer, to enhance security. It uses Bayesian Networks (BNs) to dynamically filter untrusted/unsecure federation clients. This approach presents a solution for increasing the security and robustness of FL systems, considering also privacy and performance aspects.

**Keywords:** Federated Learning · Cloud Computing · Edge Computing · Data Privacy · Bayesian Ensemble Learning

## 1 Introduction

Modern and pervasive Machine Learning (ML)-based applications require a lot of data to collect, often by Internet of Things (IoT) sensing and monitoring systems. The final aim — in particular in edge computing — is to train a ML model in the cloud and then to transfer the model to the edge again.

This simple mechanism has to face some issues as the presence of privacy-related restriction on the transfer of the data from the edge to the cloud. The scientific literature detects some mechanisms to cope with this problem, among them: homomorphic encryption [19], Federated Learning (FL) [22] and blockchains [9]. Among these approaches, only federated learning can cope with the limited computational power and energy autonomy of the embedded devices constituting edge-based systems. This notwithstanding, due to its distributed nature, FL algorithms allow attackers to mine both the privacy of data and the integrity of trained models.

The aim of this paper is to define a method able to cope with both security and sustainability ML in the edge computing environments. Such an aim is


L. Barolli (Ed.): AINA 2024, LNDECT 203, pp. 297–306, 2024.
https://doi.org/10.1007/978-3-031-57931-8_29

pursued by combining FL and Ensemble Learning (EL) — another widespread ML technique, based on the combination of different trained models.

More concretely, the objectives of this paper are:

- to propose an approach combining FL and EL, able to cope with security and performance issues;
- to define a technique based on the adoption of Bayesian Networks (BNs) dynamically filtering untrusted federation clients;
- to overcome the problem of returning an ensemble model to the edge by integrating existing model averaging techniques.

The paper is based on the following previous studies: in [10] a mechanism based on BNs evaluating the trustworthiness of several off-the-shelf sensors is introduced; in [6], a variant of the same mechanism is applied to a special case of ensemble methods, Random Forests (RFs), introducing the Reputation Oriented Random Forest (RORF) method. The present paper frames the basic BN-based techniques into a wider FL-based approach for edge computing. Another important original contribution of the paper is the introduction of an extra layer with respect to the classical FL schemas; this layer enables the mitigation of security issues, while the integration with existing model averaging approaches guarantees the generation of a small and lightweight (global) model.

The rest of the paper is organized as follows: Sect. 2 clearly states the problem, describing the security issues to deal with. Section 3 reviews the scientific literature. Section 4 describes the overall approach and supporting architecture. Section 5 gives technical details and describes the process of (local) learning, (global) merging and transmitting the trained model back to the edge. Section 6 ends the paper, discussing current development and future research steps.

## 2 Motivation

FL has attracted in these years the interest of scholars and practitioners. With the increasing diversification and complexity of the local nodes concerns as mutual trust, efficiency, and convergence quality have become an essential prerequisite, while in practical applications, FL needs to solve the heterogeneity of devices in storage, computing and communication capabilities in different local application environments. In [18], several requirements are identified as essential to guarantee a sustainable edge-level ML-environment. Among them:

- keeping small the model size, by using as few trainable parameters as possible;
- keeping low the inference time and energy, by minimizing the number of computations.

In addition to the operational requirements, FL is affected by more security issues than traditional centralised approaches due to the increase in the attack surface[13]. In this paper, a survey of the possible security attacks and countermeasures in FL is proposed. Potential security attacks scenarios are considered, referring to each stage in which a FL system is acting. They can be summarized as follows:

- *privacy inference attacks*: they treat the privacy of the federation participant by inferring sensitive-information; by intercepting the communication between the federates and the server, an attacker can sniff the $\Delta W$ sent by the federated to the server and infer, using reconstruction techniques as in [11], information about the data used by the federated to train the proposed model update;
- *poisoning attacks*: they try to alter the model update $\Delta W$ sent by each federate; it attempts to alter the model update itself (model poisoning) or the data used by the federated to compute the model update (data poisoning);
- *evasion*: aiming to cheat trained model with noisy input data, able to generate erroneous values, very far from the non-noised input.

In [13], the poisoning attack has Robustness Aggregation as a countermeasure. The approach presented in this paper can be framed in this context: in [13], one of the founding techniques of the Robustness Aggregation is weighted averaging, or more complex approaches, clustering the model updates and excluding the updates that have a large impact on the trained model (e.g., see [8]).

Let us now formalise the problem by supposing $n$ working devices, each of whose has its own local data. For the sake of simplicity, let us consider a simple classification problem where, according to $k$ features $f_1, f_2, \ldots, f_k$, a sample is classified in $m$ possible classes $c_1, c_2, \ldots, c_m$. Hence, the model considered in this paper is such a classifier, which is trained locally with own data by each federated and that the server should improve according to the different "update requests"coming from the participants.

## 3 Related Work

FL offers a promising collaborative approach that allows many participants to jointly construct a shared model without revealing their individual training data. A critical challenge in this context is to ensure adversarial robustness, where malicious participants cannot undermine the overall integrity of the system [1,2,16,17]. An effective aggregation strategy is critical to the security of FL systems, helping to mitigate the risks of tampering, adversarial attacks, and privacy breaches during the collaborative learning process.

Several approaches exist in the literature to improve the security of FL systems. An example of FL framework was proposed in [14]. This system prioritizes privacy by incorporating homomorphic encryption and provides a mechanism for the server to penalize malicious participants by effectively extracting logarithmic function gradient data. An adaptive federated aggregation approach was adopted to mitigate poisoning attacks. This method assesses the reliability of users using coordinated medians as a reference point and dynamically adjusts the weights of their gradients accordingly. Malicious gradients were filtered using a logarithmic function. In [3], a different aggregation algorithm was, instead, proposed. It adopts a Bayesian inference perspective by sampling global models of superior quality and combining them via a Bayesian ensemble model.

In particular, by treating each client's model as a possible global model, a distribution of global models is constructed. This distribution allows for an ensemble of Bayesian models by aggregating the results of multiple global models.

On the other hand, reputation is also critical to improving the robustness and reliability of federated systems by helping to detect malicious participants. A reputation-aware hierarchical aggregation framework was introduced in [15]. A reputation-based method to select client updates for aggregation was used. It aims to reduce unnecessary local update exchanges. The FLClient's reputation score is modelled as a function of the quality of the update of its local model and the quantity of its dataset contributed in the current round. The quality function is measured in terms of accuracy against the test data.

The reputation approach in the FL architecture in [21] involves, instead, evaluating their contributions by examining the similarity of their uploaded gradients through vector comparison. In [20], the reputation score was determined by three metrics that assessed the comparison between the performance of the local model at the current iteration, the performance of the temporary global model generated from the current iteration, and the performance of the global model from the last iteration.

## 4 The Proposed Architecture

Figure 1 depicts the schema proposed in this work. The main idea behind this work is to introduce an extra layer inside the federation server that pre-processes the local changes to bring to the model.

The first stage from the left is the *Federated Layer*, which contains n federates. Each federated trains the model on the base of its own *Local Data* and, according to the FL general principles, a federated is the only entity knowing such a data. When local models are trained, a federated sends to the central server the difference between the global model ($\Delta W_i$).

As stated in Sect. 1, this architecture introduces a pre-processing stage at the server, named *Federation Filter*, whose main objective is to exclude malicious/-failing federated from the final stage of *Model Averaging*. At the centre of the entire approach, there is the idea that the Federation Filter determines and uses a dynamic index of the trustworthiness of each federated: the index is named in this paper $\rho_i$. We consider two different temporal steps: $k$ is the current step and $k-1$ that is the previous one[1]. The first task is to consider the *Global Model* at the previous step ($GM^{k-1}$) and to add it to the model changes proposed by each federated. Through Eq. 1, the local models are built back and sent to the *Ensemble Learning* module. The underlying idea is to consider a weak learner for each federated learner "proposing" a model.

$$M_i^k = GM^{k-1} + \Delta W_i^k \, \forall i \in \{0, \ldots, n\} \quad (1)$$

Each model is then sent to the Bayesian Ensemble Learning (BEL) module. As an output, the BEL module generates a current reliability information, sent

[1] For the sake of the readability, where omitted in the figure, we refer to the k-th step.

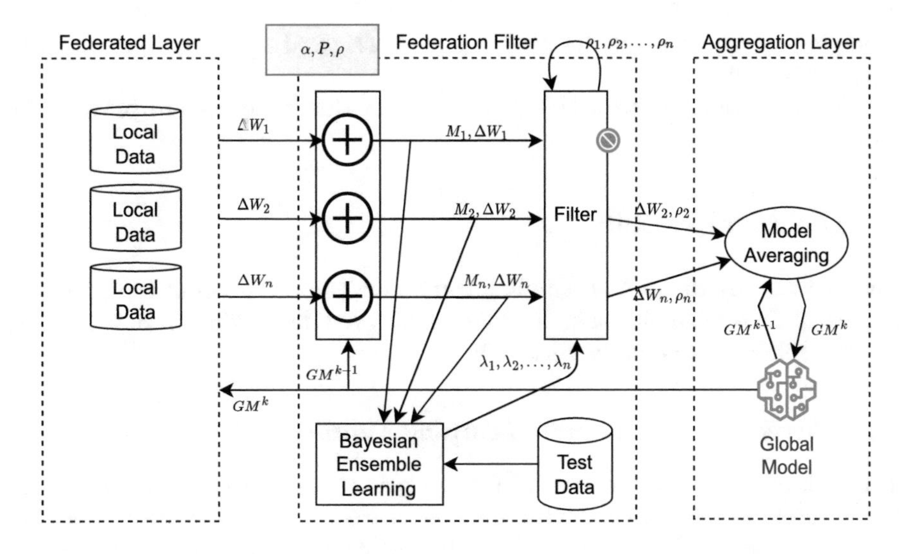


**Fig. 1.** The overall approach

to the Filter module, which can be summarised by a list of temporary reliability parameters $\lambda_1, \lambda_2, \ldots, \lambda_n$, determining the degree of concordance of the decisions between each weak learner and the ensemble model on the base of the Test Data.

The Filter module has two objectives: (1) to update the reputational indices $\rho_i$ in time, according to the values of such variables at the previous stages and from the results of the BEL module elaboration; (2) to pass to the Federation Server stage only the deltas of the model and the estimations of the reputation indices of those federated that are considered trustworthy.

The behaviour of the Filter module is affected by three global parameters:

- $\alpha$: this is a learning factor which determines how much the current reputation is affected by the history (i.e., the previous reputation) and how much by the current learner's performance (i.e., the $\lambda_i$ parameter); $\alpha \in ]0, 1]$);
- $P$: this is a threshold of the reputation used to filter or not the learner $P \in ]0, 1]$;
- $\rho$: this is the minimum reachable reputation for a learner ($\rho \in ]0, 1[ \wedge \rho < P$). When $\rho$ is zero, a non-trustable learner cannot become trusty any more; hence, the lower is $\rho$, the longer is the period of time it takes to reconsider the learner trustable.

The final stage of this architecture is the Federation Server, which oversees the concrete merging of the model updates coming from the federated. As a secondary objective of the paper is to reuse existing architectures, proposing a novel approach for model averaging is not in the scope of the paper; existing averaging mechanisms can be used, e.g., the weighted aggregation mechanism reported in [12].

Further details on both the Filter and the BEL modules are provided in Sect. 5.

# 5 Technical Details

This section has the aim of detailing some interesting aspects of the proposed approach. Two main elements are considered: the BEL module in Subsect. 5.1, and the Filter module in Subsect. 5.2.

## 5.1 The Bayesian Ensemble Learning Module

The task of the generation and of the analysis of ensemble model is based on the property of BNs of combining different learners, according to the Decision Support System (DSS) paradigm [4,5]. In particular, considering the formalisation expressed in Sect. 2, while the model managed by the federated classifies $k$ features on in $m$ classes, the BN ensemble model manages the output of such classifiers having both as input and output the $m$ classes.

The Bayesian approach used in this paper extends the one reported in [7]; the main difference between the approach presented in this paper and the one reported in the cited paper starts from a shift in the scope of the model itself. While the model proposed in [7] has the final aim to generate the ensemble model, here such a model is "used only" to evaluate the reliability parameters $\lambda_1, \lambda_2, \ldots, \lambda_n$. There are two main phases: **building** of the BN ensemble model, and its **analysis**. The first phase starts from the BN model structure reported in Fig. 2 and performs according to the Python-like code reported in Listing 1.1.

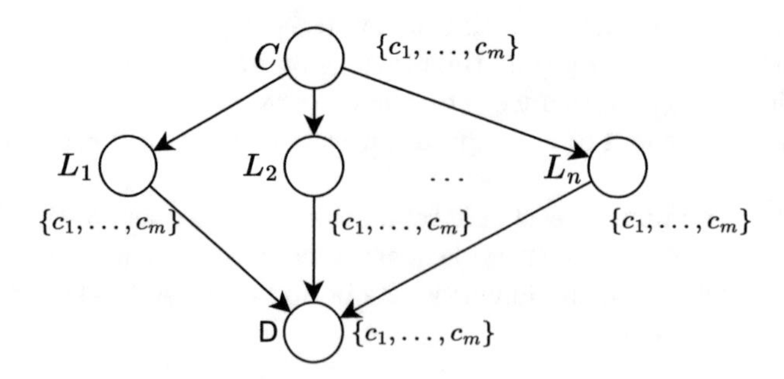


**Fig. 2.** The Bayesian Ensemble Model: model structure

**Listing 1.1.** BN ensemble model construction

```
1 def bembuilding(federates,testdata):
2     bn = BNEnsembleModel()
3     for i in range(0,len(federates)):
4         m = federates[i].model()
5         cm = m.test(testdata)
6         bn.populate_LCPT(i,cm)
7     bn.populate_DCPT()
8     return bn
```

At line 2, an instance of the BN model depicted in Fig. 2 is generated. From line 3 to 6 a loop over all the federated performs the following: at line 4, the model $M_i^k$ computed from Eq. 1 is retrieved; such model is tested with Test Data at line 5 and at line 6 the confusion matrix (i.e., `cm`) is used to populate the Conditional Probability Table (CPT) of the $L_i$ node of the BN model, according to the CPT structure reported in Table 1. At line 7, the CPT node of the `D` node is populated, according to the structure reported in Table 2. This CPT implements a majority voting logic, giving the maximum output to the most frequent class present in the row. For the sake of the concreteness, the tables report the case of three federates and two classes: a different case can be easily built by extending the reported example..

**Table 1.** CPT of $L_i$ node

| $C$ | $c_1$ | $c_2$ |
|---|---|---|
| $c_1$ | $1 - \epsilon_{1,2}$ | $\epsilon_{1,2}$ |
| $c_2$ | $\epsilon_{2,1}$ | $1 - \epsilon_{2,1}$ |

**Table 2.** CPT of $D$ node

| $L_1$ | $L_2$ | $L_3$ | $c_1$ | $c_2$ |
|---|---|---|---|---|
| $c_1$ | $c_1$ | $c_1$ | 1 | 0 |
| $c_1$ | $c_1$ | $c_2$ | 1 | 0 |
| $c_1$ | $c_2$ | $c_1$ | 1 | 0 |
| $c_1$ | $c_2$ | $c_2$ | 0 | 1 |
| $c_2$ | $c_1$ | $c_1$ | 1 | 0 |
| $c_2$ | $c_1$ | $c_2$ | 0 | 1 |
| $c_2$ | $c_2$ | $c_1$ | 0 | 1 |
| $c_2$ | $c_2$ | $c_2$ | 0 | 1 |

Once the model is built, the analysis phase is run. By using the a-posteriori analysis of BNs, the `D` and `C` are observed to assume the different values from $c_1$ to $c_m$; for each of these values, the distributions of $L_1$ to $L_n$ are computed: the probability of error (i.e., a possible $L_i$ value different from the observed one in $C$ and $D$) contributes to the computation of $\lambda_i$. More in the details, $\forall i \in \{1, \dots, n\}$, $\lambda_i$ is the average of all the error probabilities over all the observations on the set of the classes.

### 5.2 The Filter Module

The two actions performed by the Filter module are detailed in the following.

*Update of the Reputational Indices.* Starting from the reliability indices given by the BEL module, i.e. $\lambda_1, \lambda_2, \ldots, \lambda_n$, the Filter module updates the reputational indices $\rho_1, \rho_2, \ldots, \rho_n$. The formula ruling such an update also uses two of the three global parameters $\alpha$ and $\rho$. Equation 2 reports the law of reputation updating, which represents a variant of the law already published in [10],

$$temp = \rho_i^{k-1} \cdot (1 + (2\lambda_i - 1) \cdot \alpha) \quad (2)$$

while Eq. 3 normalises the value of the new reputation inside the $[\rho, 1]$ range.

$$\rho_i^k = min(1, max(temp, \rho)) \quad (3)$$

As stated in Subsect. 5.1, $\lambda_i$ spans between 0 and 1 (i.e., for respectively fully non-trustable and fully trustable learners). When $\lambda_i = 0$, Eq. 2 becomes $temp = \rho_i^{k-1} \cdot (1 - \alpha)$, while in the case of $\lambda_i = 1$, Eq. 2 becomes $temp = \rho_i^{k-1} \cdot (1 + \alpha)$. Both the cases reflect an evolution of the $\rho_i$ parameter that considers the previous value, enriched by the current learner performance. The speed of the evolution of the reputation parameter can be changed by tuning the $\alpha$ parameter. All these equations are valid $\forall i \in \{1, \ldots, n\}$.

*Filtering Federated Learners.* Given the variable framework described above, the filtering of the learners is quite straightforward. This paper proposes a very basic filtering mechanism, even if more complex policies can be adopted. The filtering is based on the construction of the predicate described in Eq. 4, which considers valid the federates whose reputation is greater than the threshold $P$.

$$\pi : \mathbb{N}_n \longrightarrow \mathbb{B} \mid \pi(i) = (\rho_i^k \geq P),\ \forall f \in \mathbb{N}_n \quad (4)$$

Further mechanisms can be considered — e.g., more complex outlier detection algorithms — which are out of the scope of this paper.

## 6 Discussion and Conclusions

EL is a powerful technique that combines the strengths of multiple models to achieve higher accuracy and better generalization. The future of FL promises to enable large-scale machine learning applications while preserving data privacy and security.

The methodology proposed in the paper, which combines FL with EL to address security and privacy challenges in a distributed environment, represents a potential advance in the field, although still in an early-stage form.

The introduction of the Federation Filter represents a crucial step in improving the security of FL systems. By dynamically assessing the trustworthiness of federated learners using BNs, the system gains the capability to filter out potentially malicious participants. This is particularly relevant in edge computing, where resource-constrained devices are more susceptible to adversarial attacks and situations in which FL shows all its benefits.

The proposed updating mechanism for reputation indices, which incorporates both historical reputation and current performance, adds a level of sophistication to the FL system. The flexibility introduced by parameters such as $\alpha$, $P$, and $\rho$ allows the system's response to different reliability scenarios to be fine-tuned. The filtering mechanism, while basic, provides a solid foundation for subsequent more complex detection policies.

The integration of EL and FL increases security and contributes to the sustainability of the entire system. EL's ability to combine several trained models gives robustness against individual vulnerabilities and adversary attacks. This combination ensures that the generated global model is not only secure, but also maintains a small and lightweight profile, which is critical for edge devices with limited resources. Using BNs, the system adds insight into the reliability of federated learners. This approach is particularly effective in defeating adversary attempts to compromise the integrity of the FL system.

In conclusion, the proposed method of combining FL with EL, incorporating a federation filter and using BNs, provides a holistic solution to the security and sustainability challenges inherent in edge computing environments. The approach leverages the strengths of FL to fit the limitations of edge devices while introducing new mechanisms to strengthen security.

Dynamic updating of reputation indices, driven by both historical data and actual performance, ensures adaptability to changing conditions and increases system resilience. The integration of EL ensures that the overall model remains secure and resource efficient, in line with the needs of edge computing.

Although the method discussed represents a promising solution, ongoing research and development is critical. Implementations and validations are needed to evaluate the scalability, effectiveness, and adaptability of the proposed approach. Future works may consider implementation and testing of the proposed architecture through the use of BN- and FL-specific frameworks. Finally, a case study to comprehensively validate the proposed approach also could be implemented in the future. Overall, the work presented represents a first step toward creating a more secure and resilient FL system.

## References


1. Agrawal, S.,et al.: Federated learning for intrusion detection system: concepts, challenges and future directions. Comput. Commun. (2022)
2. Campanile, L., Marrone, S., Marulli, F., Verde, L.: Challenges and trends in federated learning for well-being and healthcare. Procedia Comput. Sci. **207**, 1144–1153 (2022)
3. Chen, H.Y., Chao, W.L.: Fedbe: making Bayesian model ensemble applicable to federated learning. arXiv preprint arXiv:2009.01974 (2020)
4. Chen, W.H., Wang, W.H.A.: A Bayesian network based decision support system, pp. 865–872 (2014). https://doi.org/10.14809/faim.2014.0865
5. Cooper, G.F., Herskovits, E.: A Bayesian method for the induction of probabilistic networks from data. Mach. Learn. **9**(4), 309–347 (1992). https://doi.org/10.1023/A:1022649401552

6. De Biase, M., Marulli, F., Verde, L., Marrone, S.: Improving classification trustworthiness in random forests, pp. 563–568 (2021). https://doi.org/10.1109/CSR51186.2021.9527939
7. De Stefano, C., D'Elia, C., Marcelli, A., di Freca, A.S.: Using Bayesian network for combining classifiers. In: 14th International Conference on Image Analysis and Processing (ICIAP 2007), pp. 73–80 (2007). https://doi.org/10.1109/ICIAP.2007.4362760
8. Fang, M., Cao, X., Jia, J., Gong, N.Z.: Local model poisoning attacks to byzantine-robust federated learning, pp. 1623–1640 (2020)
9. Fernández-Caramés, T.M., Fraga-Lamas, P.: A review on the use of blockchain for the internet of things. IEEE Access **6**, 32979–33001 (2018). https://doi.org/10.1109/ACCESS.2018.2842685
10. Flammini, F., Marrone, S., Nardone, R., Caporuscio, M., D'Angelo, M.: Safety integrity through self-adaptation for multi-sensor event detection: methodology and case-study. Futur. Gener. Comput. Syst. **112**, 965–981 (2020). https://doi.org/10.1016/j.future.2020.06.036
11. Haim, N., Yehudai, G., Vardi, G., Shamir, O., Irani, M.: Reconstructing training data from trained neural networks, vol. 35 (2022)
12. Li, Z., Lin, T., Shang, X., Wu, C.: Revisiting weighted aggregation in federated learning with neural networks, vol. 202, pp. 20412–20451 (2023)
13. Liu, P., Xu, X., Wang, W.: Threats, attacks and defenses to federated learning: issues, taxonomy and perspectives. Cybersecurity **5**(1) (2022). https://doi.org/10.1186/s42400-021-00105-6
14. Liu, X., Li, H., Xu, G., Chen, Z., Huang, X., Lu, R.: Privacy-enhanced federated learning against poisoning adversaries. IEEE Trans. Inf. Forensics Secur. **16**, 4574–4588 (2021)
15. Panigrahi, M., Bharti, S., Sharma, A.: A reputation-aware hierarchical aggregation framework for federated learning. Comput. Electr. Eng. **111**, 108900 (2023)
16. Paragliola, G.: A federated learning-based approach to recognize subjects at a high risk of hypertension in a non-stationary scenario. Inf. Sci. **622**, 16–33 (2023)
17. Paragliola, G., Ribino, P., Ullah, Z.: A federated learning approach to support the decision-making process for ICU patients in a European telemedicine network. J. Sens. Actuator Netw. **12**(6), 78 (2023)
18. Sarwar Murshed, M., Murphy, C., Hou, D., Khan, N., Ananthanarayanan, G., Hussain, F.: Machine learning at the network edge: a survey. ACM Comput. Surv. **54**(8) (2022). https://doi.org/10.1145/3469029
19. van Dijk, M., Gentry, C., Halevi, S., Vaikuntanathan, V.: Fully homomorphic encryption over the integers. In: Gilbert, H. (ed.) EUROCRYPT 2010. LNCS, vol. 6110, pp. 24–43. Springer, Heidelberg (2010). https://doi.org/10.1007/978-3-642-13190-5_2
20. Wang, Y., Kantarci, B.: Reputation-enabled federated learning model aggregation in mobile platforms. In: 2021-IEEE International Conference on Communications, pp. 1–6 (2021)
21. Xu, X., Lyu, L.: A reputation mechanism is all you need: collaborative fairness and adversarial robustness in federated learning. arXiv:2011.10464 (2020)
22. Yang, Q., Liu, Y., Chen, T., Tong, Y.: Federated machine learning: concept and applications. ACM Trans. Intell. Syst. Technol. **10**(2) (2019). https://doi.org/10.1145/3298981

# Situation Awareness in the Cloud-Edge Continuum

Giuseppe D'Aniello[1(✉)], Matteo Gaeta[1], Francesco Flammini[2,3], and Giancarlo Fortino[4]

[1] Department of Information Engineering, Electrical Engineering and Applied Mathematics, University of Salerno, Fisciano, Italy
gidaniello@unisa.it

[2] IDSIA USI-SUPSI, University of Applied Sciences and Arts of Southern Switzerland, Lugano, Switzerland

[3] School of Innovation, Design and Engineering, Mälardalen University, Västerås, Sweden

[4] Department of Informatics, Modeling, Electronics and Systems, University of Calabria, Rende, Italy

**Abstract.** Edge Computing is becoming a promising computing paradigm that addresses the limitations of cloud computing concerning latency, autonomy, and costs. To facilitate more intelligent applications made possible by the Edge Computing paradigm, it is essential to integrate intelligence and adaptability into devices located at the network's edge. The paper explores the potential integration of Situation Awareness (SA) capabilities into the Cloud-Edge continuum. This integration aims to empower smarter applications while effectively managing challenges related to low latency, high autonomy, and cost-effective solutions. Within our illustrative example in healthcare, we showcase how the proposed SA cloud-edge continuum architecture enables efficient data processing and decision-making.

## 1 Introduction

Recent technological advancements have had a tremendous impact across various sectors of our digital world. We find ourselves surrounded by hundreds of smart sensors and autonomous intelligent systems [1]. However, this proliferation often does not make us more informed; instead, the vast amount of data overwhelms us, making it increasingly challenging to understand what is happening around us. Consequently, making accurate and timely decisions in critical domains such as healthcare, cybersecurity, and military has become more challenging.

While the increased autonomy of intelligent systems provides significant and indispensable support for handling domains with dynamics too rapid for direct human intervention, it's crucial that these systems allow human operators to act as effective supervisors.

Human operators must comprehend the autonomous system's state to interact with it effectively (human-autonomy teaming [13]). Otherwise, there is a


L. Barolli (Ed.): AINA 2024, LNDECT 203, pp. 307–316, 2024.
https://doi.org/10.1007/978-3-031-57931-8_30

risk of being out of the command and control loop of the system. This scenario could be perilous, especially when dealing with unforeseen situations beyond the capabilities of the autonomous system. Simultaneously, the autonomous system itself must comprehend the ongoing environment in which it operates. It needs a comprehensive representation of the situation and the ability to anticipate future occurrences to act both actively and proactively [8]. This capability of both human and artificial agents to understand the current situation and to be able to anticipate future status of the environment and act according to it is defined as Situation Awareness (SA).

SA has been defined by Endsley as a three-level model: i) perception of elements, ii) comprehension of their current states, iii) projection of future statuses [10]. SA is considered a critical prerequisite for dynamic decision-making and the performance of actions in operational and tactical tasks, as making correct decisions is challenging without first having a good awareness of what is happening in the environment. The sharing of an understanding of the current situations between human and artificial agents is fundamental for human-autonomy teaming and ensures that both humans and machines maintain command and control. This is particularly critical in navigating and responding to unforeseen challenges.

The exponential increase in the number of sensors and autonomous intelligent systems calls for new models and architectures for processing the obtained information. It's widely acknowledged that transmitting data from its source to a centralized data center leads to increased communication overhead, higher bandwidth usage, and privacy concerns. Relying solely on cloud processing proves inefficient for real-time systems like health monitoring, autonomous driving, and smart city frameworks. To address this, there's a growing need to perform data computation locally, as much as possible at the network's edge, leveraging Edge Intelligence [2] instead of solely relying on centralized data centers. This involves deploying AI near edge devices capable of handling complex tasks and adapting swiftly to dynamic scenarios. Presently, microcontrollers possess sufficient power to make autonomous decisions based on sensor-derived data, analyze information, transmit to low-latency actuators, and transfer summarized data to the cloud. Integrating intelligence into sensor nodes represents the progression towards achieving an enhanced awareness at the edge [1], and only when needed, to use the cloud data center to perform processing and analysis on aggregated data, with higher latency but also with higher computational power.

This paper introduces an approach for SA in the Cloud-Edge Continuum, with the goal to improve the efficiency of those applications that orchestrate different devices scattered around the network, while also paying attention to the cost and efficiency of these devices.

To demonstrate the feasibility of our approach, we present an illustrative example in the domain of the health-care, in order to better monitoring patients and facilitate situation-aware telemedicine applications.

## 2 Related Works

The survey by V. Barbuto et al. [2] highlights the growing trend of Edge Intelligence (EI) and its increasing presence in recent literature, leading to a surge in systematic studies within this domain [16]. This evolution is an inherent part in the evolution of the specific EI methods [11,20] and in many cases the broader EI concepts [18,19,22]. An interesting conceptual framework is proposed by F. Flammini in [12], outlining a cloud-edge continuum architecture (which also incorporates fog-level computing) for Cyber-Physical Systems (CPS). Specifically, this continuum is employed as the underlying architecture to support digital twins (DT) for real-time modeling and performability prediction across various domains of CPS. Other key studies in the field, such as those carried out by Zhou et al. [22] and Deng et al. [6], have been seminal in posing insights not only on what the essential elements of EI were but also its potential applications. By noting the large footprints the works have had in citations, one can infer these insinuations are only growing relevant to everyday contexts.

The incorporation of Situation Awareness (SA) in EI, particularly in edge computing argues to appear as an emerging area of interest. It is highly significant in real-time processing environments and those that engage quick decision-making. In the context of edge computing, the role of the SA becomes pronounced in that it helps in the processing of data closer to the source, thereby reducing latency and enhancing response times in critical situations such as military operations [2]. This development in EI, more so as viewed from the perspective of SA, is a big stride marking that intelligent systems need to be not only technologically advanced but also should show an increased level of contextual awareness and responsiveness. Moreover, the convergence of cloud computing and SA is emerging as a powerful alliance, offering a rich source of insights for EI.

For example, the Samsara architecture [17] exemplifies this trend showing how cloud computing is improved in efficiency and responsiveness by SA. Such developments bode increased value of context-aware and adaptable systems in an increasing array of application domains spans from security-sensitive military environments to general-purpose cloud computing platforms. Most approaches to SA limit themselves to edge or cloud computing with a very distinct lack of cloud-edge continuum spanning methods.

Furthermore, most research efforts in this field do not comprehensively address all stages of Endsley's SA model [10]. This observation underscores the necessity for more integrative methods in forthcoming Edge Intelligence research to fully encapsulate Perception, Comprehension, and Projection-the three critical phases of Endsley's SA model.

## 3 An Approach for Situation Awareness in the Cloud-Edge Continuum

In the Cloud-Edge continuum [7], multiple levels of data processing and transmission can be distinguished. Initially, the sensors collect data and can perform basic processing tasks, such as signal conditioning, data cleaning, and data fusion. Subsequently, the processed data is transmitted to edge devices situated in the edge layer. These devices enable processing in close proximity to the sensors, leveraging machine learning and data analytics tasks as well [2]. Finally, the edge devices transmit the aggregated and filtered data to the cloud. Here, all the data originating from the sensor networks and edge layer can be integrated, providing human operators with a comprehensive overview of the environment.

To advance intelligence at the edge and enhance device autonomy, we integrate SA capabilities into various levels of the Cloud-Edge Continuum. The proposed conceptual architecture for the SA-enabled Cloud-Edge Continuum is illustrated in Fig. 1. At the foundational level, a sensor-equipped smart environment generates data for monitoring both the environment itself and the behavior of individuals within it. These sensors transmit raw data to the edge-level devices.

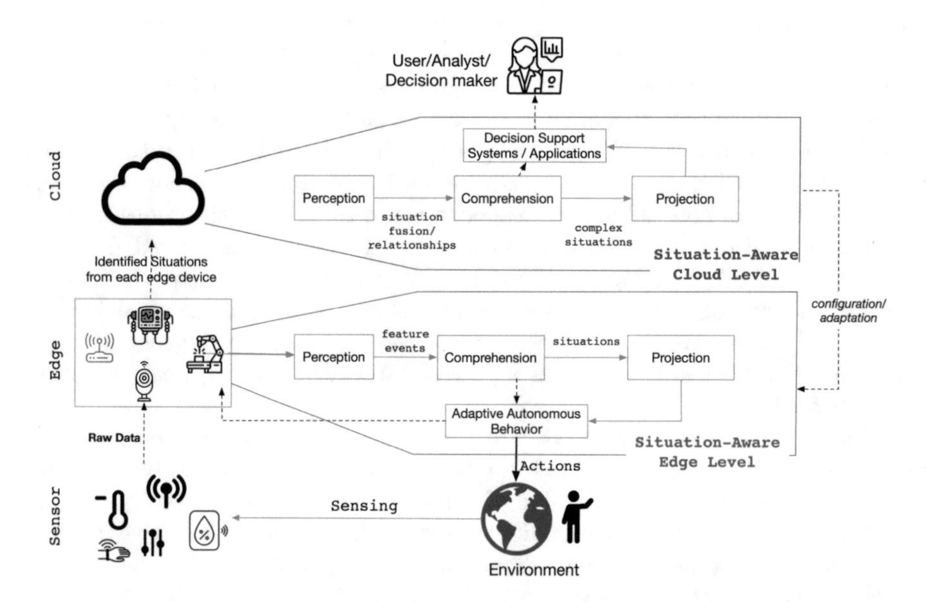


**Fig. 1.** Conceptual Architecture of the Situation Awareness enabled Cloud-Edge Continuum

Within the edge level, each device is equipped with a SA loop, drawing inspiration from Endsley's well-established SA model [10]. Specifically, the devices receive raw data from monitored sensors and undergo processing at the perception level to perform tasks such as data cleaning, harmonization, and fusion. This step identifies basic features and simple events. Subsequently, these features and events are processed at the comprehension level, whose objective is to identify the situation using specific techniques such as machine learning classifiers [21]. Situations are identified concerning a model of available situations managed by the edge device. Finally, the current situation is utilized to forecast the future state of the environment at the projection level. This predictive task holds significance as it enables proactive actions by the edge device considering plausible future environmental states. The identified and projected situations govern the adaptation of the smart edge device's behavior: the device selects actions to be executed (via its actuators or interfaces) to alter the environment's state. This modification initiates a new cycle of SA for the device, as fresh data is acquired from the sensors. Due to the execution of the SA loop onboard the edge device, actions can be performed quasi-real-time, ensuring low-latency [14]. Moreover, this approach aids in preserving data privacy and security. Only data authorized for transmission by the edge device will be sent to the cloud data center. To minimize data volume, the device may transmit representations of identified situations. These representations are abstract, compact, synthetic summaries of the raw sensor data collected by the edge device, encapsulating domain-relevant concepts [5].

The cloud data center initiates an additional SA loop involving human operators, analysts, and decision-makers concerned with supervising and managing the monitored environment. Specifically, at the perception level, various situations sent by individual edge devices are fused, integrated, and harmonized. At the comprehension level, intricate relationships among the atomic situations from the edge level are analyzed to discern a comprehensive global situation, presenting an overarching view of the environment's status. Finally, predictive models at the projection level forecast potential trends and evolutions in the situation, facilitating timely and proactive decision-making.

The identified and projected situations are pivotal in supporting situation-aware decision support systems and data analytics dashboards, providing analysts and decision-makers with an encompassing view. Additionally, at the cloud level, the identified global situations can influence the configuration adjustments of individual edge devices as required.

## 4 Illustrative Example

A case study in the healthcare domain [4,15] can be helpful to demonstrate the application of the proposed conceptual framework in monitoring patients to facilitate situation-aware telemedicine applications.

Building upon the conceptual framework mentioned in Sect. 3, this paper introduces an illustrative example, particularly inspired by the research presented in the study carried out by G. D'Aniello et al. [9]. This example demonstrates the practical application of SA in a real-world scenario, leveraging the principles and methods discussed in the aforementioned study.

In the context of smart wearable computing systems, SA plays a pivotal role in enhancing the responsiveness and adaptability of these devices. The study focuses on the use of Machine Learning (ML) techniques and Context Space Theory (CST) to identify and understand varying situations in a dynamic environment. Originating from the work of Boytsov et al. [3], CST theory enables the system to understand and classify the context based on various environmental and situational parameters.

These methodologies align seamlessly with our proposed framework, particularly in processing and interpreting data from wearable devices at the edge level.

The example scenario, depicted in Fig. 2, involves the deployment of smart wearable devices in a healthcare setting. These devices specifically are smartwatches and smartphones, equipped with a suite of sensors, in particular accelerometers, gyroscopes, and heart rate monitors, in order to continuously collect and transmit patient health parameters. The sensory data undergoes an initial processing phase at the edge (e.g., on the user smartphone), where:

- The **Sensing** stage collects data through deployed sensors in the environment.
- The **Perception** stage involves data collection and cleaning, alongside feature extraction from raw sensor outputs.
- At the **Comprehension** stage, the context is recognized, activities are identified, and the situation is understood, using machine learning (ML) models pre-trained at the Cloud level, as we will explore later.
- The **Projection** stage includes predicting future patient health events, enabling proactive medical intervention.

The collected data is synthesized into actionable insights, represented by high-level daily life activities (e.g., sitting, standing, walking) and situations (e.g., watching TV, working, eating). This data is then transmitted to the cloud layer for comprehensive analysis. In this phase, advanced processing occurs according to Endsley's SA model, involving:

- In the **Perception** phase, data aggregation and event correlation from multiple users.
- In the **Comprehension** phase, the update of the ML models that will be used at the edge level for situation identification is performed. Moreover, this level supports online applications (e.g., telemedicine) by providing user with information regarding complex situations, trend analysis, monitoring of multiple users.
- In the **Projection** phase, where data mining and rule mining techniques are applied on the recognized situations to support situation projection and predictive analysis.

One of the advantage of this approach lies in the edge-layer processing, as shown in the Fig. 2. By handling data preprocessing locally, it minimizes network traffic and reduces congestion. Notably, the role of the cloud extends to refining models on the edge based on collective patient data, enhancing the system's comprehension and projection capabilities, and ensuring that care remains both individualized and informed by broader health trends.

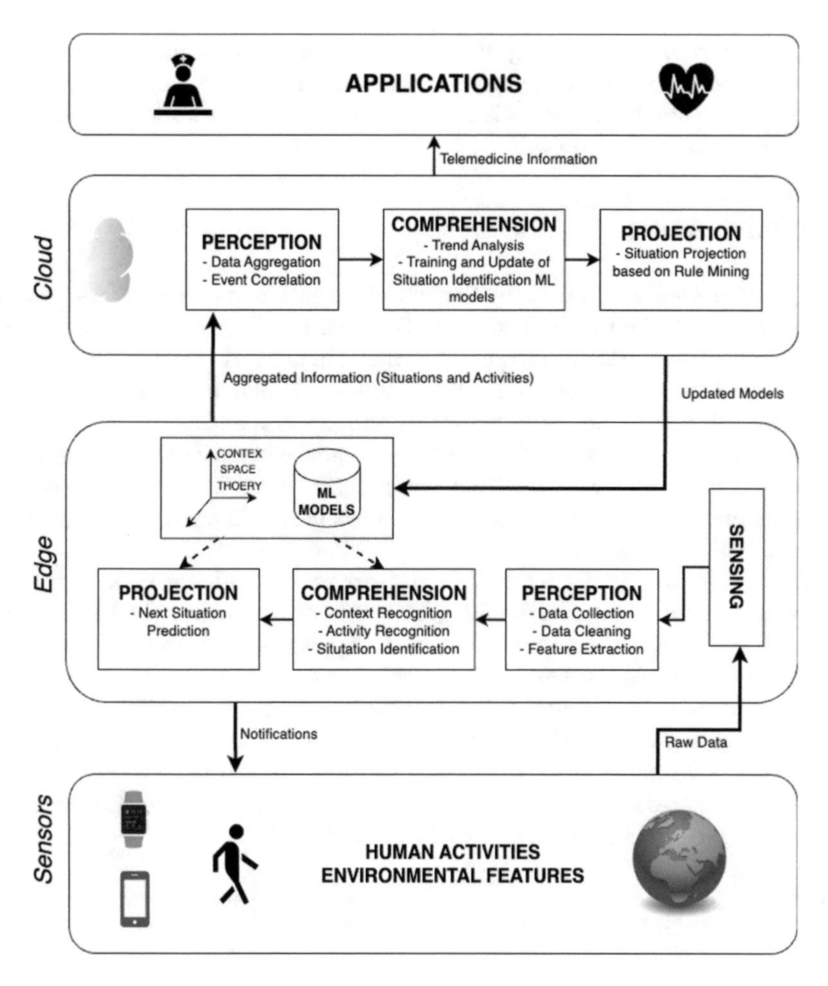


**Fig. 2.** Schematic Overview of the SA-Enhanced Patient Health Monitoring System in the Cloud-Edge Continuum

On the other hand, the integration of SA within the edge and in the layers ensures timely responses to patient needs and enhances the efficiency of healthcare services. This holistic approach, facilitated by the Cloud-Edge Continuum, empowers healthcare professionals to make informed decisions and provide targeted care, exemplifying the effective use of SA in healthcare applications. For instance, a doctor using an app for monitoring the heart rate of patients, leveraging this approach can more accurately evaluate ECG data. By correlating heartbeat anomalies with the patient's activities and situations, the doctor (or an intelligent system) can discern normal from abnormal patterns. For example, an elevated heartbeat while walking may be deemed normal, whereas a sudden increase during rest could indicate an issue. This context-rich insight empowers healthcare professionals to provide more targeted and informed care.

Lastly, the localized processing of data at the edge layer significantly enhances patient privacy. By processing sensitive health information on the device itself, only essential summarized data is transmitted over the network, minimizing the exposure of raw patient data.

## 5 Conclusions

This paper presented a novel integration of Situation Awareness (SA) into the Cloud-Edge continuum, specifically applied to a healthcare monitoring context. Our approach demonstrates significant benefits in network efficiency by localizing data processing at the edge, reducing bandwidth usage and latency. Having an intelligent edge layer not only preserves the privacy and security of sensitive health data but also allows for real-time responsiveness critical in healthcare scenarios. The cloud's role, far from being diminished, is enhanced as it serves to update and refine the computational models based on aggregated data across patients. This synergy between cloud and edge computing facilitates a more comprehensive and predictive healthcare service by enabling personalized and context-aware interventions. Looking forward, the proposed SA-based Cloud-Edge approach holds the potential for broad application beyond healthcare, into domains where rapid decision-making and adaptability are paramount. Future work will focus on extending this framework to other domains including IoT, cybersecurity, logistics, exploring the scalability of the model and its effectiveness in different environmental contexts. We will explore the balance between edge processing and cloud analytics to optimize the efficiency and effectiveness of intelligent systems in an increasingly connected world.

**Acknowledgements.** The authors would like to thank Luca Aliberti for his contribution to the definition of the illustrative example.

## References

1. Al-Saedi, A.A., Boeva, V., Casalicchio, E., Exner, P.: Context-aware edge-based AI models for wireless sensor networks. An overview. Sensors **22**(15) (2022). https://doi.org/10.3390/s22155544
2. Barbuto, V., Savaglio, C., Chen, M., Fortino, G.: Disclosing edge intelligence: a systematic meta-survey. Big Data Cogn. Comput. **7**(1) (2023). https://doi.org/10.3390/bdcc7010044
3. Boytsov, A., Zaslavsky, A., Synnes, K.: Extending context spaces theory by predicting run-time context. In: Balandin, S., Moltchanov, D., Koucheryavy, Y. (eds.) NEW2AN/ruSMART -2009. LNCS, vol. 5764, pp. 8–21. Springer, Heidelberg (2009). https://doi.org/10.1007/978-3-642-04190-7_2
4. Casillo, M., Cecere, L., Colace, F., Lorusso, A., Marongiu, F., Santaniello, D.: Internet of things in SPA medicine: a general framework to improve user treatments. In: Proceedings - 2023 IEEE International Conference on Smart Computing, SMARTCOMP 2023, pp. 309–313 (2023). https://doi.org/10.1109/SMARTCOMP58114.2023.00077

5. D'Aniello, G., Gravina, R., Gaeta, M., Fortino, G.: Situation-aware sensor-based wearable computing systems: a reference architecture-driven review. IEEE Sens. J. **22**(14), 13,853–13,863 (2022). https://doi.org/10.1109/JSEN.2022.3180902
6. Deng, S., Zhao, H., Fang, W., Yin, J., Dustdar, S., Zomaya, A.Y.: Edge intelligence: the confluence of edge computing and artificial intelligence. IEEE Internet Things J. **7**(8), 7457–7469 (2020)
7. Di Martino, B., Venticinque, S., Esposito, A., D'Angelo, S.: A methodology based on computational patterns for offloading of big data applications on cloud-edge platforms. Futur. Internet **12**(2) (2020). https://doi.org/10.3390/fi12020028
8. D'Aniello, G.: Fuzzy logic for situation awareness: a systematic review. J. Ambient Intell. Humaniz. Comput. **14**(4), 4419–4438 (2023). https://doi.org/10.1007/s12652-023-04560-6
9. D'Aniello, G., Gaeta, M., Gravina, R., Li, Q., Rehman, Z.U., Fortino, G.: Situation identification in smart wearable computing systems based on machine learning and context space theory. Inf. Fusion **104**, 102,197 (2024)
10. Endsley, M.R.: Toward a theory of situation awareness in dynamic systems. Hum. Factors **37**(1), 32–64 (1995). https://doi.org/10.1518/001872095779049543
11. Feng, R., Feng, X.: Robot edge intelligence and data survey. In: 2021 IEEE International Conference on Dependable, Autonomic and Secure Computing, pp. 843–848. IEEE (2021)
12. Flammini, F.: Digital twins as run-time predictive models for the resilience of cyber-physical systems: a conceptual framework. Philos. Trans. R. Soc. A: Math. Phys. Eng. Sci. **379**(2207) (2021). https://doi.org/10.1098/rsta.2020.0369
13. Flammini, F., Alcaraz, C., Bellini, E., Marrone, S., Lopez, J., Bondavalli, A.: Towards trustworthy autonomous systems: taxonomies and future perspectives. IEEE Trans. Emerg. Top. Comput. 1–13 (2022). https://doi.org/10.1109/TETC.2022.3227113
14. Greco, A., Saggese, A., Vento, M., Vigilante, V.: Vehicles detection for smart roads applications on board of smart cameras: a comparative analysis. IEEE Trans. Intell. Transp. Syst. **23**(7), 8077–8089 (2022). https://doi.org/10.1109/TITS.2021.3075749
15. Greco, L., Percannella, G., Ritrovato, P., Tortorella, F., Vento, M.: Trends in IoT based solutions for health care: moving AI to the edge. Pattern Recognit. Lett. **135**, 346–353 (2020). https://doi.org/10.1016/j.patrec.2020.05.016
16. Guo, B., Zhang, D., Wang, Z.: Living with internet of things: the emergence of embedded intelligence. In: 2011 International Conference on Internet of Things and 4th International Conference on Cyber Physical and Social Computing, pp. 297–304. IEEE (2011)
17. Neves, V., Pilla, M., Yamin, A., Pilla, L.: Samsara architecture: exploring situation awareness in cloud computing management. In: 2018 Symposium on High Performance Computing Systems (WSCAD), pp. 84–92. IEEE (2018)
18. Peltonen, E., et al.: The many faces of edge intelligence. IEEE Access **10**, 104,769–104,782 (2022)
19. Ramya, R., Ramamoorthy, S.: Survey on edge intelligence in IoT-based computing platform. In: Hu, YC., Tiwari, S., Trivedi, M.C., Mishra, K.K. (eds.) Ambient Communications and Computer Systems. LNNS, vol. 356, pp. 549–561. Springer, Singapore (2022). https://doi.org/10.1007/978-981-16-7952-0_52
20. Xu, S., Qian, Y., Hu, R.Q.: Edge intelligence assisted gateway defense in cyber security. IEEE Netw. **34**(6), 14–19 (2020)

21. Ye, J., Dobson, S., McKeever, S.: Situation identification techniques in pervasive computing: a review. Pervasive Mob. Comput. **8**(1), 36–66 (2012). https://doi.org/10.1016/j.pmcj.2011.01.004
22. Zhou, Z., Chen, X., Li, E., Zeng, L., Luo, K., Zhang, J.: Edge intelligence: paving the last mile of artificial intelligence with edge computing. Proc. IEEE **107**(8), 1738–1762 (2019)

# XSS-Unearth: A Tool for Forensics Analysis of XSS Attacks

Davide Alfieri, Massimo Ficco, Michele Mastroianni(✉), and Francesco Palmieri

Dipartimento di Informatica, Università degli Studi di Salerno, Via Giovanni Paolo II, 132, 84084 Fisciano, SA, Italy
mmastroianni@unisa.it

**Abstract.** One of the most common application layer attacks which also involves cloud systems is called Cross-Site Scripting (XSS), and Those attacks link a victim to the internet by stealing their cookies and other private information. XSS attacks are in the third place the OWASP Top10 of most critical web application threats, and the literature shows a number of tools implemented by scholars and professionals in order to mitigate this particular threat; despite that, there are no tools intended to extract forensics information after a successful or unsuccessful attack.

In this paper is proposed a forensic tool, XSS-Unearth, developed for detection of forensics traces in case of XSS attacks. Using an approach based on a XSS payload list, the tool proposed can be helpful in order to produce digital evidence about XSS attacks. Furthermore, this tool may be successfully integrated with other well-known forensic tools, such as Plaso, to support further analysis.

**Keywords:** Cross Site Scripting (XSS) · forensics analysis · Brute force XSS · injection attacks · blacklist

## 1 Introduction

Due to the pervasiveness of the new ICT technologies, such as cloud and fog systems, and new methods in the field of Artificial Intelligence and Machine Learning, the number of connected devices has been steadily rising in recent years, which has given attackers abundance of new opportunities to attack them. While security professionals work to prevent intrusions, digital forensics specialists are crucial in determining the extent of the harm and the source of the attack [1]. The outcomes of the digital forensics could possibly be regarded as a useful resource for security experts to face the future intrusion in ICT systems [10]. One of the most common application layer attacks which also involves cloud systems is called Cross-Site Scripting (XSS), and Those attacks link a victim to the internet by stealing their cookies and other private information. XSS attacks are in the third place the OWASP Top10 of most critical web application threats, and the literature shows a number of tools implemented by scholars and professionals in order to mitigate


L. Barolli (Ed.): AINA 2024, LNDECT 203, pp. 317–326, 2024.
https://doi.org/10.1007/978-3-031-57931-8_31

this particular threat; despite that, there are no tools intended to extract forensics information after a successful or unsuccessful attack.

In this paper is proposed a forensic tool, XSS-Unearth, devoted to the detection of forensics traces in case of XSS attacks. Furthermore, this tool may be successfully integrated with other well-known forensic tools, such as *Plaso* [5] to help further analysis. The tool has been released on *Github* platform[1].

The paper is organized as follows: in the next Section are listed some related work, in Sect. 3 are discussed the approach used, the design and the implementation of XSS-Unearth, in the Sect. 4 is proposed a case study with the related results, while in the Sect. 5 is shown the possible integration of the proposed tool in other well-known frameworks. The last section is devoted to conclusion and future work.

## 2 Related Work

There are there definitely few contributions in scientific literature regarding gathering of forensics data in case of XSS attacks. A quick search on Scopus database finds only twelve papers, of which only eight are really related to the topic of the research.

In 2007, Wang et al. [21] proposed a scheme to collect evidence after suffering XSS attacks from network systems and the management strategies to prevent XSS attacked from network intrusions; in the paper are also discussed the primary sources of forensics data.

In Suteva et al. [18,19] is proposed a technique called *post-mortem computer forensic analysis* on both victim and attacker machine to find digital evidences useful to identify the XSS attack, using logfiles as the main sources of information.

Kumar et al. presented a XSS Digital Forensics Investigation Model (XSDFIM), including twelve phases, that will help in the admissibility of the gathered evidences [12][11]. Hooshmand et al. [6] propose *D-ForenRIA*, a forensic tool aimed to automatically trace the user activities in Rich Internet Applications (RIAs); the tool needs only the full HTTP traces of the sessions.

Taha et al. [20] proposed an approach based on Python functions and regular expressions to evaluate the efficiency of web pages and prevent XSS attacks. Bhushan [2] discusses the various forensics inspection models presented in scientific literature and proposed a new model specifically suited for XSS attacks.

At the best of our knowledge, in scientific literature are not documented general-purpose tools aimed to examine computer data to automatically gather and/or discover relevant information to be used in a forensics analysis.

## 3 Design and Implementation of XSS-Unearth

The NIST 800-86 [8] outlines the essential phases of a forensic investigation process. The phases defined are the following:

[1] https://github.com/drak3-4598/xss_unearth.

- **Collection**: identification and collection of data from prospective sources;
- **Examination** examine the data previously collected, that is assessing and extracting the pertinent pieces of information from the gathered data;
- **Analysis** study and analyze the data to draw conclusions from it;
- **Reporting** obtaining, arranging, and presenting the data obtained during the previous phase.

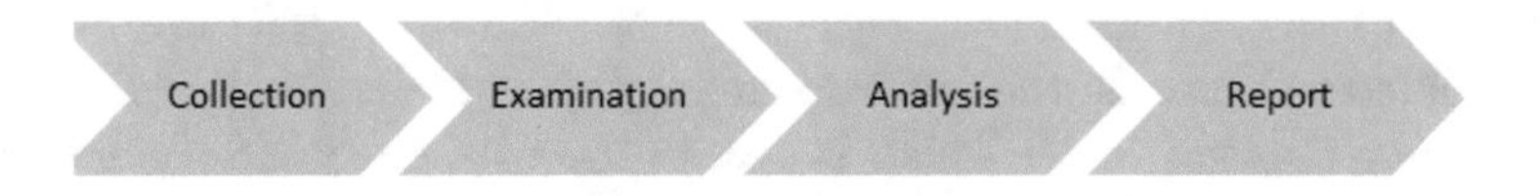


**Fig. 1.** The four phases of forensics process (adapted from [14]).

The time sequence of the four phases is drawn in Fig. 1. XSS-Unearth is meant to be used in the *Analysis* phase, in which the useful data have been collected and examined. The analysis phase aims to determine individuals, locations, objects, and events. It also establishes the connection between the information gathered in order to define conclusions and provide evidence.

From a forensics perspective, the pertinent data and the related artifact in which those data could be located in the case of XSS attacks have been identified by Wang et al. [21], and summarized in Table 1.

**Table 1.** Evidence finding

| Artifact | Description | Evidence |
|---|---|---|
| HTML Tags | Tag Scripts | Potential malicious code |
| | Tag attributes | |
| Register files | Bash history | Activity (sequences of commands typed by the user) |
| | Browser history | Suspicious URLs |
| Log files | Web server | Attacker IP address |
| | | URLs of compromised web pages |
| | | URLs of compromised web pages |
| | | HTTP requests (GET, POST) |
| | | HTTP headers |
| | Database | Injected scripts in stored data |

So, a key role in the forensics analysis is played by computer log files but, although reading through logs by hand is technically feasible, it is far more typical to employ analytical tools and utilities to scan the logs in an effective, heuristic fashion, mostly because log files are often unmanageable due to their huge length. Moreover, because of the vast amount of data gathered from the logs, it is essential to decrease the dimensionality of the features in order to effectively investigate attacks [13].

An effective approach to gathering evidence while reducing the amount of data listed in log files could be suggested by some tools implemented to detect XSS vulnerabilities using a *Brute Force* approach [4], such as BruteXSS[2] and XSSer[3]. In those tools, a dataset containing known malicious code patterns is used in order to generate XSS attack patterns to verify web application vulnerabilities.

The dataset chosen for the tool implementation is *Cross Site Scripting (XSS) Vulnerability Payload List*[4], which has been used by many XSS vulnerability detection tools and contains 1292 different attacks patterns; the dataset has also been successfully used by scholars to implement various XSS scanning and detection tools [9,15].

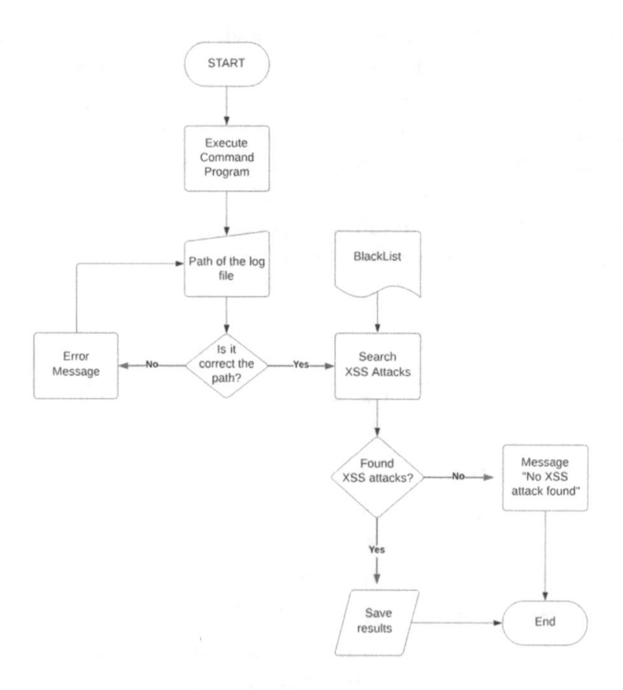


**Fig. 2.** Flowchart Diagram of XSS-Unearth.

The tool implementation is outlined in Fig. 2. XSS-Unearth takes as inputs a series of logfiles, for instance Apache and MySQL logfiles:

- **access_log.log**: is a registry of Apache Tomcat in which all HTTP requests between client and server are recorded.

[2] https://github.com/shawarkhanethicalhacker/BruteXSS-1.
[3] https://tools.kali.org/web-applications/xsser.
[4] https://github.com/payloadbox/xss-payload-list.

- **general_query.log**: is a MySQL registry where all queries are recorded.

Another input file is a *Blacklist* containing the previous identified XSS payloads list. For the implementation the Python programming language version 3.11 has been used, and the following libraries/modules:

- **os**: for handling I/O operations
- **re**: for regular expressions
- **datetime**: for datetime handling
- **urllib.parse**: for parsing URLs

## 4 Case Study

To test the tool, a reference architecture and a test web application have been implemented and a case study has been implemented (Fig. 3). A web application with a small database for the handling data. An attacker discovers some XSS vulnerabilities in it, injects the XSS payload that is stored in the database. When the payload is executed, the attacker is able to capture sensitive data.

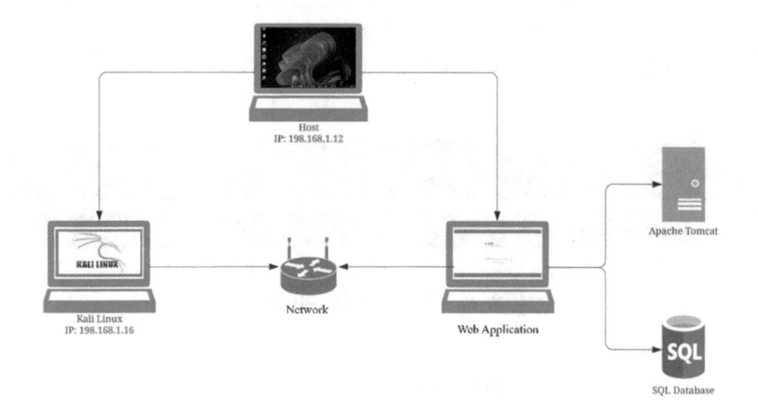


**Fig. 3.** Execution environment of case study

The PC that acts as host for both the web application and Kali Linux has the following characteristics:

- Windows 11 Home 64 bit operating system
- Intel Core (TM) i7-8750H CPU @ 2.20 GHz (12 CPUs), 2.2 GHz
- RAM 16384 MB

To execute the attacks Kali Linux has been chosen, a GNU/Linux distribution based on Debian, designed for computer forensics and computer security. We will use Kali Linux version Linux 6.1.0-kali5-amd64. The web application has been specially designed *vulnerable-by-design* to run attacks. This is a web application where registered users can enter some data. The technologies used for the implementation are Java Spring Boot for the back-end, Thymeleaf for the front-end and MySQL for a small database in which the data. The server hosting the application is Apache Tomcat/10.1.11 on port 8080. Finally, regarding communication, Kali Linux and the application communicate via a virtual network adapter in bridge mode.

In order to test our tool in this context, we have assumed the following hypothetical scenario: an air monitoring network has been implemented using 36 fixed stations and 5 mobile laboratories. The stations analyze the pollutants prescribed by D.Lgs. 155/2010 according to their positions: in general, the pollutants are oxides of nitrogen, carbon monoxide, particulates (P.M. both 10 and 2.5), ozone, benzene and sulfur dioxide. The data are collected hourly and then validated using national guidelines [3].

On 31 August 2023, an attacker discovers some XSS vulnerabilities in the web application and decides to execute these attacks to compromise functionality and data. After reporting, the investigation begins.

So these are the steps to take:

1. **First Step**: Identification of the full path of the log files
   - *C:\..\LogFiles\tomcat\logs\access_log.31-08-2023.log*
   - *C:\..\MySQLServer8.0\Data\general_query.log*
2. **Second Step**: Start the tool and perform analysis on files (Fig. 4)

```
PS C:\Users\dadea\Desktop\xss_unearth\xss_unearth\xss_unearth> python xss_unearth.py
Enter the full path of the log file you want to analyze: C:\Users\dadea\Desktop\Log Files\tomcat\logs\access_log.31-08-2023.log
```

```
PS C:\Users\dadea\Desktop\xss_unearth\xss_unearth\xss_unearth> python xss_unearth.py
Enter the full path of the log file you want to analyze: C:\ProgramData\MySQL\MySQL Server 8.0\Data\general_query.log
```

**Fig. 4.** Performing analysis

3. **Third Step**: Output results (Fig. 5).

```
Found 39 potential XSS attacks:

Results saved to 'Results\xss_results_2023-09-06_13-03.log and Results\summary_2023-09-06_13-03.log
```

```
Found 60 potential XSS attacks:

Results saved to 'Results\xss_results_2023-09-06_13-11.log and Results\summary_2023-09-06_13-11.log
```

**Fig. 5.** Output of analysis

We have seen that the tool performs the analysis correctly and saves the results in two files. Here, we will treat only the file named "summary", while the other one will be discussed in Sect. 5. The summary file contains the following information: total number of potential attacks found, the log lines in which they were tracked and the contents of the lines. Figure 6 shows the analysis summary file on the access_log file:

**Fig. 6.** File summary of access_log.log

We can observe that the forensic investigator from this file can collect the following evidence:

- The **number of attacks** that hit the web application are **39**;
- Detected attacks indicate attempts by the attacker to **execute commands** within HTML comments (e.g. Line 85), **query XSLT** to get information about the XSLT provider and get information about the server (e.g. Riga 247), **run a JavaScript script** within web pages (e.g. Riga 153).
- The server responded more with a **200 success message**.
- The attacks were actually executed on **31/08/2023** in a range of time from **11:34 to 11:37**.
- The attacks come from the IP address: **198.168.1.16**.

Now we examine the general_query summary file shown in Fig. 7:

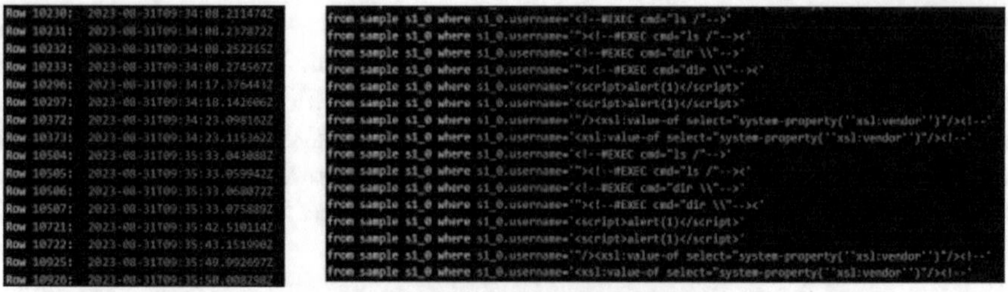

**Fig. 7.** File summary of general_query.log

- The **number of attacks** that hit the web application are **60** (it takes into account that the file is unique and therefore also keeps track of queries made in the past);
- **Malicious SQL queries** were performed to get database information, run malicious scripts
- These executions are recorded on **31/08/2023** in a time interval **09:34 to 9:35** (MySQL uses a different time zone than Apache)

Thus, we can deduce that the tool has actually managed to correctly detect the XSS attack searching within the log files. You can see that the analysis phase has been greatly simplified: in fact, going from a file of about two thousand lines to one of just forty, it was easier to highlight and collect the evidence.

## 5 Integration with Plaso and Other Forensics Tools

The tool implemented is not only meant to perform forensic analysis, but also to support other forensic tools that have different purposes. In general, using more forensic tools and techniques in an investigation is extremely effective and beneficial because you can confirm the findings made by an instrument, increasing the reliability of the evidence collected, maximise the retrieval of information from digital devices, even if data is altered or destroyed, conduct cross-analysis to identify links and relationships between evidence, and maintain a strict chain of evidence custody [7].

Application layer incidents, such as XSS attacks, can be detected inspecting a number of logs, including the network, access, authentication, and other logs, as well as in the numerous log file traces that are kept on the Apache server. In a forensics perspective, the aim of the log analysis is to identify application layer attacks [16].

As an example of the integration of XSS-Unearth with other forensics tools, it has been chosen to use the output of the proposed tool to help timeline. The Plaso framework[5] is a collection of tools, based on Python, designed to help investigators with their timeline research by automatically parsing various log

[5] https://github.com/log2timeline/plaso.

files and artifacts and creating a super timeline in a simple and straightforward manner [17].

XSS-Unearth may contribute enormously to Plaso's work performing the analysis on the log file, highlighting the suspect lines (containing the XSS scripts), and then defines a very precise context. If Plaso goes to process the data with this extract we are quite certain that he can extract the events in the defined context and determine what actions were performed before, during and after the accident.

## 6 Conclusion and Future Work

In this paper has been described design and implementation of XSS-Unearth, a forensics tool suited for server-side analysis of logfiles in order to detect forensics traces of XSS attacks. The use of this new tool may be of great help in investigations and integrates effectively with other forensics framework/tools such as Plaso, etc.

Future work will be oriented to implement integration of XSS.Unearth with other well-known forensic tools such as Autopsy.

## References

1. Barbierato, E., Gribaudo, M., Iacono, M., et al.: A performance modeling language for big data architectures. In: ECMS, pp. 511–517 (2013). https://doi.org/10.7148/2013-0511
2. Bhushan, S.: A novel digital forensic inspection model for XSS attack. In: Kumar, R., Ahn, C.W., Sharma, T.K., Verma, O.P., Agarwal, A. (eds.) Soft Computing: Theories and Applications, pp. 747–759. Springer, Singapore (2022). https://doi.org/10.1007/978-981-19-0707-4_68
3. Campanile, L., Cantiello, P., Iacono, M., Lotito, R., Marulli, F., Mastroianni, M.: Applying machine learning to weather and pollution data analysis for a better management of local areas: the case of Napoli, Italy. In: International Conference on Internet of Things, Big Data and Security (2021). https://api.semanticscholar.org/CorpusID:235259306
4. Caturano, F., Perrone, G., Romano, S.P.: Discovering reflected cross-site scripting vulnerabilities using a multiobjective reinforcement learning environment. Comput. Secur. **103**, 102204 (2021) https://doi.org/10.1016/j.cose.2021.102204. https://www.sciencedirect.com/science/article/pii/S0167404821000286
5. Debinski, M., Breitinger, F., Mohan, P.: Timeline2GUI: a Log2Timeline CSV parser and training scenarios. Digit. Investig. **28**, 34–43 (2019) https://doi.org/10.1016/j.diin.2018.12.004. https://www.sciencedirect.com/science/article/pii/S1742287618303232
6. Hooshmand, S., et al.: D-ForenRIA: distributed reconstruction of user-interactions for rich internet applications. In: Proceedings of the 25th International Conference Companion on World Wide Web, WWW 2016, pp. 211–214. Companion, International World Wide Web Conferences Steering Committee, Republic and Canton of Geneva, CHE (2016). https://doi.org/10.1145/2872518.2890547

7. Javed, A.R., Ahmed, W., Alazab, M., Jalil, Z., Kifayat, K., Gadekallu, T.R.: A comprehensive survey on computer forensics: state-of-the-art, tools, techniques, challenges, and future directions. IEEE Access **10**, 11065–11089 (2022). https://doi.org/10.1109/ACCESS.2022.3142508
8. Kent, K., Chevalier, S., Grance, T.: Guide to integrating forensic techniques into incident. Technical report, 800-86 (2006)
9. Khazal, I.F., Hussain, M.A.: Server side method to detect and prevent stored XSS attack. Iraqi J. Electric. Electron. Eng. **17**(2) (2021)
10. Khodayarseresht, E., Majumdar, S.: Digital forensics for emerging technologies: present and future. In: Innovations in Digital Forensics, pp. 1–11. World Scientific (2023)
11. Kumar, S., Pathak, S., Singh, J.: A comprehensive study of XSS attack and the digital forensic models to gather the evidence. ECS Trans. **107**(1), 7153 (2022). https://doi.org/10.1149/10701.7153ecst
12. Kumar, S., Pathak, S., Singh, J.: An enhanced digital forensic investigation framework for XSS attack. J. Discrete Math. Sci. Crypt. **25**(4), 1009–1018 (2022). https://doi.org/10.1080/09720529.2022.2072424
13. Lin, X.: Introductory Computer Forensics: A Hands-on Practical Approach. Springer, Cham (2018). https://doi.org/10.1007/978-3-030-00581-8
14. Maratsi, M.I., Popov, O., Alexopoulos, C., Charalabidis, Y.: Ethical and legal aspects of digital forensics algorithms: the case of digital evidence acquisition. In: Proceedings of the 15th International Conference on Theory and Practice of Electronic Governance, ICEGOV 2022, pp. 32–40. Association for Computing Machinery, New York, NY, USA (2022). https://doi.org/10.1145/3560107.3560114
15. Perumal, S., Sujatha P, K.: Stacking ensemble-based XSS attack detection strategy using classification algorithms. In: 2021 6th International Conference on Communication and Electronics Systems (ICCES), pp. 897–901 (2021). https://doi.org/10.1109/ICCES51350.2021.9489177
16. Sree, T.R., Bhanu, S.M.S.: Data collection techniques for forensic investigation in cloud. In: Shetty, B.S.K., Pavanchand Shetty, H. (eds.) Digital Forensic Science, chap. 6. IntechOpen, Rijeka (2020). https://doi.org/10.5772/intechopen.82013
17. Studiawan, H., Hasan, M.F., Pratomo, B.A.: Rule-based entity recognition for forensic timeline. In: 2023 Conference on Information Communications Technology and Society (ICTAS), pp. 1–6 (2023). https://doi.org/10.1109/ICTAS56421.2023.10082742
18. Suteva, N., Mileva, A., Loleski, M.: Computer forensic analisys of some web attacks. In: World Congress on Internet Security (WorldCIS-2014), pp. 42–47 (2014). https://doi.org/10.1109/WorldCIS.2014.7028164
19. Suteva, N., Mileva, A., Loleski, M.: Finding forensic evidence for several web attacks. Int. J. Internet Technol. Secured Trans. **6**, 64–78 (2015)
20. Taha, T.A., Karabatak, M.: A proposed approach for preventing cross-site scripting. In: 2018 6th International Symposium on Digital Forensic and Security (ISDFS), pp. 1–4 (2018). https://api.semanticscholar.org/CorpusID:19189439
21. Wang, S.J., Chang, Y.H., Chiang, W.Y., Juang, W.S.: Investigations in cross-site script on web-systems gathering digital evidence against cyber-intrusions. Future Gener. Commun. Network. (FGCN 2007) **2**, 125–129 (2007). https://api.semanticscholar.org/CorpusID:687227

# MUSA: A Platform for Data-Intensive Services in Edge-Cloud Continuum

Marco Anisetti[1], Claudio A. Ardagna[1], Massimo Banzi[2], Filippo Berto[1](✉), Ruslan Bondaruc[1], Ernesto Damiani[1], Alessandro Pedretti[1], Arianna Pisati[1], and Antonio Retico[3]

[1] Università degli Studi di Milano, Milan, Italy
{marco.anisetti,claudioa.ardagna,filippo.berto,ruslan.bondaruc, ernesto.damiani,alessandro.pedretti,arianna.pisati}@unimi.it

[2] Telecom Italia, Rome, Italy
massimo.banzi@telecomitalia.it

[3] Almaviva, Rome, Italy
a.retico@almaviva.it

**Abstract.** In the rapidly evolving landscape of modern applications, the Edge-Cloud Continuum emerges as a pivotal paradigm, promising unprecedented flexibility and efficiency in service deployments. As the demand for low-latency, high-throughput applications intensifies, the Continuum provides a dynamic framework, enabling the distribution of computational tasks between centralized cloud servers and decentralized edge devices. However, the transition from traditional cloud-centric models to the Continuum introduces complexities that necessitate careful consideration. Besides development, Continuum applications call for a placement process with the aim to allocate services to the best suitable deployment node, according to application requirements and nodes capabilities. Furthermore, controlling non-functional properties within the Cloud-Edge Continuum and balancing trade-offs between performance, reliability, and security becomes increasingly intricate in this distributed architecture. This paper addresses the above challenges proposing MUSA, a deployment platform for data-intensive workflows of services integrating continuous non-functional properties verification.

## 1 Introduction

The advent of Edge Computing has reshaped application deployment, offering improved support for real-time applications and better service placement to fulfill Quality of Service (QoS) requests. Despite physical separation, the edge relies heavily on the cloud for resource-intensive tasks, accentuated by the rise of machine learning (ML) and distributed approaches like federated learning (FL) [1]. This scenario resulted in a novel computational paradigm, the Edge-Cloud Continuum, which merges cloud and edge strengths, delivering scalability, flexibility, and mobility.

Handling services in edge-cloud continuum rises a number of challenges, both technological and administrative, also considering that over the last decade the IT architectures have rapidly evolved from predominantly holistic paradigms to increasingly fragmented and very specialized service offerings. This architectures more and more often


L. Barolli (Ed.): AINA 2024, LNDECT 203, pp. 327–337, 2024.
https://doi.org/10.1007/978-3-031-57931-8_32

spawn outside the boundaries of single organizations and become part of complex Business to Business (B2B) ecosystems. From the technology perspective, this has brought to the production of tools and platforms that are very specialized for a specific aspect or goal in data and service management.

Our paper presents the initial results of Project Multilayered Urban Sustainability Action (MUSA)[1], an innovation ecosystem funded by NRRP – Next Generation Europe, a platform for handling data-intensive workflows operating in Edge-Cloud continuum and continuously monitored to guarantee non-functional behavior such as privacy and security. The MUSA platform unprecedented ambition is to encompass, organize and serve a vast set of tools and technologies, and to convey it into a single and well-governed service platform. The MUSA platform is inspired by key motivations directly descending from the MUSA ambitions, data strategy, and regulatory constraints, which can be translated into well-defined architectural principles and drivers, as follows:

- An holistic, innovative, and long-term sustainable digital architecture for the storage and safe exchange of big data for life sciences.
- Big Data for the development and sharing of new technologies in life sciences and medicine research.
- Development of technologies and customizable tools for continuous monitoring, well-being, and health.
- Develop, implement, and sustain technological innovation in health.
- Enable economic impact of structured collection of big data in life science.

The MUSA architecture development methodological approach is based on alternate rounds of software selection and pilot realizations, with a constant monitoring and tracking of architecture decisions. Sector best practices and gradually consolidating requirements from research use cases are the input for realizing the MUSA Cloud Data Hub Architecture (Sect. 4), supporting continuous adaptation to the evolving context of MUSA research activities.

The contribution of this paper is threefold: *i)* a new notion of Edge-Cloud Continuum empowered by 5G MEC nodes (Sect. 3), *ii)* a service framework for managing

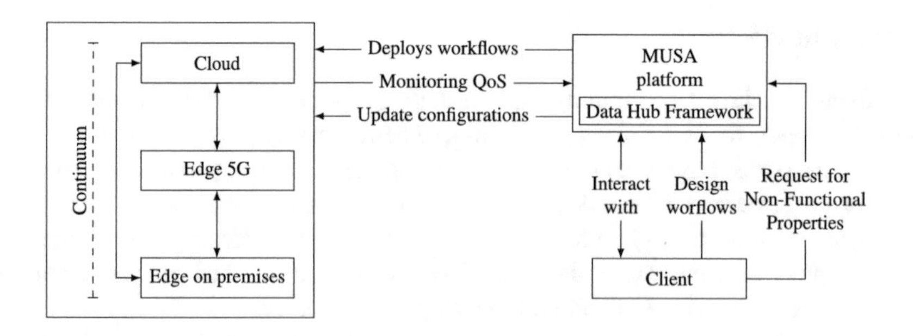


**Fig. 1.** MUSA's Edge Continuum workflow deployment.

[1] https://musascarl.it/.

workflow execution on such advanced Edge-Cloud Continuum (Sect. 4), *iii)* proof of concept pilots showing the utility and feasibility of MUSA (Sect. 6).

## 2 Related Work

The Continuum aims to decouple resource management and deployment configuration from application execution. This is achieved by abstracting the deployment infrastructure, allowing developers to define applications in terms of functional capabilities and desired QoS properties [2]. The assumption is that non-functional properties can be ensured through suitable deployment configurations without modifying application logic [3]. For example, in a Continuum, an application's services can be deployed for low latency and privacy in data gathering (on-premises or Edge), and high computational power for data modeling (Cloud). This contrasts with Cloud-only or Edge-only scenarios, where similar requirements may demand application logic modification.

Despite the Edge-Cloud Continuum's potential, research into its full exploitation is in its infancy. Existing solutions, mainly in serverless computation, fall short for complex Continuum applications, lacking a non-functional property-aware approach [4]. Some works [5–8] focus on the Edge-Cloud Continuum scenario instead. Nevertheless, only the architecture presented in [7] stands out for its ability to seamlessly deploy applications along the Continuum, being fully independent of specific technology and cloud service provider (CSP) involvement. Most works [6,7,9–12] take QoS requirements and constraints into account, either fully or partially. Notably, only [6,9] offer a comprehensive approach to modeling both applications and the environment. Unfortunately, none of the existing solutions address the life-cycle management of deployment, adapting to changes in context. In summary, each solution has its drawbacks, such as limited application scenarios, dependence on specific technologies, or inadequate composition models.

## 3 Edge-Cloud Continuum Digital Platform Enabled by 5G Network

Figure 3 shows the MUSA Edge-Cloud Continuum platform comprising three main elements: a public Cloud platform, a 5G-enabled edge computing platform, and an on-premises edge platform. The MUSA platform manages these elements through a transparent deployment framework, leveraging their unique features to provide an improved experience for service workflows in a uniform management solution. MUSA will also provide a set of managed cloud services, in the form of Software as a Service, using the same continuum infrastructure. A client can interact with the MUSA platform to design and deploy service workflows made of custom services and platform services. Among the available services, MUSA offers a Data HUB Framework to support advanced storage of data for data intensive workflows. In addition, clients can annotate their workflows with non-functional requirements to be satisfied. The MUSA platform will then instruments the workflows and select deployment to fulfill these requirements.

The main difference between MUSA's continuum infrastructure and those previously proposed in the literature are *i)* the presence of a 5G-enabled edge platform, *ii)* the possibility of deploying non-functional-aware workflows of services, *iii)* the intensive data processing carried out by such workflows deployed on the continuum. In order to obtain such infrastructure, Internet Service Providers (ISP) and Telco providers collaborate with MUSA, providing both network and computation resources with fine-grained control. The MUSA platform can then manage these resources by deploying services on the ISP's servers or Telco provider's Mobile Edge Computing nodes [13, 14], allowing them access through the ISP/Telco providers' private network. This solution offers a compromise between cloud and on-premises solutions, providing high availability and access to hardware resources while also ensuring stronger guarantees on the privacy and security of the data. The private nature of 5G's network slices enables the creation of a private high-performance network link between the on-premises infrastructure and the 5G edge platform.

MUSA 5G edge nodes will play a fundamental role for building an effective continuum, providing standardized high reliability infrastructure for service delivery and execution. The MUSA deployment is distributed across 5G Edge, on-premises edge and Cloud computing nodes, exploiting the capabilities provided by the 5G core network, such as low latency, network slicing and computation locality. MUSA platform is grounded on the capability of 5G Core Network to *i)* set up private connections having advanced non functional properties and *ii)* push the computation to the Edge MEC, as close as possible to the end users, *iii)* automatically adapting its configuration to ensure the target QoS levels. The MUSA Edge 5G provides automated management of services in virtualized or containerized environments. These are exposed and integrated in the 5G network as network services, allowing to transparently deploy them in the mobile service provider's data center. This solution enables several use cases where end-users may prefer CSP-managed services or platforms but have strict privacy or performance requirements that limit them to on-premises hosting. MUSA's edge-continuum solution offers the security and privacy advantages of on-premises hosting while retaining the scalability and management delegation of cloud platforms.

## 4 MUSA Cloud Data Hub Framework

The MUSA Cloud Data Hub is a service framework with the ambitious goal of providing the ecosystem for an effective and safe exchange of data and services related to life science coming from research activities, biomedical data analysis, and diagnostics monitoring of human behaviors. The framework is at the basis of the MUSA platform and allows to support the management of service workflows over the Edge-Cloud Continuum.

The Service Framework, represented at high-level in Fig. 2, is based on a multi-layered service infrastructure that features: i) A catalog of added-value IT services designed to fit the domain business cases, ii) Technical services for the seamless distributed deployment of added-value services across the Continuum, iii) Technical services for the enrollment, integration and certification of new services in the MUSA Cloud Hub. The ecosystem will be based on an innovative distributed edge-cloud network by federating a central cloud platform, 5G edge-nodes and local nodes made available by MUSA partners.

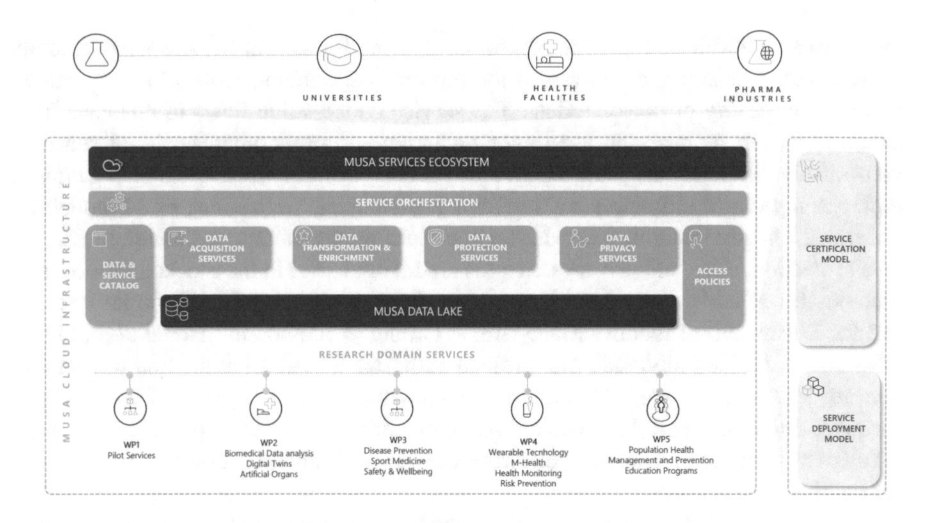


**Fig. 2.** MUSA Cloud Data Hub: high-level view.

The Service Framework is practically realized by a real-life platform model, the MUSA Cloud Platform (MCP). The MCP is a durable, scalable, cost-effective, and flexible infrastructure for data processing, storage, and analytics featuring all the applications and technical components suitable for satisfying the different stages of data and service lifecycle.

The MCP is implemented giving preference to Open Source packages, choice that offers characteristics of transparency, security, cost-effectiveness, and interoperability while fostering innovation through experimentation and rapid development, and minimizing the risk of technological lock-in. The software selection we use to implement MCP is compliant to the ISO/IEC 25010[2] and 25012[3] standards, defining the quality characteristics of software products and their use. Namely, the products selected for the platform stack are evaluated in terms of their characteristics of 1) Functional Suitability; 2) Performance efficiency; 3) Compatibility; 4) Usability; 5) Reliability; 6) Security; 7) Maintainability 8) Portability. Specifically with respect to the latter category, and for the sake of guaranteeing the fulfillment of the MUSA requirement of straightforward deployment of data/service pipelines across the continuum, the components are selected also based on the possibility for them to be deployed in a containerized and orchestrated environment.

From the functional point of view a seamless integration with the EDGE for the realization of the Cloud-EDGE Continuum is assured by the *Multi-Protocol Ingestion* services, funneling e.g. data from the EDGE into the Big Data Layer through data pipelines built over a *lambda* architecture [15]. The inbound data acquisition/enrichment/serving flow from EDGE to cloud is of course only one of the supported patterns. Among

[2] ISO - ISO/IEC 25010:2011—Systems and software engineering - Systems and software Quality Requirements and Evaluation (SQuaRE)—System and software quality models.

[3] ISO - ISO/IEC 25012:2008—Software engineering - Software product Quality Requirements and Evaluation (SQuaRE)—Data quality model.

the Advanced Services, for instance, the Local Data Protection Service is specifically designed for research and analysis in the bio-medical domain, remarkably delicate in terms of sensibility of personal data. The service added-value flows in this case from Cloud to EDGE by enabling local anonymization and pseudonymization of sensible, non-transferable data-sets with techniques ranging from *homomorphic encryption* to *differential privacy*, ultimately making them available e.g. for training of AI models by Federated Machine Learning Services. Other cutting-edge features directed to address data protection, trust, and privacy concerns are recognizable in the blockchain-powered modules for reliable, non-alterable, non-repudiable Consent/Policy/Contract tracking and for Decentralized Identity Management aiming to provide individual digital identities and to enable individual, fine-grained authorization, accounting, and traceability both for MUSA managed users and of anonymous users by 3rd-party MUSA business partners. Enhanced by a passwordless authentication system, this approach puts users in full control of their credentials, ensuring secure and seamless access without the need for traditional passwords.

The MUSA services are made available to be composed into the service workflows to the user communities and partners for the realization of virtuous ecosystems by multiple integration patterns and interfaces leveraging both on API integration technologies and web solutions, with a particular focus being dedicated to ensure, in any operational circumstances, the traceability, certification, and non-repudiation of user actions across a natively distributed deployment.

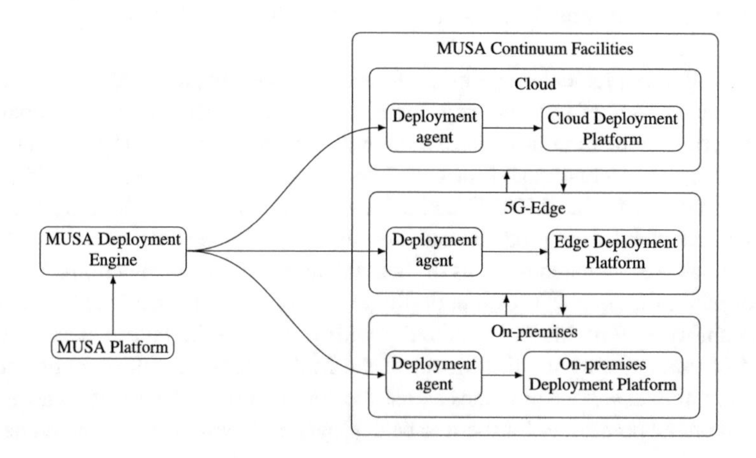


**Fig. 3.** Deployment Architecture for Edge-Cloud Continuum.

## 5 MUSA Deployment Methodology

In this section, we put forward a deployment architecture and process within the Continuum, focusing on ensuring QoS for service workflows and detailing the collaboration between the MUSA Deployment Engine, Facilities, and Deployment Agents. We also discuss how Cloud Service Providers play a crucial role in supporting this integration.

### 5.1 Deployment Architecture

We propose a deployment architecture designed to ensure a seamless Quality of Service (QoS) and adherence to constraints during the execution of a service workflow within the Continuum, as depicted in Fig. 3. This architecture grounds on the MUSA Deployment Engine service, a central orchestrator that interacts with Deployment Facilities, including cloud platforms, telco-edge, and on-premises nodes, through Deployment Agents. Clients engage with the Deployment Engine via the MUSA Platform, which allows to submit service workflows along with QoS and constraints specifications. Agents, in turn, receive from the Engine a deployment recipe that holds the deployment instructions on what services to deploy, which are the target facilities, what configuration parameters to use, and what communication links are needed between services.

In this setup, the MUSA Deployment Engine is tasked with selecting the optimal deployment configuration. It implements a deployment algorithm/strategy that deploys services where all requirements are satisfied and, in addition, can optimize some metrics given by the CSP (e.g., cost efficiency, load balancing, resource utilization, user proximity). The implementation of an optimal deployment strategy is beyond the scope of this work, but many approaches can be followed inspired by the state-of-the-art in the service placement domain, namely heuristic, optimization and machine learning techniques. Generating a deployment configuration involves also gathering information about the capabilities of the Deployment Facilities, which are matched with requirements to find eligible deployments. Capabilities are provided by the CSP for each facility it owns and are expressed according to a vocabulary shared by all CSPs.

Subsequently to the configuration generation, deployment instructions are issued to the Deployment Agents. These Agents are crucial in establishing a deployment continuum across the facilities, guiding the service deployment process and ensuring integration that is aware of QoS and constraints. To support this integration, the Cloud Service Provider (CSP) provides the Deployment Agents with access to essential resources and services, such as resource management for deployment and security services. This can include access to non-functional certificates or authentication services.

The architecture also accommodates changes in QoS or budget constraints, allowing for the redeployment of the workflow with updated parameters. This redeployment capability is not only useful for adapting to changes in requirements but also facilitates service migrations and other changes that might occur post-deployment.

### 5.2 Deployment Process

Building on the architecture shown in Fig. 1, we presents a methodology for deploying services in the Edge-Cloud Continuum. This process starts with a client creating a service composition workflow and adding Quality of Service (QoS) requirements and constraints, resulting in what we call an Annotated Service Composition Template. This template is a graphical representation of the services a client wishes to deploy, outlining how these services interact and their execution parameters. It's essentially a directed graph with services as nodes and interactions (like data and control exchanges) as edges.

Service providers then label their facilities with appropriate properties and provide access points (hooks) for Deployment Agents, creating an Annotated Deployment Facilities Graph. This is a directed graph representing each facility as a vertex, connected by links. These links indicate how facilities can communicate, whether through the internet or private channels. The client sends this template to the MUSA Deployment Engine, which triggers the deployment matching. This is the process of finding the best way to deploy a service workflow in the Continuum, considering the QoS requirements and constraints. It involves comparing the Annotated Service Composition Template and the Annotated Continuum Facilities Graph to find a suitable deployment solution. The solution must meet all requirements and constraints for the services and their interactions. If no suitable solution is found, it means deployment is not possible under the given conditions. If there are multiple solutions, they are ranked according to the CSP's policy, like minimizing operational costs.

This process generates Deployment Recipes for service workflow deployment across the Edge-Cloud Continuum. Such recipes provide detailed instructions on how to deploy the service and integrate it with the necessary components of the facility. They cover aspects like service configuration, required support components, and their integration modes. This methodology ensures that all QoS requirements and constraints are met while aligning with the CSP's internal policies. However, the specifics of generating these deployment recipes from the selected deployment solution are beyond this paper's scope and will be explored in future work.

## 6 Proof of Concept

We present a first proof of concept of our platform in operation, focusing on a real use case on molecular docking.

### 6.1 Use Case: Molecular Docking

Given the large extent of information on ligands and targets involved in diseases available in public databases, increasing efforts have been made on the application of in silico strategies over the last decades. Indeed, it has been demonstrated that in silico approaches provide novel valuable opportunities for drug discovery and development [16]. In this regard, molecular docking has become among one of the widely used computational approaches to repurpose compounds towards novel therapeutic targets [17]. This structure-based method [16] can be used to predict the foremost binding modes of a given ligand within the binding site of a target protein. The docking process embraces two main steps first, sampling conformations of the ligand in the active site of the protein, then ranking these conformations through a scoring function [18, 19]. Docking can be applied not only in drug discovery, but also in reverse screening techniques, which allow predicting the biological targets of a molecule of interest, representing a valuable approach for computational target fishing and profiling [20]. Although good biological activity against an intended target is necessary, a drug candidate needs to pass additional tests and clinical trials for toxicity, side effects, bioavailability, and efficacy [21]. Since some of these tests may be performed by in silico approaches as the

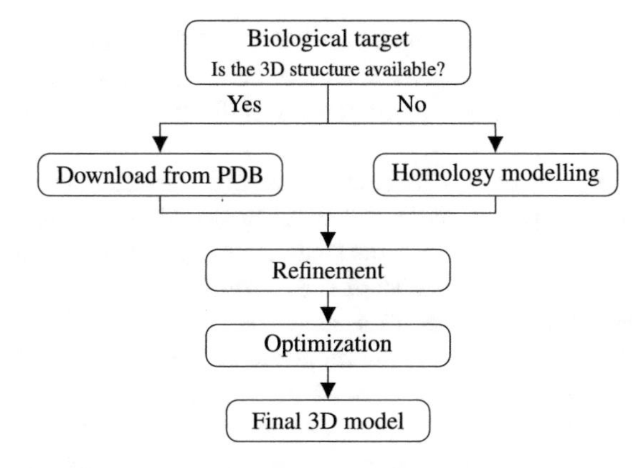


**Fig. 4.** Protein preparation workflow.

docking techniques, such methods may be therefore regarded as a tool for a low-cost and fast-speed testing in the early stages of drug development [22]. A large database of 3D structures of biologically relevant proteins is required in order to perform the reverse screening which consists in multiple molecular docking calculations of a given query molecule with each the binding site of each identified protein. In details, 485 proteins were collected as drug targets, namely proteins associated with a specific disease and for which either an approved drug or a drug under clinical trial is available. The structures were downloaded from the Protein Data Bank (PDB) following two main criteria: (I) the presence of a ligand in the active site and (II) the quality of the experimental model in term of resolution. In case of homomultimeric proteins, only a single monomer was selected according to their quality and structure completeness. The structures have then been prepared by removing crystallization chemicals and non-relevant co-crystallized molecules, filling the unresolved gaps and loops by modeling them (if required) by homology techniques. If no structure matching the required criteria was available, they have been either entirely modeled through homology modeling when the similarity with a resolved structure was greater than 0.8 or built through ab initio deep learning-based algorithms. Finally, the models have been completed by adding the hydrogens, ionized at physiological pH (7.4) and minimized by Molecular Mechanics (MM) algorithms. A schematic representation of such workflow is shown in Fig. 1. As a last thing we identified the binding site's coordinates for ten structures that have been selected to test the workflow execution.

### 6.2 Use Case Design and Deployment

To design the molecular docking pipeline, the customer, who is an expert in molecular design, access the MUSA platform, picks the services needed for the process (some of them are customer's provided services containerized by MUSA), combines them in a workflow, and selects for each service a set of non-functional requirements to be satisfied. After submission, the MUSA platform generates a suitable deployment by matching the customer's requirements with the capabilities provided by the MUSA facilities associated to the customer, and assigns each service to the Deployment Agent running on the matched facility. The Agents are then in charge of providing the services with the required resources, to set the environment to fulfill the requirements, and finally to run the services.

Given the peculiarity of this molecular docking workflow and the requirements (i.e., computational efficiency), molecular docking workflow was executed entirely on Cloud. As depicted in Fig. 4 the workflow takes as input a molecule and a set of proteins for docking, and encompasses three steps (i.e., format verification, format conversion, and docking). The first step checks if the molecule is given in a 3D format. If not, the second step is performed to convert the molecule in 3D. Otherwise, the last step, i.e. docking, is executed in parallel for each of the specified proteins in the input. Each step was containerized and the workflow was designed on AWS Step Functions This implementation streamlines the execution of the molecular docking process, making it a robust and efficient tool for virtual screening in drug discovery.

## 7 Conclusions

With the fast evolution of modern applications, controlling non-functional properties without impacting on application logic is vital. This work aims at proposing MUSA a platform for services workflow management in the Edge-Cloud Continuum, empowered by 5G technologies and including a Cloud Data Hub as a services catalog.

**Acknowledgment.** This work is partly supported by the project MUSA – Multilayered Urban Sustainability Action – project, funded by the European Union – NextGenerationEU, under the National Recovery and Resilience Plan (NRRP) Mission 4 Component 2 Investment Line 1.5: Strengthening of research structures and creation of R&D "innovation ecosystems", set up of "territorial leaders in R&D" (CUP G43C22001370007, Code ECS00000037). It is also partially supported by Università degli Studi di Milano via the program "piano sostegno alla ricerca" and "One Health Action Hub: University Task Force for the resilience of territorial ecosystems", – PSR 2021 – GSA – Linea 6.

## References

1. Liu, L., Zhang, J., Song, S., Letaief, K.B.: Client-edge-cloud hierarchical federated learning. In: ICC 2020 - 2020 IEEE International Conference on Communications (ICC), pp. 1–6 (2020)
2. Bittencourt, L., et al.: The internet of things, fog and cloud continuum: integration and challenges. Internet Things **3–4**, 134–155 (2018)
3. Anisetti, M., Berto, F., Bondaruc, R.: QoS-aware deployment of service compositions in 5G-empowered edge-cloud continuum. In: 2023 IEEE 16th International Conference on Cloud Computing (CLOUD), pp. 471–478 (2023)
4. Shafiei, H., Khonsari, A., Mousavi, P.: Serverless computing: a survey of opportunities, challenges, and applications. ACM Comput. Surv. **54**(11s) (2022)
5. Fu, K., Zhang, W., Chen, Q., Zeng, D., Guo, M.: Adaptive resource efficient microservice deployment in cloud-edge continuum. IEEE Trans. Parallel Distrib. Syst. **33**(8), 1825–1840 (2022)

6. Orive, A., Agirre, A., Truong, H.-L., Sarachaga, I., Marcos, M.: Quality of service aware orchestration for cloud-edge continuum applications. Sensors **22**(5), 1755 (2022)
7. Casola, V., Benedictis, A.D., Martino, S.D., Mazzocca, N., Starace, L.L.L.: Security-aware deployment optimization of cloud-edge systems in industrial IoT. IEEE Internet Things J. **8**(16), 12 724–12 733 (2021)
8. Nastic, S., Raith, P., Furutanpey, A., Pusztai, T., Dustdar, S.: A serverless computing fabric for edge & cloud. In: 2022 IEEE 4th International Conference on Cognitive Machine Intelligence (CogMI), pp. 1–12 (2022)
9. Brogi, A., Forti, S.: QoS-aware deployment of IoT applications through the fog. IEEE Internet Things J. **4**(5), 1185–1192 (2017)
10. Akhtar, N., Raza, A., Ishakian, V., Matta, I.: COSE: configuring serverless functions using statistical learning. In: IEEE INFOCOM 2020 - IEEE Conference on Computer Communications, pp. 129–138 (2020). ISSN: 2641-9874
11. Anisetti, M., Ardagna, C.A., Damiani, E., Gaudenzi, F., Jeon, G.: Cost-effective deployment of certified cloud composite services. J. Parallel Distrib. Comput. **135**, 203–218 (2020)
12. Quenum, J.G., Josua, J.: Multi-cloud serverless function composition. In: Proceedings of the 14th IEEE/ACM International Conference on Utility and Cloud Computing, ser. UCC 2021, pp. 1–10. Association for Computing Machinery, New York, NY, USA (2021)
13. Anisetti, M., Berto, F., Banzi, M.: Orchestration of data-intensive pipeline in 5G-enabled edge continuum. In: 2022 IEEE World Congress on Services (SERVICES), pp. 2–10 (2022)
14. Ranaweera, P., Jurcut, A., Liyanage, M.: MEC-enabled 5G Use cases: a survey on security vulnerabilities and countermeasures. ACM Comput. Surv. **54**(9), 186:1–186:37 (2021). https://dl.acm.org/doi/10.1145/3474552
15. Marz, N., Warren, J.: Big Data: Principles and Best Practices of Scalable Real-Time Data Systems. Manning Publications Co. (2015)
16. Barneh, F., Jafari, M., Mirzaie, M.: Updates on drug-target network; facilitating polypharmacology and data integration by growth of DrugBank database. Briefings Bioinf. **17**(6), 1070–1080 (2015)
17. Pushpakom, S., et al.: Drug repurposing: progress, challenges and recommendations. Nat. Rev. Drug Discovery **18**(1), 41–58 (2019)
18. Halperin, I., Ma, B., Wolfson, H., Nussinov, R.: Principles of docking: an overview of search algorithms and a guide to scoring functions. Proteins: Struct., Funct., Bioinf. **47**(4), 409–443 (2002)
19. Kitchen, D.B., Decornez, H., Furr, J.R., Bajorath, J.: Docking and scoring in virtual screening for drug discovery: methods and applications. Nat. Rev. Drug Discovery **3**(11), 935–949 (2004)
20. Chen, Y., Zhi, D.: Ligand-protein inverse docking and its potential use in the computer search of protein targets of a small molecule. Proteins: Struct., Funct., Bioinf. **43**(2), 217–226 (2001)
21. DiMasi, J.A., Bryant, N.R., Lasagna, L.: New drug development in the United States from 1963 to 1990. Clin. Pharmacol. Ther. **50**(5–1), 471–486 (1991)
22. Azzaoui, K., et al.: Modeling promiscuity based on in vitro safety pharmacology profiling data. ChemMedChem **2**(6), 874–880 (2007)

# Towards a Patterns Driven Cloud Edge Continuum Architecture for eHealth

Alessia Sabia[1(✉)] and Beniamino Di Martino[1,2,3]

[1] Department of Engineering, Università degli Studi della Campania "Luigi Vanvitelli", via Roma 4, 81031 Aversa, Italy
{alessia.sabia, beniamino.dimartino}@unicampus.it

[2] Department of Computer Science and Information Engineering, ASIA University, Taichung, Taiwan

[3] Department of Computer Science, University of Vienna, Vienna, Austria

**Abstract.** Nowadays different software applications rely on server cloud to host their resources and provide their services. The advent of cloud computing has revolutionized software development and maintenance processes but it also introduced vulnerabilities due to network connectivity and data storage, especially in real-time scenarios and sensitive sectors such as healthcare. These concerns have raised the necessity of a new kind of system capable of providing a solution to these problem, leading to the development of edge-based systems. Such systems aim to pre-process data near its source, optimizing bandwidth usage, ensuring data privacy and enabling timely data processing. This paper presents a Cloud Edge Architecture Patterns-based system for a Real-Time Drug Response Monitoring. The system architecture captures biometric data via IoT devices, processes it locally, and uses Federated Learning for global insights, exploiting Federated Patterns for model improvements. This approach not only enhances data privacy but also provides pharmaceutical insights and paves the way for advancements in precision medicine.

## 1 Introduction

Cloud Computing is a term used to indicate a set of resources and services that are made available at a low cost and in a flexible way. The advent of Cloud Computing has in fact brought numerous benefits [5], especially in contexts where the required resources need to be immediately available and scalable without suffering excessive costs and time [9]. The architecture of Cloud Computing is a layered architecture that is implemented using data centers in which thousand of nodes are stacked together. Basically, the system is composed of three layers: the Resource layer, which consists of physical or virtualized resources; the Platform layer containing application capability components and computing frameworks and the Application Layer which includes components such as analytical, transactional and interactive components [12]. We can indeeed distinguish between Public, Private, Community and Hybrid Cloud; the main difference between


L. Barolli (Ed.): AINA 2024, LNDECT 203, pp. 338–346, 2024.
https://doi.org/10.1007/978-3-031-57931-8_33

them lies in accessibility. In Public Cloud anyone can access to the allocated resources since users access and share the same infrastructure; in Private Cloud resources are restricted to the organizations and are not shared with other external parties.

However, several concerns arise using Cloud Computing solutions as there's nothing that can completely guarantee the protection of data stored in the Cloud [3]. These issues are of particular importance in context such as healthcare where the privacy of data can compromise the usability of the system itself. Furthermore, systems functionalities are highly dependent on the network connection quality; in fact a mere disconnection can immediately disrupt access to services. Another important issue arises due to IoT devices, as the presence of large amount of devices that produce a huge amount of data increases problems related to data flows and network bandwidth [10]. All these vulnerabilities are of extremely importance in real-time environments, where data transmission have to be immediate in order to have a reliable service. The advent of Edge Computing introduces the possibility to process data as close as possible to its source. Unlike Cloud systems, in Edge Computing data is first pre-processed at Edge Nodes, in order to filter only relevant data; subsequently the data is forwarded to the central Cloud Server for computationally intensive tasks [7]. This approach has several benefits: it optimizes the bandwidth usage of IoT devices, guarantees efficient data transmission and ensure data privacy since raw data is processed at Edge Nodes. The aim of this paper is to introduce a pattern-driven architecture for a Real-Time Drug Response Monitoring System, showing how Cloud Edge and Federated Learning architectural patterns can help in the development of systems relevant also for sensitive context such as healthcare. In the following sections, we first introduce the primary design patterns employed in Cloud Edge systems, then a case study focused on a real-time system for drug response monitoring is presented, demonstrating how Cloud Edge and Federated Learning patterns can be exploited to realize the system.

## 2 State of the Art

In the context of software engineering, a design pattern is described as a reusable solution to a recognized issue. Once a particular problem is identified, design patterns assist in preventing the implementation of new solutions that could lead to errors. For the development of Cloud Edge systems, various design patterns are defined at different levels; we can distinguish between Deployment Patterns, Computational Patterns, and Architectural Patterns [2,6,8]. Deployment Patterns describe how the software is deployed on the target architecture, Computational Patterns detail how computation is organized among the various nodes, whereas Architectural Patterns outline how nodes should be organized within the system.

Among the **Deployment Patterns**, we find the *Canary Deployment* and the *Blue-Green Deployment* [17]. Specifically, in the *Canary Deployment* Fig. 1, the deployment occurs on a limited set of servers referred to as *Canary*, where the functionality introduced into the system is tested. In the *Blue-Green Deployment*, different servers are utilized, upon which a router switches, and each of them contains various versions of the software.

Regarding **Architectural Patterns**, we can find the *Multi-tier* pattern that is defined as an architecture in which various components are organized into tiers or layers [15]. Among the Architectural Patterns for Cloud-Edge systems in the *Multi-tier* context, we find the *Edge One Tier Architecture Pattern*, Fig. 2.

This pattern consists of a single-layer architecture, made up of a bottom layer containing IoT devices connected to Edge Nodes through a simple WLAN connection. The *Edge Two Tier Architecture Pattern* in Fig. 3, similarly to the previous pattern, is composed of a bottom layer that contains IoT devices; the second layer contains Edge Nodes connected to IoT devices through a WLAN connection on one side, while on the other side they are connected to a central Cloud server through a WAN connection. Lastly, we have the *Edge Two Tier Architecture Pattern with Orchestrator*, which is an extension of the Edge Two Tier pattern that includes an Edge Orchestrator responsible for managing the workload among the various Edge Nodes in the network.

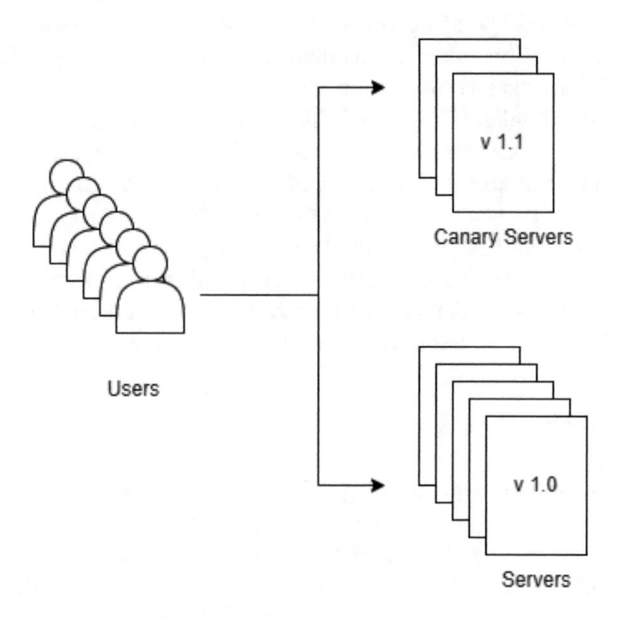


**Fig. 1.** Canary Deployment

## 3 The Proposed Architecture

In [4], Eysenbach states that

> eHealth is an emerging field in the intersection of medical informatics, public health and business, referring to health services and information

delivered or enhanced through the Internet and related technologies. In a broader sense, the term characterizes not only a technical development, but also a state-of-mind, a way of thinking, an attitude, and a commitment for networked, global thinking, to improve health care locally, regionally, and worldwide by using information and communication technology.

Hence, actualizing a system in the healthcare context requires essential pre-requisites that traditional Cloud Computing can't satisfy, as highlighted in [1], due to issues and limitations including security, data privacy, and transmission issues. On the other hand, Edge Computing offers solutions to overcome the limitations of Cloud approaches, allowing the development of systems that satisfy eHealth domain specific requirements. In this paper, we propose for a pattern-driven architecture based on the Cloud Edge Continuum for the creation of a Real-Time Drug Response Monitoring System for patients.

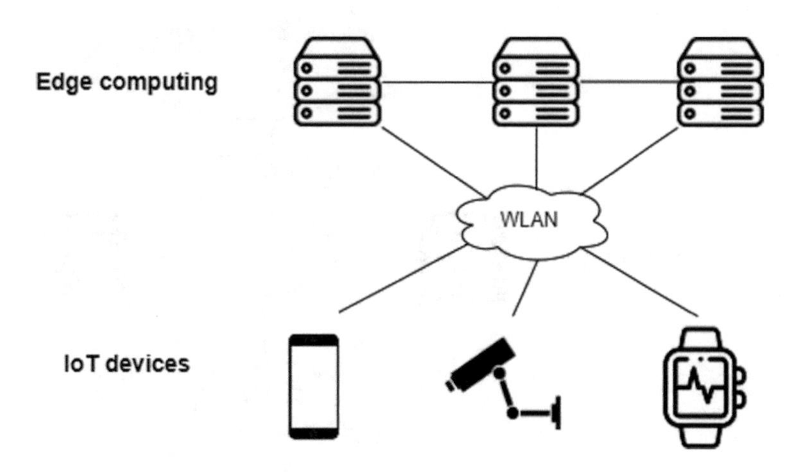


**Fig. 2.** Edge One Tier Architecture Pattern

### 3.1 Motivation

An Adverse Drug Reaction is an unexpected reaction of the body to the administration of a medicine. ADRs can be difficult to predict, as their causes can be linked to several factors such as the composition of the medicine, the dosage or unknown reactions. In some cases an ADR can lead to morbidity or even mortality [14]. We can distinguish between two categories of ADRs, Type A ADRs are predictable and dose related, while Type B reactions cannot be explained on the basis of drug composition [16]. Therefore, the possibility of being able to immediately monitor and detect an ADR in a patient can bring various benefits associated with the improvement of the entire healthcare system. Firstly, if doctors could be promptly notified in the event that a patient, who had started a drug treatment, manifested an ADR, then the treatment or dosage could be immediately modified, thus avoiding more disastrous consequences. Furthermore, the collected and anonymized data could be useful for pharmaceutical industries to monitor and analyze, on a large scale, the effects of medicines on the market. To manage a system that can achieve the purposes set out, it is necessary to guarantee data privacy and security, as well as efficient and real-time data transmission. In this context, the best solution for creating such a system falls on Cloud Edge architectures. In particular, through the use of Cloud Edge and Federated Learning, it is possible to develop an efficient and distributed system that ensures compliance with all the constraints imposed by the eHealth paradigm. In the next section, it will be shown how Cloud Edge and Federated Learning Patterns can help in the development of the Real-Time Drug Response Monitoring System.

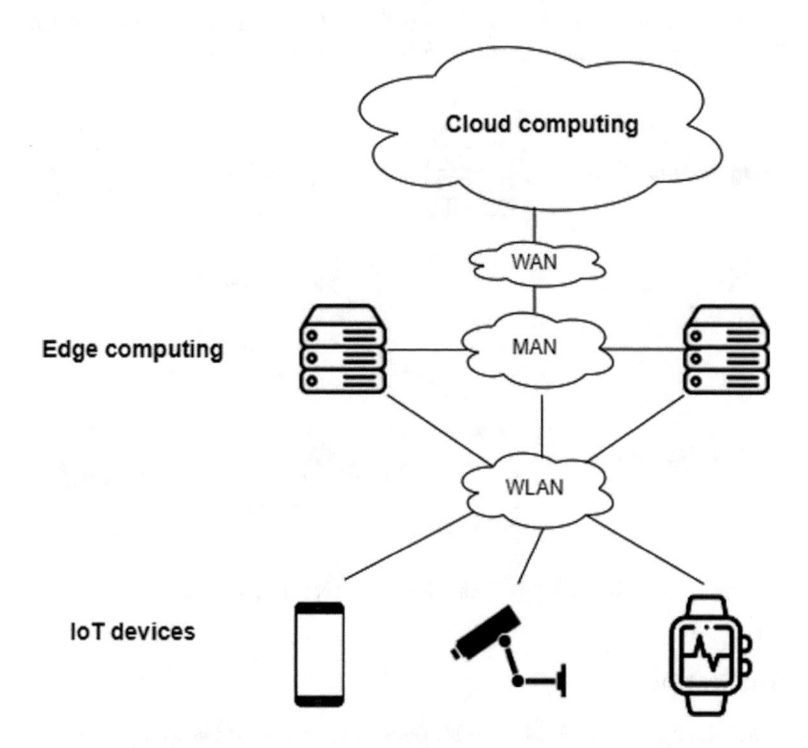


**Fig. 3.** Edge Two Tier Architecture Pattern.

### 3.2 Cloud Edge Based Solution

The purpose of the system is to gather data from patients to detect signals of ADR, alert physicians for timely intervention, and at the same time make data available for the pharmaceutical industry to gain insights. The Real-Time Drug Response Monitoring System relies on a Cloud Edge architecture and uses the main Cloud Edge architectural patterns. The central architecture of the system is based on the "Edge Two Tier with Orchestrator" pattern, as illustrated in

Fig. 5. The bottom level of this structure is composed of IoT devices, such as smartwatches. These devices are designed to capture biometric parameters from users, including blood pressure, body temperature, and blood oxygen levels. The gathered data is then transmitted via a WLAN connection to an Access Point linked to an Edge Node. Inside the Edge Nodes, the data undergoes to noise removal and filtering processes, to consider only the relevant data for subsequent processing. At this point, the filtered data is sent to the central server, after appropriate anonymization, to be made available for the pharmaceutical industries which will be able to perform large-scale analytics activities on the collected data. Simultaneously, the data is used to train the local Federated Learning model, which aims to identify anomalies in the patient's biometric parameters. It is crucial to observe that the data housed within the Edge Nodes is directly accessible to healthcare professionals. This accessibility allows physicians to timely identify patients who exhibit ADRs and intervene to change drug or drug posology. In the next stage, the central Cloud server, hosting the global Federated Learning model, receives two types of data from the Edge Nodes: the data, previously anonymized as mentioned before, and the updated parameters of the Federated Learning model trained using the Federated Averaging algorithm. Specifically, the gradients computed by each Edge Node during the local learning process are transmitted.

The necessity to ensure the preservation of data privacy, gives rise to the need to use a model that guarantees adherence to these constraints. Federated learning has been used frequently in the medical context, as attested by the work of [11]. In fact, the use of federated learning guarantees, through distributed training, the avoidance of exchanging sensitive data along the network, thus promoting data privacy [13]. Furthermore, Federated Learning offers advantages such as:

- Reduction of network traffic: Since in Federated Learning the model is trained in a distributed manner on local devices rather than centrally on a server, only the model weights are sent to the central server for updates, while raw data remains within the Edge Nodes. Indeed during the model update, only a limited number of devices are selected. This reduces network traffic since, only essential information is transferred, enhancing the efficiency of network resource utilization and avoiding the bottleneck effect due to limited network bandwidth.
- Preservation of privacy: The training data stays on local devices and the model is updated without raw data leaving the devices; only model parameters are sent to the Cloud Server. This approach guarantees user privacy, as there is no sensitive data transmission and any data detail remains local.
- Model convergence: Since model training is realized among different devices in a distributed manner, and during the training, only a limited set of devices is selected for updates, this process leads to a more effective and faster model convergence.
- Minimization of the risk of overfitting: The peculiarity of Federated Learning, where the data on which the model is trained comes from distinct devices, and

therefore, the data presents heterogeneity, offers the advantage of reducing the risk of overfitting the model.

In the context of this system, we can leverage specific patterns of Federated Learning to optimize data exchange between models, especially in scenarios where the data exhibits non-IID (Non-Independently and Identically Distributed) characteristics.

Firstly, the *Message Compressor Pattern* serves to enhance data transmission by facilitating compression at the local level before transmitting data to the global model. This strategic compression mitigates bottleneck effects arising from the constrained bandwidth of IoT devices. By reducing the volume of data sent, this pattern efficiently addresses the limitations imposed by device bandwidth, ensuring a smoother flow of information between local and global models.

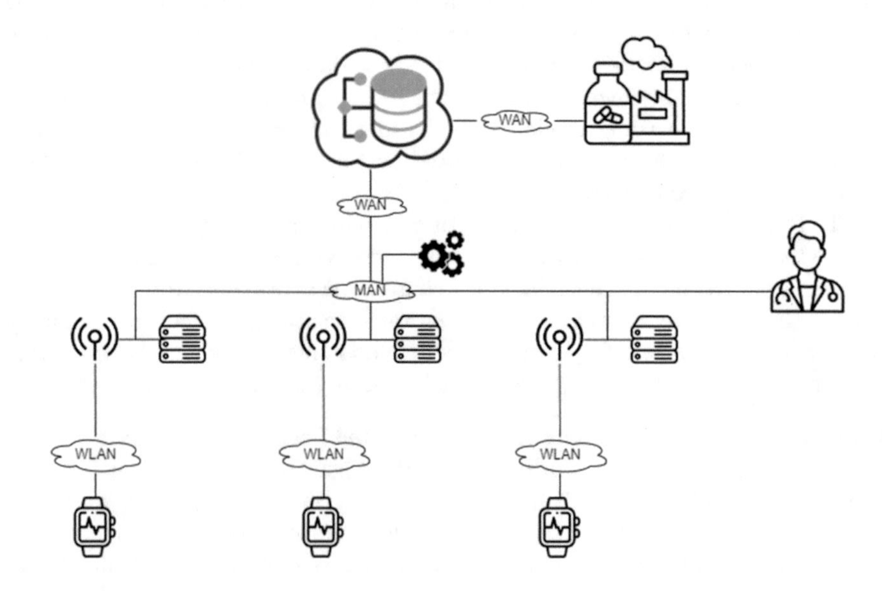


**Fig. 4.** Drug Monitoring System Architecture

Secondly, the *Client Cluster Pattern* plays a pivotal role in managing non-IID data. This pattern involves the grouping of diverse clients based on specific features. By clustering clients with similar characteristics, the system can effectively handle the challenges posed by non-IID data distribution. This not only aids in optimizing model convergence but also contributes to improved overall performance. The organized grouping facilitated by the Client Cluster Pattern fosters a more efficient collaboration among clients, ultimately enhancing the convergence and efficacy of the final model. The application of Federated Learning patterns is illustrated in Fig. 4. In order to explore a potential implementation of the system, the following services have been identified:

- For an efficient transmission of data from IoT devices to Edge Nodes:
  - AWS IoT Core, ensuring secure and scalable connectivity for IoT devices, it only requires utilization within the AWS ecosystem but has no limitations on IoT device compatibility.
  - Azure IoT Hub specific to Microsoft Azure and compatible with any IoT device.
  - Google Cloud IoT Core specific to Google Cloud Platform and compatible with any IoT device.
- For Edge Nodes implementation:
  - AWS Greengrass that can be integrated within the AWS ecosystem.
  - Azure IoT Edge which is compatible with the Microsoft Azure ecosystem.
  - Google Cloud Edge IoT for the Google Cloud System.

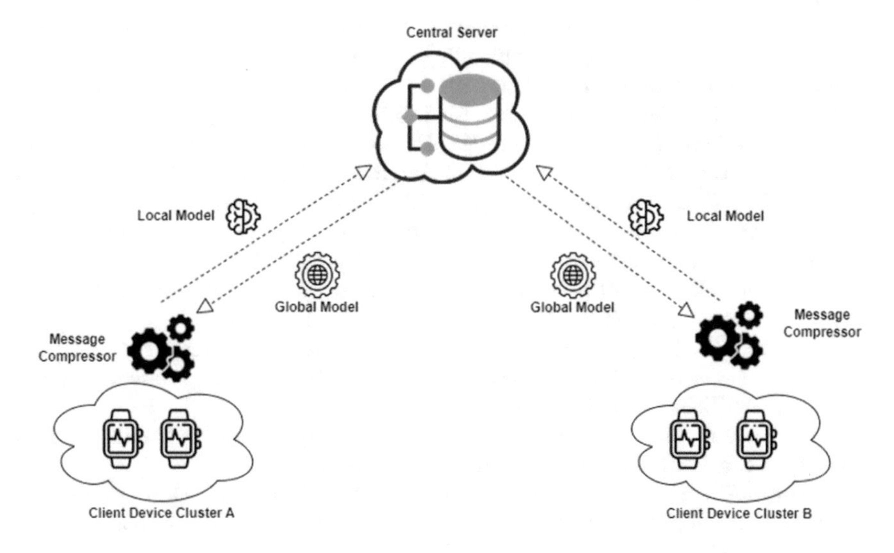


**Fig. 5.** Federated Learning patterns

## 4 Conclusions

The integration of Edge and Cloud technologies within the Real-Time Drug Response Monitoring System marks a transformative leap forward in the realm of medical data management and analysis. Through the strategic utilization of Federated Learning, the system not only addresses issues related to data privacy but also provides pharmaceutical entities with the possibility of realize insights on a large scale. This synergistic interplay between IoT devices and advanced computational architectures not only promises but delivers timely interventions for patients, ultimately enhancing the overall landscape of healthcare outcomes.

The proposed system highlight how it is possible to use design patterns and in particular Cloud Edge and Federated Learning patterns to realize a system capable of a transformative impact of technology in patient-centric healthcare solutions.

**Acknowledgments.** Alessia Sabia is a PhD student enrolled in the National PhD program in Artificial Intelligence, XXXIX cycle, course on Health and life sciences, organized by Università Campus Bio-Medico di Roma.

## References

1. Bahrami, M., Singhal, M.: A dynamic cloud computing platform for eHealth systems. In: 2015 17th International Conference on E-health Networking, Application and Services (HealthCom), pp. 435–438 (2015)
2. Di Martino, B., Esposito, A.: Applying patterns to support deployment in cloud-edge environments: a case study. In: Barolli, L., Woungang, I., Enokido, T. (eds.) AINA 2021. LNNS, vol. 227, pp. 139–148. Springer, Cham (2021). https://doi.org/10.1007/978-3-030-75078-7_15
3. Di Martino, B., Venticinque, S., Esposito, A., D'Angelo, S.: A methodology based on computational patterns for offloading of big data applications on cloud-edge platforms. Future Internet **12**(2) (2020)
4. Eysenbach, G.: What is e-health? J. Med. Internet Res. **3**(2), E20 (2001)
5. Jin, H., Ibrahim, S., Bell, T., Gao, W., Huang, D., Wu, S.: Cloud types and services. In: Furht, B., Escalante, A. (eds.) Handbook of Cloud Computing. Springer, Boston, MA (2010). https://doi.org/10.1007/978-1-4419-6524-0_14
6. Di Martino, B., Esposito, A.: A tool for mapping and editing of cloud patterns: the semantic cloud patterns editor. Stud. Inf. Control **27**(1), 117–126 (2018)
7. Di Martino, B., Esposito, A.: An overview of reference architectures for cloud continuum interoperability. IEEE Comput. Soc. **2**(3) (2022)
8. Di Martino, B., Esposito, A., Cretella, G.: Semantic representation of cloud patterns and services with automated reasoning to support cloud application portability. IEEE Trans. Cloud Compu. **5**(4), 765–779 (2017)
9. Di Martino, B., Esposito, A., Damiani, E.: Towards AI-powered multiple cloud management. IEEE Internet Comput. **23**(1), 64–71 (2019)
10. Sadeeq, M.M., Abdulkareem, N.M., Zeebaree, S.R.M., Ahmed, D.M., Sami, A.S., Zebari, R.R.: IoT and cloud computing issues, challenges and opportunities: a review. Qubahan Acad. J. **1**(2), 1–7 (2021)
11. Pezzullo, G.J., Esposito, A., di Martino, B.: Federated learning of predictive models from real data on diabetic patients. In: Barolli, L. (ed.) Advanced Information Networking and Applications, vol. 655, pp. 80–89. Springer, Cham (2023). https://doi.org/10.1007/978-3-031-28694-0_8
12. Qian, L., Luo, Z., Du, Y., Guo, L.: Cloud computing: an overview. In: Jaatun, M.G., Zhao, G., Rong, C. (eds.) CloudCom 2009. LNCS, vol. 5931, pp. 626–631. Springer, Heidelberg (2009). https://doi.org/10.1007/978-3-642-10665-1_63
13. Rieke, N., Hancox, J., Li, W., et al.: The future of digital health with federated learning. npj Digit. Med. **3**, 119 (2020)
14. Schatz, S., Weber, R.J.: Adverse drug reactions. Pharmacy Pract. **1**(1) (2015)
15. Schuldt, H.: Multi-tier architecture. In: Liu, L., Özsu, M.T. (eds.) Encyclopedia of Database Systems, pp. 1862–1865. Springer, Boston (2009). https://doi.org/10.1007/978-0-387-39940-9_652
16. Shibbiru, T., Tadesse, F.: Adverse drug reactions: an overview. History **1**(0) (2016)
17. Yoder, J.W., Aguiar, A., Merson, P., Washizaki, H.: Deployment patterns for confidences. In: HILLSIDE Proceedings of the 8th Asian Conference on Pattern Languages of Programming, pp. 1–9, March 2019

# Architectural Patterns for Software Design Problem-Solving in the Implementation of Federated Learning Structures Within the E-Health Sector

Beniamino di Martino[1,2,3], Domenico Di Sivo[4], and Antonio Esposito[1](✉)

[1] Department of Engineering, University of Campania "Luigi Vanvitelli", Caserta, Italy
{beniamino.dimartino,antonio.esposito}@unicampania.it
[2] Department of Computer Science and Information Engineering, Asia University, Taichung, Taiwan
[3] Department of Computer Science, University of Vienna, Vienna, Austria
[4] Università Campus Bio-Medico, Rome, Italy
domenico.disivo@unicampus.it

**Abstract.** Current research in artificial intelligence, including Machine Learning and Deep Learning, is driving innovation in various fields. In the healthcare sector, where considerable amounts of data are used for studies, early diagnosis and disease monitoring, the importance of addressing security and privacy issues is clear. Cloud Edge and Federated Learning, a more privacy-focused approach, allow algorithms to be trained without actual data exchange, using decentralized models. Recent studies show that prototypes trained with Federated Learning and Cloud Edge paradigms achieve reliable performance, generating robust models while preserving security and privacy. This study proposes a review focused on specific Patterns that, when applied to the design of Cloud Edge architectures within the healthcare sector, solve multiple problems, exploring challenges, implications and potential in this context. In particular, for each of the described Patterns, this paper will focus on the specific problem they face and present the actual solution, specifically in association with Federated Learning approaches.

## 1 Introduction

The surge in IoT devices, including smartphones and smartwatches, has led to substantial Edge data accumulation due to advancements in sensor and communication technology [1]. Machine Learning and Deep Learning highlight the crucial importance of this Edge data across applications such as predictive text input, personalized recommendations, intrusion detection, AI-driven diagnostics in healthcare [2], and traffic pattern analysis in smart transportation.


L. Barolli (Ed.): AINA 2024, LNDECT 203, pp. 347–356, 2024.
https://doi.org/10.1007/978-3-031-57931-8_34

Concerns about information security and privacy have prompted privacy protection laws like GDPR and the Consumer Privacy Bill of Rights, making traditional centralized data processing impractical. Federated Learning, a distributed approach, addresses challenges by keeping raw training data on the user's side, transmitting only local model parameters to a central manager [1]. This model manager conducts local model aggregation, providing a solution for secure training across multiple data pools.

The shift from centralized Machine Learning to distributed Federated Learning mirrors the evolution of computing architecture from centralized to distributed models. Cloud computing faces challenges like computational limitations and communication inefficiencies, leading to the rise of Edge Computing. This shift processes data anywhere on the path from its generation to the Cloud, reducing latency for real-time applications.

This article focuses on two distributed strategies: Federated Learning and Cloud-Edge Collaborative Computing Architectures. It emphasizes the deployment of Federated Learning within the Cloud-Edge architecture, recognizing it as crucial for future web services. The paper introduces collaborative learning mechanisms and identifies key patterns addressing challenges in data privacy, communication protocols, and system scalability.

## 2 Federated Learning in E-Health Sector

Federated Learning encounters challenges in delivering personalized services within the Internet of Things (IoT), primarily due to the constraints of a global model that captures only the statistical characteristics of various IoT devices. This limitation poses difficulties in adapting to diverse personal styles, such as interpreting identical body weight and height measurements differently based on varied living environments and external factors when predicting diseases. Moreover, issues arise from the heterogeneity in computing resources and network conditions among Federated Learning devices. Recognizing the significance of personalized Federated Learning models, recent advancements have been made, particularly in the domain of smart healthcare. A groundbreaking study presented in [3] addresses critical challenges associated with personalized Federated Learning for healthcare services. This study emphasizes natural variations in physical characteristics among individuals and introduces a Cloud Edge in Federated Learning architecture named FedHome for in-home healthcare services. FedHome trains local models at home on the network Edge, with the Cloud handling global model aggregation. The shared model addresses diverse datasets but may underperform for specific users. For personalized in-home health monitoring, users integrate the global model with their health data, allowing in-home IoT devices to download and train personalized models. While FedHome excels in training accuracy and model size on a human activity recognition dataset, it faces limitations due to data distribution imbalances. Addressing label heterogeneity [4], devices employ lightweight CNN models, keeping learning parameters locally to personalize the pain estimation model and enhance data privacy. Evaluation of the UNBC-McMaster Shoulder Pain dataset shows that personalized Federated Learning with privacy measures outperforms conventional Federated Learning, reducing the risk of data breaches by not sharing all updates with the aggregation server. Figure 1 illustrates how a Cloud-Edge architecture can support Federated Learning in e-Health contexts.

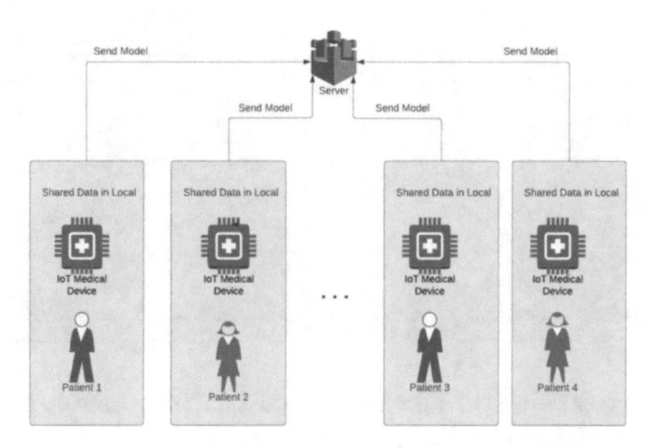


**Fig. 1.** A cloud-edge architecture for Federated Learning enabled for health monitoring services.

### 2.1 Two-Tier Edge Architecture

A simplified architecture for Federated Learning can be achieved by using two different architectural tiers. Figure 2 describes such a solution. The designed two-tier Edge computing system architecture incorporates computational resources across various levels and networks (LAN, MAN, WAN). The first tier involves the Edge device layer, including mobile and IoT devices communicating with the local Edge server through WLAN, the second tier comprises global Cloud servers providing services via WAN. Edge servers deliver computing services within WLAN coverage, and the Cloud server caters to remote users. Nearby Edge servers can assist users in neighbouring LANs, and tasks are scheduled to under-utilized servers or the Cloud when resources are insufficient. In a health sector scenario, multiple strategically deployed Edge servers handle tasks generated by patients' mobile devices, adapting to varying user densities in different locations throughout the day. Such locations can be categorized into different attractiveness levels based on device density[5]. Due to the high device density and excessive task offloading, LANs may experience congestion in both network and computational resources. To address this issue, the local Edge server must offload tasks beyond its capacity to nearby Edge servers or the Cloud server. However, optimizing task scheduling to achieve desired objectives poses a challenging problem.

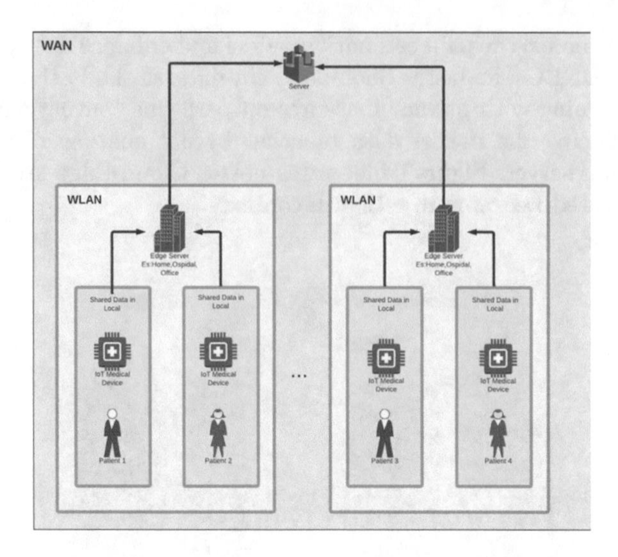


**Fig. 2.** A Cloud-Edge architecture for Federated Learning enabled for health monitoring services designed with a two-tier architecture.

# 3 Application of Selected Patterns to E-Health Cloud Edge Architectures

In software engineering, an architectural Pattern represents a reusable solution to problems that commonly arise within specific contexts in software Architectural design [6]. The Patterns presented in this paper have been carefully chosen from a large list to tackle the specific and recurring challenges associated with Federated Learning and Cloud Edge systems. The paper also explores how to optimally combine information from different sources. To provide a clearer understanding of how these Patterns function, we define the lifecycle of a model within a Federated Learning system. Each Pattern presented is associated with a specific state transition in this lifecycle, offering a clear and structured framework for the practical implementation of these Patterns.

## 3.1 Pattern for Client Registration

**Problem:** The challenge lies in effectively monitoring and identifying dishonest, failed, or dropout nodes in Federated Learning systems. This issue is crucial for the security of both the central server and client devices against potential adversarial threats. Ensuring system integrity in Federated Learning depends on accurate tracking of participating nodes. Dishonest or failed nodes can introduce vulnerabilities, compromising security, while dropout nodes disrupt model training and aggregation. To address these concerns, an efficient tracking mechanism is essential to identify instances of dishonesty, failures, or dropouts, providing transparency into client-device interactions. This tracking layer is fundamental for prompt detection and mitigation of potential security breaches. Additionally, achieving synchronization in the model training process across client devices in each aggregation round adds complexity. Without detailed records of connection and training information, achieving alignment becomes challenging, leading to discrepancies in training data and hindering overall model performance. In summary, the challenge underscores the need for robust tracking mechanisms in Federated Learning systems. Addressing this challenge enhances security and ensures effective alignment of the model training process, contributing to the success and reliability of Federated Learning implementations.

**Solution:** A client registry systematically captures and stores comprehensive details about client devices engaged with the system since their initial connection. The recorded information encompasses essential data such as the device ID, timestamps indicating connection initiation and termination, and specifics regarding device resources, including computational capabilities, communication parameters, power consumption, and storage capacity. Access to the client registry is subject to restrictions defined by the agreement established between the central server and the participating client devices. Known registry Patterns: IBM Party Stack [7], SIEMENS Industrial Metadata Management [8].

Figure 3 describes the implementation of the Client Registration Pattern in an e-health application.

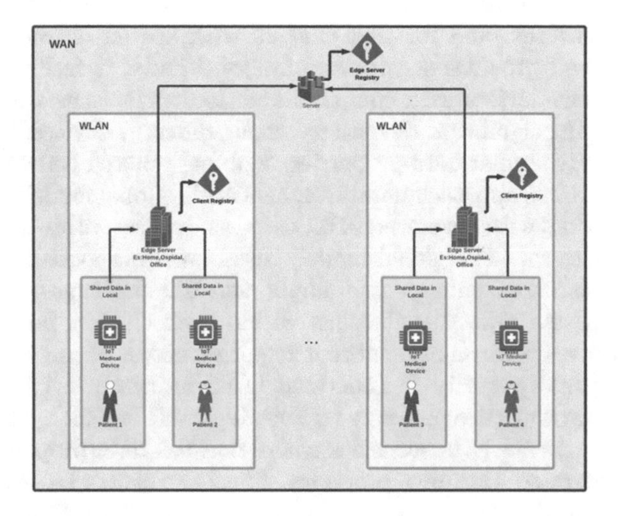


**Fig. 3.** A Cloud-Edge architecture for Federated Learning enabled for health monitoring services designed with client registry Pattern implemented.

### 3.2 Client Cluster Pattern

**Problem:** Federated Learning models created under non-IID data properties are perceived as less generalized. The primary reason behind this lies in the absence of adequately representative data labels originating from the client devices. In such scenarios, the distribution of data across different clients deviates significantly, leading to challenges in creating a model that can effectively generalize to diverse and unseen data points. The lack of a consistent and well-distributed set of data labels hinders the model's ability to learn Patterns that are applicable across various instances. Additionally, the local models developed on individual client devices may exhibit significant divergence from one another. This divergence can be attributed to the varying characteristics of the data available on different devices. As each client device may possess a unique subset of data, the local models trained on these subsets may capture distinct Patterns and trends. Consequently, when these locally trained models are aggregated to form a global model, the disparities among them can lead to reduced model coherence and hinder overall performance. Addressing the challenges associated with non-IID data is crucial for enhancing the generalization capabilities of Federated Learning models and ensuring their effectiveness across diverse datasets.

**Solution:** Clustering client devices based on properties such as data distribution, feature similarities, and gradient loss is an effective strategy to tackle challenges related to non-IID data. This approach involves grouping devices with similar data patterns, and forming clusters with shared characteristics. This clustering method provides several advantages. Firstly, it facilitates the identification of client devices with comparable data distributions within the same cluster, addressing significant deviations among different devices. Clusters categorize clients with similar data properties, enabling tailored Federated Learning for each cluster. This approach notably enhances the global model's performance for client networks with severe non-IID data, as devices within a cluster share similar data patterns. The global model, based on aggregated knowledge from these clusters, is better equipped to handle non-IID challenges, mitigating the impact of disparate data distributions within each cluster. Importantly, this approach improves performance without requiring access to raw local data, preserving privacy and security in Federated Learning systems. Clustering client devices based on properties proves to be a strategic solution for optimizing model performance, particularly in scenarios where non-IID data properties challenge traditional Federated Learning processes. Figure 4 reports an example of the use of the Cluster Pattern. Known cluster Patterns: Iterative Federated Clustering Algorithm (IFCA) [9], Clustered Federated Learning (CFL) [10], SIEMENS Industrial Metadata Management [8].

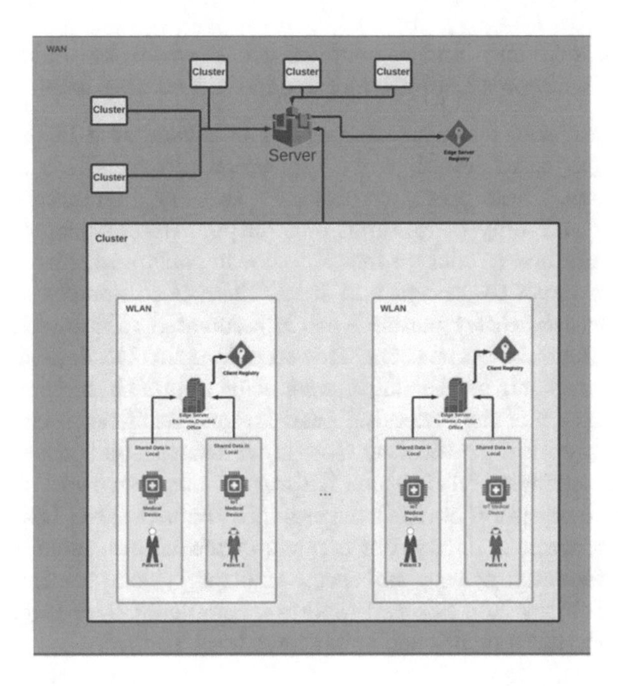


**Fig. 4.** A cloud-edge architecture for Federated Learning enabled for health monitoring services designed with client registry and cluster Pattern implemented.

### 3.3 Secure Aggregator Pattern

**Problem:** The absence of robust security measures leaves Federated Learning systems vulnerable to specific types of security threats, notably the honest-but-curious and active adversary threats. These threats can compromise the integrity, privacy, and effectiveness of Federated Learning processes. Let's delve into the implications of these security concerns: Honest-But-Curious Threats: An honest but curious adversary in Federated Learning strictly follows the protocol but aims to gather extra information. Without proper security measures, these adversaries may exploit the process, risking the exposure of sensitive data and compromising the confidentiality of user information and proprietary models.

**Active Adversary Threats:** An active adversary in Federated Learning is malicious, seeking to manipulate the process. This poses risks such as biased models, disrupted communication, compromised updates, and privacy breaches. The absence of security measures raises concerns about model reliability, accuracy, and potential privacy issues.

**Privacy Concerns:** Both adversary types pose privacy risks, threatening sensitive information in Federated Learning. Inadequate security may result in privacy violations, hindering collaboration benefits. Addressing this requires advanced cryptographic techniques, secure protocols, and privacy-preserving mechanisms like homomorphic encryption, secure multi-party computation, and differential privacy. Clear guidelines and regulations are essential for building trust and promoting the widespread adoption of Federated Learning technologies.

**Solution:** The role of a security aggregator in a Federated Learning system is pivotal, managing secure multiparty computation protocols for model exchanges and aggregations. These protocols provide a security guarantee, ensuring that each party is aware only of its input and output. Homomorphic encryption is employed as a method to encrypt models, allowing only authorized client devices and the central server to decrypt and access them. Additionally, pairwise masking and differential privacy methods are implemented to diminish model interpretability for unauthorized parties. This technique involves introducing noise to parameters or gradients or utilizing a generalized approach, further fortifying the security and privacy of the Federated Learning process. Known cluster Patterns: SecAgg [11] a practical protocol by Google for secure aggregation in the Federated Learning settings; HybridAlpha [12] serves as a framework responsible for overseeing the integration of client devices into the Federated Learning process. The security operations encompass a range of techniques, including functional encryption, differential privacy and secure multiparty computation, and TensorFlow Privacy Library [13]. Figure 5 reports a complete Cloud-Edge architecture where all the Patterns mentioned so far have been applied.

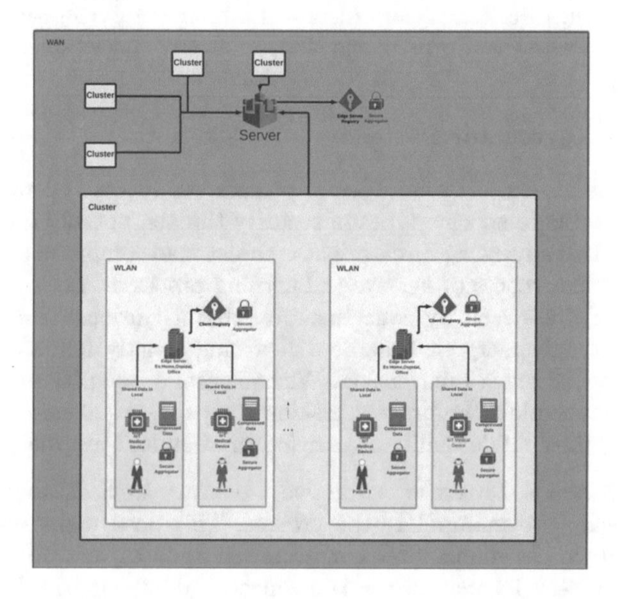


**Fig. 5.** A cloud-edge architecture for Federated Learning enabled for health monitoring services designed with client registry, cluster Pattern and Secure aggregator Pattern implemented.

## 4 Conclusion

In conclusion, Federated Learning, a pioneering distributed Machine Learning paradigm, prioritizes data privacy through intelligent IoT and smart mobile device resource utilization. Despite this commitment, adopting Federated Learning presents substantial architectural challenges in orchestrating complex distributed systems. This paper contributes by presenting three meticulously aligned Federated Learning architectural patterns, offering valuable guidance for architects designing robust systems. These patterns address challenges associated with the distributed nature of Federated Learning, laying a foundation for more effective and secure implementations. Future research will delve deeper into architectural designs, focusing on enhancing trust in Federated Learning frameworks. This work aims to pioneer innovative solutions, anticipating and adapting to emerging trends in the evolving landscape of privacy concerns and distributed Machine Learning. Through ongoing exploration and refinement, the goal is to strengthen Federated Learning, making it more resilient, trustworthy, and adaptable to the demands of a privacy-centric and distributed Machine Learning environment.

**Acknowledgements.** Domenico Di Sivo is a PhD student enrolled in the National PhD in Artificial Intelligence, XXXVIII cycle, course on Health and life sciences, organized by "Università Campus Bio-Medico di Roma".

## References

1. Sandhu, A.K.: Big data with cloud computing: discussions and challenges. Big Data Mining Anal. **5**(1), 32–40 (2021)
2. Di Martino, B., Esposito, A., Pezzullo, G.J., Weng, T.-H.: Evaluating machine and deep learning techniques in predicting blood sugar levels within the e-health domain. Connect. Sci. **35**(1), 2279900 (2023)
3. Wu, Q., Chen, X., Zhou, Z., Zhang, J.: FedHome: cloud-edge based personalized federated learning for in-home health monitoring. IEEE Trans. Mob. Comput. **21**(8), 2818–2832 (2020)
4. Gudur, G.K., Balaji, B.S., Perepu, S.K.: Resource-constrained federated learning with heterogeneous labels and models. arXiv preprint: arXiv:2011.03206 (2020)
5. Wan, J., Jiang, C.: Reinforcement learning-based workload scheduling for edge computing. PREPRINT (Version 1). Research Square (2021)
6. Beck, K.: Using pattern languages for object-oriented programs. In: OOPSLA 1987 Workshop on the Specification and Design for Object-Oriented Programming (1987)
7. Ludwig, H., et al.: IBM federated learning: an enterprise framework white paper v0. 1. arXiv preprint: arXiv:2007.10987 (2020)
8. "Siemens industrial metadata management," https://documentation.mindsphere.io/resources/html/asset-manager/en-US/index.html. Accessed 01 Jan 2024
9. Ghosh, A., Chung, J., Yin, D., Ramchandran, K.: An efficient framework for clustered federated learning. IEEE Trans. Inf. Theory **68**(12), 8076–8091 (2022)

10. Sattler, F., Müller, K.-R., Samek, W.: Clustered federated learning: model-agnostic distributed multitask optimization under privacy constraints. IEEE Trans. Neural Netw. Learn. Syst. **32**(8), 3710–3722 (2020)
11. Bonawitz, K., et al.: Practical secure aggregation for privacy-preserving machine learning. In: proceedings of the ACM SIGSAC Conference on Computer and Communications Security 2017, pp. 1175–1191 (2017)
12. Xu, R., Baracaldo, N., Zhou, Y., Anwar, A., Ludwig, H.: Hybridalpha: an efficient approach for privacy-preserving federated learning. In: Proceedings of the 12th ACM Workshop on Artificial Intelligence and Security, pp. 13–23 (2019)
13. TensorFlow privacy library. https://github.com/tensorflow/privacy/. Accessed 01 Jan 2024

# Integration of Landsat and Sentinel Data for Localized High-Resolution Monitoring of the Territory

Gholamreza Nikravesh, Raffaele Persico(✉), Alfonso Senatore, and Giuseppe Mendicino

Department of Environmental Engineering (DIAm), University of Calabria, Arcavacata, Italy
raffaele.persico@unical.it

**Abstract.** In this contribution, we propose a joined use of optical satellite data achieved from Sentinel and Landsat platforms in order to achieve localized monitoring of some areas of interest with regard to the drought. This integration of data allows enhanced spatial and temporal resolutions to be fulfilled. In fact, for the quoted satellites, the pixel size is of the order of 20 or 30 m depending on the radiometer, and the interpolation of the temporal series allows an average time interval between two subsequent satellite images of the order of 2.5 days. A supervised multiband thresholding algorithm has been implemented for the cloud masking because only almost cloud-free images provide information of interest, which, in our case, is based on the NDVI spectral index.

**Keywords:** Calabria Region · Landsat and Sentinel data · NDVI Index · Drought

## 1 Introduction

The increasing availability of open-access satellite data at medium resolution makes it possible to monitor phenomena of interest for environmental protection, such as drought, fires, floods, coastal erosion and so on. In this framework, the integration of Landsat/OLI and Sentinel/MSI data is an issue of interest for the recent literature [1–5], even more increased after the recent launch of the Landsat 9 platform on September $27^{th}$, 2021. In fact, the two radiometers OLI and MSI work on several similar bands in the frequency range between the visible and the short-wave infrared (SWIR). Also, the available geometrical resolution is similar in the two cases (30 m for Landsat/OLI and 20 m for Sentinel/MSI in the bands of our interest). The temporal resolutions are different because the Landsat (8 or 9) platforms observe the same scene after eight days, whereas the Sentinel (2A and 2B) platforms observe the same scene after five days. Both platforms have Sun-synchronous orbits, i.e. they illuminate the same scene at the same time they did at their previous passage, but this time is not the same for Sentinel and


L. Barolli (Ed.): AINA 2024, LNDECT 203, pp. 357–367, 2024.
https://doi.org/10.1007/978-3-031-57931-8_35

Landsat data. Also, the comprehensive scene is not the same. The integration of Landsat and Sentinel data improves the average time rate of the data, which becomes on average of the order of 2.5 days (of course, the time step between two consecutive satellite images is not constant vs. time when joining Sentinel and Landsat data) driving it to values not so far from those achievable from moderate resolution radiometers, as e.g. SLSTR (the radiometer mounted on Sentinel 3 platforms) or MODIS, for which two images per days are guaranteed but the pixel size is of the order of 250 m. However, the integration of the MSI and OLI data creates some problems. In fact, the technical differences between the radiometers, the slightly different bands and the different spatial resolution drive to different evaluations of the NDVI [6].

In this contribution, we will present results based on the NDVI spectral index. The area of interest (Fig. 1) is part of an agricultural company whose name is Terzeria, located in the Sibari Plain (Calabria, Southern Italy, Lat. 39.7763 °N, Lon. 16.4485 °E), extending over about 450 hectares, divided into 77 sub-areas characterized by different uses of the soil, i.e. different cultivations, some of which stable (rice, fruit trees) and some other seasonally rotating.

**Fig. 1.** The area of the Terzeria Farm in the Sibari Plain, Italy.

# 2 Method and Materials

## 2.1 Cloud Masking

We have preliminarily implemented an algorithm for cloud masking based on multiband thresholding. In fact, the reflectance of the clouds is customarily higher than that of the reference (cloud-free) scenario in any optical band ranging from the visible to the short wave infrared SWIR (wavelengths ranging about from 0.4 to 1.6 $\mu$m), with the exception of the areas covered with snow. However, even if the distinction between clouds and snow is, in general, an issue [6], in the area of interest the snow is an extremely rare phenomenon, and its occurrence can be neglected. With regard to cloud masking, there are several algorithms (see, e.g. the review paper [7] and references therein). We have avoided statistical algorithms based on a reference cloud-free scenario [7]. In fact, in the case of our interest, the cloud-free scenario would not be constant vs. time due to the seasonal variation of the state of the vegetation, the alternation of the cultivations and to the customary agricultural activities (irrigation, pruning, ploughing, harvesting, etc.). Therefore, following we have started with a supervised classification based an a multithreshold algorithm [8]. In particular, we have set seven initial thresholds on seven optical bands (ultra-blue, blue, green, red, near-infrared, SWIR1 and SWIR2) available both from the Landsat/OLI and Sentinel/MSI radiometers. The algorithm classifies as cloudy any pixel where the reflectance overcomes the threshold in each the seven bands. The value of the threshold, band per band, has been initially achieved from the direct observation of some images, and then it has been refined by an algorithm that has tested automatically many vectors of thresholds, randomly chosen in a neighbour of the first vector of thresholds. Each trial value of the vector of thresholds was temporarily considered optimal and replaced the previously store values if the corresponding evaluation of the cloud coverage was closer to those found in the metadata of the satellite images. We have looked in particular for an optimized seasonal vector of thresholds referred to the summer period of the year 2023.

This algorithm, however, was calibrated over the entire satellite scene, i.e. a territory much larger than the area of interest. Therefore, this global optimization was exploited as starting point for a further supervised heuristic optimization at the local level, namely at the spatial scale of the Terzeria farm. After this step, a spatial filtering based on the two-dimensional Fourier Transform has been set up. In particular, we have set an ideal spatial filter saving the spatial frequencies of the cloud mask in a circle of radius 0.01 $m^{-1}$ and erasing the spatial frequencies outside it. This has the effect to reduce further on the clouds smaller than 100 m in any directions, statistically deemed to be artifacts, while preserving larger clouds, statistically estimated to be authentic clouds. With reference to each to the 77 sub-areas constituting the Terzeria farm, the images with cloud coverage smaller than five per cent are retained, whereas the other ones are discharged. The fact that the decision is done at very local level makes the results more reliable and allows to save more information with respect to a cloud-coverage threshold fixed on the entire Terzeria farm. In fact, of course we might have some sub-areas covered by clouds and some other cloud free.

### 2.2 Interpolation of MSI and OLI Data

The monitoring of drought conditions is performed by means of the normalized differential vegetation index (NDVI), spatially averaged on the sub-area of interest. Beyond the spatial average value, we have also calculated the standard deviation for each sub-area. In fact, being each sub-area characterized essentially by the same cultivation, in the absence of pathological phenomena, any sub-area is expected to show a quite homogeneous reflectivity pixel by pixel. Consequently, the standard deviation is expected to be relatively low and, in particular, lower than the average value, with the exception of the days where the average NDVI assumes very low values. The time series of the Landsat data have been inserted within that of the Sentinel data. In order to do this, we have corrected the bias observed between the NDVI values provided by both platforms. The bias is due to the different spectral sensitivity of the radiometers and to the slight differences among the homologous bands in relationship to the MSI and to the OLI radiometers (indeed, some minimal differences exist even between the bands of the MSI radiometers mounted on Sentinel 2A and Sentinel 2B). In Fig. 2 (upper panel), we show in particular the values of NDVI achieved in one of the sub-areas of the Terzeria farm (the behaviour is similar in any other sub-area) for each day when an almost-cloud-free image (coverage up to five per cent) Sentinel/MSI or Landsat/OLI was available for that sub-area during the summer 2023. As can be seen, the values achieved by Landsat/OLI are, on average, lower than those achieved by Sentinel/MSI. In particular, this happens in the three days where we have had the passage of both a Sentinel 2 and a Landsat platform (these days are June $8^{th}$, July $7^{th}$ and August $27^{th}$).
In particular, in these three days, there is a bias quite constant between the NDVI evaluated from the Sentinel platform and that evaluated from the Landsat platform. Such a bias cannot be attributed (if not quite marginally) to a physical difference of the scene occurred in the time interval (a few hours) between the passage of the two satellites, neither it can be due to, e.g. the different solar irradiation or the different view angles of the satellites with respect to the scene. In particular, even if these factors could affect the reflectance, the NDVI (like most spectral indices) amortizes their effects due to the fact that it is defined by means of a ratio. In particular, as well known, the NDVI index is given by:

$$NDVI = \frac{R_{NIR} - R_{RED}}{R_{NIR} + R_{RED}} \tag{1}$$

where $R_{NIR}$ is the reflectivity of the considered pixel in the near-infrared band (ranging from 0.854 to 0.874 µm for Sentinel 2A and 2B and from 0.85 to 0.88 µm in Landsat 8 and 9) and $R_{RED}$ is the reflectivity in the band of red (ranging from 0.65 to 0.68 µm in Sentinel 2A and 2B and from 0.64 to 0.67 $\mu$m in Landsat 8 and 9). Since the effects of the sunray incidence are the same at the numerator and denominator, their effect is implicitly mitigated, as well as the effects of the view angle of the satellite. We have exploited the data retrieved in the three said days in order to extract the bias, provided by the algebraic average of the three visible (and quite similar) biases. This bias has been added to all the Landsat data (gathered in any date), keeping the Sentinel data unchanged. The result achieved after this correction is shown in the lower panel of Fig. 2.

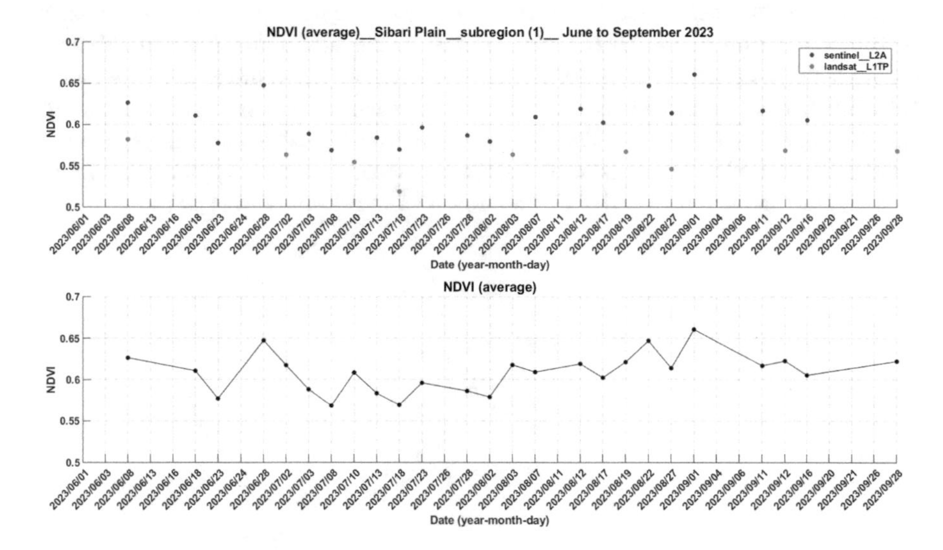


**Fig. 2.** Upper panel: Sentinel 2 (A and B) and Landsat (8 and 9) NDVI index throughout the summer period 2023 in one of the sub-areas of the Terzeria farm. Lower panel: the joined graph accounting for both radiometers after correction of the bias of the Landsat data.

## 3 Results

In the following, we provide some results representing the spatially averaged NDVI and its standard deviation vs. time during the summer period of 2023. The sub-area is put into evidence for each graph. In particular, among the 77 sub-areas, we have chosen the records no. 1, 5, 26, 64 and 77. For each of them, both the spatially averaged NDVI and the corresponding standard deviation are shown.

It can be seen that, as expected, the standard deviation in most cases is smaller than the average value, confirming indirectly the substantial homogeneity of the soil use in each sub-area, regardless of the extension of the sub-region or the particular use of the soil in each sub-area. Figures 3, 4, 5, 6 and 7 provide some results worthy of further investigation, such as the progressive decrease of the NDVI from June 8 up to July 7 in sub-area 64 and from June 18 up to July 7 in sub-region 77. Future investigation will be aimed at understanding the causes of this behaviour as well as other anomalies not shown here.

Finally, the data have been also associated with regard to the kind of cultivation. In fact, some of the sub-areas can also be associated to the same culture, at least in a specific period. In Fig. 8 we have joined together all the results relative to the cultivation of the rise in the summer 2023, distinguishing the sub-areas through the colours of the curves. We can observe an analogous behaviour of the NDVI, that in all the relevant sub-areas reaches its maximum value at the end of august. We also see that in the subregion 68 the standard deviation is meaningfully higher than elsewhere from the second half of August on. We will investigate further on in order to understand the reason of this datum.

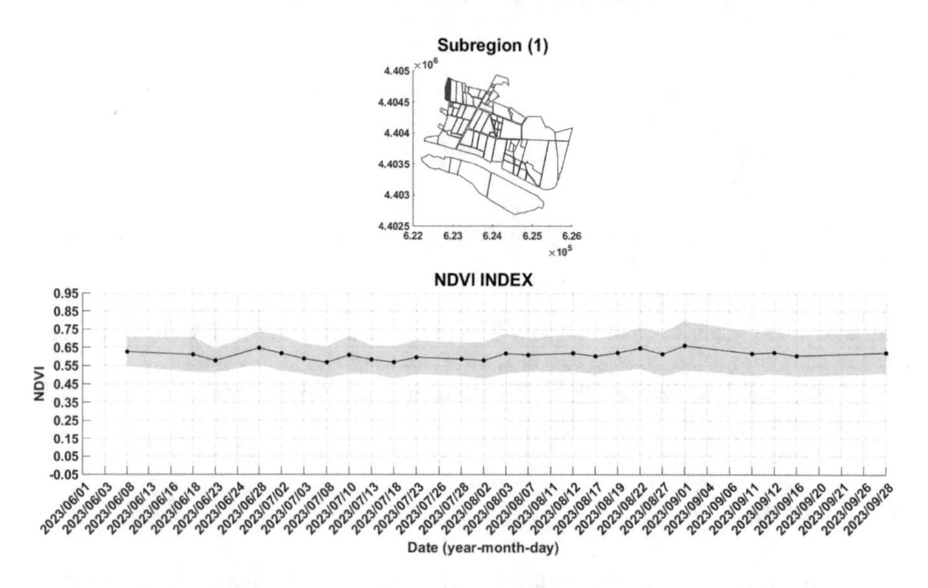


**Fig. 3.** Spatially average NDVI and standard deviation during the summer period of 2023. Record no. 1.

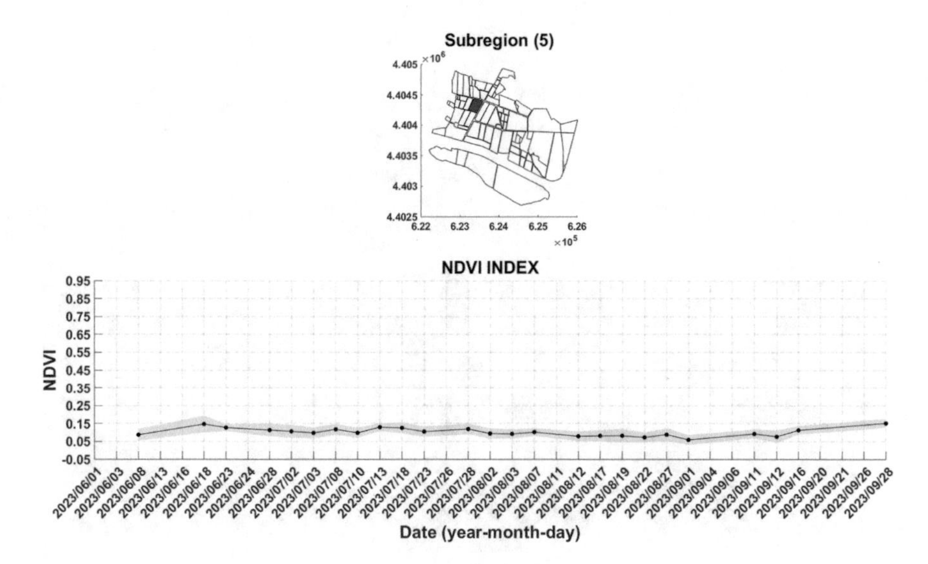


**Fig. 4.** Spatially average NDVI and standard deviation during the summer period of 2023. Record no. 5

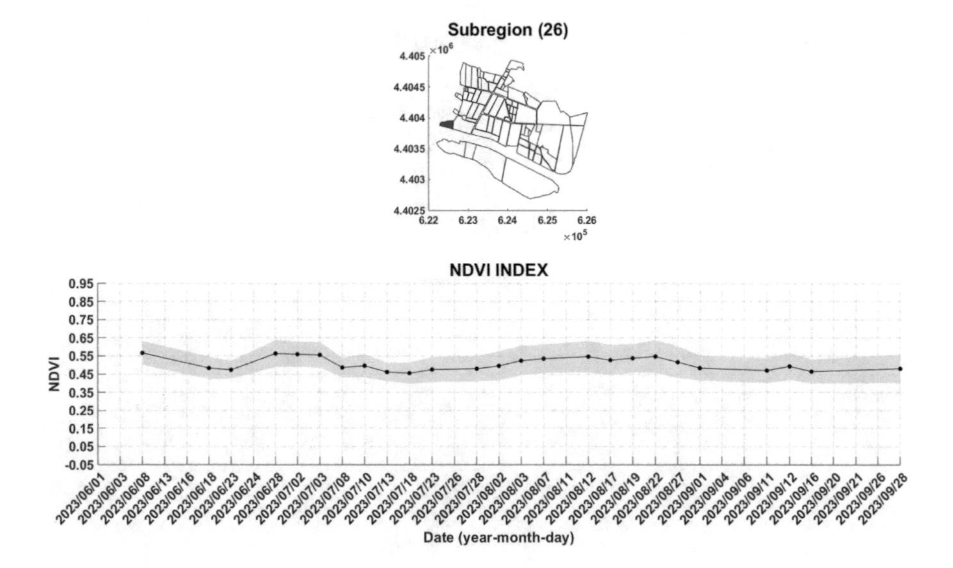


**Fig. 5.** Spatially average NDVI and standard deviation during the summer period of 2023. Record no. 26

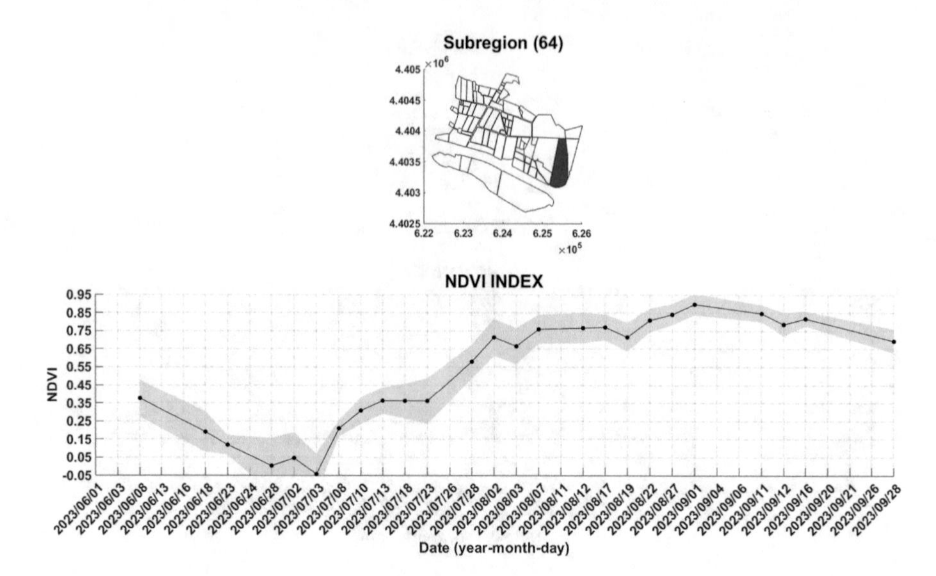


**Fig. 6.** Spatially average NDVI and standard deviation during the summer period of 2023. Record no. 64

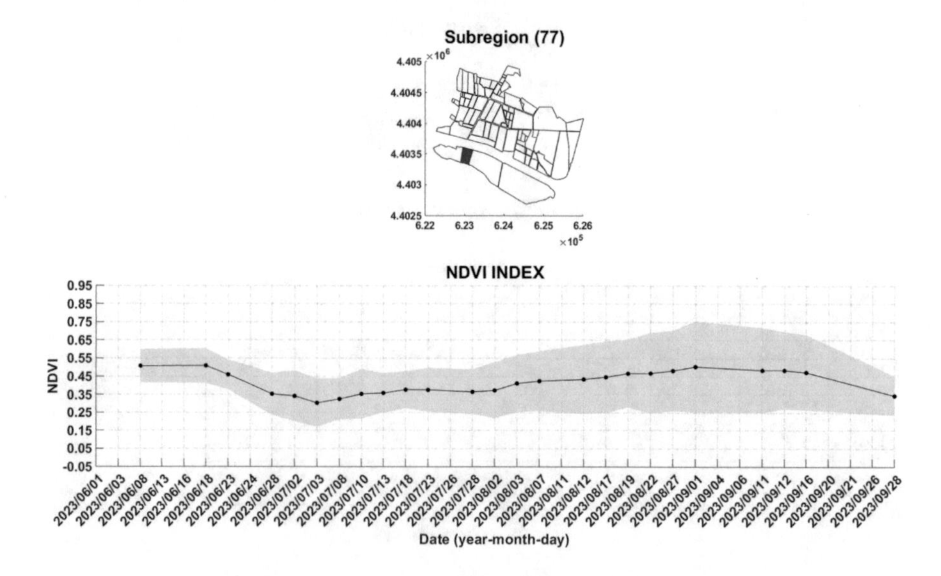


**Fig. 7.** Spatially average NDVI and standard deviation during the summer period of 2023. Record no. 77

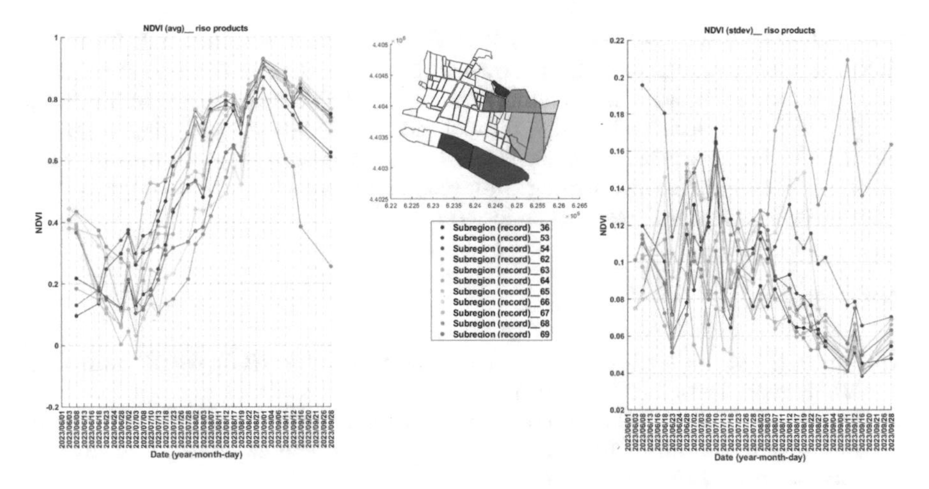


**Fig. 8.** Average NDVI (left hand side) and standard deviation referred to the cultivation of the rise.

## 4 Conclusions and Future Developments

We have presented a procedure for the monitoring of the drought conditions in an area exploited for industrial agriculture in Southern Italy, showing the main features of an algorithm aimed at examining the behaviour vs. time of the NDVI index averaged at the local level (a few hectares) and of its standard deviation achieved on the same areas.

The algorithm also contemplates the possibility of clicking on the points of the graphs opening in this way the relative real colour images. This is scheduled in order to offer the possibility of a last check regarding anomalous values of the average NDVI, so that the final user will be enabled to be definitively sure that they are not due to clouds escaped to the cloud masking algorithm.

Further developments will be to monitor also other previous seasons and enable the final user to exploit the satellite data (integrated with measurements at ground level) in order to make decisions economically convenient for the Terzeria company, mitigating in particular the local effects of climate change by means of technologically supported decisions with regard to the irrigation, the choice of the cultivations and their rotation, the fertilization and so on. Remote sensing monitoring will be part of a more comprehensive drought monitoring system [9] involving geophysical prospecting [10–13] in order to assess the shallow geological scenario and investigate possible correlations between it and the drought conditions. Moreover, a monitoring network concerning meteorological, hydrological and hydrogeological variables will be displaced in order to calculate real-time drought indices (e.g., SPI, SPEI or GRI [14]). Finally, several forecasting systems will be exploited, from the short-range to seasonal models [15], in order to forecast drought onset and other relevant hydrometeorological events like heat waves, cold spells or heavy precipitation.

**Acknowledgements.** This work was funded by the Next Generation EU - Italian NRRP, Mission 4, Component 2, Investment 1.5, call for the creation and strengthening of 'Innovation Ecosystems', building 'Territorial R&D Leaders' (Directorial Decree n. 2021/3277) - project Tech4You - Technologies for climate change adaptation and quality of life improvement, n. ECS0000009. This work reflects only the authors' views and opinions, neither the Ministry for University and Research nor the European Commission can be considered responsible for them.

## References

1. Franch, B., et al.: A method for Landsat and Sentinel 2 (HLS) BRDF normalization. Remote Sens. **11**, 632 (2019)
2. Griffiths, P., Nendel, C., Hostert, P.: Intra-annual reflectance composites from Sentinel-2 and Landsat for national-scale crop and land cover mapping. Remote Sens. Environ. **220**, 135–151 (2019)
3. Roy, D.P., Yan, L.: Robust Landsat-based crop time series modelling. Remote Sens. Environ. **238**, 110810 (2018)
4. Schwieder, M., Leitão, P.J., da Cunha Bustamante, M.M., Ferreira, L.G., Rabe, A., Hostert, P.: Mapping Brazilian savanna vegetation gradients with Landsat time series. Int. J. Appl. Earth Obs. Geoinf. **52**, 361–370 (2016)
5. Calcagno, G., Mendicino, G., Monacelli, G., Senatore, A., Versace, P.: Distributed estimation of actual evapotranspiration through remote sensing techniques. In: Rossi, G., Vega, T., Bonaccorso, B. (Eds.) Methods and Tools for Drought Analysis and Management, Series: Water Science and Technology Library, Vol. 62, pp. 125–148. Springer, Dordrecht, The Netherlands (2007). https://doi.org/10.1007/978-1-4020-5924-7_7
6. Tiede, D., Sudmanns, M., Augustin, H., Baraldi, A.: Investigating ESA Sentinel-2 products' systematic cloud cover overestimation in very high altitude areas. Remote Sens. Environ. **252**, 112163 (2021). https://doi.org/10.1016/j.rse.2020.112163
7. Chaves, M.E.D., Picoli, M.C.A., Sanches, I.D.: Recent applications of Landsat 8/OLI and Sentinel-2/MSI for land use and land cover mapping: a systematic review. Remote Sens. **12**, 3062 (2020). https://doi.org/10.3390/rs12183062
8. Qiu, S., Zhu, Z.: He, B: Fmask 4.0: improved cloud and cloud shadow detection in Landsats 4–8 and Sentinel-2 imagery. Remote Sens. Environ. **231**, 111205 (2019). https://doi.org/10.1016/j.rse.2019.05.024
9. Mendicino, G., Versace, P.: Integrated drought watch system: a case study in Southern Italy. Water Resour. Manage **21**, 1409–1428 (2007). https://doi.org/10.1007/s11269-006-9091-6
10. Catapano, I., Crocco, L., Persico, R., Pieraccini, M., Soldovieri, F.: Linear and nonlinear microwave tomography approaches for subsurface prospecting: validation on real data. IEEE Trans. Antennas Wirel. Propag. Lett. **5**, 49–53 (2006)
11. Pierri, R., Leone, G., Soldovieri, F., Persico, R.: Electromagnetic inversion for subsurface applications under the distorted Born approximation. Nuovo Cimento. **24C**(2), 245–261 (2001)
12. Calia, A., Leucci, G., Lettieri, M.T., Matera, L., Persico, R., Sileo, M.: The mosaic of the crypt of St. Nicholas in Bari (Italy): integrated GPR and laboratory diagnostic study. J. Archaeol. Sci. **40**(12), 4162–4169 (2013)
13. Persico, R., et al.: GPR investigations at St. John's co-cathedral in Valletta. Near Surf. Geophys. **17**(3), 213–229 (2019). https://doi.org/10.1002/nsg.12046

14. Mendicino, G., Senatore, A., Versace, P.: A groundwater resource index (GRI) for drought monitoring and forecasting in a Mediterranean climate. J. Hydrol. **357**(3–4), 282–302 (2008). https://www.sciencedirect.com/science/article/pii/S0022169408002321
15. Furnari, L., Magnusson, L., Mendicino, G., Senatore, A.: Fully coupled high-resolution medium-range forecasts: Evaluation of the hydrometeorological impact in an ensemble framework. Hydrol. Process. **36**(2), e14503 (2022). https://doi.org/10.1002/hyp.14503

# Personalized Federated Learning in Edge-Cloud Continuum for Privacy-Preserving Health Informatics: Opportunities and Challenges

Mario Bochicchio[1,2(✉)] and Sileshi Nibret Zeleke[1]

[1] Department of Computer Science, Università Degli Studi di Bari Aldo Moro, Bari, Italy
{mario.bochicchio,sileshi.zeleke}@uniba.it

[2] Digital Health National Lab, CINI - Consorzio Interuniversitario Nazionale per l'Informatica, Roma, Italy

**Abstract.** The growing interest in advanced data-intensive models for healthcare applications presents several challenges and opportunities. Federated learning (FL) emerges as an attractive solution to allow decentralized nodes to collectively train shared machine learning models without the need of transmitting sensitive data to a central database. This can safeguard privacy while effectively leveraging the distributed computational resources available in the cloud-edge continuum. In health informatics, the need for robust privacy-preserving mechanisms is paramount, especially when the nodes of the FL system are associated with datasets from individual patients, as opposed to the case of databases that include many patients (such as those available from hospitals). This need becomes particularly significant when addressing diagnoses and predictive analytics in personalized medicine, precision medicine, risk stratification, and longitudinal monitoring. We explore the applications of FL frameworks in the context of cloud-edge in healthcare. We identify real-world settings to assess the benefits and challenges of personalized federated learning. These include data imbalance issues, usability, promoting replicability, improving security, minimizing environmental impact (greenness), and optimizing overall efficiency.

## 1 Introduction

The rapid development of artificial intelligence (AI) models and Internet of Medical Things (IoMT) brings an opportunity for healthcare applications spanning from daily activity monitoring for continual change to disease diagnosis and decision support for healthcare providers. The IoMT devices are edge devices with limited computational power and storage. To apply AI for healthcare implementation, the need for high-quality and quantity data is always at the corner. Especially in the health sector, not only finding these data but protecting the privacy and security of available data is always a problem. To overcome this issue, the FL approach comes into play. FL has gained much attention in healthcare applications in recent years, some applications are medical imaging [1], COVID-19 detection [2, 3], clinical decision support [4, 5], in-home activity monitoring [6], and others. Healthcare applications widely use horizontal federated learning since


L. Barolli (Ed.): AINA 2024, LNDECT 203, pp. 368–378, 2024.
https://doi.org/10.1007/978-3-031-57931-8_36

all the parties share the same feature space but have different data samples [7]. In HFL, the training step starts with participants downloading the latest model from the server individually and training the model based on the local data. Then, participants send the trained model parameters to the server. Finally, the server aggregates the gradients of each user and returns the updated model to each participant to update the local models. Recently, the availability of wearable and smart devices presents an opportunity to look for data availability. But still using wearable data and protecting privacy under the umbrella of different data protection regulations is challenging [8]. One of the main challenges in FL application in healthcare is the heterogeny of data, as different subjects of the federation have unique characteristics in terms of gender, behavior, previous medical history, age, living condition, and others that can affect the possibility to detect specific health condition. To provide personalized services to IoMT users, the conventional FL faces several challenges. The first one is that the global model only captures the statistical characteristics of different IoMT devices, but it is hard to provide distinct personal styles. To overcome this challenge, a recent trend is to make the FL more fine-grained. This concept brings the idea of personalized federated learning (PFL) into play. This study is more interested in the concept of an on-device-based model for individual pre-frail monitoring systems and explores the challenges and characteristics of such approaches in health. Figure 1 shows that PFL allows for the development of AI models that can offer personalized treatment and monitoring at the edge level, with the global model being trained on diverse data sources.

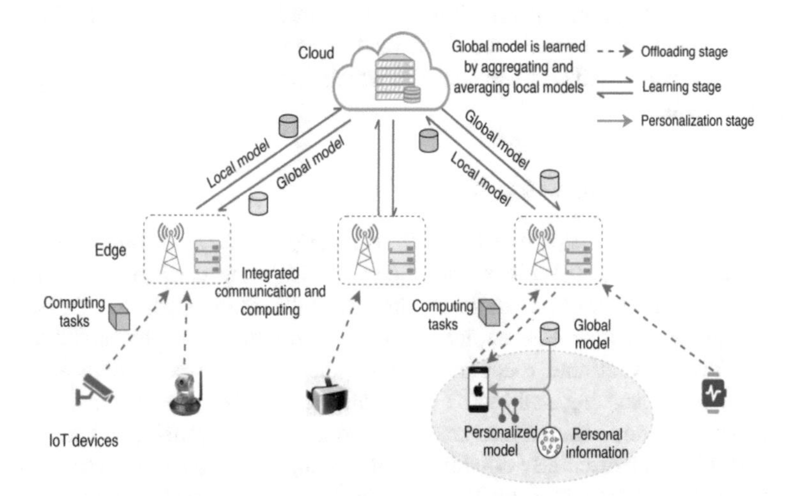


**Fig. 1.** Personalized federated learning framework for intelligent IoT applications [9]

In this context, the fusion of privacy preservation and federated learning to comply with data regulations, such as the General Data Protection Regulation (GDPR) presents both challenges and promising opportunities. FL is a transformative paradigm in health informatics by allowing researchers, individuals, and practitioners to collaborate, mitigating the risks associated with centralized data repositories and facilitating advancements in predictive analytics, disease diagnosis, personalized treatment recommendations, and long-term monitoring.

**Motivation:** The application of FL in health informatics has significantly increased over the years. A major issue is that conventional AI models may perform good on controlled data sources, but still perform poorly on samples coming from clinical practice in a real-life medical setting. However, most studies have concentrated on institutional settings where health institutions, owning databases on multiple patients, serve as nodes of the FL system. In this study, we will explore the challenges and opportunities of PFL in the edge-cloud implementation scenario to answer the following research questions (RQ).

- **RQ1.** What is the current state of FL and PFL in healthcare?
- **RQ2.** In the healthcare domain, what are the challenges of PFL implementation?
- **RQ3**. What are the opportunities for PFL in healthcare applications in terms of privacy preservation and cloud-edge continuum?

## 2 Background and Related Work

The studies reviewed in Sect. 2.1 favor the application of FL in health care through the federation of different medical institutions, each participating with its own patient database, to train a global model. Each database includes data about the individuals diagnosed, treated and/or monitored at the institutions.

Section 2.2 is focused on FL systems in which federation members are represented by single patients.

### 2.1 Federated Learning and Healthcare

The study [8] proposed adaptive FL to derive the generalization bound of the mixture of local and global models to find the optimal mixing parameter and propose a communication-efficient optimization method for collaborative learning. In FedHealth [10], a framework for wearable healthcare for activity recognition achieved better results relative to the conventional machine learning model. It performs data aggregation through federated learning and builds personalized models through transfer learning, achieving accurate and personalized healthcare without compromising privacy and security. FedBN [11], is particularly effective in addressing the problem of feature shift non-independently identical data (non-IID) datasets, such as different sensors in wearable devices or medical imaging. [13] proposed self-supervised FL for magnetic resonance imaging to explore the advantage of FL on non-fully labeled data. Researchers in [12] propose an FL-based methodology to support the diagnosis of acute neurological symptoms such as severe headache or loss of consciousness through X-ray scan imaging in smart healthcare units. Each hospital runs a CNN-based DenseNet1212 model that can support feature propagation, encourage feature reuse, and minimize the number of neural parameters to train the X-ray image dataset.

The feasibility of FL in medical imaging has been investigated via a real-world experiment conducted at the University of Pennsylvania and 19 other institutions worldwide in the collaborative healthcare project [13]. Particularly, the Intel company has provided support to the FL-health project by leveraging the capabilities of Intel Xeon Scalable processors for running FL functions at hospitals and the cloud server. In addition, Nvidia corporation has established a platform for deploying on-devices for smart and wearable devices to handle medical image and video processing.

## 2.2 Personalized Federated Learning for Healthcare

The study [14], focuses on a multi-device-environment, where each user can have multiple devices with different processing power. They propose personalized federated learning for scenarios where users can have sufficient data on one of their edge devices. In PfedHN [15], the authors proposed personalized models for users using Hyper-networks. To train personalized models for multiple clients with different data distributions. Hypernetworks are one of the widely used meta-learning techniques that output the weights of another target network that performs the learning task. However, the empirical evaluation of the study is limited to a few personalized federated learning challenges and standard benchmarks, which may not fully capture the performance of the approach in real-world scenarios. The study [16] trained a model that learns the similarity between clients based on the statistics of the layers' outputs of the pre-trained network while preserving the specificity of each client with different local batch normalization.

The work in [17] developed a personalized FL framework to predict pain using face images. Each device employs a lightweight convolutional neural network model and keeps the learning parameters and updates of the last layer locally to personalize the local pain estimation model and increases the protection of data privacy against potential adversaries. Despite the potential application of the technique in the health field, the study failed to investigate the challenges of the data imbalance effect.

FedHome [6] proposed an edge-cloud FL architecture for in-home remote health monitoring services, where each local model is trained at home located at the network edge and the cloud is responsible for global model aggregation. The training on the cloud server mainly relies on the distributed datasets, which may be different among different homes. The shared model only captures the common features of all health users, and thus it may perform poorly on a particular user. As a result, to achieve personalized in-home health monitoring, each user integrates the trained global model with his health data. However, the drawback of the study is the convergence of the proposed FL algorithm is still an open problem. In IoT applications, hardware heterogeneity is another major challenge that requires users to have personalized models. However, a non-significant number of literature has focused on the effect of differences in user processing power on the overall performance.

However, understanding and identifying the challenges and opportunities of personalized aspects in edge-cloud scenarios is open research. Thus, this study presents a survey on the application of PFL in the medical domain besides identifying key challenges and opportunities for PFL.

# 3 Challenges and Opportunities

## 3.1 Challenges in Personalized Federated Learning

In this section, we will discuss the challenges that arise because of the nature of the PFL technique and data-related challenges in privacy-preserving healthcare applications.

**Data Heterogeneity.** To achieve a desirable training performance in PFL-based healthcare systems, a critical problem to be addressed is the non-IIDness of the medical data, which potentially makes the FL training divergent in the training. For example, an individual can wear the sensors for a longer period for a certain type of monitoring and data collection than other individuals. The current situation is that the data samples lack independence and are dispersed uniformly. It causes an increased bias in the model [18]. By intelligently choosing the participant, it can reduce the time of communication in FL and counterbalance the bias introduced by data. However, the data are not always IID type. Without addressing this non-IIDness issue, training would suffer greatly in quality or possibly diverge. To address the problem, solutions such as providing an additional subset of datasets to allocate equitably among clients [18], can be a suitable solution. Another intriguing technique is to implement the feature shift across diverse clients [11]. To evaluate non-IID data in the FL-based smart healthcare industry, quantitative criteria are required, such as standard deviation, precision, and accuracy about label/feature distribution skew and homogenous partitions [19].

**Concept Drift Problem.** Given a PFL system's ability to learn from non-stationary data in dynamic real-world systems, we can anticipate changes in the underlying data distributions over time this phenomenon is known as concept drift. Learning in the presence of concept drift often involves drift detection, drift understanding, and drift adaptation [20]. Little research has yet been done on this problem, but this kind of situation is quite frequent in real life and poses new challenges. The work [21] presents a method called concept-drift-aware federated averaging to study the problem of concept drift in FL. To detect drift in the global model, two timestamps are defined. The difference is identified using two data storage for short-term and long-term memory. It remains an open direction to leverage existing drift detection and adaptation algorithms to improve learning on dynamic real-world data in PFL systems.

**Data Quantity and Feature Distribution Skew.** FL clients maintain local data of varying quantities, with some clients holding significantly more data than others. Data size skew is pervasive in real-world environments due to diverse usage patterns across FL clients. To simulate data size, data from an imbalanced dataset can be used directly without re-sampling and could affect the model performance, but it can be used to mimic a real-world scenario. The feature distribution may vary across clients, while the conditional distribution of target given some features is the same across clients. For example, in health monitoring applications, the distributions of user activity data vary considerably according to their habits and lifestyle patterns [6, 10]. This implies feature representation across the clients' data points and training procedure will bring a challenge to overcome for efficient application of PFL in applications like healthcare. In addition, data and label noise in personalized scenarios need attention for efficient application.

**Health Dataset Benchmarks.** Realistic datasets are critical to the progress of this field. To enable effective PFL, datasets with more modalities, such as audio, video, and sensor signals for a wider range of real-world AI applications, are required [22]. Medical dataset quantity and quality have frequently hampered the creation of a robust solution to the FL algorithm. The dataset utilized in FL studies may greatly vary depending on the study goal. For example, some datasets concentrate on medical image classification and segmentation, while others concentrate on network communication performance. However, there is still a lack of benchmark datasets, especially in the medical field. Thus, a reliable benchmark is required to assess the performance of the FL that employs several medical data sources. Benchmarking PFL for healthcare presents various hurdles and a set of existing issues that researchers must address.

Most researchers evaluate PFL algorithms in a single non-IID scenario by leveraging an already pre-partitioned public dataset or by partitioning a public dataset to meet the intended non-IID environment [23]. For a more accurate comparison, the research community must gain a better grasp of the many non-IID circumstances in real-world federated learning to mimic actual non-IID settings. Possible circumstances include temporal shifts in data distribution. Such attempts initial greater collaboration among researchers and industry practitioners.

**Communication Efficiency and Device Heterogeneity.** FL networks can consist of many contributing model updates, in which communication becomes a key bottleneck. Multi-participant situations are unavoidable when attempting to create high-performance models. In this situation, network communication is many orders of magnitude slower than local processing because of constrained resources, such as bandwidth, energy, and power. To address this challenge, we can take advantage of two major features, with the first being the reduction of the total number of communication rounds. The second is reducing the size of the messages communicated in each round [24]. FL can be designed to perform client-side filtering, reduce the frequency of updates, and peer-to-peer learning [25]. However, an efficient communication protocol for frequent updates of parameters for healthcare sensitivity and privacy-preserving mechanisms is still an open problem for PFL.

In practice, healthcare systems based on edge-cloud continuum involves a significant number of participating IoMT devices. With IoMT, the fast and unpredictable changes in wireless channels pose another significant barrier to the secure exchange of learning updates [26]. Differences in device hardware capabilities may result in varying storage and computing capacities, inevitably leading to device heterogeneity. For example, one client could be a smartphone and another a smartwatch. Smartphones have more storage and computing capacity than smartwatches, resulting in device heterogeneity. As a result, during federated communication, smartwatches may have prolonged local runtimes, leading to dropped or lost connections, which can affect system efficiency and stability.

**AI Architecture and Evaluation Metrics Challenges.** In the presence of heterogeneity, federated neural architectures are particularly sensitive to hyperparameter selections and may hence experience poor learning performance if not carefully calibrated [27]. As a preliminary study, Neural Architecture Search (NAS) [28] is a promising technique that can help PFL reduce human design effort by optimizing model architecture based on specified scenarios. It will be especially useful for parameter decoupling and knowledge distillation-based PFL approaches. The FL model architecture should also fit the underlying data distribution well. But Finding a general solution to address the non-IID learning degradation problems in PFL. In addition, reproducibility and standardization in PFL can make it difficult to achieve crucial comparison outcomes from different investigations.

We can categorize the challenges related to evaluation metrics into model performance, system performance, and trustworthiness [7]. Evaluating performance in PFL applications is largely concerned with accuracy advances in model performance. However, addressing the costs of achieving PFL is also necessary. When pursuing accurate models, there are frequent trade-offs in terms of system scalability, as well as communication and computing overheads. When pursuing accurate models, it is important to consider healthcare-specific metrics like patient outcome metrics, concordance index, sensitivity, and specificity in applying PFL models.

**Explainability.** The explainability of models is important in applications like healthcare but has not yet been vetted in the FL studies. Because of the volume and variability of dispersed datasets, attaining explainability in PFL presents unique and challenging issues. Striving for FL model explainability may also be associated with potential privacy risks from inadvertent data leakage, as demonstrated where certain gradient-based explanation methods are prone to privacy leakage [29]. Future research could focus on developing a framework that strikes both explainability and privacy.

**Usability, Client Participation, and User Characteristics.** In the PFL context, clients may have different levels of participation in training rounds. Some clients may contribute more frequently or with larger datasets. Benchmarking personalized federated learning needs to consider scenarios where the number of participating clients and their data contributions can vary. Ensuring user engagement in the learning process and providing models that are usable and beneficial for individual users are also existing challenges in PFL. Models need to be interpretable and user-friendly while still maintaining privacy. In addition, users' preferences and characteristics may change over time. In this sense, PFL systems should account for dynamic changes in user behavior and be adaptable to evolving user characteristics.

**Privacy and Security Challenges.** With any edge-cloud architecture, PFL is also prone to different cybersecurity challenges. We can categorize the cyber challenges into attack occurrence scenarios, such as the transmission medium, data, and model. In a medium attack, the target of the attacker is the link between the local models and the aggregation model. For example, man-in-the-middle attacks, eavesdropping, jamming, and replay attacks are some of the conventional attacks encountered [30]. In a data poisoning attack, the attacker alters the data on which the local model is trained to affect the global model decision and global performance. In the same way, model poisoning affects the local model to cause an error in the global model [31].

### 3.2 Opportunities of PFL

In the ever-changing healthcare landscape, the need for personalized and patient-centric solutions is growing. It emerges as a game-changing concept that smoothly mixes personalized healthcare with the critical necessity to maintain patient confidentiality.

**Predictive Decision for Personalized Medicine.** PFL transforms predictive analytics in healthcare by integrating global model weights with individualized data generated at the edge. Individual devices in this decentralized method train models using locally stored longitudinal health data, enabling the development of predictive analytics models suited to specific patients. PFL enables the estimate of sickness risks, prediction of treatment outcomes, and prediction of likely side effects based on personalized health histories. Edge-level data processing protects privacy by storing sensitive information on local devices, making PFL a new and privacy-conscious framework for enhancing individualized therapy and prediction in healthcare [32].

**Adaptive Health Recommendations and Treatment Plans.** Individuals participating in the federation are unique in different aspects of health and physical state. PFL brought an opportunity that enables the creation of models that can personalize treatment plans based on individual characteristics. This can lead to more effective and targeted interventions, especially in areas like cancer treatment or chronic disease management.

**Remote Monitoring and Alert.** FL makes it easier to create models for remote patient monitoring, which is important for preserving data privacy and real-time alerts for patients with chronic conditions who require timely medical attention. This is especially useful for ongoing monitoring of chronic illnesses and early diagnosis of potential health problems for pre-frail people. By allowing models to be trained on individual devices rather than centralizing sensitive patient data, PFL ensures that personal health information remains private throughout the remote monitoring process. This method not only improves the efficiency of continuous healthcare monitoring but also demonstrates a dedication to protecting privacy, making PFL a promising framework for increasing remote monitoring while being privacy-conscious [6].

**Inclusive Cooperation.** Aside from algorithmic issues, future PFL research could focus on encouraging collaboration among self-interested data owners. For example, data owners with personalized FL models may need to collaborate by sharing their model parameters with other appropriate data owners to respond to changes in the learning challenge over time in dynamic real-world applications [33]. Incentive mechanism design is an interesting study area for this objective. Game theory, pricing, and auction processes can be used to create appropriate incentive schemes to encourage the development of open collaborative PFL systems [34].

Generally, the main challenges are the non-IID nature of medical data, leading to biases in models, and the concept drift problem in dynamic real-world data, requiring further exploration. Issues related to data quantity, feature distribution, and the absence of realistic health dataset benchmarks are also obstacles to the sustainability of such systems. Communication overheads, device heterogeneity, and challenges in AI architecture selection, particularly in the context of edge-cloud continuum networks, are important research directions for efficient implementations. Evaluation metrics and the need for model explainability pose additional hurdles, with implications for system scalability and user trust. Conversely, the opportunities for PFL in healthcare, such as predictive decision-making for personalized medicine, adaptive health recommendations, remote monitoring, and fostering inclusive cooperation among data owners, encourage the studies and applications in an edge-cloud continuum. These opportunities signify the potential for PFL to revolutionize healthcare by offering tailored treatment plans, improving patient monitoring, and creating collaborative efforts in dynamic real-world applications.

## 4 Conclusion

In this study, we present the opportunities and challenges of PFL in cloud-edge environment for healthcare applications. The rise in the implementation of privacy preservation issues and user data protection regulations brings an alarming concern about how to implement intelligent IoMT systems. Even though PFL offers many opportunities, like providing a robust model from heterogeneous learning, privacy preservation, personalized treatment, remote monitoring, and others, it has many challenges that slow down the implementation progress. This study presented the most influential challenges and opportunities of PFL in the healthcare domain to answer our initial research questions.

In the future, the authors would like to implement PFL for real-world applications in edge-cloud scenarios in the health monitoring domain besides investigating the performance in terms of communication, sensitivity, and specificity besides user acceptability.

**Acknowledgment.** This research is supported by the European Union – Next Generation EU, in the context of the National Recovery and Resilience Plan Investment", Project Age-It (Ageing Well in an Ageing Society).

## References

1. Adnan, M., Kalra, S., Cresswell, J.C., Taylor, G.W., Tizhoosh, H.R.: Federated learning and differential privacy for medical image analysis. Sci. Rep. **12**, 1953 (2022). https://doi.org/10.1038/s41598-022-05539-7
2. Dayan, I., Roth, H.R., Zhong, A., et al.: Federated learning for predicting clinical outcomes in patients with COVID-19. Nat. Med. **27**, 1735–1743 (2021). https://doi.org/10.1038/s41591-021-01506-3
3. Zhang, W., et al.: Dynamic-fusion-based federated learning for COVID-19 detection. IEEE Internet Things J. **8**, 15884–15891 (2021). https://doi.org/10.1109/JIOT.2021.3056185
4. Xue, Z., et al.: A resource-constrained and privacy-preserving edge-computing-enabled clinical decision system: a federated reinforcement learning approach. IEEE Internet Things J. **8**, 9122–9138 (2021). https://doi.org/10.1109/JIOT.2021.3057653
5. Lim, W.Y.B., et al.: Dynamic contract design for federated learning in smart healthcare applications. IEEE Internet Things J. **8**, 16853–16862 (2021). https://doi.org/10.1109/JIOT.2020.3033806
6. Wu, Q., Chen, X., Zhou, Z., Zhang, J.: FedHome: cloud-edge based personalized federated learning for in-home health monitoring. http://arxiv.org/abs/2012.07450 (2020)
7. Tan, A.Z., Yu, H., Cui, L., Yang, Q.: Towards personalized federated learning. IEEE Trans. Neural Netw. Learning Syst. **34**, 9587–9603 (2023). https://doi.org/10.1109/TNNLS.2022.3160699
8. Ge, Y., Zhou, Y., Jia, L.: Adaptive personalized federated learning with one-shot screening. IEEE Internet Things J. (2024). https://doi.org/10.1109/JIOT.2023.3346900

9. Wu, Q., He, K., Chen, X.: Personalized federated learning for intelligent IoT applications: a cloud-edge based framework. IEEE Open J. Comput. Soc. **1**, 35–44 (2020). https://doi.org/10.1109/OJCS.2020.2993259
10. Chen, Y., Wang, J., Yu, C., Gao, W., Qin, X.: FedHealth: a federated transfer learning framework for wearable healthcare. http://arxiv.org/abs/1907.09173 (2021)
11. Li, X., Jiang, M., Zhang, X., Kamp, M., Dou, Q.: FedBN: federated learning on non-iid features via local batch normalization. http://arxiv.org/abs/2102.07623 (2021)
12. Srivastava, U.C., Singh, A., Kumar, D.K.S.: Intracranial hemorrhage detection using neural network based methods with federated learning. http://arxiv.org/abs/2005.08644 (2022)
13. Unknown: Federated learning: protecting data at the source. Intel Labs (2023). https://www.intel.com/content/www/us/en/research/news/federated-learning-protecting-data-at-the-source.html. Accessed 25 Jan 2024
14. Cho, H., Mathur, A., Kawsar, F.: FLAME: federated learning across multi-device environments. In: Proceedings of ACM Interaction Mobile Wearable Ubiquitous Technology, vol. 6, pp. 1–29 (2022). https://doi.org/10.1145/3550289
15. Shamsian, A., Navon, A., Fetaya, E., Chechik, G.: personalized federated learning using hypernetworks. http://arxiv.org/abs/2103.04628 (2021)
16. Lu, W., et al.: Personalized federated learning with adaptive batchnorm for healthcare. http://arxiv.org/abs/2112.00734 (2022)
17. Rudovic, O., et al.: Personalized federated deep learning for pain estimation from face images. http://arxiv.org/abs/2101.04800 (2021)
18. Zhao, Y., Li, M., Lai, L., Suda, N., Civin, D., Chandra, V.: Federated learning with non-IID data (2018). https://doi.org/10.48550/arXiv.1806.00582
19. Li, Q., Diao, Y., Chen, Q., He, B.: Federated learning on non-IID data silos: an experimental study. http://arxiv.org/abs/2102.02079 (2021)
20. Lu, J., Liu, A., Dong, F., Gu, F., Gama, J., Zhang, G.: Learning under concept drift: a review. IEEE Trans. Knowl. Data Eng. (2018). https://doi.org/10.1109/TKDE.2018.2876857
21. Casado, F.E., Lema, D., Criado, M.F., Iglesias, R., Regueiro, C.V., Barro, S.: Concept drift detection and adaptation for federated and continual learning. Multimed Tools Appl. **81**, 3397–3419 (2022). https://doi.org/10.1007/s11042-021-11219-x
22. Matsuda, K., Sasaki, Y., Xiao, C., Onizuka, M.: Benchmark for personalized federated learning. IEEE Open J. Comput. Soc. **5**, 2–13 (2024). https://doi.org/10.1109/OJCS.2023.3332351
23. Yang, L., Huang, J., Lin, W., Cao, J.: Personalized federated learning on non-IID Data via group-based meta-learning. ACM Trans. Knowl. Discov. Data **17**, 1–20 (2023). https://doi.org/10.1145/3558005
24. Li, T., Sahu, A.K., Talwalkar, A., Smith, V.: Federated learning: challenges, methods, and future directions. IEEE Signal Process. Mag. **37**, 50–60 (2020). https://doi.org/10.1109/MSP.2020.2975749
25. Xu, J., Glicksberg, B.S., Su, C., Walker, P., Bian, J., Wang, F.: Federated learning for healthcare informatics. http://arxiv.org/abs/1911.06270, (2020)
26. Li, J., et al.: A federated learning based privacy-preserving smart healthcare system. IEEE Trans. Ind. Inf. **18**, 2021–2031 (2022). https://doi.org/10.1109/TII.2021.3098010
27. Li, X., Huang, K., Yang, W., Wang, S., Zhang, Z.: On the convergence of FedAvg on Non-IID data. http://arxiv.org/abs/1907.02189 (2020)
28. Zhu, H., Zhang, H., Jin, Y.: From federated learning to federated neural architecture search: a survey. http://arxiv.org/abs/2009.05868 (2020)
29. Shokri, R., Strobel, M., Zick, Y.: On the privacy risks of model explanations. http://arxiv.org/abs/1907.00164 (2021)

30. Yaacoub, J.-P.A., Noura, H.N., Salman, O.: Security of federated learning with IoT systems: issues, limitations, challenges, and solutions. Internet Things Cyber-Phys. Syst. **3**, 155–179 (2023). https://doi.org/10.1016/j.iotcps.2023.04.001
31. Mammen, P.M.: Federated learning: opportunities and challenges. http://arxiv.org/abs/2101.05428 (2021)
32. Thwal, C.M., Thar, K., Tun, Y.L., Hong, C.S.: Attention on personalized clinical decision support system: federated learning approach. In: 2021 IEEE International Conference on Big Data and Smart Computing (BigComp), pp. 141–147. IEEE, Jeju Island (2021). https://doi.org/10.1109/BigComp51126.2021.00035
33. Zheng, S., Cao, Y., Yoshikawa, M., Li, H., Yan, Q.: FL-Market: trading private models in federated learning. In: 2022 IEEE International Conference on Big Data (Big Data), pp. 1525–1534 (2022). https://doi.org/10.1109/BigData55660.2022.10020232
34. Zhan, Y., Zhang, J., Hong, Z., Wu, L., Li, P., Guo, S.: A survey of incentive mechanism design for federated learning. IEEE Trans. Emerg. Topics Comput. (2021). https://doi.org/10.1109/TETC.2021.3063517

# Examining University Students and Teachers' Behavioral Intention to Upgrade Blended Learning Using an Extended TAM Model

Rund Fareed Mahafdah[1(✉)], Seifeddine Bouallegue[2], and Ridha Bouallegue[1]

[1] Innov'COM Laboratory, High School of Communications (Sup'COM), University of Carthage, Carthage, Tunisia
{rund.mahafdah,ridha.bouallegue}@supcom.tn

[2] University of Doha for Science and Technology, Doha, Qatar
seifeddine.bouallegue@udst.edu.qa

**Abstract.** Various new revolutions are observed with the changing face of learning paradigms in the education industry due to the impact of the global pandemic. This study reveals the intentions of students and academicians towards improving e-learning systems. The Technology Acceptance Model (TAM) is used to identify the acceptability and expected improvements by them. The model identifies certain factors and categorizes them under various hypothesis parameters. A survey was conducted to collect opinions from multiple respondents, consisting of two main parts. The initial part includes questions for academic staff (16 questions), while the second part includes 18 questions for students. The questions focus on the current e-learning system, its adaptability, and the expected improvements at the next level. The data collected can be used to assess e-learning systems and their issues. The integration of blended learning and IoT devices can improve current features. Data was collected for two months, from March to April 2023. The survey was distributed in Middle Eastern universities, and over twenty universities responded. The staff submitted 1080 responses, and approximately 740 university students participated in the second part.

**Keywords:** Blended learning · Traditional learning · Artificial intelligence · IoT · Integrated learning

## 1 Introduction

A nation's growth depends on its education system. Teaching methods have changed significantly in the past 50 years. The use of different teaching methods has many advantages over traditional methods. Education researchers use various models to study and analyze teaching and learning. Given its long history, the final-level teaching and learning system needs a complete overhaul. Students expect schools and universities to adapt to changing educational landscapes.


L. Barolli (Ed.): AINA 2024, LNDECT 203, pp. 379–391, 2024.
https://doi.org/10.1007/978-3-031-57931-8_37

Numerous polls have been conducted to improve education. [1] Modern methods may need improvement. Use technology to implement and integrate these unique technical ideas into current systems. The educational system is being enhanced in various ways, including classroom teaching [2], mobile learning [3], learning management systems (LMS) [4], virtual classrooms [5], blended learning [6], and smart classrooms [7]. Classroom teaching is considered the most traditional method, but it is becoming increasingly popular due to its participative and efficient transfer of information. Mobile learning is becoming increasingly important for young minds, and many students choose this approach. Students and teachers are adopting learning management systems (LMS) due to their unique capabilities, with Middle Eastern higher education institutions showing a higher adoption rate. Virtual classrooms allow for easy exploration of subjects and seamless collaboration between instructors and students, promoting joint efforts on projects and assignments. Blended learning is becoming the norm in education, utilizing mobile learning, virtual tutoring, mobile devices, and LMS. Smart classrooms are being deployed at all educational levels to engage students and make instruction more enjoyable using technology and intelligent equipment. Companies are creating interactive and engaging smart classroom software to grab students' attention and track them without human intervention.

## 2 Problem Identification

More students and teachers are using blended learning to improve comprehension. This mixed-learning model is hard. Integrating specialized methods to improve education and value-added services is debatable. Learning prediction was advised and prediction-based systems are expected to grow in this research project. Computational pacing streamlines student learning and assessment. [8–11] Further case studies examined innovative interactive learning methods that outperform traditional ones. Smart methods, sensor-based devices with IoT capabilities, and cutting-edge tools simplify use case education [12–15].

The study examines user acceptance of new methods. The research always targets online learners, the main education technology users. Internet surveys fit the study's tech focus. Comments determine whether staff and faculty use modern technology and teaching methods. Staff believe in an improved blended and e-learning system, according to multiple questions. Most of the 1080 Middle East staff respondents were Saudi. We answered questions with thoughts and comments. Cronbach's alpha values are used to determine which elements can be trusted for accurate results based on presented information. The scale rated faculty responses. Mapping fuzzy values to numbers helped identify scale. IBM SPSS analyzes statistical data. Universities have adjusted processes to accommodate diverse student input, resulting in 74 responses using Google Forms.

This method is accessible for both techies and non-techies, making it suitable for a wide range of participants. The survey uses simple instructions and weights, and Cronbach's alpha determines factor dependability. The study also identified learning technique improvements and statistically tests every category's questions for reliability. The findings highlight the importance of diverse student input in academic research.

## 3 Using the TAM Model for the Measuring of Intentions

The project's most important step was categorizing intention-detecting aspects. Three staff and three student subsections make up the grouped areas. Six sections were inspected. A Google form offers 16 questions regarding staff goals and 18 on students' goals. The subsections were grouped and aggregated to evaluate if there was a link between respondents' inputs (Fig. 1).

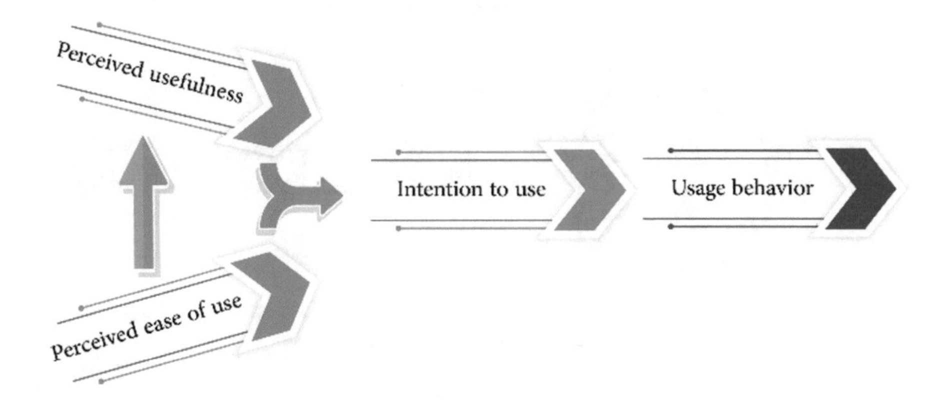


**Fig. 1.** TAM Model for User Acceptance.

Davis' 1989 technology acceptance paradigm defines student and teacher goals in this study. Existing e-learning needs help to spread. The research includes the Technology Acceptance Model (TAM) perceived usefulness survey. This approach can examine user groups' goals for improved technology. Many aspects are divided into components in the study. Each section impacts consumer acceptance, intentions to use, utility, ease of use, etc. Table 1 below lists this model's hypotheses.

Based on these predictions and observations, the model predicts whether Middle Eastern students and professors would accept the new e-learning system design (Fig. 2).

Table 2 describes the method used to gather and categorize data from various respondents by characteristics to provide a comprehensive overview of the Technology Acceptance Model hypothesis. Categories and parameters are used to determine user intent on the survey's acceptance of improved e-learning models. The study aims to discover software users' goals and support Table 1's hypothesis.

**Table 1.** Hypothesis

| H1: H5 | Hypothesis |
|---|---|
| H1 | Perceived ease of use will have significant impact on teachers and students with the new system |
| H2 | Perceived usefulness will have a significant impact on intentions to use the updated model by students and teachers |
| H3 | Attitude to use the new model have a positive impact by students and teachers |
| H4 | User behavior to the new model have a significant impact based on the inputs |
| H5 | Adaptability of the new model will have a positive effect on students and teachers |

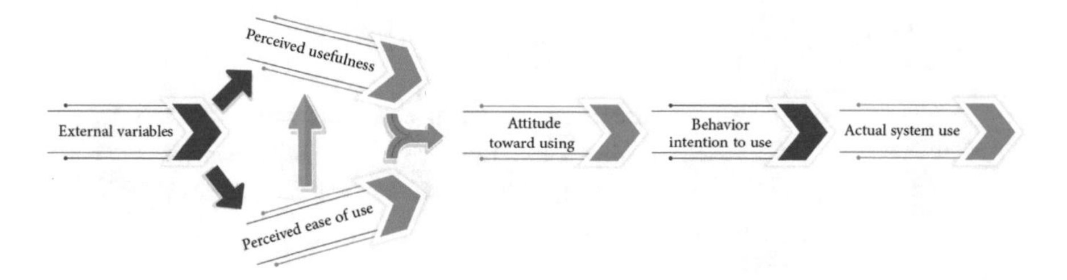


**Fig. 2.** Hypothesis Factors Mapping

**Table 2.** Approach Information

| Subject | Subject Computer science, e-learning |
|---|---|
| Specific area | Blended learning, future, e-learning. |
| Type of data | Tables. |
| Method of collecting data | The data is collected using online Google forms. |
| Data format | Raw data was analyzed data, and finally filtered |
| Parameters for data collection | Data was collected from 1080 participants from March to April of 2023 for the Staff Domain |
| Description of data collected | Data is collected from various respondents in Saudi Arabia via email and what's app messages |
| Data source | Around twenty universities participated in providing inputs |
| Related information | Author's name: Rund Fareed Mahafdah |

### 3.1 Value of the Data

The data is valuable for consciousness research and e-learning system development. This extensive field-specific content targets technology-savvy people from a variety of academic backgrounds. Global online learning communities, LMS and software providers, and e-learning scholars seeking publication opportunities may benefit. This information is important for evaluating educational institutions and organizations, assisting in their assessment processes, and discovering incremental changes and user objectives in current technology. This dataset is useful for understanding e-learning and related activities beyond education and training.

### 3.2 Factors Towards Staff Intentions

Several factors affect staff e-learning enhancement intentions. Because school staff and teachers need to improve traditional classroom methods, the e-learning system must be accepted. E-learning proficiency must be assessed in teachers. Students, instructors, and modern e-learning technologies must be compatible to advance teaching methods beyond traditional roles. Combining blended learning with e-learning, which uses software-driven learning and other technologies, may improve education. Staff attitudes toward blended learning in the current educational system must be assessed. SPSS analyzes four questions to evaluate its value. Enhancing MOOCs with AI requires user-friendly teaching methods. AI can make e-learning systems more dynamic than traditional ones. Staff use of AI to achieve e-learning goals is revealed by AIELS' reliability statistics, evaluated through four SPSS questions. However, IoT devices integrated with AI for e-learning raise concerns, emphasizing the need for careful consideration.

Table 3 shows ELS (which targets the current e-learning system), BL (which emphasizes blended learning), and AIELS (staff comments on AI-based e-learning services).

### 3.3 Factors Towards Student Intentions

The ELSS factor measures student interest in e-learning systems and their willingness to use them for effective education. Students answer five questions to assess their e-learning platform engagement. The second part of the study examines students' views on e-learning AI. This section has five questions about using this technology. IBM SPSS measures students' awareness, interest, and readiness to adopt AI in education to assess this method's reliability. Third, students prefer traditional learning over computer-aided methods. This factor examines students' views on e-learning system flexibility in the context of global educational reform, using SPSS to assess reliability. The tables of questions used by staff and student respondents to evaluate the various elements in the study are shown in the following tables.

**Table 3.** Staff questions

| Factor | Question |
|---|---|
| ELS | Are you satisfied with the present E-Learning System? |
| | What rating would you give the existing E-Learning System that you use? |
| | Do you think there is a need for the change/upgrade of the E-Learning System? |
| | Is the present E-Learning System efficient enough for the teaching and learning procedure? |
| | Are the MOOC Courses sufficient for teaching over the E-Learning System? |
| BL | Does your E-Learning System Support Blended Learning Options? |
| | Can Blended Learning be used over the Traditional Learning Methodology of E-Learning System? |
| | Do you think that the use of IoT can be more helpful in providing Blended learning to modern E-Learning Systems? |
| | Can we replace human evaluation procedures with the help of a Software Driven System? |
| | Which one is better: Blended learning, Traditional Learning, machine-based learning, or Online Learning? |
| AIELS | Do you think we can measure a student's cognitive skills using Artificial Intelligence? |
| | Do you think we can measure the Intelligence Quotient of a student using Artificial Intelligence? |
| | Can we use social media as a Learning Platform for teaching? |
| | Can the software-based systems be used to evaluate the Lab Learning/practical-based learning for students? |
| | Is Sentiment Analysis Possible Using Artificial Intelligence? |
| | Do you think that AI can improve the Teaching methodology for staff members? |

Table 4 shows that students' opinions are influenced by are ELSS (it represents the present e-learning system, which student uses), ELS-AI (this factor focuses on students' perception towards artificial intelligence-based e-learning system), and TL (which represents the traditional learning model that various universities and schools have used since ancient time), the traditional learning model used by universities and schools since ancient times. Student opinions on the proposed changes to the e-learning system are covered in all questions.

**Table 4.** Students Questions

| Factor | Question |
|---|---|
| ELSS | Are they compatible with the present e-learning system? |
| | Do you use the e-learning system 100% of the time without failure? |
| | Are you engaged all the time in the e-learning system without any other work? |
| | Is your e-learning system sufficient to learn your course completely? |
| | How many times do you use the e-learning system in one day? e |
| | Are you able to understand your teacher during the classroom with the help of online learning? |
| ELS-AI | Is your e-learning system able to identify your emotions while you are taking a class? |
| | Is your e-learning system able to identify your emotions while you are taking a class? |
| | Do you think blended learning is important compared to the traditional e-learning system? |
| | Are online studies better than the traditional mode of teaching and learning? |
| | Do you think that an e-learning system is able to check your intelligence quotient? |
| TL | Which is better, traditional learning or e-learning? |
| | Do you think you can roll back to traditional learning? |
| | Rate your e-learning system on a scale of one to five |
| | How much is your comfort level with your present e-learning system? |
| | Rate your experience with the e-learning system? |

Table 5 lists this model's hypotheses and examines factors supporting them.:

The study evaluates students and teachers' attitudes towards e-learning through blended learning and artificial intelligence. It uses various variables to assess their intentions and adapts to changing educational circumstances. The study uses the technology acceptance model to determine staff and student behavioral intentions. It also examines stakeholders' perspectives and expectations, demonstrating their willingness to accept and commit to e-learning with blended learning and AI. The study emphasizes the importance of system reliability in modern educational technology.

**Table 5.** Students Opinion Factors

| HYPOTHESIS H1:H5 | FACTORS THAT ARE ANALYZED TO SUPPORT THE HYPOTHESIS |
|---|---|
| H1 | TL |
| H2 | BL |
| H3 | AIELS, ELSAI |
| H4 | BL, TL |
| H5 | ELSS, ELS |

# 4 Discussion and Analysis of Faculty Intentions

## 4.1 Statistical Modelling Data - E-Learning System Acceptability

The 1,089-record analysis was substantial. There are a total of 1089 valid cases without any exclusions, accounting for 100% of the total cases. List-wise deletion, which considers all variables in the procedure, was implemented without excluding any cases. This study's accurate data-gathering ensures reliability and credibility for subsequent analyses and results. Staff members strongly support enhancing the existing e-learning system, as evidenced by a 0.765 Cronbach's reliability coefficient and a 0.752 Cronbach's alpha.

Table 6 shows the relationships between variables and faculty and staff queries. Many observations support the idea that blended learning technology should improve and replace the current e-learning system.

**Table 6.** Inter-item correlation matrix for e-Learning system items

| Factors | ELS1 | ELS2 | ELS3 | ELS4 | ELS5 |
|---|---|---|---|---|---|
| ELS1 | 1.000 | 0.802 | 0.382 | 0.382 | −0.047 |
| ELS2 | 0.802 | 1.000 | 0.325 | 0.325 | 0.045 |
| ELS3 | 0.203 | 0.115 | 1.000 | −0.062 | 0.023 |
| ELS4 | 0.382 | 0.325 | 1.000 | 1.000 | 0.063 |
| ELS5 | −0.047 | 0.045 | 0.023 | 0.063 | 1.000 |

## 4.2 Blended Learning Integration

Examining the second component can determine if blended and e-learning are appropriate. Reliability statistics based on a score of Cronbach's alphas based on standardized items is 0.545, where the number of items equals 4, indicating that many people support improving e-learning paradigms with blended learning and IoT devices. Additionally, staff cannot simultaneously use other Thing devices or blended learning methods. These strategies are rarely used but are effective in e-learning and future student training. The calculation of scale statistics, which include a mean of 7.39, a variance of 4.020, and a standard deviation of 2.005, closely corresponds to the staff responses regarding blended learning. These statistics are based on a dataset consisting of four items. The mean value acts as a comprehensive representation of all inquiries and factors related to blended learning, showing a significant correlation with staff perceptions or opinions. Additionally, the relatively low standard deviation indicates a higher level of acceptability and consistency in the responses received from different respondents regarding blended learning. This highlights a coherent consensus among staff members.

### 4.3 Artificial Intelligence Acceptability

Artificial Intelligence Acceptability reliability statistics show a Cronbach's Alpha of 0.556, indicating moderate internal consistency in survey responses. For future consideration, student emotional and sentimental intelligence assessments are suggested. The computed Alpha factor supports artificial intelligence as a catalyst for education technology. The mean scale value is 9.93, indicating respondents' positive view of AI's significant benefits in online education. The high demand for online teaching and learning tools supports its widespread use.

## 5 Discussion and Analysis of Student Intention

### 5.1 Affinity of Students Towards E-Learning

740 students completed the e-learning implementation survey. All students understood and answered questions correctly. After analyzing survey respondents' generous responses, Cronbach's Alpha is calculated. The high-reliability coefficient of 0.712, calculated using Cronbach's Alpha Standardized 0.703 with 5 Items, indicates a consistent response to well-researched questions. This method may attract students, and the survey model's unique characteristics can personalize e-learning experiences. Based on five items, the statistical numbers showed a mean of 14.95, a variance of 24.598, and a standard deviation of 4.960. This information is easy to manage and interpret due to the high mean and low standard deviation. The elevated mean indicates a strong dataset central tendency, while the reduced standard deviation indicates less variability around it. Thus, these statistical characteristics facilitate the ease of managing and drawing conclusions from this dataset.

### 5.2 Artificial Intelligence - Students Perception

Based on the reliability statistics computed here, including Cronbach's Alpha of 0.728 and Cronbach's Alpha based on standardized items of 0.744, derived from a dataset comprising five items, student feedback suggests positive perceptions regarding the current e-learning system's flexibility. Additionally, respondents propose potential enhancements through the integration of blended learning approaches and IoT devices. The mention of Cronbach's Alpha value of 0.728 by the respondents indicates a level of confidence in the reliability of the data used for their assessments and feedback on the e-learning system (Table 7).

**Table 7.** Statistical summary of item characteristics in dataset

| - | Mean | Min | Max | Range | Max/ Min | Variance | Items |
|---|---|---|---|---|---|---|---|
| Item Means | 2.324 | 1.703 | 3.459 | 1.757 | 2.032 | 0.461 | 5 |
| Item Variances | 2.475 | 1.779 | 3.740 | 1.961 | 2.102 | 0.644 | 5 |
| Inter-Item Covariances | 0.864 | 0.385 | 1.445 | 1.060 | 3.750 | 0.115 | 5 |
| Inter-Item Correlations | 0.368 | 0.147 | 0.715 | 0.568 | 4.867 | 0.026 | 5 |

The correlation between several items and their variance lowers values. E-learning systems, blended learning, and Internet of Things-based gadgets can help many students, according to many students. The dataset's scale statistics reveal a mean of 11.62, a variance of 29.654, and a standard deviation of 5.446. These values, derived from five items, indicate a moderate to higher level across the assessed elements. Additionally, the relatively wider dispersion of responses suggests variability within the dataset. Given the circumstances, the difference between ideals and student answers is acceptable. A high variance indicates good second-factor adaptation for all items in this factor analysis.

### 5.3 Traditional Learning Affinity Analysis

The reliability statistics displayed, with Cronbach's Alpha at 0.240 and Cronbach's Alpha based on standardized items at 0.344, computed from a dataset of four items, indicate remarkably low values concerning the acceptance of conventional learning and learner affinity. These findings suggest a lack of consensus among respondents regarding traditional teaching methods, portraying a diminished popularity among learners. Consequently, this observation emphasizes a shift in student preferences towards more openness and acceptance of modern, non-traditional teaching methodologies (Table 8).

**Table 8.** Descriptive statistics and variability of e-Learning system items

| - | Mean | Min | Max | Range | Max/ Min | Variance | Items |
|---|---|---|---|---|---|---|---|
| Item Means | 3.416 | 2.919 | 4.108 | 1.189 | 1.407 | 0.249 | 4 |
| Item Variances | 1.477 | 1.090 | 1.856 | 0.766 | 1.702 | 0.115 | 4 |
| Inter-Item Covariances | 0.108 | −0.695 | 0.971 | 1.666 | −1.398 | 0.619 | 4 |
| Inter-Item Correlations | 0.116 | −0.444 | 0.683 | 1.127 | −1.538 | 0.297 | 4 |

## 6 Results and Discussion

Based on the analysis of responses from various respondents, this study recommends updating the educational paradigm. During pandemics, educational approaches change (Table 9):

**Table 9.** Reliability assessment of factors influencing staff and student intentions in e-Learning systems

| S. NO. | FACTOR | CRONBACH VALUE |
|---|---|---|
| Staff Intentions | | |
| 1. | ELS | 0.765 |
| 2. | BI | 0.5434 |
| 3. | AIELS | 0.556 |
| Student Intentions | | |
| 4. | ELSS | 0.712 |
| 5. | ELSAI | 0.728 |
| 6. | TL | 0.240 |

The study encompasses five hypotheses (H1 to H5) and their associated factors and results. Hypotheses H1 and H2, regarding 'TL' and 'BL', respectively, are supported. Additionally, H3 explores 'AIELS' and 'ELSAI,' H4 includes 'TL' and 'BL,' and H5 involves 'ELS' and 'ELSS,' all showing supported outcomes. These findings suggest a strong alignment between the specified factors and their positive reception based on the Cronbach Alpha values. The questionnaire employed in this research is notably parameter-focused, gathering responses from multiple respondents. It is evident that the study's analyses heavily rely on data from a multitude of respondents, affirming the reliability of conclusions drawn regarding the hypotheses and their associated factors.

## 7 Conclusion and Future Work

The research program aims to enhance e-learning management systems by integrating IoT devices. The study examines educators' and students' views on this change, using massive survey data analysis. The research supports blended learning and suggests that e-learning system designers should integrate AI and machines to revolutionize education. This study improved e-learning but has limitations. Consider that academic fields and institutions may limit generalizability. The study's assumption of uniform technical readiness may restrict its usefulness due to educational institutions' varied infrastructure. The study may only apply to some cultures because it may only account for some cultural factors affecting technology acceptability. To advance, research should examine these limits. E-learning can be improved by considering disciplinary and institutional differences, adapting plans to different technology infrastructures, and extensive cultural assessments. Many academic and artistic fields can benefit from research.

## References

1. Djamolovhich, S.N.: Improvement of education system management based on modern management approaches (2019)
2. Jabbarova, A.: The Importance of the Teaching Method-Theory and its Application (2020)
3. Lin, X.F., et al.: Chinese undergraduate students' perceptions of mobile learning: conceptions, learning profiles, and approaches (2019)
4. A. Khan, R., Qudrat-Ullah, H.: Adoption of LMS in Higher Educational Institutions of the Middle East. ASTI, Springer, Cham (2021). https://doi.org/10.1007/978-3-030-50112-9
5. Moser, L., Seufert, S., Guggemos, J.: Institutional levers to implement digital learning methods in the classroom. Use Case: i-MOOC to foster information literacy in upper secondary education in Switzerland. In: European Association for Research on Learning and Instruction (EARLI) SIG 6-7 Conference (2020)
6. Alhramelah, A., Alshahrani, H.: Saudi graduate student acceptance of blended learning courses based upon the unified theory of acceptance and use of technology. Aust. Educ. Comput. **35**, 1–22 (2020)
7. Mehmood, R., et al.: UTiLearn: a personalized ubiquitous teaching and learning system for smart societies (2017)
8. Chen, X., et al.: Learning to stop while learning to predict. In: International Conference on Machine Learning. PMLR (2020)
9. Rashidov, A.S.: Interactive methods in teaching mathematics: CASE STUDY method. Sci. Res. **3**, 21–24 (2020)
10. Martinez, P.: Machine learning applications in educational settings: current developments and future prospects. Comput. Educ. **128**, 143–156 (2022)
11. Jin, X., et al.: Epidemiological, clinical and virological characteristics of 74 cases of coronavirus-infected disease 2019 (COVID-19) with gastrointestinal symptoms. Gut **69**(6), 1002–1009 (2020)
12. Fernández-Caramés, T.M., Fraga-Lamas, P.J.S.: Teaching and learning IoT cybersecurity and vulnerability assessment with Shodan through practical use cases. Sens. J. **20**(11), 3048 (2020)

13. Smith, J.A., Johnson, K.L.: Exploring IoT security and vulnerability assessment using real-world scenarios. Cybersecur. J. **25**(5), 1123–1140 (2021)
14. Pillai, R., et al.: Students' adoption of AI-based teacher-bots (T-bots) for learning in higher education. Inf. Technol. People (2023). https://doi.org/10.1108/ITP-02-2021-0152. (ahead-of-print)
15. Holmes, et al.: Artificial intelligence in education. Globethics Publications (2023)

# Applying Artificial Intelligence in the E-Learning Field: Review Article

Rund Fareed Mahafdah[1](✉), Seifeddine Bouallegue[2], and Ridha Bouallegue[1]

[1] Innov'COM Laboratory, High School of Communications (Sup'COM), University of Carthage, Carthage, Tunisia
{rund.mahafdah,ridha.bouallegue}@supcom.tn

[2] University of Doha for Science and Technology, Doha, Qatar
seifeddine.bouallegue@udst.edu.qa

**Abstract.** Artificial intelligence has succeeded in creating a massive technological revolution in the education sector especially in e-learning, by improving the overall learning experience for all educational institutions and their staff. In this paper, we will conduct a review study about how artificial intelligence enhances and helps the eLearning field by applying many techniques. These techniques are adaptive learning, intelligent tutoring systems, automated grading and assessment, automated administration, chatbots and virtual assistants, speech recognition, virtual labs and simulations, content recommendation, predictive analytics, and improving language learning. The findings indicate that the use of these techniques based on artificial intelligence can improve the education process. Also, it handles the challenges and obstacles that learners and students face when using the eLearning system.

**Keywords:** Artificial intelligence · E-Learning · Adaptive learning · Chatbots

## 1 Introduction

AI mimics human cognition with technology. Usually, it means a machine can act like a smart human. AI involves theoretically and practically designing and building computer systems that can perform human tasks like visual perception, speech recognition, decision-making, and language translation [1,2]. Modern societies use "AI" more. Autopilots, telemedicine, chat bots, Big Data, smart cities, smart homes, automated surveillance, AI in weapons, online and traditional education, industry, and cyber justice are improving and affecting human life [2,3].

E-learning, also known as online or electronic learning, involves learning through technology and media. Electronic learning is acquiring knowledge and skills. E-learning uses the Internet to let students access educational resources remotely and at their convenience. Online courses, degrees, and programs dominate e-learning. E-learning is widespread, as discussed in our previous articles [4,5].


L. Barolli (Ed.): AINA 2024, LNDECT 203, pp. 392–403, 2024.
https://doi.org/10.1007/978-3-031-57931-8_38

Traditional methods are inferior to online learning. Learning can be self-paced or at a venue. Despite geographical barriers, e-learning is cost-effective and efficient. These and other benefits explain electronic learning's growth. By 2025, global e-learning revenue will reach $325 billion, up from $107 billion in 2015 [5,6]. Artificial intelligence has transformed education, especially e-learning, by improving student and staff learning. E-learning can be improved by adaptive learning, intelligent tutoring, automated grading and assessment, administration, chatbots and virtual assistants, speech recognition, virtual labs and simulations, content recommendation, predictive analytics, and language learning [7,8].

This article reviews previous articles on AI in e-learning and its improvement. This review article reminder is organized: Sect. 1 presents the brief introduction about this study. Section 2 provides a background on how AI improves eLearning. AI in eLearning articles are summarized in Sect. 3. AI in eLearning is compared to traditional methods in Sect. 4. eLearning AI challenges and limitations and ethical considerations are discussed in Sect. 5. The article concludes with future work in Sect. 6. Article concludes with future work suggestions in Sect. 7.

## 2 Background

In this section, we present the different ways to help the eLearning system based on AI technique. But we will focus on the Adaptive Learning and Chatbots.

### 2.1 Adaptive Learning

Teaching and learning are adaptive. Technology and analytics help the teacher adapt to the changing curriculum. This strategy lets teachers evaluate students' learning, identify their needs, and use customized instruction to improve learning [9].

Adaptive learning tailors education to each student. Individualized instruction helps students learn and achieve goals. In adaptive learning, instructors and students use LMS. Data analysis with this technology helps educators understand student needs and traits [9,10].

Real-time adaptive learning software adjusts lessons and practice activities to student proficiency. This improves academic performance. Real-time adaptive learning software analyzes student progress-related concepts and skills. Certain software packages distinguish between "performance data" and "engagement" data to measure student engagement [11]. The latter includes logins and activity duration. This adaptive learning tool uses AI and machine learning to analyze student academic paths and achievements. The analysis shows that the software offers current students personalized review or practice based on its algorithms [12].

Each student using adaptive learning courseware will follow a different path through the material. One student will learn the topic, while another will be given additional materials. An additional student will receive a thorough evaluation and few quiz-based practice problems [13]. Adaptive learning possesses many advantages [14]:

1. It allows students to study at their own pace, skipping over familiar concepts and focusing on slower concepts.
2. Increase student engagement by tailoring lessons and activities to their needs and readiness.
3. This tool helps educators and administrators provide targeted support by identifying specific students or course sections that require additional attention.
4. This tool allows faculty and administrators to actively participate in continuous improvement by comparing data across semesters.
5. The exploitation of this resource offers a cost-effective alternative to the procurement of expensive textbooks.

### 2.2 Chatbots Technique

Bots chat with users to mimic human interaction. Modern chatbots use NLP to answer user questions automatically. Chatbots streamline user information retrieval by answering questions. Voice, text, or a mix of both eliminates manual examination and intervention [15]. Smart home speakers, SMS, WhatsApp, Facebook Messenger, and Slack use chatbots. Recent chatbots, called "intelligent virtual assistants" or "virtual agents," understand real conversations using complex language models. Additionally, they can automate relevant actions. Offices are using Siri and Amazon Alexa to help customers and employees. These mechanisms allow chatbots to communicate with users [16,17].

Chatbots, initially designed as interactive FAQs, now offer pre-programmed responses to some questions. Despite their limited natural language understanding, they allow users to select basic terms and phrases for discussion, addressing complex queries and answering unintended questions. [18].

Better chatbot algorithms can handle complex rules-based programming and NLP. This technology allows consumers to ask questions casually and engagingly. A new chatbot anticipates user queries using contextual awareness and machine learning. Human language data improves chatbots [19]. Modern AI chatbots use NLU to interpret user input freely. This ability overcomes typos and language translation errors. Advanced AI tools determine the chatbot's "intent" from user input using semantic content [20]. Conversational AI responds properly. AI uses machine learning and deep learning to expand its user query and answer database. LLMs boost chatbot deployment and customer satisfaction [20,21].

## 3 Literature Review

This section describes some previous papers that are related to how AI can help and improve the eLearning process by using several techniques.

### 3.1 Adaptive Learning Technique

Table 1 shows the summarization of the previous articles related to this study for the Adaptive Learning Technique.

Alshmrany et al. [22] proposed CNN-LFD for learning style prediction. Automated adaptive e-learning predicts learning style and counts. The database received student data after user ID verification. The questionnaire score, login credentials (session ID, learner ID, and course ID), and login time (location and session ID) are extracted. Recorded user training. CNN-LFD predicted sequential/global, feature-based active/reflective, sensing/intuitive, visual/verbal learning styles. Experimental results improved learning style prediction and the CNN-LFD has 97.09% of accuracy, 94.76% of specificity, 92.12% of sensitivity, and 97.56% of precision.

An adaptive e-learning paradigm by Arsovic et al. [23] improved learning and decision-making. The students' model, adaption module, expert data analysis and decision-making system, learning materials repository, and instructional approach database were included. Learning management tailored to each student's background and style. Clear ability to adapt course content, organization, and assessment to student performance. The popular open-source LMS improved usability and deployment with the idea. This higher education case study showed how the model adapts to different learning formats. Giving one group of students a non-adaptive and one an adaptive e-learning course tested the idea. The adaptive model customized teaching and learning, improved student performance across multiple dimensions, and enabled continuous development.

Khan et al. [24] examined how AI and big data affect E-learning system integration to improve Saudi university and institution education. They sampled 290 college students and the data is analyzed using SPSS/SmartPLS 3. AI and big data can help higher education teachers and students learn, especially during an epidemic, says a study. Students, teachers, and institutions benefit from this study's resource access, task efficiency, and customization. It will improve student-teacher relations.

Liu et al. [25] tested 128 first-year students on adaptive learning. Math, biology, chemistry, and information literacy were fixed. Student learning in four curriculum areas was studied using mixed methods. Also examined were student perceptions and experiences with adaptive systems and how student characteristics affected intervention decisions. The adaptive learning intervention improved chemistry. Other three were unaffected. The only student trait that predicted engagement was math anxiety. Although time and design limited student achievement, they had fun. Results favored adaptive learning.

According to two Chinese experiments, YiXue is an adaptive learning system that provided by Li et al. [26]. YiXue was tested against two other adaptive learning platforms in an after-school English language arts class. This platform has six features:

1. The detailed learning path knowledge map was hierarchical.
2. Diagnostic pre-assessment.
3. Automated differentiated instruction matches student abilities.

4. Excellent educational materials.
5. Giving students immediate feedback and detailed explanations.
6. Teacher help.

The study found that YiXue improved student learning efficiency over other platforms. Surveys show treatment group students are happier at school.

A personalized adaptive e-learning system based on students' learning style and initial knowledge was evaluated by Köhler et al. [27]. Computer network at Yogyakarta public vocational secondary school had 62 students. Researchers gave students the USE Questionnaire to evaluate system usability. A 30-question survey evaluates usefulness, usability, learning, and satisfaction. Each question had four Likert scale answers. Students initially liked the adaptive e-learning system's cross-domain usability. The multiple linear regression showed that usefulness, usability, and learning affect satisfaction. Finally, regression showed that usefulness and ease of use affect satisfaction but not learning.

In-service math, science, and English teachers received extensive training in a vibrant educational setting. Technical Teaching Content Elmaadaway et al. knowledge support [29]. 173 educators were in 83 adaptive learning and 90 Zoom groups. TPACK questionnaires were given before and after the trial. Experimental TPACK beat control. All teachers thought adaptive learning made tech-heavy lessons. A study suggests dynamic learning environments to improve in-service teachers' TPACK. This text discussed adaptive learning environment issues and research.

Rishard et al. [29] created "Adaptivo" a personalized E-learning system for different learning styles and skill levels. To improve efficacy and interactivity, Adaptivo considered learner characteristics like time availability, online interactions, and learning duration. The system then tailors education to each student's learning style and knowledge. Testing adaptive learning on students. Students enjoyed teaching. When content matches learning style and prior knowledge, grades rise.

Mirata et al. [30] say higher education instructors struggle with adaptive learning. The four-stage Delphi design emphasized adaptive learning obstacles. The empirical study included experts from Switzerland and South Africa, two institutions with different organizational and socioeconomic contexts. Multiple contexts helped researchers understand adaptive learning. The study found eight categories and three dimensions-technical, teaching and learning, and organizational. Their study found significant university difficulty and ranking differences. South Africa and Switzerland differ due to socioeconomic factors and university organizational contexts like type, instructional methods, and implementation stage. IT advised higher education leaders and project implementers on adaptive learning. Infrastructure, institutional support, and dedication are needed.

**Table 1.** The Summarization of The Previous Articles

| Ref | Objective | Challenges | Platform | Results |
|---|---|---|---|---|
| [22] | They proposed an algorithm called CNN-LFD (Convolutional Neural Network-based Levy Flight Distribution) for the purpose of predicting learning styles | – | CNN-LFD | The proposed CNN-LFD algorithm outperforms alternative methods with 97.09% accuracy, 94.76% specificity, 92.12% sensitivity, and 97.56% precision. |
| [23] | They introduced adaptive e-learning to customize learning and improve decision-making. | – | adaptive model | The adaptive model can improve learning performance, tailor teaching and learning approaches, and enable continuous development. |
| [24] | They examined how artificial intelligence and big data affect E-learning system integration to improve educational opportunities for Saudi Arabian students in various institutions and universities. | – | AI and big data | AI and big data can help educators and students in higher education institutions facilitate teaching and learning, especially during an epidemic. |
| [25] | They examined the effects of an adaptive learning intervention on first-year pharmacy professional degree students (n = 128). | – | mixed methods design | The results emphasize the importance of design principles in adaptive learning. |
| [26] | The YiXue adaptive learning platform was assessed. | – | YiXue | YiXue improved student learning efficiency compared to other platforms. |
| [27] | A personalized adaptive e-learning system based on students' learning style and initial knowledge was evaluated. | – | personalized adaptive e-learning system | Students initially liked the adaptive e-learning system's usability across all domains |
| [28] | They developed a dynamic educational environment in which in-service math, science, and English teachers receive comprehensive and ongoing support for each TPACK component. | – | TPACK | All educators believed adaptive learning training helped create technology-rich lesson plans. |
| [29] | Their personalized adaptable E-learning system "Adaptivo" tailors learning to each learner's learning style and level of expertise. | efficacy and interactivity of the learning process | Adaptive | They reported that students are satisfied with the teaching method. |
| [30] | They examined the barriers to integrating adaptive learning principles into higher education instruction. | Infrastructure, institutional dedication, support, and resources. | Delphi | The four challenges are: infrastructure, institutional dedication, support, and resources |

### 3.2 Chatbots Technique

Tables 2 shows the summarization of the previous articles related to this study for the Chatbots Technique.

Making a prototype A chatbot for university students in various courses was described by Colace et al. [31]. The main focus has been designing an architectural framework and model that efficiently manages communication and gives learners accurate answers. Created a query-recognition system. The system answers students using NLP and domain ontologies. Experiments validated the model after implementation.

WU et al. [32] studied online learning with chatbots. Testing their emotional isolation and disconnection reduction ability. We created an E-Learning assistant chatbot for evaluation. The chatbot's NLP framework uses retrieval-based and QANet models. A two-model hybrid chatbot improves E-Learning platforms. Their chatbot's response context emphasizes instruction and informal conversation, making it more human. The experimental study and questionnaire suggest chatbots could reduce E-Learning user isolation and disengagement. Chatbots outperform E-Learning platform teacher counseling. An AI-powered e-learning chatbot by Janati et al. [33] helped students interact and recommend content. A preliminary analysis of e-learning multimedia content yielded the most information for these goals. A new NLP keyword extraction method was developed. Then they suggested keyword-based multimedia indexing. Online messaging networks use multimedia indexing for Chatbots. Keywords simplify multimedia presentation in the proposed method. Comparisons were made to literature-based methods. Keywords improve multimedia indexing and representation, says this study. Their Chatbot retrieved information quickly using indexed multimedia content. Chatbots accelerated response and satisfied learners.

With an educational chatbot, Baha et al. [34] taught secondary school students Logo using NLP and deep learning. Moroccan public schools tried the chatbot. Many Souss Massa Regional Center for Education and Training Professions instructors backed this project. They split 109 people into three groups: a control group using conventional methods and two experimental groups using digital content and chatbots. Initial research suggested chatbots could help students learn. Promote self-directed learning, reduce stress, optimize time, and motivate. In technologically advanced classrooms, AI improves student engagement, academic performance, and interactions.

A Telegram, Facebook Messenger, and Line chatbot for student inquiries was described by Lee et al. [35]. This chatbot obeyed and answered natural language questions. When professors upload course data to an online database, the chatbot can answer content and logistics questions like the class schedule. Logging in lets the chatbot tailor responses to student profiles like class schedules and score distribution. A survey of 10 undergraduate computer science students found that their chatbot reduced online tutoring workload. E-learning systems benefit from this chatbot.

**Table 2.** Previous Articles Summarization - Chatbots Papers

| Ref | Objective | Challenges | Platform | Results |
|---|---|---|---|---|
| [31] | Thy offered educational assistance to university students | – | Learning Bot | After the model was implemented, experiments showed its practicality and efficacy. |
| [32] | They facilitating users' learning on electronic learning platforms | – | QANet model | The chatbot outperforms the E-Learning platform's teacher counselling function. |
| [33] | TThey facilitate learner interaction and recommend e-learning content based on their needs. | Response time | e-learning AI-powered Chatbot | Chatbot reduced response time and met learner needs. |
| [34] | They used a practical chatbot to teach secondary school students Logo, an educational programming language. | – | – | AI technologies improve student engagement, academic performance, and student interactions in technologically advanced classrooms. |
| [35] | They delivered timely responses to student enquiries on many popular social platforms, including Telegram, Facebook Messenger, and Line | – | Learning Chatbot | They found that integrating this chatbot into e-learning systems is useful. |

## 4 Comparative Analysis

In this section, we compare AI applications in eLearning with traditional methods that include the following aspects: learner outcomes, engagement, and cost-effectiveness, to highlight AI's advantages and drawbacks. Table 3 shows the comparison between AI technology in eLearning and traditional education methods.

**Table 3.** Comparing AI technology in eLearning and traditional education methods

| Criteria | AI in eLearning | Traditional Methods |
|---|---|---|
| Learner Outcomes | - AI can assess learner performance to determine task difficulty.<br>- Learner can spot and fix mistakes immediately.<br>- Learning outcomes improve based on performance. | This type of student takes time to receive feedback and may have received it recently. |
| Engagement | Interactive simulations and virtual reality create an engaging AI learning environment. | Books and lectures make it hard for students to find the interactive elements that leaders use to make the environment more engaging. |
| Cost-Effectiveness | AI needs a high initial investment to develop this technology in eLearning. | - Traditional methods require classrooms to teach. Continuous maintenance and operational costs are needed.<br>- Teachers need high pay. |
| Conclusion | AI in personalized learning and its ability to provide prompt feedback may improve learner outcomes compared to traditional instructional methods. | AI technology may be expensive to start, but its scalability makes it a cost-effective solution over time, especially for large eLearning projects |

## 5 Challenges and Limitations

AI integration in eLearning presents challenges such as data privacy, the digital divide, and the need for human oversight. Data privacy concerns necessitate edtech companies to prioritize security and adhere to data protection, security, and intellectual property laws. AI systems require high processing power and solid infrastructure to efficiently digest large amounts of data. Many businesses use outdated technology, which is unsuitable for AI deployment. To optimize AI systems, organizations must invest in advanced infrastructure, tools, and applications. Educator training is essential for optimizing these systems to benefit students.

Integration into existing systems is more complex than installing plugins on an LMS. Assessing storage capacity, processors, and infrastructure is crucial, as is staff training to use new tools and identify algorithm malfunctions. Collaboration with AI-savvy vendors can help overcome these challenges. Cost is another challenge, but upgrading IT infrastructure and training staff can help reduce costs. Affordable training programs or free apps can help reduce costs.

# 6 Ethical Considerations

## 6.1 Algorithmic Bias

Algorithmic bias in AI, particularly eLearning systems, is the occurrence of unjust and biassed results that arise from the use of algorithms. Unintentional bias can occur at several phases of the development and implementation of AI systems.

## 6.2 Student Privacy Issues

The incorporation of AI in the field of eLearning necessitates careful deliberation of the significant consequences pertaining to the safeguarding of student privacy. The utmost significance is in the careful implementation of protocols for gathering, preserving, and utilising student data, with a specific emphasis on protecting privacy and complying with applicable regulations.

# 7 Conclusion and Future Work

AI in eLearning is thoroughly examined in this study. Adaptive learning, intelligent tutoring systems, automated grading and assessment, automated administration, chatbots and virtual assistants, speech recognition, virtual labs and simulations, content suggestion, predictive analytics, and language learning enhancement are examined. The findings suggest that artificial intelligence could improve education and solve eLearning system issues. AI can improve eLearning by personalizing learning, increasing productivity, and providing insightful student performance evaluations. eLearning has seen several AI advances and patterns, including customized learning trajectories: AI can analyze student data like learning styles, preferences, and performance history to create customized learning trajectories. This allows adaptive learning experiences tailored to each student's needs and learning pace. Intelligent Tutoring Systems: AI-driven tutoring systems can provide immediate feedback, explanations, and resources. These systems can adapt to students' learning styles and provide targeted support. There are plans to conduct a thorough examination of all artificial intelligence (AI) methods in the future.

# References


1. Albahra, S., et al.: Artificial intelligence and machine learning overview in pathology & laboratory medicine: a general review of data preprocessing and basic supervised concepts. In: Seminars in Diagnostic Pathology, vol. 40, no. 2, pp. 71-87. WB Saunders (2023)
2. Xu, Y., et al.: Artificial intelligence: a powerful paradigm for scientific research. Innovation **2**(4) (2021)
3. Zhao, S., Blaabjerg, F., Wang, H.: An overview of artificial intelligence applications for power electronics. IEEE Trans. Power Electron. **36**(4), 4633–4658 (2020)

4. Kulkarni, P. V., Rai, S., & Kale, R. (2020, January). Recommender system in elearning: a survey. In Proceeding of International Conference on Computational Science and Applications: ICCSA 2019 (pp. 119-126). Singapore: Springer Singapore
5. Yeung, C.L., Zhou, L., Armatas, C.: An overview of benchmarks regarding quality assurance for elearning in higher education. In: 2019 IEEE Conference on e-Learning, e-Management & e-Services (IC3e), pp. 1-6. IEEE (2019)
6. Amarneh, B.M., Alshurideh, M.T., Al Kurdi, B.H., Obeidat, Z.: The Impact of COVID-19 on E-learning: advantages and challenges. In: Hassanien, A.E., et al. (eds.) Proceedings of the International Conference on Artificial Intelligence and Computer Vision (AICV2021). Advances in Intelligent Systems and Computing, vol. 1377, pp. 75-89. Springer, Cham (2021). https://doi.org/10.1007/978-3-030-76346-6_8
7. Tang, K.Y., Chang, C.Y., Hwang, G.J.: Trends in artificial intelligence-supported e-learning: a systematic review and co-citation network analysis (1998–2019). Interact. Learn. Environ., 1-19 (2021)
8. Soltani, A., Izquierdo, A.: Adaptive learning under expected and unexpected uncertainty. Nat. Rev. Neurosci. **20**(10), 635–644 (2019)
9. Liu, L., Jiang, H., He, P., Chen, W., Liu, X., Gao, J., Han, J.: On the variance of the adaptive learning rate and beyond. arXiv preprint: arXiv:1908.03265 (2019)
10. Imhof, C., Bergamin, P., McGarrity, S.: Implementation of adaptive learning systems: current state and potential. Online Teach. Learn. High. Educ., 93-115 (2020)
11. Wang, S., et al.: When adaptive learning is effective learning: comparison of an adaptive learning system to teacher-led instruction. Interact. Learn. Environ. **31**(2), 793–803 (2023)
12. Osadcha, K., Osadchyi, V., Chemerys, H., Chorna, A.: The review of the adaptive learning systems for the formation of individual educational trajectory (2020)
13. Xu, Z., Dai, A.M., Kemp, J., Metz, L.: Learning an adaptive learning rate schedule. arXiv preprint: arXiv:1909.09712 (2019)
14. Zhang, Y., Liu, X., Bao, F., Chi, J., Zhang, C., Liu, P.: Particle swarm optimization with adaptive learning strategy. Knowl.-Based Syst. **196**, 105789 (2020)
15. Adamopoulou, E., Moussiades, L.: An overview of chatbot technology. In: Maglogiannis, I., Iliadis, L., Pimenidis, E. (eds.) Artificial Intelligence Applications and Innovations. IFIP Advances in Information and Communication Technology, vol. 584, pp. 373–383. Springer, Cham (2020). https://doi.org/10.1007/978-3-030-49186-4_31
16. Skjuve, M., Følstad, A., Fostervold, K.I., Brandtzaeg, P.B.: My chatbot companion-a study of human-chatbot relationships. Int. J. Hum Comput Stud. **149**, 102601 (2021)
17. Amiri, P., Karahanna, E.: Chatbot use cases in the COVID-19 public health response. J. Am. Med. Inform. Assoc. **29**(5), 1000–1010 (2022)
18. Følstad, A., et al.: Future directions for chatbot research: an interdisciplinary research agenda. Computing **103**(12), 2915–2942 (2020)
19. Taecharungroj, V.: "What Can ChatGPT Do?" Analyzing early reactions to the innovative AI Chatbot on Twitter. Big Data Cogn. Comput. **7**(1), 35 (2023)
20. King, M.R.: The future of AI in medicine: a perspective from a Chatbot. Ann. Biomed. Eng. **51**(2), 291–295 (2023)
21. Przegalinska, A., Ciechanowski, L., Stroz, A., Gloor, P., Mazurek, G.: In bot we trust: a new methodology of chatbot performance measures. Bus. Horiz. **62**(6), 785–797 (2019)

22. Alshmrany, S.: Adaptive learning style prediction in e-learning environment using levy flight distribution based CNN model. Clust. Comput. **25**(1), 523–536 (2022)
23. Arsovic, B., Stefanovic, N.: E-learning based on the adaptive learning model: case study in Serbia. Sādhanā **45**(1), 266 (2020)
24. Khan, M.A., Khojah, M., Vivek.: Artificial intelligence and big data: the advent of new pedagogy in the adaptive e-learning system in the higher educational institutions of Saudi Arabia. Educ. Res. Int. **2022**, 1–10 (2022)
25. Liu, M., McKelroy, E., Corliss, S.B., Carrigan, J.: Investigating the effect of an adaptive learning intervention on students' learning. Educ. Tech. Res. Dev. **65**, 1605–1625 (2017)
26. Li, H., Cui, W., Xu, Z., Zhu, Z., Feng, M.: Yixue adaptive learning system and its promise on improving student learning. In CSEDU (2), pp. 45-52 (2018)
27. Hariyanto, D., Triyono, M., Köhler, T.: Usability evaluation of personalized adaptive e-learning system using use questionnaire. Knowl. Manage. E-Learn. **12**(1), 85–105 (2020)
28. Elmaadaway, M.A.N., Abouelenein, Y.A.M.: In-service teachers' TPACK development through an adaptive e-learning environment (ALE). Educ. Inf. Technol. **28**(7), 8273–8298 (2023)
29. Rishard, M.A.M., et al.: Adaptivo: a personalized adaptive E-learning system based on learning styles and prior knowledge. In: 2022 Seventh International Conference on Informatics and Computing (ICIC), pp. 1-9. IEEE (2022)
30. Mirata, V., Hirt, F., Bergamin, P., van der Westhuizen, C.: Challenges and contexts in establishing adaptive learning in higher education: findings from a Delphi study. Int. J. Educ. Technol. High. Educ. **17**, 1–25 (2020)
31. Colace, F., De Santo, M., Lombardi, M., Pascale, F., Pietrosanto, A., Lemma, S.: Chatbot for e-learning: a case of study. Int. J. Mech. Eng. Robot. Res. **7**(5), 528–533 (2018)
32. Wu, E.H.K., Lin, C.H., Ou, Y.Y., Liu, C.Z., Wang, W.K., Chao, C.Y.: Advantages and constraints of a hybrid model K-12 E-Learning assistant chatbot. IEEE Access **8**, 77788–77801 (2020)
33. El Janati, S., Maach, A., El Ghanami, D.: Adaptive e-learning AI-powered chatbot based on multimedia indexing. Int. J. Adv. Comput. Sci. Appl. **11**(12) (2020)
34. Ait Baha, T., El Hajji, M., Es-Saady, Y., Fadili, H.: The impact of educational chatbot on student learning experience. Educ. Inf. Technol., 1-24 (2023)
35. Lee, L.K., Fung, Y.C., Pun, Y.W., Wong, K.K., Yu, M.T.Y., Wu, N.I.: Using a multiplatform chatbot as an online tutor in a university course. In: 2020 International Symposium on Educational Technology (ISET), pp. 53-56. IEEE (2020)

# Impact of IoT Adoption on Supply Chain Management and Supplier Relationships: Systematic Review

Adil Hammadi(✉), Torsten Reiners, and Beenish Husnain

Curtin University, Perth, Australia
{a.hammadi,t.reiners}@curtin.edu.au,
beenish.husnain@postgrad.curtin.edu.au

**Abstract.** Digital technologies, such as the Internet of Things (IoT), Big Data Analytics, Cloud Computing, 3D Printing, Artificial Intelligence, Machine Learning, and Blockchain, rapidly transform industries. IoT, in particular, has revolutionised the world and impacted our daily lives. Despite this, there is still a gap in the literature regarding a comprehensive explanation of the drivers and barriers to IoT adoption in supply chains, specifically from the perspective of Supply Chain Management (SCM) and Supplier Relationship Management (SRM). Moreover, the literature surrounding SRM and IoT adoption is relatively scarce and scattered. Thus, to address this gap and support future research, we conducted a systematic review of the literature, which will be presented in this paper. The review intends to create a body of understanding to extend the research by linking it to supplier relationship management theories.

## 1 Introduction

Industry 4.0 is seen as an evolution for supply chains by integrating innovative technology like Big Data, Cloud Computing, Artificial Intelligence, 3D Printing, Robotics, Blockchain and Machine Learning [1]. Cyber-physical systems support the monitoring and decision-making in global, real-world supply chains by having access to real-time data and information, creating a digital twin of the supply chain network to gain complete visibility and, in addition to that, the potential for agility, speed, and intelligent decisions to support a customer-centric, responsive, end-to-end supply chain being mutually beneficial for all stakeholders [2–4]. Further, the transparency of the supply chain and real-time information mitigates the risk of impact by global events and geopolitical situations.

Traditional supply chains used to comprise different companies for production and distribution. Supply chain management (SCM) functions, such as manufacturing, procurement, supply, storage, and distribution, were fragmented and scattered [1]. The proliferation and acceptance of digital technologies have transformed supply chain functions towards integrated and centralised functions [1]. Supply chain integration provides better information sharing and visibility regarding end-to-end tracking of the flow of goods and services [1, 5, 6].


L. Barolli (Ed.): AINA 2024, LNDECT 203, pp. 404–414, 2024.
https://doi.org/10.1007/978-3-031-57931-8_39

Cyber-physical systems in the digital supply chain depend on real-world, real-time, and high-detailed data from the supply chain network, including all stakeholders and logistics processes. The Internet of Things (IoT) represents technology that interconnects all processes and entities in the supply chain to collect continuous massive data utilising specialised intelligent sensor networks as a data source for digital twins and big data analytics [7]. IoT uses sensors for environmental monitoring, accuracy measurements in quality control, and logging and controlling to detect abnormal events like vibrations or shifts to the tolerance boundary. This allows automatic alerts and controls to implement actions like exchanging parts or maintenance as early as possible to prevent more severe damage or loss. In the IoT-based innovative and digital SCM process, IoT integrates all functions, such as suppliers, manufacturers, customers, storage, inventory management, retail, and distribution [7].

Despite the potential benefits of implementing IoT in SCM, its adoption is limited for Supplier Relationship Management (SRM), according to a study by [8]. The general focus of studies is on IoT applications downstream in the supply chain, with the impact of IoT on SRM being mostly ignored or considered less relevant. This knowledge gap has been identified as a future research area in the same study. Additionally, [3] stated in 2019 that limited research on IoT applications addresses the entire supply chain management process.

In this research, we conduct a systematic literature review to identify the main barriers and drivers of IoT adoption in SRM. Specifically, this research seeks to address the following questions:

RQ1: What are the key barriers and drivers of IoT adoption in SRM, considering technological and organisational factors?

RQ2: How does IoT adoption impact key aspects of SRM, such as information visibility, collaboration, performance monitoring, and risk management?

Answering these questions will provide valuable insights for practitioners and academics in understanding the leverage of this technology in SRM practices and contribute to a better understanding of IoT's potential in enhancing supplier relationships. The literature review enables academics to progress with further research on the theory integration and future models to support comprehensive IoT adoption across the end-to-end supply chain. The main contribution of this paper is presenting the results of this systematic review in the form of tables and figures illustrating "drivers of IoT adoption" and "barriers of IoT adoption" in SCM and SRM. The review intends to create a body of understanding to extend the research by linking it to supplier relationship management theories.

## 2 IoT, SCM and SRM

Supply chains operate in a very dynamic and everchanging environment because of economic (inflation, energy cost, commodity prices), social (political instability, unrest), and natural (natural disasters, floods) factors [3]. This creates vulnerability to various threats and risks at every level [3]. Modern end-to-end supply chains must incorporate this by being agile and possessing the velocity to address untoward challenges and gain an advantage towards conventional supply chains [5].

Tadejko [9] stated that end-to-end real-time visibility, tracking, and traceability are essential to mitigate and handle challenges. Ghabak and Seetharaman [10] claim that early implementation of mitigation measures prevents supply chain disruptions. Agile supply chains seem to have become the minimum requirement these days as they proactively address risks and threats and improve the effectiveness of the supply chain management process.

Agility can be achieved by digitalising the supply chain using IoT technology, which provides end-to-end visibility across the whole supply chain [11]. Agility reflects the proactiveness of an organisation by swiftly and efficiently dealing with the rapid changes and fluctuations in the global supply chains [11, 12]. According to Ghabak and Seetharaman [10], IoT, in conjunction with Big Data analysis and other Industry 4.0 concepts, increases the effectiveness of supply chain management by providing a decision support system and improving agility and visibility.

All stakeholders in agile supply chains, for example, suppliers, buyers, customers, manufacturers, warehousing, and logistics, should be interlinked for effective and timely decision-making [10]. In this regard, RFID technology, as part of IoT systems, is an established technology to collect real-time data to inform and support decision-making processes by buyers, suppliers, manufacturers, distributors, and retailers [13].

IoT makes the process of information sharing and dissemination smooth, efficient, and effective in global supply chains and consequently eliminates the demand variability concerns [14]; in particular, counteracting the "bullwhip" effect by removing information asymmetries between stakeholders to make accurate and informed decisions in support of the supply chain rather than local firms [14, 15].

### 2.1 Significance of Research

Supplier relationship management is an essential function of supply chain management that cannot be ignored, as suppliers play a crucial role in the success of supply chain management [16], and Alhyari [12] claim that buyer-supplier relationships are pretty valuable in supply chain management from a "relational view" perspective. That is why the researchers will examine the literature to highlight the impact of IoT adoption in terms of benefits (drivers) and bottlenecks (barriers). A secondary aim of the review is to set a snapshot of the current state of research to examine the implications of IoT adoptions through the lens of established supply chain theories in future research, i.e., looking into the Transaction Cost Economics, Agency Theory, Resource-based Theory, Relational View Theory, Power Position Theory, Dynamic Capabilities Theory.

## 3 Systematic Literature Review Process

To answer the research questions depicted in the introduction, we conducted a systematic literature review (SLR) using the modified "PICO" method (population, interest, context); see [17] and [18] on the SLR process. The SLR is about reviewing, analysing, and collecting all information about a specific effect [19].

In systematic reviews, literature is evaluated for evidence with the help of a set of clear, to-the-point, and predefined research criteria to find the answer to the research question. This minimises the bias and provides reliable results for decision-making [20]. The Boolean operators AND, OR, and NOT were used with the keywords to combine, expand, and filter out the relevant and irrelevant material [18]. The search term was designed using critical terms from the research question and the Boolean operators, as indicated below.

("Drivers of Internet of Things" OR "Barriers of Internet of Things" OR "Internet of Things") AND ("Supplier Relationship Management" OR "Buyer supplier relationships" OR "Supply Chain Management" OR "Supply Chain Management Performance")

We selected the following databases for the initial search for relevant material: *Science Direct, Google Scholar, Emerald, Wiley Online Library, IEEE Xplore, Taylor & Francis, Sage Journal, and SpringerLink*. These databases combine the most relevant sources for materials related to SCM, Supplier Relationship Management (SRM), and the Internet of Things (IoT). The period for the search was defined from 2014 to 2023 to cover the significant period of IoT development in supply chains.

Figure 1 shows an overview of the search. The initial 6156 articles were first filtered by subject focus to 4300 remaining articles as well as the exclusion of articles (down to 3235) based on specific keywords that set the context and content of the paper, e.g., digital currencies, bibliometrics, coronaviruses, hypotheses, neural networks, and agricultural production Minor reduction of the number of articles resulted from excluding duplicates and non-English articles. The next step was an in-depth filtering of articles by validating the titles and abstract for relevance, i.e., having the focus on the keywords in the context of the research questions. We examined the articles by skimming, scanning, reading superficially, and, in some cases, reading thoroughly by analysing the content [21]. The final subset consists of 80 articles.

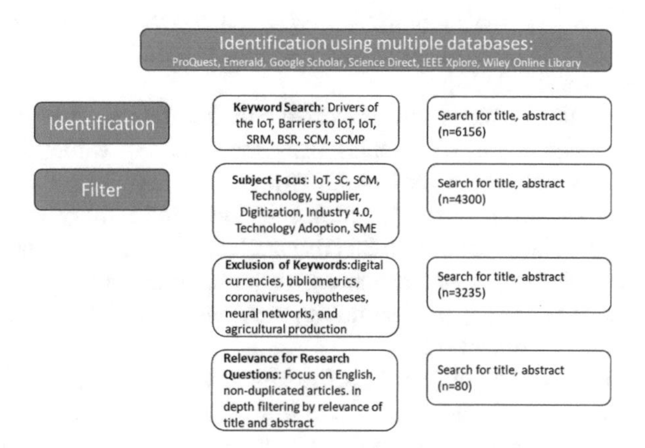


**Fig. 1.** Literature review process (self-design)

# 4 Results

The impact of IoT adoption on the supply chain can be considered positive or negative. The positive impacts result in higher revenue, reduced cost, waste minimisation, and increased productivity and can bring satisfaction to supply chain partners [3, 22, 23]. Therefore, "positive impacts" can be regarded as the "drivers" [22]. "Barriers" are the "negative impacts", which are the challenges and risks associated with the implementation of IoT or the negative influence linked with the adoption of technology [3, 22, 23]. Table 1 provides the references to drivers of IoT in SCM and SRM, while Table 2 provides the references to barriers.

## 4.1 Key Drivers of IoT Adoption in SRM (RQ1)

**From an IoT perspective, supply chain visibility** is an organisational capability to collect and assess the information to achieve the corporate objective [24]. End-to-end visibility in the supply chain operations, the ability of remote visual operation on production lines, and real-time information sharing regarding the outsourced operation are compelling factors for IoT adoption to achieve enhanced visibility [1]. It also includes sharing information with suppliers.

**Location tracking** is a significant enabler of IoT adoption [24]. IoT supports tracking every step in the manufacturing process [25]. Components tagged with IoT sensors can be located and tracked throughout the manufacturing and assembly [1, 26]. IoT is used to precisely and accurately track, discover, and trace goods and services in real-time in conjunction with GPS and GIS over the network in logistics and warehouses [6, 24, 26]. This results in the efficient movement of goods. 90% of logistics companies have implemented IoT technology [27], creating **logistical efficiency**.

According to Ali, Ashraf [22], IoT provides a large amount of information and data on the various objects connected to the network. When used in conjunction with other technologies, such as cloud computing and Big Data analytics, IoT enables precise and swift **decision-making** [22, 28].

Velocity in **communication and effective coordination** is another driver of the adoption of IoT technology. IoT enables real-time reporting within and outside the organisation in the supply chain network [2, 28].

IoT enhances the **organisation's efficiency and productivity** and brings more revenue [6, 28]. Being innovative is another driver in the adoption of digital technologies.

**Customer satisfaction** is also becoming very important and is a critical driver in adopting digital technology [6]. IoT technology is essential in addressing customer needs and meeting customer demands [2].

Predictive analysis of **equipment maintenance** with the help of Big Data while using information collected from IoT devices is another driver of IoT adoption [6].

Significant labour **costs can be reduced** by deploying RFID tags in warehouse operations [29]. Ahmed, Kalsoom [24] mention that IoT can reduce 40% to 70% of **inventory costs** by deploying RFID tags. IoT-enabled cameras are being deployed to quantify the number of items in warehouse bins [1]. RFID tags can significantly improve material loading and unloading time [29].

**Table 1.** IoT Drivers in SCM

| Drivers | Publications |
|---|---|
| Visibility | [1, 3, 5, 7, 8, 16, 24, 26, 28, 30–35] |
| Traceability (tracking) | [1, 3, 5–9, 22, 24–27, 29–31, 36–38] |
| Logistical efficiency | [3, 5–7, 9, 24, 26, 27, 29, 36] |
| Decision making | [3, 5, 7, 22, 28, 39] |
| Communication | [2, 5, 6, 8, 14, 28, 39–41] |
| Organisational efficiency | [1–3, 6, 7, 9, 16, 28, 35, 39, 42–44] |
| Customer satisfaction | [1, 2, 6–8, 22, 26, 27, 33, 34] |
| Maintenance excellence | [1, 6, 7, 26, 31, 35, 40] |
| Cost saving | [1, 3, 6, 7, 16, 24, 29, 31, 33, 35, 39, 42] |
| Efficiency warehousing | [1–3, 6, 7, 9, 24–27, 31, 45] |
| Inventory management | [1, 3, 6, 24–27, 29, 31, 36, 45] |

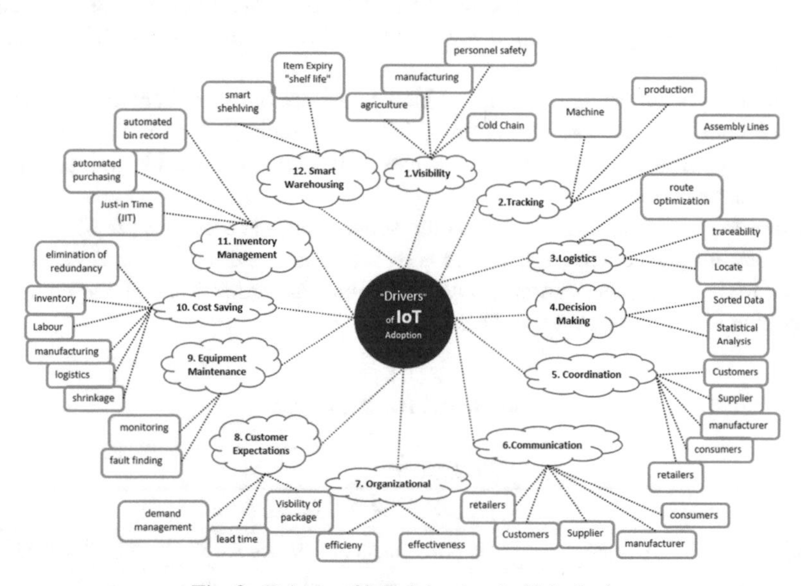


**Fig. 2.** Drivers of IoT Adoption (self-design)

## 4.2 Key Barriers of IoT Adoption in SRM (RQ1)

There are several barriers or concerns of IoT adoption in SCM and SRM. One of the significant barriers is **cost**, especially related to expensive RFID tags, IoT sensors, and the required infrastructure. Manufacturing companies, particularly SME suppliers, may face considerable pressure due to the high cost of IoT deployment. The return on investment (ROI) sometimes outweighs the benefits of IoT adoption.[14, 16, 29, 46].

The risk of **security breaches** may cause supply chain managers to hold the adoption of digital technologies [28, 35]. It has been emphasised that supply chains may collapse and face disastrous outcomes if hackers penetrate the systems and make slight algorithm changes, especially where machine-to-machine transactions are being made [23, 39].

Tajima [29] treats the consumer's data privacy as one of the significant barriers to implementing RFID tags. Pal [5] mentions that RFID technology is vulnerable to being taken over by malicious actors, posing a risk to data privacy and consumer liberties.

**IoT technology** is quite complex and requires a communication network, a centralised data storage repository, and the deployment of sensor devices [33]. The existing firewalls, proxy settings, routers, and other hardware systems organisations use in routine typically do not work for IoT-type mobile network devices. Companies require new hardware and different firewalls for the application of IoT [28, 47]. IoT generates a large amount of data, which requires a repository for storage, data management capabilities, and analytical technology to gain the advantage of the retrieved data [9, 35], i.e. if it contains significant amounts of redundant information without benefits [33].

Implementation of IoT will affect the **organisation's policies** [6, 23]. Understanding what type of technology will align with the organisation's strategic objectives is also very important [6, 28]. Why is the technology needed? Who will be the beneficiary? The systems ' concept selection, design, and architecture must be carefully evaluated and selected [6].

**Scalability** is an important aspect which cannot be ignored. According to Sutikno and Thalmann [33], the main problem of IoT is scalability, which is the ability of the system to support the addition of more devices and increase bandwidth, storage capacity, and processing power.

**Standardisation** of IoT devices in the supply chain is essential [23, 24]. Without standardisation, the concept of a unified network cannot be established. Moreover, considering IoT's global application and supply chain standardisation's global nature, it is challenging and impossible to achieve [23].

**Table 2.** IoT Barriers in SCM

| Barriers | Publications |
|---|---|
| Investment concerns | [7, 14, 23, 28, 29, 33, 35, 40, 46] |
| Security & regulatory concerns | [3, 7, 9, 23, 24, 28, 30, 33, 35, 37, 39, 40, 42, 48–52] |
| Privacy concerns | [3, 5, 7, 9, 23, 28–30, 33, 39, 42, 47, 48, 50–52] |
| Technology concerns | [9, 23, 24, 28, 33, 47, 49, 52] |
| Data management concerns | [6, 8, 9, 33, 35, 49] |
| Organisational concerns | [1, 6, 23, 28, 33, 35, 49, 52, 53] |
| Scalability concerns | [1, 9, 11, 23, 33, 37, 49] |
| Standardisation concerns | [3, 7, 9, 23, 24] |

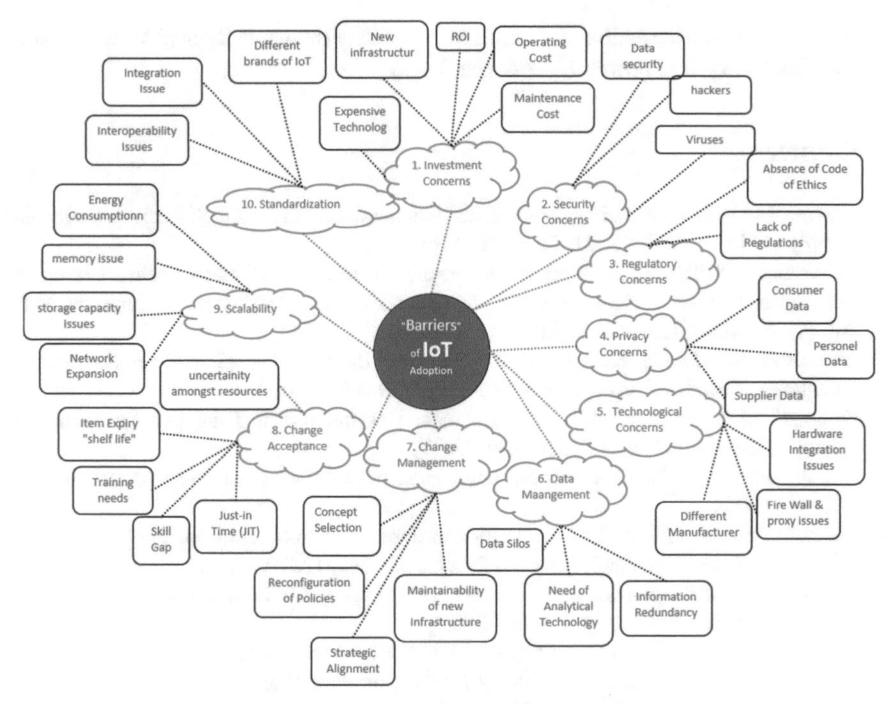


**Fig. 3.** IoT Barriers in SCM (self-design)

### 4.3 Impact of IoT Adoption on Key SRM Aspects (RQ2)

Effective strategies for developing good relationships with suppliers can significantly improve the overall performance of supply chains. These strategies can enhance efficiency, collaboration, risk management, and innovation [54]. Supplier relationships and supply chain management performance concerning IoT can be analysed and examined regarding agility, visibility, collaboration, and smartness/intelligence using dynamic capabilities [10–12].

## 5 Discussion and Future Work

Upon thoroughly examining the literature, Figs. 2 and 3 have revealed several drivers and barriers. Notably, this study has found that digital technology serves as a **dynamic capability for firms**, providing them with increased visibility, agility, and flexibility. Buyers and suppliers can exchange real-time information using IoT technology, allowing them to predict market dynamics and quickly respond to changes. However, the early adoption of IoT technology by large buyers may create an atmosphere of "**opportunism**," causing small suppliers to delay their adoption of digital technologies or choose not to adopt them.

This research has the potential to be expanded upon by exploring the impact of IoT on the adoption of SRM from the perspective of supplier relationship management theories, including transaction cost economics, agency theory, relational view, dynamic capabilities theory, and power dependence theory.

## References


1. Attaran, M.: Digital technology enablers and their implications for supply chain management. Supply Chain Forum Int. J. **21**, 158–172 (2020)
2. Ernawan, Y., Wijaya, C., Indriati, F.: Application of Internet of Things to ensure improved performance within the internal supply chain unit of an organization. Int. J. Commun. Netw. Inf. Secur. **14**(1s), 121–135 (2022)
3. Ben-Daya, M., Hassini, E., Bahroun, Z.: Internet of things and supply chain management: a literature review. Int. J. Prod. Res. **57**(15–16), 4719–4742 (2019)
4. Rahman, M.A., Asyhari, A.T.: The emergence of Internet of Things (IoT): connecting anything, anywhere. Computers **8**(2), 40 (2019)
5. Pal, K.: Internet of Things impact on supply chain management. Procedia Comput. Sci. **220**, 478–485 (2023)
6. Yang, M., Fu, M., Zhang, Z.: The adoption of digital technologies in supply chains: drivers, process and impact. Technol. Forecast. Soc. Chang. **169**, 120795 (2021)
7. Khan, Y., et al.: Application of Internet of Things (IoT) in sustainable supply chain management. Sustainability **15**(1), 694 (2022)
8. Rebelo, R.M.L., Pereira, S.C.F., Queiroz, M.M.: The interplay between the Internet of things and supply chain management: challenges and opportunities based on a systematic literature review. Benchmark. Int. J. **29**(2), 683–711 (2022)
9. Tadejko, P.: Application of Internet of Things in logistics–current challenges. Ekonomia i Zarządzanie **7**(4), 54–64 (2015)
10. Ghabak, V., Seetharaman, A.: Integration of machine learning in agile supply chain management. In: 2023 15th International Conference on Computer and Automation Engineering (ICCAE). IEEE (2023)
11. Yan, J., et al.: Intelligent supply chain integration and management based on cloud of things. Int. J. Distrib. Sens. Netw. **10**(3), 624839 (2014)
12. Abdallah, A.B., Alfar, N.A., Alhyari, S.: The effect of supply chain quality management on supply chain performance: the indirect roles of supply chain agility and innovation. Int. J. Phys. Distrib. Logist. Manag. **51**(7), 785–812 (2021)
13. Li, S., Xu, L.D., Zhao, S.: The internet of things: a survey. Inf. Syst. Front. **17**, 243–259 (2015)
14. Yan, B., Huang, G.: Supply chain information transmission based on RFID and internet of things. In: 2009 ISECS International Colloquium on Computing, Communication, Control, and Management. IEEE (2009)
15. Jiang, W.: An intelligent supply chain information collaboration model based on Internet of Things and big data. IEEE Access **7**, 58324–58335 (2019)
16. Birkel, H.S., Hartmann, E.: Internet of Things–the future of managing supply chain risks. Supply Chain Manag. Int. J. **25**(5), 535–548 (2020)
17. Durach, C.F., Kembro, J., Wieland, A.: A new paradigm for systematic literature reviews in supply chain management. J. Supply Chain Manag. **53**(4), 67–85 (2017)
18. Atkinson, L.Z., Cipriani, A.: How to carry out a literature search for a systematic review: a practical guide. BJPsych Adv. **24**(2), 74–82 (2018)
19. Davis, J., et al.: Viewing systematic reviews and meta-analysis in social research through different lenses. Springerplus **3**(1), 1–9 (2014)

20. Snyder, H.: Literature review as a research methodology: an overview and guidelines. J. Bus. Res. **104**, 333–339 (2019)
21. Bowen, G.A.: Document analysis as a qualitative research method. Qual. Res. J. **9**(2), 27–40 (2009)
22. Ali, S.M., et al.: Drivers for Internet of Things (IoT) adoption in supply chains: implications for sustainability in the post-pandemic era. Comput. Ind. Eng. **183**, 109515 (2023)
23. Birkel, H.S., Hartmann, E.: Impact of IoT challenges and risks for SCM. Supply Chain Manag. Int. J. **24**(1), 39–61 (2019)
24. Ahmed, S., et al.: Towards supply chain visibility using internet of things: a dyadic analysis review. Sensors **21**(12), 4158 (2021)
25. Wang, Y., et al.: A literature review on the application of digital technology in achieving green supply chain management. Sustainability **15**(11), 8564 (2023)
26. Abdel-Basset, M., Manogaran, G., Mohamed, M.: Internet of Things (IoT) and its impact on supply chain: a framework for building smart, secure and efficient systems. Futur. Gener. Comput. Syst. **86**(9), 614–628 (2018)
27. Witkowski, K.: Internet of things, big data, industry 4.0–innovative solutions in logistics and supply chains management. Procedia Eng. **182**, 763–769 (2017)
28. Oubrahim, I., Sefiani, N.: Exploring the drivers and barriers to digital transformation adoption for sustainable supply chains: a comprehensive overview. Acta Logistica (AL) **10**(2), 305–317 (2023)
29. Tajima, M.: Strategic value of RFID in supply chain management. J. Purch. Supply Manag. **13**(4), 261–273 (2007)
30. Qiu, X., et al.: Physical assets and service sharing for IoT-enabled Supply Hub in Industrial Park (SHIP). Int. J. Prod. Econ. **159**, 4–15 (2015)
31. Soori, M., Arezoo, B., Dastres, R.: Internet of things for smart factories in industry 4.0: a review. Internet Things Cyber-Phys. Syst. **3**, 192–204 (2023)
32. Pradhan, S.K., Routroy, S.: Improving supply chain performance by Supplier Development program through enhanced visibility. Mater. Today Proc. **5**(2), 3629–3638 (2018)
33. Sutikno, T., Thalmann, D.: Insights on the internet of things: past, present, and future directions. TELKOMNIKA (Telecommun. Comput. Electron. Control) **20**(6), 1399–1420 (2022)
34. Cui, L., et al.: Improving supply chain collaboration through operational excellence approaches: an IoT perspective. Ind. Manag. Data Syst. **122**(3), 565–591 (2022)
35. Kalsoom, T., et al.: Impact of IOT on Manufacturing Industry 4.0: a new triangular systematic review. Sustainability **13**(22), 12506 (2021)
36. Lou, P., et al. Agile supply chain management over the internet of things. In: 2011 International Conference on Management and Service Science. IEEE (2011)
37. Sun, Z.-H., et al.: Potential requirements and opportunities of blockchain-based industrial IoT in supply chain: a survey. IEEE Trans. Comput. Social Syst. **9**(5), 1469–1483 (2021)
38. Fawcett, S.E., et al.: Information technology as an enabler of supply chain collaboration: a dynamic-capabilities perspective. J. Supply Chain Manag. **47**(1), 38–59 (2011)
39. Osmonbekov, T., Johnston, W.J.: Adoption of the Internet of Things technologies in business procurement: impact on organizational buying behavior. J. Bus. Ind. Market. **33**(6), 781–791 (2018)
40. Falkenreck, C., Wagner, R.: The Internet of Things-Chance and challenge in industrial business relationships. Ind. Mark. Manag. **66**(10), 181–195 (2017)
41. Kauffman, R., Pointer, L.: Impact of digital technology on velocity of B2B buyer-supplier relationship development. J. Bus. Ind. Market. **37**(7), 1515–1529 (2022)
42. Yerpude, S., Singhal, T.K.: Supplier relationship management through internet of things-a research perspective. In: 2018 International Conference On Advances in Communication and Computing Technology (ICACCT). IEEE (2018)

43. De Vass, T., Shee, H., Miah, S.J.: The effect of "Internet of Things" on supply chain integration and performance: an organisational capability perspective. Aust. J. Inf. Syst. **22** (2018)
44. Saravanan, G., et al.: Implementation of IoT in production and manufacturing: an Industry 4.0 approach. Mater. Today Proc. **51**, 2427–2430 (2022)
45. Tsang, Y.P., et al.: An Internet of Things (IoT)-based risk monitoring system for managing cold supply chain risks. Ind. Manag. Data Syst. **118**(7), 1432–1462 (2018)
46. Son, B.-G., et al.: The dark side of supply chain digitalisation: supplier-perceived digital capability asymmetry, buyer opportunism and governance. Int. J. Oper. Prod. Manag. **41**(7), 1220–1247 (2021)
47. Chanson, M., et al.: Blockchain for the IoT: privacy-preserving protection of sensor data. J. Assoc. Inf. Syst. **20**(9), 1274–1309 (2019)
48. Liu, Y., Zhang, S.: Information security and storage of Internet of Things based on block chains. Futur. Gener. Comput. Syst. **106**, 296–303 (2020)
49. Ghaffari, K., et al.: A comprehensive framework for Internet of Things development: a grounded theory study of requirements. J. Enterp. Inf. Manag. **33**(1), 23–50 (2020)
50. Nozari, H., Fallah, M., Szmelter-Jarosz, A.: A conceptual framework of green smart IoT-based supply chain management. Int. J. Res. Ind. Eng. **10**(1), 22–34 (2021)
51. Harkin, D., Mann, M., Warren, I.: Consumer IoT and its under-regulation: findings from an Australian study. Policy Internet **14**(1), 96–113 (2022)
52. Rathore, B., et al.: Identification and analysis of adoption barriers of disruptive technologies in the logistics industry. Int. J. Logist. Manag. **33**(5), 136–169 (2022)
53. Ocicka, B.: How a digital platform transforms the value proposition in purchasing and buyer-supplier relationship management (2021)
54. Teller, C., et al.: The importance of key supplier relationship management in supply chains. Int. J. Retail Distrib. Manag. **44**(2), 109–123 (2016)

# IoT Environment Based on Multi Agent System in Smart Home

Ameni Harrabi[1(✉)], Hamdi Hassen[2], Khlil Ahmad Alsulbi[3], and Meftah Zouai[4]

[1] Business Computing Department, Higher Institute of Management, University of Sousse, Sousse, Tunisia
harrabiamani1996@gmail.com

[2] Miracl lab, University of Sfax, Sfax, Tunisia
hhassen2006@yahoo.fr

[3] Department of Computers College of Engineering and Computers, Al-Qunfodah Umm Al-Qura University, Makkah, Saudi Arabia
kasulbi@uqu.edu.sa

[4] LINFI Laboratory, Mohamed khider University, Biskra, Algeria
meftah.zouai@univ-biskra.dz

**Abstract.** The ubiquity of Internet of Things (IoT) technology has automated and digitized various aspects of communication, data collection, and management. However, the primary challenge lies in establishing effective communication between devices, given the distinct properties and heterogeneity of each device. This paper addresses key challenges in the IoT domain, emphasizing the need to resolve issues such as data management, service discovery, user interfaces, service customization, and context awareness. Notably, the paramount challenge remains the communication among IoT objects.

To tackle this, we propose a solution through the implementation of a Multi-Agent System (MAS). A MAS comprises a collection of agents situated in an environment where interactions rely on specific relationships. The appeal of utilizing multi-agent systems lies in their ability to enhance robustness and efficiency, facilitate collaboration among existing systems, and address problems inherent in distributed data, knowledge, and control due to agent involvement.

Our proposed approach involves the development of an IoT environment by seamlessly integrating agents into IoT objects. Through experiments, we demonstrate that this approach holds promise as a viable solution to streamline communication among heterogeneous objects within the IoT environment.

**Keywords:** Internet of Things (IoT) · Pervasive Systems (PS) · Multi Agent System (MAS) · Smart Home

## 1 Introduction

The swift advancement of technology is making daily life more convenient. In our digital society, the integration of pervasive systems and Internet of Things (IoT) technologies is resulting in the generation of extensive and diverse data.


L. Barolli (Ed.): AINA 2024, LNDECT 203, pp. 415–426, 2024.
https://doi.org/10.1007/978-3-031-57931-8_40

Middleware architectures like the Common Object Request Broker Architecture (CORBA) and Distributed Component Object Model (DCOM) play a crucial role in facilitating communication between distributed architectures and pervasive systems.

Connected devices, endowed with intelligence, enable the collection of substantial data, supporting informed decision-making and reducing operational costs in both industrial and domestic settings [1].

Despite the numerous benefits and technological advancements, this domain presents challenges. Each device, sensor, or object possesses unique characteristics, contributing to a heterogeneous environment. Additionally, challenges such as service discovery, communication, data management, user interfaces, and service adaptation persist.

A Multi-Agent System (MAS) comprises intelligent, mobile, reactive, and proactive agents, capable of self-action and environmental interaction. To address these challenges, we propose the implementation of a MAS to enhance the IoT environment.

The complexity of IoT architecture necessitates effective control. The current study delineates the role of MAS in IoT and its applications.

This paper is structured as follows: Sect. 2 defines fundamental terms and concepts relevant to the Internet of Things and pervasive systems. In Sect. 3, we delve into the key features of MAS. Section 4 presents related work, while Sect. 5 discusses the problem statement. Our proposed approach is detailed in Sect. 6, followed by an exploration of the experimental study and validation component in Sect. 7. Finally, we conclude with a discussion of the main findings.

## 2 Pervasive Systems and Internet of Things (IoT)

One could contend that PS places a greater emphasis on the aspects of human-computer interaction, whereas IoT centers more around the interconnection of devices. Nevertheless, these two communities exhibit substantial commonalities in their interests and technical goals. They both delve into considerations beyond technology, addressing concerns like privacy, security, and ethics.

### 2.1 Pervasive Systems (PS)

Pervasive Systems(PS) or ubiquitous systems and Artificial Intelligence (AI) have broadened the role of IoT systems. PS refers to the growing trend of incorporating computing capabilities into everyday objects to maximize their communication and utility while minimizing the need for the end user to use a computer. Devices for pervasive computing are always available and connected. PS can be used with any device, anytime, anywhere, and in any data format on any network [2].

### 2.2 Interent of Things (IoT)

IoT is a technology that enables remote monitoring and connection of devices. The IoT concept has grown rapidly in recent years and it is used nowadays, is now used in different contexts, including smart homes, healthcare, and industry. IoT-integrated wireless sensor network technology enables global connectivity of smart devices with cutting-edge functions.

### 2.3 IoT Based on Middleware Architecture

The IoT basef on middleware is software that serves as a bridge between its many parts, enabling communication between them that otherwise would not be feasible. Middleware connects many, frequently sophisticated, pre-existing applications that weren't initially intended to be connected. Almost anything can be connected to the IoT and used to send and receive data across a network. By offering a layer of connectivity for the sensors as well as for the application layers that provide services and ensure effective communication between the software. The integration of devices with automatic setup features and field buses used in IoT environments, as well as the integration of numerous middleware technologies (CORBA, DCOM) [3].

## 3 Multi Agent System (MAS)

Researchers from across disciplines, including computer science and civil engineering, have given MAS a great deal of attention as a way to break down complicated problems into smaller jobs.

### 3.1 Comprehensive Knowledge of the MAS

Multi-Agent Systems (MAS) represent a distinct category within distributed intelligent systems, featuring autonomous agents existing in a decentralized environment without a globally reliable information source. According to the MAS equation, agents engage in communication to accomplish tasks [4]. Essentially, the foundational components of any MAS include its agents, environment, interaction, and organization.

### 3.2 Software Agent

An agent can be conceptualized as a dynamic entity, encompassing processes, robots, or software, situated within an environment where interactions are contingent upon specific relationships. It functions as an autonomous, intelligent, mobile, reactive, and proactive entity, capable of influencing both itself and its surroundings. Furthermore, an agent possesses the ability to learn from its past experiences and engage in rational reasoning. Various definitions of agents emerge based on application-specific characteristics, as articulated by Russell et al., who define an agent as a "flexible autonomous entity capable of perceiving the environment through connected sensors." [5].

### 3.3 MAS Environment

The MAS (Multi-Agent System) environment, serving as the task environment, delineates the realm in which an agent operates and accomplishes diverse tasks, significantly shaping the agent's behavior. In this context, we present several potential characterizations of the environment derived from this synthesis. Describing an environment in a broad manner poses challenges, as the nature and characteristics of a system's environment are deeply intertwined with the specific problem it aims to address. Consequently, there exists a notion of predicting the subsequent action, as highlighted in [6].

## 4 Related Works

This section concentrates on the various endeavors documented in the literature that aim to create an intelligent home environment, as detailed in Table 1.

**Table 1.** Related works

| Ref. | Year | Focus | Platform | Protocol for Communication |
|---|---|---|---|---|
| [7] | 2020 | An AI-controlled RESTful middleware for smart home sensors, actuators, and other devices | eVATAR | REST and JSON |
| [8] | 2021 | Framework for smart home activity identification and load monitoring proposed | IoT platform | Home area network (HAN) |
| [9] | 2022 | A review of interoperability difficulties and solutions in the IoT context | Middleware | MQTT |
| [7] | 2023 | Smart Home Load Monitoring and Activity Recognition Using IoT | Node Microcontroller, Raspberry Pi | Bluetooth or Wi-Fi |
| [10] | 2023 | This research proposes the CAMID architecture, in which a middleware for the smart home domain has been developed. Furthermore, they expanded the distribution mechanism | Middleware | MQTT, JAVA |

The preceding table provided an overview of diverse research studies within the IoT domain. Notably, there is a limited number of articles in the literature that specifically address the incorporation of an agent layer in the development of smart home applications. Instead, the predominant focus in existing literature revolves around the utilization of middleware layers like DCOM, CORBA, and others as a means of facilitating communication between objects.

## 5 Problem Statement

In the existing literature, a limited number of works address the communication dynamics among objects within the IoT environment. Despite the abundance of articles, there is a noticeable absence of discussions regarding the integration of an agent layer for the development of smart home applications. Instead, the prevailing approach in numerous articles involves the utilization of middleware layers, such as DCOM and CORBA, as instrumental mechanisms for object communication [7,8].

Communication challenges arise particularly when dealing with diverse objects, each possessing specific characteristics and heterogeneity. This diversity gives rise to complications in their communication. Many research endeavors employing middleware layers often lack concepts of intelligence and autonomy. Objects within this framework lack the ability to recognize tasks from past experiences and consequently lack the foresight to predict subsequent actions [9].

Despite the significant advancements in IoT technology, aiming to become a pivotal force in the future, the current architecture of IoT and middleware (e.g., DCOM, CORBA) is notably rigid. Communication among objects, such as sensors and devices, proves challenging due to various reasons. Primary among these challenges are issues related to service discovery, user interfaces, services adaptation, and context awareness. Another significant hurdle is the use of different platforms, each operating as a separate system. Additionally, the heterogeneity and diversity of attributes among devices contribute to the communication difficulties.

A crucial aspect is the lack of intelligence and autonomy in sensors. Each device operates in a procedural manner when prompted by another object, devoid of cognitive capacity, flexibility, real-time communication, or self-learning capabilities.

## 6 The Proposed Approach

The proposed approach involves establishing an IoT environment through the integration of Multi-Agent Systems (MAS). The utilization of MAS aims to govern and coordinate multiple devices within the IoT network, ensuring the seamless and efficient collaboration of various system components.

The concept of MAS has evolved to encompass systems comprised of multiple autonomous entities exhibiting specific qualities: each agent capable of task completion, absence of global system control, sharing and distribution of information and data, and self-sufficiency of subsystems.

Over time, advancements in MAS have enabled modern agents to perform diverse human-like functions, including learning, thinking, negotiating, self-organizing, and fostering trust among one another [12].

Building upon the agent paradigm, we propose a novel IoT architecture that enhances the system with increased autonomy, coordination, and cooperation. This involves leveraging agent functionality and implementing the approach by integrating sensors with a Raspberry Pi to facilitate communication. A smart agent is integrated into the object (sensor) to create an intelligent environment.

Our research aims to explore further aspects of MAS as a vital component of the IoT environment. We provide a detailed description of our architecture and its components, followed by an exploration of its general functionality. However, there is a recognized limitation in the cognitive capability of designing truly autonomous and intelligent IoT devices.

To address this, we introduce an agent layer, as depicted in Fig. 1, aiming to enhance the cognitive abilities of the IoT devices.

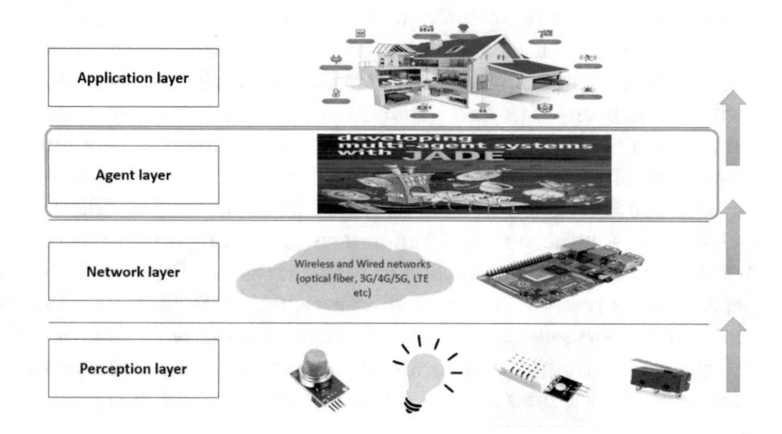


**Fig. 1.** The proposed architecture

My proposed architecture is composed of 4 layers:

- **Perception Layer:** Integrated with the Arduino platform's gas sensor, lamp sensor, door lock, and humidity/temperature sensor.
- **Network Layer:** This layer facilitates communication between sensors using a mini-card Raspberry Pi and the MQTT protocol.
- **Agent Layer:** Embedded as an agent within the IoT device, it manages the data gathered by the sensors, analyzes this data, and makes decisions.
- **Application Layer:** Our work is contained within the environment or application that constitutes a smart home.

We incorporated the agent layer into the IoT architecture to ensure the autonomy and intelligent capabilities of the system. Utilizing the Raspberry Pi, we facilitated communication among sensors and devices. Consequently, these objects now exhibit the autonomous and intelligent attributes inherent in the agent layer.

The decision-making process of an agent relies on the information received from other devices and the data collected by sensors. To facilitate communication with these devices, the Network and Data Communication layer (NDC) is imperative.

Agents process the real-time data gathered by physical layer environment sensors. As a result, a series of software-layer decisions undergo changes based on the agent's objectives, leading to varied behavior depending on the input received.

# 7 Experimental Study and Validation

In this section, we present the experimental study and validation of our IoT environment within a smart home context.

## 7.1 The IoT Setup

To experiment with our IoT environment, we utilized four sensors for intercommunication: Gas sensor, DHT22 sensor, Lamp sensor, and DICGU Micro Switch Sensor (for Door). This paper focuses on two specific examples.

**Example 1: Relay Control for Lamp.** To illustrate the capabilities of our IoT setup, we employed an ESP32 with a relay to control a lamp. This configuration provides an effective means of managing AC household lighting, as depicted in Fig. 2. **Gas Sensor:** In our setup, we employ an Arduino-connected gas sensor. These electronic devices, commonly referred to as gas sensors or gas detectors, are designed to detect and classify various types of gases, as illustrated in Fig. 3.

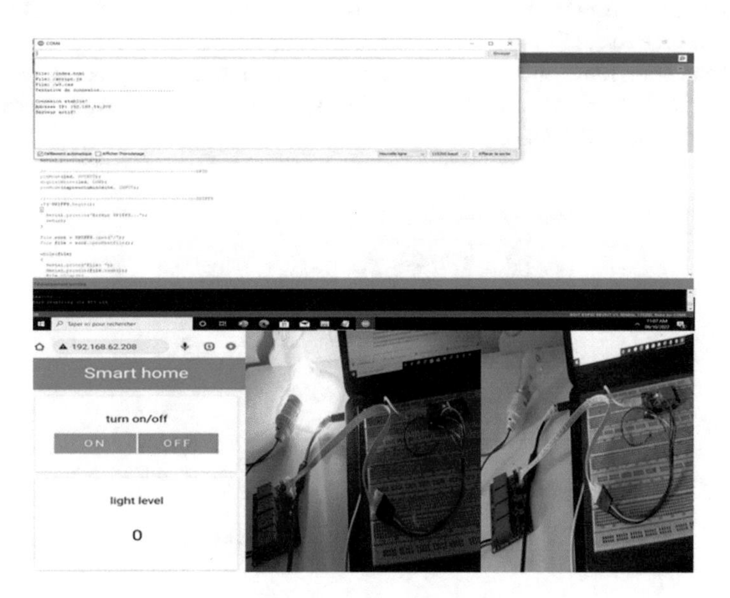


**Fig. 2.** Relay control With lamp

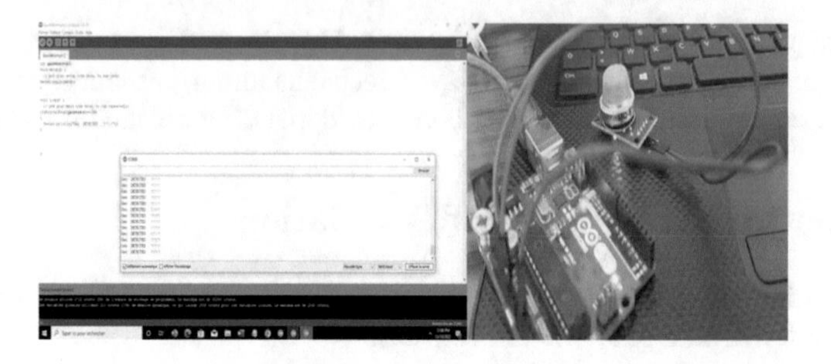

**Fig. 3.** Gas sensor

### 7.2 The MAS Environment

We established a Multi-Agent System (MAS) and a decentralized ecosystem using Eclipse. Our integration involved the Jade platform, a preferred choice among MAS researchers for deploying their agents. Jade provides a set of sophisticated Application Programming Interfaces (APIs) developed in Java, simplifying the construction of various essential components found in traditional software agents. **Create smart Home application** To commence, let's delve into the application of our choice, namely the "Eclipse IDE," a platform utilizing the Java programming language. In Fig. 4 below, our project named "Smart Home" is depicted. Subsequently, we generated all the classes central to our programming, including "AgentDHT22," "AgentLamp," "AgentGaz," "AgentDoor," and "AgentCircuitBreake."

On the other hand, after the creation of five agents, we injected each agent with specific behaviors and methods for sending and receiving messages to develop the `AgentDHT` class.

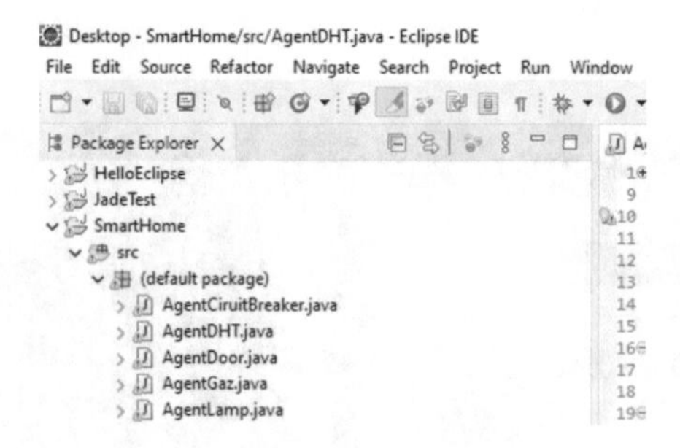


**Fig. 4.** Smart Home application

### 7.3 IoT Environment Based on MAS

In this segment, we configured the Raspberry Pi-a compact computer designed for embedded applications, equipped with a set of GPIO (General Purpose

Input/Output) pins enabling exploration of IoT and control of electronic components for physical computing. We installed the MQTT protocol on the Raspberry Pi to facilitate agent communication, enhancing connectivity between users and devices. This addition streamlines connectivity while minimizing network traffic and improving energy efficiency for IoT devices, as depicted in Fig. 5.

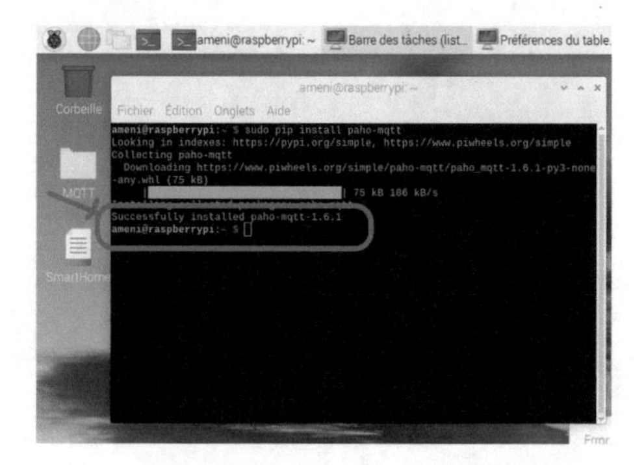


**Fig. 5.** MQTT protocol in raspberry pi

## 7.4 Validation

We showcased the interaction between agents during the installation process and their utilization of the JADE platform. Subsequently, we assessed our methodology in a real-world scenario, drawing comparisons between an older smart home lacking a Multi-Agent System (MAS) and a newer smart home that incorporates MAS. The examination covered the intricacies of agent interactions, including the transmission and reception of messages and information. The Agent Communication Language (ACL) served as the primary language facilitating communication between agents.

Furthermore, we illustrated the process of sending a message using ACL. In this scenario, the sender is the "agentGas," while the recipients are "the other agents," with the content stating "Gas leak at home," as depicted in Fig. 6.

We have elucidated the process by which an object can establish communication with another object through a sniffer. The seamless integration of the sniffer into the Jade environment proves particularly beneficial for scrutinizing agent behavior, as illustrated in Fig. 7.

**Agent Communication:** To facilitate collaboration, various IoT devices establish connections through JADE agents. Each JADE agent possesses a virtual mailbox where communications from other agents are stored. These mailboxes function as lists, organizing messages in the order of their reception. JADE messages utilize instances of the ACLMessage class from the `jade.lang.acl` pack-

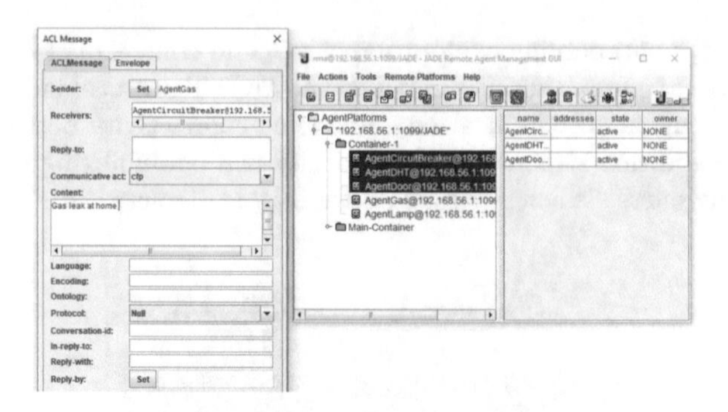


**Fig. 6.** Agent Communication language

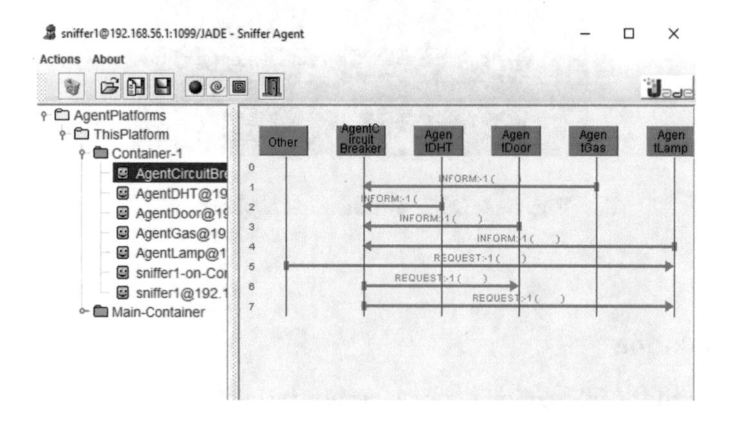


**Fig. 7.** Send request between agents with a sniffer.

age, adhering to FIPA-ACL standards. In our experimental setup, the Multi-Agent System (MAS) ensured effective communication among devices, overcoming disparities in their characteristics. Additionally, the MAS addressed challenges related to service discovery, data management, user interfaces, and service adaptation and prediction.

### 7.5 Result Discussion

To evaluate our approach, we deployed our IoT environment on two architectures: one with a Multi-Agent System (MAS) and the other without. We identified several crucial factors and features, including:

- **Communication:** Without a MAS, establishing effective communication among numerous devices with diverse properties is challenging in an IoT environment. We addressed this by embedding an agent into each object to facilitate seamless communication between them.

- **Autonomy:** In the previous IoT environment lacking MAS, autonomy was scarcely present due to a supervisor managing tasks and overseeing message exchange between objects.
- **Failure:** When a supervisor facilitates communication between objects, any failure in the supervisor can disrupt the entire communication network between objects.
- **Coordination:** In an environment without MAS, coordination is indirect. However, with MAS, coordination occurs directly through interconnected agents.
- **Prediction:** Simple agents in the absence of MAS lack the capability to predict future events.

Our findings reveal that integrating various characteristics of a multi-agent system can significantly enhance the IoT environment. The differentiation between an object (sensor) in an IoT environment without MAS and an object with an agent in MAS is illustrated in Table 2.

**Table 2.** Comparison between Single Object (IoT) and Object based on MAS

| Object (IoT) | Object with Agent (MAS) |
|---|---|
| Has a state and behavior | Has a state and behavior. |
| Communicates with other agents by sending messages | Communicates with other agents by sending messages. |
| Communication is used to activate methods of the object | Communication is used for negotiation between agents and establishing an interaction. |
| An object is obliged to perform a method if another object invokes it | An agent may refuse to execute a request from another agent. |
| Not a single automatic learning or detection concept | Ability to learn from the past and to reason rationally. |
| No intelligence | Emergence of an intelligent individual and collective |

## 8 Conclusion and Future Work

The integration of Multi-Agent Systems (MAS) with the Internet of Things (IoT) represents a significant advancement in the realm of technology. This synergy offers potent and efficient solutions to a multitude of challenges. It facilitates seamless communication between heterogeneous objects, such as sensors and devices, optimizing time and resources within a responsive framework.

Our contribution in this work involves proposing an IoT environment by incorporating an agent layer into the device. This addition imbues the device with agent attributes, including autonomy, mobility, intellect, cooperation, and organization.

Despite the widespread applicability of MAS, challenges persist, such as agent coordination and security. Additionally, creating an agent capable of executing desired actions without explicit programming, and enabling the agent to autonomously adapt its behavior to incorporate newly discovered activities, presents a complex issue. Each agent is assigned a sub-task or region within the home and engages in negotiations with others to achieve specific goals, such as comfort or energy savings.

Looking ahead, future research should shift focus from the Internet of Things to the Internet of Agents (IoA). The field of multi-agent systems has evolved, empowering agents to perform diverse tasks such as learning, reasoning, negotiation, and self-organization. Rather than programming simple objects to execute predefined methods, security challenges become prominent.

Building upon the foundation of the IoT, transitioning to the Internet of Agents becomes imperative to accommodate the rapid proliferation of devices.

## References

1. Centenaro, M., Costa, C., Granelli, F., Sacchi, C., Vangelista, E.: A survey on technologies, standards, and open challenges in satellite IoT. IEEE Commun. Surv. Tutor. **23**(3), 1693–1720 (2021)
2. TUDelft, https://research.tudelft.nl/en/publications/pervasive-ai-for-iot-applications-a-survey-on-resource-efficient. Accessed 5 Jan 2024
3. GeeksforGeeks. https://www.geeksforgeeks.org/introduction-to-pervasive-computing. Accessed 07 Jan 2024
4. Tubitak. https://www.Submitted to The Scientific & Technological Research Council of Turkey. Accessed 3 Jan 2024
5. Russell, S., Norvig, P.: Artificial Intelligence: A Modern Approach, vol. 25, p. 27. Prentice-Hall, Englewood Cliffs (1995)
6. Weyns, D., Van Dyke Parunak, H., Michel, F., Holvoet, T., Ferber, J.: Environments for multiagent systems state-of-the-art and research challenges. In: Weyns, D., Van Dyke Parunak, H., Michel, F. (eds.) E4MAS 2004. LNCS (LNAI), vol. 3374, pp. 1–47. Springer, Heidelberg (2005). https://doi.org/10.1007/978-3-540-32259-7_1
7. Franco, P., Martinez, J.M., Kim, Y.C., Ahmed, M.A.: IoT based approach for load monitoring and activity recognition in smart homes. IEEE Access **9**, 45325–45339 (2021)
8. Vahdat, H.: CAMID: architectural support of middleware for multiple-domain ubiquitous computing and IoT. J. Supercomput. **79**(1), 573–590 (2023)
9. Pazhev, G., Spasov, G., Shopov, M., Petrova, G.: On the use of blockchain technologies in smart home applications. In: IOP Conference Series: Materials Science and Engineering, vol. 1, p. 012023 (2020)
10. Albouq, S.S., Abi Sen, A.A., Almashf, N., Yamin, M., Alshanqiti, A., Bahbouh, N.M.: A survey of interoperability challenges and solutions for dealing with them in IoT environment. IEEE Access **10**, 36416–36428 (2022)
11. Cheeli, A.K, Ravi, B., Keerthi, K.: IoT based approach for load monitoring and activity recognition in smart homes. J. Positive School Psychol. 145–148 (2023)
12. ResearchOnline. https://researchonline.ljmu.ac.uk/. Accessed 10 Jan 2024

# A Review of Social Network Regulations and Mechanisms for Safeguarding Children's Privacy

Mykaele F. Abreu, Eduardo K. Viegas(✉), and Altair O. Santin

Pontifícia Universidade Católica do Paraná | Pontifical Catholic University of Parana — PUCPR, Graduate Program in Computer Science — PPGIa, Curitiba, Brazil
{mykaele.abreu,eduardo.viegas,santin}@ppgia.pucpr.br

**Abstract.** The privacy challenges of social networks and their use by children online require the assessment of legal frameworks and the implemented service provider mechanisms to identify potential gaps. This paper reviews the scientific literature on privacy mechanisms for children on social networks. First, we extensively examine existing legal regulations governing minors' privacy on these platforms. Next, we overview the current scientific literature to assess the mechanisms implemented to safeguard the privacy of minors on social networks. Our findings suggest a lack of consistency in legal frameworks for defining child privacy mechanisms. Similarly, the existing literature often fails to consider the alignment of their proposed solutions with current privacy regulations, impeding their practical implementation in real-world scenarios.

**Keywords:** Privacy · Social Network · Privacy Mechanisms · Children

## 1 Introduction

Safeguarding children's privacy is paramount due to risks like inappropriate data collection, targeted advertising, and suspicious online interactions. According to a report [7], nearly a quarter of major social media users began using these networks in early childhood, even before reaching 6 years old. Notwithstanding, ≈90% of the people aged from 9 to 17 actively maintain a profile on digital platforms. Consequently, the growing presence of children in the digital realm exposes them to the collection of personal and sensitive information, prompting considerable concerns regarding their security and privacy [10].

Over the past years, numerous laws have been enacted to safeguard personal data and ensure the protection of children. Examples include the General Data Protection Regulation (GDPR) in Europe and the General Data Protection Law (LGPD) in Brazil, which primarily focuses on personal data protection. Similarly, the Children's Online Privacy Act (COPPA) was specifically established in the United States to protect children's data [30]. In this context, the technological evolution of social networks poses several challenges in ensuring user privacy, necessitating ongoing regulatory adjustments.


L. Barolli (Ed.): AINA 2024, LNDECT 203, pp. 427–438, 2024.
https://doi.org/10.1007/978-3-031-57931-8_41

Children's privacy holds significant relevance, given the escalating use of online technologies and mobile devices among minors in recent years [28]. While this surge in children's engagement with the technological landscape has led to the routine collection and processing of data by service providers, it has concurrently exposed this group to substantial privacy risks [11]. Children can be considered vulnerable subjects due to various factors. Their vulnerability is underscored by inherent physical and psychological fragility and a limited capacity to make autonomous decisions. Special rights have been internationally and nationally acknowledged to address this vulnerable status, specifically tailored for this demographic group [1].

However, despite the existence of current regulations addressing the privacy of minors, there remains a lack of a clear definition regarding the concept of minors, whether considering the user's age or their mental capabilities. Childhood encompasses the stage of human development from birth to adolescence, marked by universally recognized characteristics like fragility and the need for special attention and care over a substantial period [22]. At the regulatory level, the Child and Adolescent Statute defines a child as a person up to the age of 12, while adolescents are those between twelve and eighteen years old, providing clarity on specific age groups. Conversely, it is noteworthy that LGPD, COPPA, and GDPR do not provide a unanimous definition of minors concerning privacy issues, leaving room for legal interpretations. The lack of a precise legal definition for the concept of minors contributes to the absence of unified and universal definitions in the realm of children's privacy on social networks [26]. Despite regulations mandating parental consent for minors to use social networks and data collection, service providers often neglect to implement compliance mechanisms. These may include measures such as photo verification of official documents, biometrics, or more robust security questions tailored to minors [20].

The absence of standardization poses challenges for global service providers, requiring them to navigate diverse regulations based on users' locations. This variability directly affects the uniform protection of children's data, as the security measures are contingent on each country's specific regulations and practices. Additionally, this lack of standardization gives rise to complications in adhering to laws, implementing specific processes, and incorporating security mechanisms to ensure compliance with the distinct legal requirements of each country. There is a scarcity of literature examining data privacy on social networks used by minors. This gap arises from the diverse nature of social media platforms, each with unique architecture, posing a challenge to unify and assess the implemented mechanisms [14]. Consequently, there is limited evidence regarding the protective mechanisms adopted by websites and applications for children, making it challenging to establish the effectiveness of existing laws. In practice, service providers often exploit legislation's lack of precise privacy definitions and the ambiguity in defining minors. Additionally, the widespread lack of transparency on many social networks hinders a proper understanding of their policies, data collection practices, and the implementation of security and privacy mechanisms [19].

In response to this gap, this paper undertakes an in-depth review of the scientific literature concerning implemented privacy mechanisms for minors to safeguard their privacy on social networks. We commence with a comprehensive review of current legal regulations governing the privacy of minors on social networks. Subsequently, we present a literature review focusing on the mechanisms in place to safeguard the privacy of minors on social networks.

## 2 Data Privacy Regulations

The imperative for personal data protection has gained prominence by enacting judicial decisions in numerous countries. Multiple data protection laws have been implemented by sharing the idea that personal data warrants distinctive legal protection. From a legal standpoint, discussions about the right to privacy ensued due to new techniques and technological tools, enabling access and disclosure of information related to an individual's private sphere. Its primary focus is to protect the data holder rather than the data itself. Consequently, any breach of an individual's personal information jeopardizes their security and integrity [10]. Data privacy involves adherence to established norms governing the collection, disclosure, and use of information, encompassing elements such as name, age, sexual orientation, race, and religious and philosophical beliefs.

Modern endeavors by numerous countries strive to safeguard privacy rights in the digital realm, leading to the continual evolution of regulations on the subject [12]. Examples of these regulations on a global scale include the General Data Protection Law (LGPD) in Brazil, the Personal and Electronic Information Privacy Act (PIPEDA) in Canada, the California Consumer Privacy Act (CCPA) in California, the APPI Amendment of 2017 (APPI) in Japan, and the General Data Protection Regulation (GDPR) in the European Union. The subsequent subsections will explore privacy-related laws relevant to this work, specifically GDPR, LGPD, and COPPA.

### 2.1 General Data Protection Regulation (GDPR)

The GDPR is legislation within the European Union that became effective in 2018, aiming to establish regulations for processing personal data. Broadly, the GDPR is designed to support individuals with greater control over their data more responsibly and transparently. It sets a standard that has influenced the approach of numerous countries worldwide.

Concerning data processing, the GDPR includes the need for specific regulatory provisions for children. More specifically, it states that children warrant special protection concerning their data due to their limited awareness of associated risks and rights. This safeguarding, especially relevant in marketing, profiling, and direct service usage, should not necessitate parental consent for preventive or counseling services directly provided to the child.

The GDPR also delineates conditions for obtaining children's consent in the context of digital services, emphasizing key aspects of data processing, such as:

- *Minimum age for consent.* Establishes a minimum age of 16 for consent, requiring parental or guardian consent for processing data of children under such age.
- *Parental consent.* Stipulates that data processing is lawful only with parental consent, necessitating the controller to verify consent diligently, leveraging available technology.
- *Clear information for children.* Data controllers are mandated to furnish children with clear and accessible information regarding the processing of their data.

Therefore, while the GDPR does not feature a dedicated section on children's data protection, it comprehensively addresses the topic, encompassing crucial provisions such as minimum age and parental consent. Notably, specific data privacy laws may exist in some European Union states.

### 2.2 General Data Protection Law (LGPD)

The LGPD is a Brazilian legislation that safeguards citizens' data privacy by establishing regulations for collecting, storing, processing, and sharing personal information. Specifically addressing children's privacy, it emphasizes processing data in the "*best interests of the minor*," aligning with the protective system for children and adolescents.

Initially, concerns arose about the necessity of parental consent for data processing. However, a statement by the National Data Protection Authority (ANPD) clarified that data processing for children and adolescents is permissible under any legal scenario outlined in the LGPD, as long as the child's interests are respected. The law outlines the importance of special measures to ensure children's privacy and data security, acknowledging their vulnerability. However, the shift away from requiring parental consent for data processing is viewed by some as a regressive step, potentially easing the responsibilities of data controllers [7].

The law also outlines requirements for processing children's data. Notably, it mandates that information on data processing must be presented in a simple, clear, and accessible manner, incorporating audiovisual resources when appropriate. This approach aims to provide necessary information for parents or legal guardians and ensure comprehension by the child, discouraging complex language and technical terms. The use of drawings, diagrams, flowcharts, videos, and other resources is recommended to enhance accessibility for both children and guardians.

### 2.3 Children's Online Privacy Protection (COPPA)

COPPA is a U.S. legislation overseen by the Federal Trade Commission, established to safeguard the online privacy of children under the age of 13. Enacted in 1998 in response to concerns about collecting children's personal information on online platforms, COPPA applies to websites and online services directed at children. The Act mandates that websites and services targeting children must

**Table 1.** Regulatory overview on privacy aspects for the child.

| Privacy Aspect | LGPD | GDPR | COPPA |
|---|---|---|---|
| Definition of personal data | ✓ | ✓ | |
| Definition of child | $\leq$12 y/o | $\leq$16 y/o | $\leq$13 y/o |
| Consent of parents or guardians | ✓ | ✓ | ✓ |
| Transparency of the use of information | ✓ | ✓ | ✓ |
| Limited Purpose | ✓ | ✓ | ✓ |
| Data Minimization | ✓ | ✓ | |
| Data collection without parental consent | ✓ | ✓ | ✓ |
| Data processing notice | ✓ | ✓ | ✓ |
| Right to Review the provided data | ✓ | ✓ | ✓ |
| Privacy-Oriented Design | ✓ | ✓ | |
| Confidentiality of Collected Information | ✓ | ✓ | ✓ |

obtain parental consent before collecting and using personal information. The legislation is structured into sections covering Definitions, General Requirements for the Protection of Children's Privacy, Prohibitions, Administration and Enforcement, Civil Sanctions, Regulations and Procedures, and Studies and Reports.

It establishes the core provisions, including the obligation for websites to notify parents and secure verifiable parental consent before collecting, using, and disclosing information from children under 13. Websites are also required to secure the information collected from children and give parents the right to review the data. It also features prohibitions on certain practices, details the role of the Federal Trade Commission, and specifies penalties for violations.

### 2.4 Discussion

The absence of adequate protection for sensitive data leaves room for potential abuse, including the misuse of information for monitoring, manipulation, and harm to children, posing both short and long-term risks. Moreover, there is an observed deficiency in legal frameworks as they often fail to consider the risks to children associated with profiling sufficiently.

Table 1 shows the regulatory overview on privacy aspects for children. It is possible to note a lack of consensus on child privacy, even deviating from the regulatory definition of minor's age. This gap may stem from legislators not fully comprehending the specific risks to children, resulting in legislation that may not be promptly updated to address emerging concerns effectively.

## 3 Privacy on Social Networks

General data protection laws primarily concentrate on safeguarding broad rights, often neglecting the specific implications for children [8]. For instance, consent requirements, privacy policies, and security measures are frequently not tailored for the child audience. Another challenge pertains to defining the *age of consent* (see Table 1). Establishing an accurate age for consent and comprehending its implications is complex, as it may vary for legal, cultural, and social reasons. This intricacy adds a layer of difficulty to addressing children's specific needs and considerations within the framework of data protection laws.

A pressing challenge involves constantly adapting laws in response to technological advances. The rapid growth of artificial intelligence, for instance, raises questions about safeguarding data in environments where autonomous algorithms process information in complex and often opaque ways. Establishing clear boundaries for the ethical and responsible use of data in such contexts presents a challenging task for policymakers and regulators.

The next subsections further explore the current literature on privacy mechanisms proposed for social networks.

### 3.1 Privacy on Social Networks

Social networks have become integral to the daily lives of millions worldwide, serving as a platform for connection, information sharing, and interaction. Notably, their utility extends beyond entertainment; companies analyze trends for personalized marketing, employers scrutinize candidate profiles, and the judiciary leverages social networks for crime-solving evidence. Social media even plays a role in influencing election outcomes [13]. This highlights that social networks accumulate diverse and comprehensive information, such as Facebook containing personal life details and LinkedIn featuring professional activities, enabling the creation of detailed individual profiles [18]. Nevertheless, the digital environment raises substantial privacy concerns, and the primary challenge in privacy solutions lies in balancing between preserving user privacy and not impeding the advantages of socializing and information sharing.

One of the main privacy problems is information leakage, which comes from users who put themselves at risk by interacting or disclosing their personal information. Still, leaks can also occur through third-party applications linked to social media accounts or even due to vulnerabilities in services provided by the social network. These information displays are of interest to different audiences, such as: (a) *Data Brokers*: sell personal information to other parties, such as banks, insurance companies, etc.; (b) *Service providers*: offer targeted services and advertisements; (c) *Criminals*: carry out social engineering, spear phishing or recovery of authentication techniques.

Concerning native privacy features, social networks usually allow users to limit access to their information, leaving it visible only to friends, for example. A user can create an account without explicitly revealing any information. This allows the user to leave their information public and/or private. However, there are privacy-related attacks that aim to infer user attributes that are incomplete or missing [10]. Several works in the literature explore attribute inference in social networks. These works can be classified into two main categories: friend-based inference and behavior-based inference [5].

Friend-based inference techniques leverage the homophily theory, positing that two friends are likelier to share the same attributes than two strangers. For instance, if most of an individual's friends on a social network are enrolled at a given university, it is plausible to infer that this individual is also a student at that university. Several studies have utilized machine learning techniques to validate this theory [23,24]. Conversely, the behavior-based inference technique utilizes the user's and similar friends' public attributes. Weinsberg *et al.* [30] proposed a method for identifying user attributes, including gender, based on the list of liked films.

In this context, having fine-grained privacy settings is essential for users to have flexibility in controlling the exposure of their information. However, this fine granularity may demand a significant cognitive effort from users, leading them to ignore and trust only default privacy settings. Therefore, several authors have proposed techniques for fine-grained privacy controls. Kruk *et al.* [14] introduced an ontology-based access control mechanism that utilizes relationships between users. This proposal employs a generic definition of relationships ("*knows*") as a trust metric. It generates rules to control a friend's access to resources based on the degree of interaction in the social network [3]. Choi *et al.* [2] advanced Kruk's work with a more refined approach, considering more granular relationships (e.g., "*worksWith*", "*isFriendOf*", "*knows*") to model the social network and access control.

Fong *et al.* [8] introduces a Role and Relationship-Based Access Control (ReBAC) model that treats relationships as polyrelational (e.g., distinguishing teacher-student relationships from parent-child relationships) and directed (e.g., distinguishing teacher-student relationships from student-teacher relationships). This model maps multiple access contexts organized in a tree-like hierarchy. When access is requested in a context, relationships from all ancestral contexts are combined with relationships in the target access context to construct a network for making authorization decisions.

Research indicates that users on social networks frequently neglect available privacy controls. For instance, over 99% of Twitter users maintain the default privacy setting, exposing their name, follower list, location, website, and biographical information. Similarly, most Facebook users retain default settings [13]. The underutilization of privacy options is primarily attributed to a non-intuitive privacy setting interface, complex privacy settings, and an inherent trust in social media.

To enhance user engagement in configuring privacy settings, providing an appropriate graphical interface that facilitates understanding and reduces configuration time is crucial. Several studies have aimed to develop such interfaces. For instance, Paul *et al.* [19] introduced Colors for Privacy Settings (C4PS), employing colors to represent different visibility levels of attributes. For example: (red) signifies visibility to no one; (blue) indicates visibility to selected friends; (yellow) denotes visibility to all friends; (gray) represents visibility to anyone. This allows users to choose the color for each attribute.

The challenge with not changing default settings is that they often tend to be more open than users would prefer [13]. Approaches have been proposed to address this issue by automatically generating more suitable default privacy settings. PriMA [28] proposed a privacy preference generator based on user profiles similar to the account owner. The PolicyMgr [26] approach employed supervised machine learning with examples of privacy policies to build classifiers that automatically generate privacy policies [6].

### 3.2 Child Privacy

Many children are growing up in a "*digital-by-default*" world, where technologies mediate interpersonal, institutional, and commercial interactions. Online interactions enable children to connect, communicate, interact, and play comfortably [15]. Privacy is crucial in supporting children's autonomy, contributing to their psychosocial development, responsibility, resilience, confidence, and critical thinking skills. Societies worldwide increasingly recognize the importance of children's privacy and safety, given that privacy violations in the digital realm can manifest in various ways. Technologies can monitor children's physical location, store their personal information and preferences, and even influence their decision-making [25].

Daniel J. Solove [27] developed a taxonomy of privacy issues, encompassing 16 privacy concerns organized into four categories: information collection, information processing, information disclosure, and intrusions. Information collection involves surveillance and monitoring, encompassing obtaining information from children in problematic ways. Information processing is associated with storing, manipulating, and using information, including aggregating information for purposes other than those initially agreed upon. Information disclosure pertains to breaches of confidentiality involving the inappropriate disclosure of information that may harm a person's reputation. Finally, invasion occurs when someone intrudes upon another person's physical, psychological, or digital space or interferes with another person's decision-making.

Eva *et al.* [25] investigated children's perceptions of privacy. The research involved 25 children aged between 10 and 11 years who participated in a workshop with tests assessing various aspects, including (1) online activities of children; (2) understanding of personal information by children; (3) types of personal information requested by applications and games; (4) acceptable online behaviors according to children; (5) risks and concerns they have in the online environment. Overall, the tests revealed that children's understanding of privacy revolves around avoiding strangers and considering their home addresses as the most confidential information. This highlights the need for efforts to increase awareness of privacy among children. Several organizations have developed materials on educating children about privacy, which are being incorporated into school curricula [1].

In developing educational materials for children, there is no consensus on whether children should participate in creating these materials. However, approaches involving games or stories are more effective [15]. For instance,

Raynes *et al.* [20] created a game called "*The Vigilantes*," aiding children in understanding information collection and its applications. Similarly, the storybook "*Superheroes*" narrates the tale of a superhero imparting lessons on personal information, online chatting, location sharing, and cyberbullying.

Apart from raising children's awareness, applications, games, websites, and social networks must comply with laws such as COPPA, GDPR, and LGPD. However, the current scenario indicates that many platforms do not adhere to legal provisions. The American Federal Trade Commission (FTC) [1] highlights misleading transparency regarding mobile application privacy, providing parents with minimal or no privacy information. An FTC study revealed that parents often struggle to determine, before downloading an app, whether it poses risks related to collecting, using, and sharing their children's personal information.

Similarly, Ilaria *et al.* [16] demonstrates that although cell phone applications are aimed at children, several applications request sensitive permissions that are inappropriate and/or expected [9]. This is critical for understanding potential legal compliance violations, as it is difficult to be clear about what information is collected and/or used. Additionally, even if an application is inactive, there is no guarantee that the application is not collecting personal information. Developer companies make free or paid applications available in stores, with the latter earning revenue through advertising, while the paid version does not. It is often more profitable for companies to have a free application with advertising than a paid one due to targeted advertising using the collected data [29].

### 3.3 Child Privacy on Social Networks

Children's privacy on social networks remains a contentious issue, especially since many platforms, including Facebook, explicitly prohibit usage by children in their terms of service. However, research revealed that millions of children under 13 use Facebook, often providing false information about their age during registration [13]. This underscores that social networks are frequently not designed with children's best interests. Platforms such as Instagram and Twitter default to keeping new account profiles public. Additionally, children may not fully comprehend that agreeing to access a social network entails consenting to collecting and sharing their personal information, including their location.

The effectiveness of identity verification systems in most social networks and online services is compromised as they rely on remote verification, easily circumvented by users. Many parents prioritize age restrictions on social networks to filter inappropriate content for their children, often overlooking privacy risks. COPPA designates the age of 13 as a pivotal point in a child's life, emphasizing their capacity for decision-making. Studies suggest that, before the age of 11, children tend to be less critical in evaluating the trustworthiness of online content [15]. For instance, an investigation involving 135 children aged 8 to 10 found that a website's dynamic features (such as animations) influenced trust perceptions, with children rating a website with animated dog images as more trustworthy than a text-only version. While children aged 8 and above can

**Table 2.** Literature overview on privacy for children on social networks.

| Work | Child Privacy | Social Net. Privacy | Legal Aspects | Eval. Conf. | New Solution |
|---|---|---|---|---|---|
| Ghazaleh and Huan [5] | | ✓ | | ✓ | ✓ |
| Kruk *et al.* [14] | | ✓ | | ✓ | ✓ |
| Choi *et al.* [2] | | ✓ | | ✓ | ✓ |
| Fong [8] | | ✓ | | ✓ | ✓ |
| Kayes and Iamnitchi [13] | | ✓ | | ✓ | ✓ |
| Paul *et al.* [19] | | ✓ | | ✓ | ✓ |
| PriMA [28] | | ✓ | | ✓ | ✓ |
| Shehab *et al.* [26] | | ✓ | | ✓ | ✓ |
| Daniel J. Solove [27] | ✓ | | | | ✓ |
| Eva *et al.* [25] | ✓ | ✓ | ✓ | ✓ | |
| Raynes [20] | ✓ | ✓ | ✓ | | ✓ |
| FTC [1] | ✓ | ✓ | ✓ | ✓ | |
| Liu *et al.* [17] | ✓ | | ✓ | ✓ | ✓ |
| Irwin *et al.* [21] | ✓ | | ✓ | ✓ | ✓ |
| Alkhalifah *et al.* [4] | ✓ | ✓ | | | ✓ |
| Rochelau and Sonia [22] | ✓ | ✓ | | | ✓ |

identify the selling intent in advertisements, they struggle to perceive persuasive intentions at this age [21].

### 3.4 Discussion

Table 2 presents the literature overview on privacy for children on social networks The observation underscores a critical gap in the current literature on children's privacy within social networks. Despite the proliferation of literature proposing innovative privacy solutions, a significant oversight is apparent in adequately addressing the legal dimensions of these proposals, particularly concerning minors. The prevailing regulatory mechanisms may not align with the proposed solutions, potentially rendering them impractical or non-compliant. This misalignment highlights a fundamental challenge where theoretical advancements in privacy protection for children on social networks face barriers in translating into effective, legally sound practices. Bridging this gap between theoretical innovation and regulatory reality is crucial for ensuring that proposed privacy measures are visionary and grounded in the legal frameworks that govern the online space for minors.

## 4 Conclusion

Children's privacy on social networks poses a multifaceted challenge in the ever-evolving digital landscape. As minors increasingly engage with online platforms, there is a pressing need to navigate the complex interplay between legal regulations, evolving norms, and the practical efficacy of existing privacy mechanisms. This paper thoroughly explores the scientific literature, delving into the mechanisms to safeguard children's privacy on social networks. The review showed a critical issue as most existing works often neglect to adequately address the legal dimensions of children's privacy, especially in social networks. This presents a notable gap, hindering the practical applicability of proposed privacy solutions within the constraints of current regulatory frameworks.

## References

1. Federal Trade Commission study (2024). https://www.ftc.gov/
2. Choi, H.C., Kruk, S.R., Grzonkowski, S., Stankiewicz, K., Davis, B., Breslin, J.: Trust models for community aware identity management. In: Proceedings of the 2006 Identity, Reference and Web Workshop, in Conjunction with WWW, vol. 1, pp. 140–154 (2006)
3. Abreu, V., Santin, A.O., Viegas, E.K., Stihler, M.: A multi-domain role activation model. In: IEEE International Conference on Communications (ICC). IEEE (2017)
4. Alkhalifah, A., Alghafis, A.: The effect of privacy concerns on children's behavior on the internet: an empirical study from the parents' perspective (2022)
5. Beigi, G., Liu, H.: A survey on privacy in social media: identification, mitigation, and applications. ACM/IMS Trans. Data Sci. **1**(1), 1–38 (2020)
6. Bulle, B.B., Santin, A.O., Viegas, E.K., dos Santos, R.R.: A host-based intrusion detection model based on OS diversity for SCADA. In: IECON 2020 Annual Conference of the IEEE Industrial Electronics Society, October 2020. IEEE (2020)
7. Cetic.br: TIC Kids Online Brasil. Technical report (2023). https://cetic.br/pt/pesquisa/kids-online/
8. Fong, P.W.: Relationship-based access control: protection model and policy language. In: Proceedings of the First ACM Conference on Data and Application Security and Privacy, February 2011, CODASPY 2011. ACM (2011)
9. Geremias, J., Viegas, E.K., Santin, A.O., Britto, A., Horchulhack, P.: Towards multi-view android malware detection through image-based deep learning. In: IEEE International Wireless Communications and Mobile Computing (IWCMC) (2022)
10. Gong, N.Z., Liu, B.: You are who you know and how you behave: attribute inference attacks via users' social friends and behaviors. In: Proceedings of the 25th USENIX Conference on Security Symposium, SEC 2016, pp. 979–995. USENIX Association, USA (2016)
11. He, J., Chu, W.W., Liu, Z.V.: Inferring privacy information from social networks. In: Mehrotra, S., Zeng, D.D., Chen, H., Thuraisingham, B., Wang, F.-Y. (eds.) ISI 2006. LNCS, vol. 3975, pp. 154–165. Springer, Heidelberg (2006). https://doi.org/10.1007/11760146_14
12. Horchulhack, P., Viegas, E.K., Santin, A.O., Ramos, F.V., Tedeschi, P.: Detection of quality of service degradation on multi-tenant containerized services. J. Netw. Comput. Appl. **224**, 103839 (2024)

13. Kayes, I., Iamnitchi, A.: Privacy and security in online social networks: a survey. Online Soc. Netw. Media **3–4**, 1–21 (2017)
14. Kruk, S.: Foam-realm: control your friends access to the resource. In: Proceedings of the First Workshop on Friend of a Friend, vol. 1 (2004)
15. Kumar, P.C., et al.: Understanding research related to designing for children's privacy and security: a document analysis. In: Proceedings of the 22nd Annual ACM Interaction Design and Children Conference, IDC 2023, June 2023. ACM (2023)
16. Liccardi, I., Pato, J., Weitzner, D.J.: Improving user choice through better mobile apps transparency and permissions analysis. J. Priv. Confidentiality **5**(2) (2014)
17. Liu, M., Wang, H., Guo, Y., Hong, J.: Identifying and analyzing the privacy of apps for kids. In: Proceedings of the 17th International Workshop on Mobile Computing Systems and Applications, February 2016, HotMobile 2016. ACM (2016)
18. Nissenbaum, H.: A contextual approach to privacy online. Daedalus **140**(4), 32–48 (2011)
19. Paul, T., Stopczynski, M., Puscher, D., Volkamer, M., Strufe, T.: C4PS - helping Facebookers manage their privacy settings. In: Aberer, K., Flache, A., Jager, W., Liu, L., Tang, J., Guéret, C. (eds.) SocInfo 2012. LNCS, vol. 7710, pp. 188–201. Springer, Heidelberg (2012). https://doi.org/10.1007/978-3-642-35386-4_15
20. Raynes-Goldie, K., Allen, M.: Gaming privacy: a Canadian case study of a children's co-created privacy literacy game. Surveill. Soc. **12**(3), 414–426 (2014)
21. Reyes, I., Wiesekera, P., Razaghpanah, A., Joel Reardon, N.V.R., Egelman, S., Kreibich, C.: "Is Our Children's Apps Learning?" Automatically Detecting COPPA Violations (2017)
22. Rocheleau, J.N., Chiasson, S.: Privacy and safety on social networking sites: autistic and non-autistic teenagers' attitudes and behaviors. ACM Trans. Comput. Hum. Interact. **29**(1), 1–39 (2022)
23. dos Santos, R.R., Viegas, E.K., Santin, A.O., Tedeschi, P.: Federated learning for reliable model updates in network-based intrusion detection. Comput. Secur. **133**, 103413 (2023)
24. dos Santos, R.R., Viegas, E.K., Santin, A.O., Cogo, V.V.: Reinforcement learning for intrusion detection: more model longness and fewer updates. IEEE Trans. Netw. Serv. Manage. **20**(2), 2040–2055 (2023)
25. Schomakers, E.M., Biermann, H., Ziefle, M.: Users' preferences for smart home automation - investigating aspects of privacy and trust. Telematics Inform. **64**, 101689 (2021)
26. Shehab, M., Cheek, G., Touati, H., Squicciarini, A.C., Cheng, P.C.: User centric policy management in online social networks. In: 2010 IEEE International Symposium on Policies for Distributed Systems and Networks. IEEE (2010)
27. Solove, D.J.: Understanding Privacy (2022)
28. Squicciarini, A.C., Paci, F., Sundareswaran, S.: Prima: a comprehensive approach to privacy protection in social network sites. Ann. Telecommun. **69**(1–2), 21–36 (2013)
29. Techcrunch: How free apps can make more money than paid apps. Technical report (2012). https://techcrunch.com/2012/08/26/how-free-apps-can-make-more-money-than-paid-apps/
30. Weinsberg, U., Bhagat, S., Ioannidis, S., Taft, N.: BlurMe: inferring and obfuscating user gender based on ratings. In: Proceedings of the sixth ACM Conference on Recommender Systems, RecSys 2012, September 2012. ACM (2012)

# Research on E-commerce Customer Segmentation Based on the K-means++ algorithm

Zhang Jing(✉)

Department of Information and Communication, Police Officers College of CAPF, Chengdu, Sichuan, China
30145002@qq.com

**Abstract.** In a long-term study, the author found that the attention and loyalty of e-commerce customers are very important factors for e-commerce customers to maintain and maintain. Therefore, on the customer value matrix AF of the traditional customer segmentation model, which represents the existing value, This article has added 2 variables that represent the value-added-potential of e-commerce customers, namely, the total clicks C representing the customer attention and the customer hold time H representing the customer loyalty, and constructs the AFCH e-commerce customer segmentation. The AFCH customer segmentation model is tested by K-means, SOM + K-means and K-means++ respectively. The error square and SSE were used as the algorithm's accurate measurement standard. The experimental effect found that the clustering results of the three algorithms were similar, The accuracy of K-Means++ algorithm is higher than that of the other two. Finally, this paper gives the experimental results of AFCH customer segmentation for an e-commerce enterprise by K-Means++ algorithm, which provides decision support for the customer management and specific marketing measures of e-commerce enterprises.

## 1 Introduction

With the rapid development of e-commerce in recent years, the phenomenon of nationwide online shopping is becoming increasingly popular. According to data from the Italian financial analysis website Finaria, there were 2.4 billion people shopping online globally in 2017, this number had already reached 3.4 billion by 2020, and the website predicts that the number of online shoppers will reach 4.9 billion by 2025. The increase of online shoppers has driven a surge in global e-commerce revenue. According to data from Statista, the global e-commerce revenue reached 2 trillion euros in 2021, it reached 23 trillion euros in 2022 with a 12% year-on-year increase. It is estimated that global e-commerce revenue will reach around 28 trillion euros by 2025. A large number of online shoppers and considerable e-commerce business income have spawned a wide range of e-commerce enterprises. Finaria's data shows that the competition between e-commerce companies is extremely fierce. Despite the thriving development of large retail e-commerce platforms such as Amazon and AliExpress, but many independent


L. Barolli (Ed.): AINA 2024, LNDECT 203, pp. 439–446, 2024.
https://doi.org/10.1007/978-3-031-57931-8_42

retail e-commerce businesses are facing greater challenges, as it is a market with a shopping cart abandonment rate of 75%. In the era of "customer economy", in such intense competition, If e-commerce wants to win survival, E-commerce wants to win survival, occupy a favorable market position, obtain sustainable development, effectively explore, manage and distinguish customer resources and carry out targeted marketing is the key to win.

## 2 Proposal of AFCH E-commerce Customer Segmentation Model

### 2.1 Existed E-commerce Customer Segmentation Model

Currently, most e-commerce customer segmentation models are either improvements on traditional customer segmentation models or incorporate certain business feature variables into traditional customer segmentation models [2–6]. This article is no exception. Here we mainly introduce two existing e-commerce customer segmentation models related to this article.

(1) Improving the RFM e-commerce customer segmentation model.
 In the reference [4], Shi Yuxue, a graduate student from Harbin Institute of Technology, incorporated the traditional RFM model to measure the current value (CV) of e-commerce customers. Additionally, the model included customer activity (total user actions, number of products clicked, number of product types clicked) and customer purchase intention (number of products favorited, number of product types favorited, number of product types added to cart) as potential value (PV), thus constructing an evaluation system for e-commerce customer value.
(2) AFCS e-commerce customer value model.
 In reference [5], the author of this article was inspired by the improvement of the improving RFM customer segmentation model in (1). However, it has the drawback of multicollinearity between the variables of purchase frequency and total purchase amount during the same period.Therefore, customer value matrix model AF which is the improved model of RFM is adopted as the current value of e-commerce customers, customer activity C (total clicks) and customer purchase intention S (product number collected and the types of products added to the shopping cart by users) are still taken as the potential value of customers, constructed AFCS customer value subdivision model.

### 2.2 Proposal of AFCH E-commerce Customer Value Model

AFCS e-commerce customer model of in (2) has duplicate calculations for the segmentation variables C and S. Therefore, this paper will merge these 2 variables into a total clicks C (C = E-commerce web page clicks + number of favorite items + number of items added to cart), representing customer attention, which is an important value-added-potential factor of e-commerce customers. Through the author's long-term tracking and research of data from a certain e-commerce platform, it has been found that customer loyalty can reflect the stability of customer relationships and effectively evaluate the quality of customer relationships. Therefore, this article believes that customer loyalty is another important factor in value-added-potential of e-commerce customers. Through the analysis of transaction records of e-commerce customers, we found that customer hold time is closely related to customer loyalty. Therefore, this article proposes to use the average purchase amount A to represent customer contribution, purchase frequency F to represent customer satisfaction, total click C count to represent customer attention, and customer hold time H to represent customer loyalty as measures of current customer value and value-added-potential, constructing an AFCH e-commerce customer value model as shown in Fig. 1.

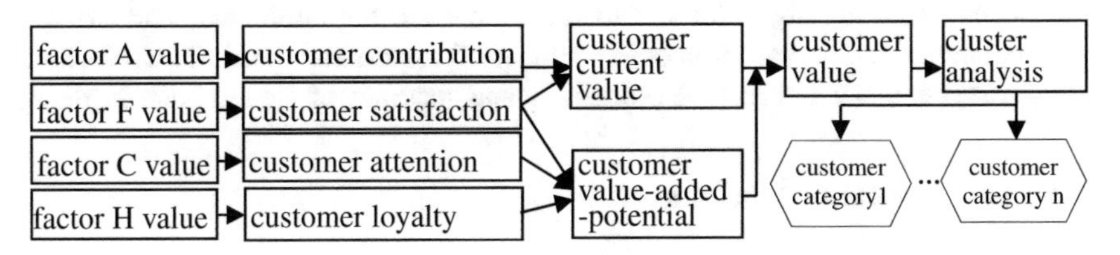


**Fig. 1.** AFCH e-commerce customer value model

## 3 Selection of Clustering Algorithm

The clustering algorithm reflects the principle of "birds of a feather flock together, and people flock together into groups". That is, a data set was divided into different classes or clusters according to a certain standard. The similarity of data objects within the same cluster and the difference of data objects in different clusters are both as large as possible. This article mainly discusses the typical partition-based clustering algorithm k-means and its derivative or combined algorithms.

### 3.1 K-means Algorithm

K-means algorithm steps [1–3, 5] flow is as follows:

(1) Assuming the sample size is M, k points were randomly selected and used as the cluster centers
(2) Calculated the distance from each point to the k cluster centers, then the point is assigned to the nearest cluster center, formed k clusters.
(3) The average of all objects in each cluster was recalculated and used as the new cluster center.
(4) Repeat (2) (3) steps to minimize the loss function corresponding to the clustering result.

Sum of Squared Errors (SSE) is the loss function as shown formula (1), which is the calculation of the sum of the squared distances of each sample to its assigned cluster centroid.

$$\mathrm{SSE} = \sum_{i=1}^{k} \sum_{p \in C_i} |p - m_i|^2 \tag{1}$$

In formula (1), p is a point in the clustering space, $m_i$ is the average value of $C_i$ cluster.

Clustering distance calculation selected Euclidean Function as shown in following formula (2).

$$d(i, j) = \sqrt{|x_{i1} - x_{j1}|^2 + |x_{i2} - x_{j2}|^2 + \cdots + |x_{ip} - x_{jp}|^2} \tag{2}$$

In formula (2), i and j are two p-dimensional data objects($x = (x_{i1}, x_{i2}, \ldots, x_{ip})$ $j = (x_{j1}, x_{j2}, \ldots, x_{jp})$).

K-means clustering algorithm has the advantages of high efficiency and low complexity, but the subjective selection of the number of clusters and the random selection of initial cluster centers can affect the objectivity and accuracy of the final cluster results.

### 3.2 SOM + K-means Algorithm

In order to overcome the limitations of k-means, the author of this paper attempted the SOM + K-Means algorithm [4, 6] for clustering. For details refer to reference 4, which is only briefly described below.

SOM + K-Means clustering algorithm can be divided into two steps.

Step 1: Input the experimental data into the SOM network and execute the SOM algorithm to continuously update the weights of the neurons in the competitive layer until the specified number of iterations is reached. End the algorithm and output the clustering results, calculate the cluster centers and the number of clusters after clustering.

Step 2: Take the cluster center and the number of clusters in step 1 as the initial cluster center and the initial K value of the K-means algorithm, respectively, execute the K-means algorithm until convergence, and output the final clustering results.

The experiment proves that SOM + K-means algorithm indeed overcome the disadvantage of the subjective selection of the number of clusters in K-means, So the accuracy improves more than the k-means. However, in the case of a large amount of data, with an increase in the number of learning times, the learning effect of the SOM network decreases.

### 3.3 K-means++ algorithm

Due to the limitations of K-means and SOM + K-Means, this paper attempts to use K-means++ clustering algorithm [7, 8].

K-means++ algorithm proposed by D. Arthur is an improvement of K-means algorithm for selecting initial centroids. K-Means++ algorithm uses the roulette wheel selection method for selecting initial centroid points:it starts by randomly selecting an initial centroid point, and then calculates the shortest distance D(X) between each data point and this centroid,and then calculates the probability of each sample being chosen as the next cluster center. Following formula (3) is the probability calculation formula.

$$G(x) = \frac{D(x)^2}{\sum_{x \in X} D(x)^2} \tag{3}$$

The larger the D(X) value is, the higher the probability this sample is selected, Then the next cluster center is selected using the roulette method according to the G(x) value.

The clustering centers selected through this method are relatively dispersed. K-Means algorithm shows that when the initial cluster points are more dispersed, the clustering effect is better and the speed is faster. Once the initial cluster centers have been selected, the rest of the steps are the same as the traditional K-means algorithm.

## 4 Simulation Experiment

The present study conducted an experiment on the data of 5000 customers from a retail e-commerce platform over a period of 4 years. The data is preprocessed according to the data mining process such as cleaning, integrating, transforming, discretizing and so on.Subsequently, then clustered the preprocessed data using K-means, SOM + K-Means and K-means++ algorithms in PyCharm Community Edition. The results were then compared and analyzed for validation.

### 4.1 Cluster Experimental Results of Three Algorithms of AFCH Model

(1) K-means

When K takes the values 2, 3, 4, 5, the clustering results of K-means algorithm for AFCH model are shown in Figs. 2, 3, 4, 5 respectively.

From Fig. 1 to Fig. 5, it can be seen that when K = 2 and K = 3, there is already a trend of clustering into 4 clusters. When K = 4, the distribution of samples into 4 clusters is very obvious, and when K = 5, although the distribution of samples in clusters is slightly chaotic, there is still a trend of 4 clusters. As can be seen, the clustering performance is best when K = 4.

(2) SOM + K-means

For the same samples in (1), when we implement the SOM + K-means algorithm, we obtain the AFCH model clustering result as shown in the Fig. 6. From Fig. 6, it can be observed that the clustering results of the AFCH model are essentially consistent with the K-means algorithm with K = 4.

(3) K-means++

For the same sample in (1), we implemented the K-means++ algorithm to get AFCH results as shown in the Fig. 7.

The experimental results from (1) (2) (3) show that the k-means algorithm (K = 4) produces similar clustering results to the SOM + K-means algorithm and the K-means++ algorithm, all three algorithms divide the samples into 4 clusters.

### 4.2 Evaluation Indicators

The clustering algorithm generally uses the SSE value as the evaluation metric for algorithm accuracy, which is calculated as shown in formula (1). A smaller SSE indicates

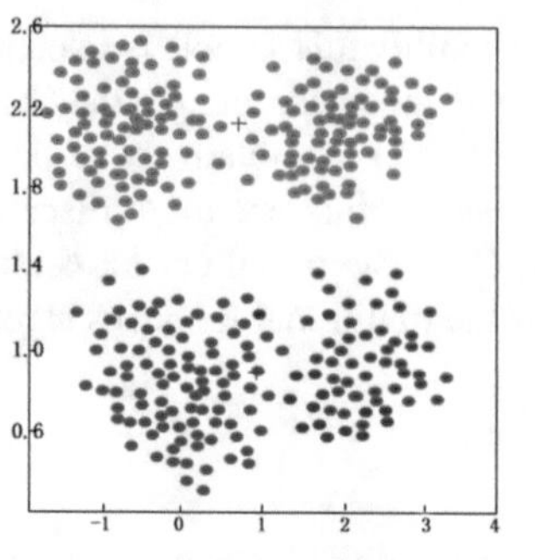


**Fig. 2.** K-means algorithm results (K = 2)

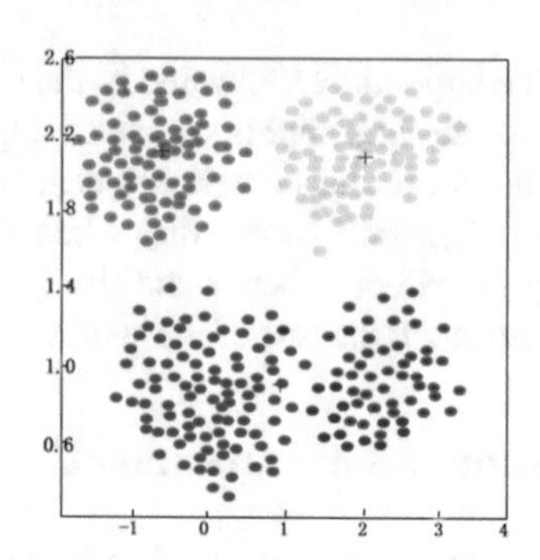


**Fig. 3.** K-means algorithm results (K = 3)

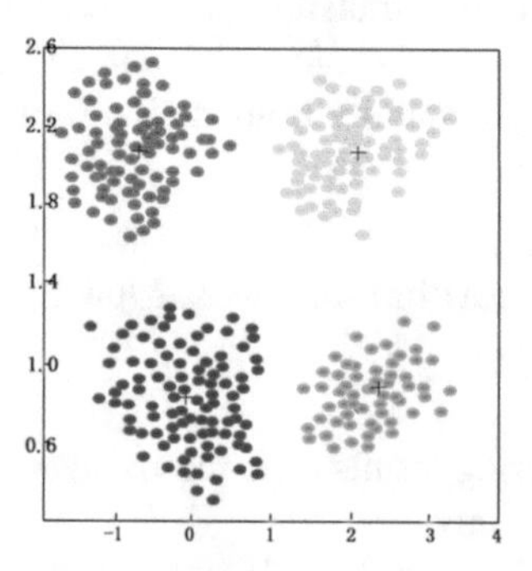


**Fig. 4.** K-means algorithm results (K = 4)

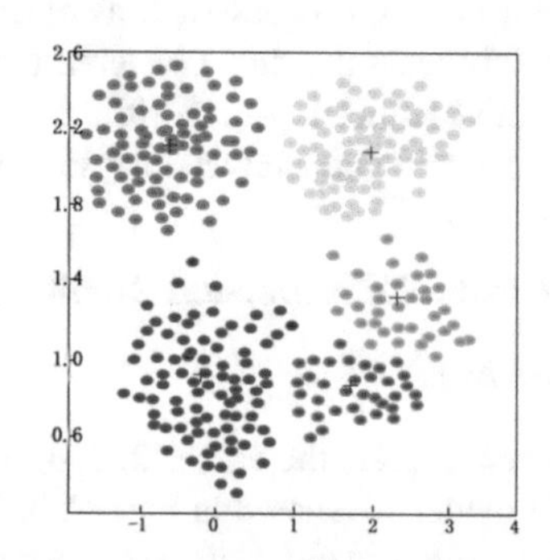


**Fig. 5.** K-means algorithm results (K = 5)

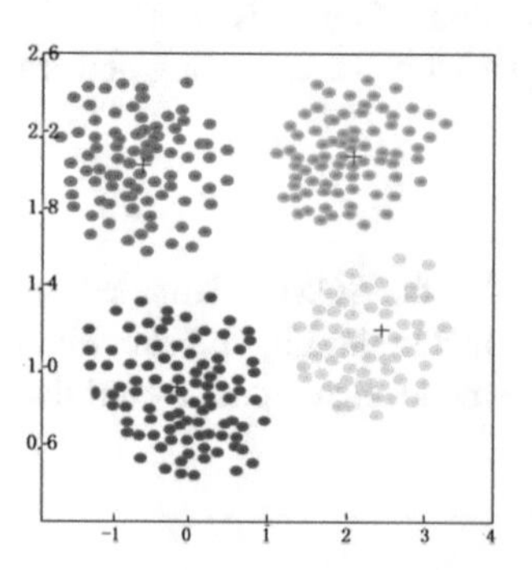


**Fig. 6.** SOM + K-means algorithm results

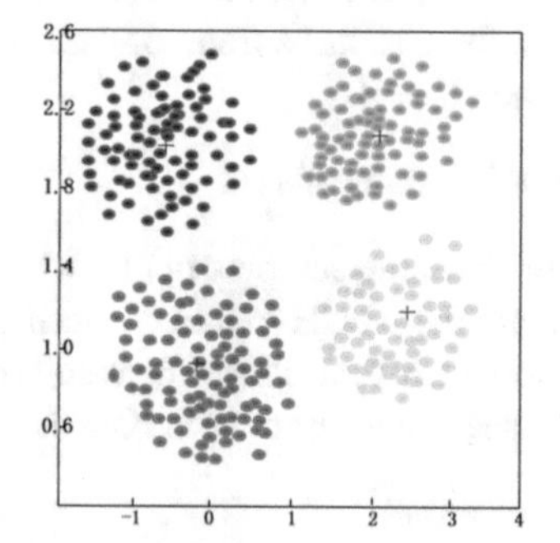


**Fig. 7.** K-means++ algorithm results

lower error and higher accuracy. The SSE values of the test data for the AFCH model using these three algorithms in three experiments are shown in Table 1.

As can be observed from Table 1: $SSE_{K\text{-means}} << SSE_{SOM+K\text{-means}} < SSE_{K\text{-means++}}$,that is, the K-means accuracy is much less than the other two algorithms SOM + K-means and K-means ++, SOM + K-means accuracy is slightly lower than the K-means++. However, As we gradually increase the training data, we find that the K-means++ clustering takes less time than SOM + K-means, When the data set exceeds

**Table 1.** AFCH's test SSE values of three algorithms

| Number | K-means | SOM + K-means | K-means++ |
|---|---|---|---|
| 1 | 1987 | 125 | 120 |
| 2 | 388 | 130 | 125 |
| 3 | 824 | 130 | 127 |
| avg. | 1066.33 | 129.33 | 124 |

1,000,000 records, the K-means++ algorithm takes about 0.5 s less time to cluster than the SOM + K_means algorithm.

### 4.3 Experimental Results

In this paper, after K-means++ clustering and subdivision of AFCH e-commerce customer model, the analysis results of e-commerce customer value of each category are shown in Table 2.

**Table 2.** AFCH e-commerce customer clustering results and value analysis table

| Classification | C1 | C2 | C3 | C4 |
|---|---|---|---|---|
| A value | 0.42 | 0.37 | 0.24 | 0.42 |
| F value | 0.09 | 0.11 | 0.05 | 0.33 |
| C value | 0.10 | 0.66 | 0.08 | 0.61 |
| H value | 0.13 | 0.71 | 0.04 | 0.84 |
| AFCH value | 0.74 | 1.87 | 0.41 | 2.20 |
| A + F value | 0.51 | 0.48 | 0.29 | 0.85 |
| F + C + H value | 0.32 | 1.48 | 0.17 | 1.78 |
| Customer value | CV↑<br>VAP↓ | CV↓<br>VAP↑ | CV↓<br>VAP↓ | CV↑<br>VAP↑ |
| Customer type | Silver customer | Copper customers | Iron customers | Gold customers |

In Table 2, the AFCH model is clustered into 4 clusters (C1 ~ C4), and we provide the clustering results for the A value (customer contribution), F value (customer satisfaction), C value (customer attention), and H value (customer loyalty). We also calculate the A + F value and F + C + H value of the clustering results. The A + F and F + C + H values accurately reflect the current value and the value-added-potential of the AFCH model customer clusters(In Table 2, CV is current value, VAP is value-added-potential, ↑represents high,↓represents low). Then, according to the results of A + F and F + C + H in descending order, the customer segments is divided into gold customers, silver customers, copper customers and iron customers. The clustered customer value matrix is shown in Fig. 8 below.

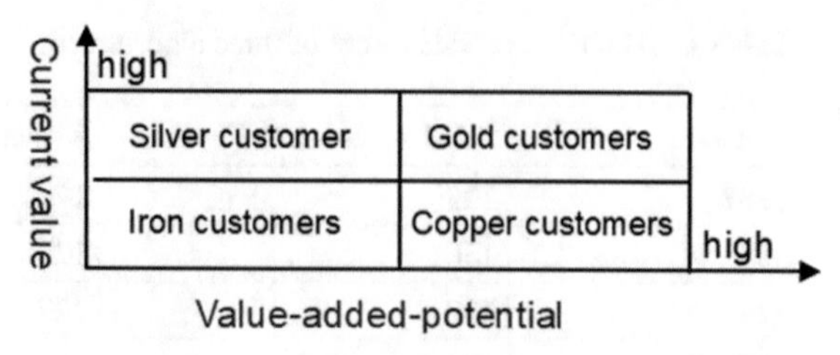


**Fig. 8.** AFCH Customer Value Matrix

## 5 Conclusion

On the base on the traditional customer value matrix, this article introduces two characteristic variables of customer value-added-potential: customer attention (customer total clicks) and customer loyalty (customer hold time), and constructs the AFCH e-commerce customer model. Then AFCH model was clustered with K-means, SOM + K-means, K-means++, and their clustering results were compared to conclude that K-means++ is superior. At last the K-means++ clustering result customer value matrix is presented, which provides a decision basis for e-commerce enterprises to implement targeted marketing and retention strategies for customers.

## References

1. Tan, P.N., Steinback, M., Kumar, V.: Introduction to Data Mining, vol. 8, pp. 305–319. China Machine Press, BeiJing (2019)
2. Kicova, E., Kral, P., Janoskova, K.: Proposal for brand's communication strategy developed on customer segmentation based on psychological factors and decision-making speed in purchasing: case of the automotive industry. Econ. Cult. **15**(1), 5–14 (2018)
3. Shi, Y.: Research on Identifying Customer Value in E-commerce Based on Improved RFM Model, pp. 19–20. Harbin Institute of Technology, Heilongjiang (2021)
4. Zhang, J., Li, J.: Research on E-commerce customer value segmentation model based on network. In: EIDWT-2023, pp. 118–128. Springer, Cham (2023). https://doi.org/10.1007/978-3-031-26281-4_12
5. Na, W.: Personalized Recommendation Based on user Behavior on an e-Commerce Platform, vol. 6, pp. 17–24. Lanzhou University of Finance and Economics, Gansu (2021)
6. Lin, H., Ji, Z.: Breast cancer prediction based on K-Means and SOM hybrid algorithm. In: Proceedings of 2020 2nd International Conference on Computer Modeling,Simulation and Algorithm (CMSA 2020), vol. 6, p. 21 (2020)
7. Han, X., Geng, M., Fan, Y.: BP neural network based on k-means++ and optimized by the genetic algorithm. In: Proceedings of the 42nd Chinese Control Conference, vol. 15, pp. 7–24 (2023)
8. Ni, W., Chu, W.: Classification of abnormal travel passengers in rail transit based on K-means++. Logist. Eng. Manag. **5** (2023)

# Early Ransomware Detection System Based on Network Behavior

Hamdi Abu-Helo(✉) and Huthaifa Ashqar

Arab American University, Ramallah, Palestine
h.abuhilo@student.aaup.edu, huthaifa.ashqar@aaup.edu

**Abstract.** Computer malware has been growing at a rapid way in recent years, with ransomware emerging as a particularly potent menace. Numerous victims, including companies, hospitals, and individuals, have suffered large financial losses as a result of the fast spread of ransomware. Due to the fact that they frequently rely on discovering infections, conventional techniques of ransomware detection have mostly proven ineffectual. Utilizing network behavior analysis for preemptive identification of ransomware incidents offers a more effective solution to this problem. This study examines network behavisor with an emphasis on ransomware, using Cerber ransomware as a case study to record and extract important information from infected host devices, a specialized testbed was built. Our model, which used the K-Nearest Neighbors (KNN) technique, attained a remarkable accuracy rate of 99.5%. This study represents a significant step towards preemptively identifying ransomware incidents based on network behavior analysis, offering valuable insights and potential solutions to combat this pervasive cyber threat.

## 1 Introduction

Malware attacks have become increasingly prevalent in targeting computer systems [1]. These malicious activities primarily occur through online channels, such as email phishing, rogue websites, and drive-by software downloads. Different types of malwares, including viruses, trojan horses, worms, and ransomware, pose significant threats to individuals and organizations [2]. Among these, ransomware has gained particular prominence, exploiting a business model known as Ransomware as a Service (RaaS) that allows even non-technical individuals to carry out attacks and demand ransoms [3]. This introductory paragraph highlights the growing use of malware, particularly ransomware, as a means for cyber attackers to exploit vulnerable systems.

### 1.1 Impact and Characteristics of Ransomware

Ransomware operates through various mechanisms, typically involving the locking and encryption of victims' devices, followed by ransom demands [4]. These attacks are facilitated by Command and Control (C&C) servers, which allow attackers to control the spread and impact of the malware. Disturbingly, statistics indicate that ransomware


L. Barolli (Ed.): AINA 2024, LNDECT 203, pp. 447–458, 2024.
https://doi.org/10.1007/978-3-031-57931-8_43

attacks affected over 51% of enterprises in 2019 [5]. Notably, even large-scale organizations have fallen victim to ransomware, such as the Colonial Pipeline Company, which paid approximately $4.4 million to recover encrypted data after a ransomware incident in May 2021 [6]. The interconnected nature of the Internet and cloud computing has amplified the propagation of ransomware through various communication protocols, enabling adversaries to employ diverse attack tactics.

### 1.2 Types and Cryptosystems of Ransomware

Ransomware can be categorized into two main types based on its behavior. The first and most common type is crypto-ransomware, which encrypts victims' data and files, rendering them inaccessible without the decryption key [7]. The second type is locker ransomware, which locks victims out of their computers, preventing normal usage [7]. Within the crypto-ransomware category, there are different varieties based on the cryptosystems employed [8]. Symmetrical cryptosystem ransomware employs a single key for both encryption and decryption, using algorithms like DES. This presents victims with the possibility of recovering their files through techniques such as memory scanning or reverse engineering to uncover the secret key. Asymmetric cryptosystem ransomware, on the other hand, encrypts files using a public key that only the attacker possesses, making it nearly impossible for victims to retrieve their files without paying the ransom. This type often uses more computational resources due to the encryption method, employing algorithms like RSA.

### 1.3 Case Study: Cerber Ransomware

One prominent ransomware strain, Cerber, serves as an ideal case study due to its heavy reliance on external communications, making it susceptible to monitoring and analysis. By examining Cerber's characteristics and behavior, valuable insights can be gained to enhance our understanding of ransomware operations and develop effective countermeasures.

## 2 Literature Review

These publications examine several methods for ransomware detection, including feature extraction methods and supervised learning algorithms. The articles also emphasize the significance of employing various datasets for machine learning model training and testing, as well as the requirement for ongoing upgrades to stay up with increasing ransomware threats. Overall, these articles give useful recommendations for further research in this field and significant insights into the possibilities of machine learning for ransomware detection.

To categorize system activity as either ransomware or benign, the authors in [9] utilize a Random Forest (RF) classifier and an Artificial Neural Network (ANN) classifier. A set of features that were gathered from a dynamic investigation of system activity are used to train the classifiers. The findings demonstrate that in terms of accuracy, precision, and recall, the RF classifier performs better than the ANN classifier. The proposed approach's usage of machine learning techniques shows how useful these methods are for creating precise and effective ransomware detection systems. The study focuses on a two-stage strategy that necessitates machine learning and dynamic analysis techniques, which may be costly in terms of time and processing. This makes the method less scalable, especially for complex systems.

The authors in [5] suggests a machine learning-based method for spotting crypto-ransomware intrusions in scenarios including file-sharing networks and encrypted traffic. To analyze network data and spot potential ransomware attacks, the method combines deep learning models and feature extraction methods. The findings demonstrate that the suggested method successfully detects ransomware attacks with high rates of accuracy. The proposed strategy is only evaluated in the study using a small number of ransomware variants and network conditions. For further research to assess the approach's efficacy and generalizability, a wider variety of ransomware attacks and network setups must be used.

By recreating the events and examining packets, the author in [8] was able to determine when the hostname and IP address were first infected. He was also able to discover how the Cerber ransomware entered the victim's device. Only the IP addresses and domain names that the ransomware virus used were found through analysis; no behavioral activity was extracted.

The author in [3] uses a honeypot and an automated method for real-time malware analysis to dynamically investigate the CryptoWall ransomware network behavior. Four virtual computers are employed in an environment called Maltester for this purpose. Some features of the CryptoWall ransomware, including some proxy servers, the protocols used, and some servers' hardcoded addresses, were able to be extracted by the author. Although the output was good, most addresses and proxy servers are easily changeable, so it is not behavioral even for this type of ransomware.

The authors in [10] provide a live forensic examination in a simulated environment of the Conti ransomware attack on a computer network. The study makes use of a specially created virtual environment to simulate the attack and assess how well live forensic procedures for spotting and addressing the ransomware attack work. The study also highlights a number of crucial characteristics that the Conti malware utilized to avoid detection and carry out the attack. These features involve manipulating Windows Management Instrumentation (WMI) and using Process Injection to run the ransomware payload. In order to avoid being discovered by conventional antivirus software, the authors also mention the usage of specialized encryption methods and obfuscation strategies. The investigation is restricted to a single ransomware version and a particular network setup, which might not be indicative of other attack situations.

The study in [11] suggested a model named EldeRan for the dynamic analysis and categorization of ransomware. EldeRan monitors a variety of actions taken by applications during the initial installation process. The findings were promising, but some ransomware still has the issue of lying dormant for an extended period of time before it begins to encrypt the victim's device.

## 3 Traffic Analysis and Features Extraction

To gather valuable insights into the network behavior of Cerber ransomware, our study focused on analyzing its network traffic and extracting a diverse range of useful features and to detect the malicious packets so we will be able label them on the final dataset. Cerber ransomware is predominantly distributed through spam emails that contain malicious file attachments. Once the attachment is opened, a downloader script is executed, which retrieves and installs Cerber's executable file from a URL on the victim's computer. Once fully installed, Cerber establishes communication with its Command and Control (C&C) server(s) to exchange encryption keys.

During the external communication with the C&C servers, we meticulously extracted a comprehensive set of features. These features were categorized into four sections during the analysis phase based on their characteristics:

- **Behavioral Features:** These features capture distinct patterns and behaviors exhibited by the ransomware, enabling differentiation from normal network traffic.
- **Non-Behavioral Features:** This category encompasses features that are not unique to Cerber ransomware and may be observed in regular network traffic as well.
- **Detectable Features:** These features allow for the detection of ransomware presence, but they do not specifically pertain to the behavioral aspects of malicious traffic. They can still be analyzed within a mixed traffic context.
- **Non-Detectable Features:** These features are exclusively observed in malicious traffic and provide strong indications of ransomware activity.

To ensure comprehensive data collection, we infected a victim device with Cerber ransomware and monitored its network traffic for one hour. Given that contact with the ransomware servers (C&C) occurs during the initial encryption and encryption key exchange phases, our focus was to identify malicious traffic patterns within the early time periods and compare them with the rest of the traffic during the hour-long observation period. Through rigorous experimentation and data analysis, the following subsections will provide a detailed exposition of these distinct feature characteristics.

By examining Cerber ransomware's network traffic and extracting various feature categories, this research contributes to a deeper understanding of its behavior and aids in the development of effective countermeasures.

1. TCP Traffic

- **RESET Connections**

It can be seen that Cerber's traffic contains some TCP reset (RST, ACK) packets that are used to abnormally terminate the malicious TCP connections. When contrasted with the normal traffic, we can figure out some normal traffic that is similar to malicious traffic. The reset connections for malicious traffic and normal traffic are depicted in Fig. 1.

Based on what is shown in Fig. 1, the RST feature would be classified as a detectable / non-behavioral feature.

Note that the red dots symbolize malicious traffic, while the blue dots symbolize normal traffic, and this is the same for all figures in the next analysis. It should be

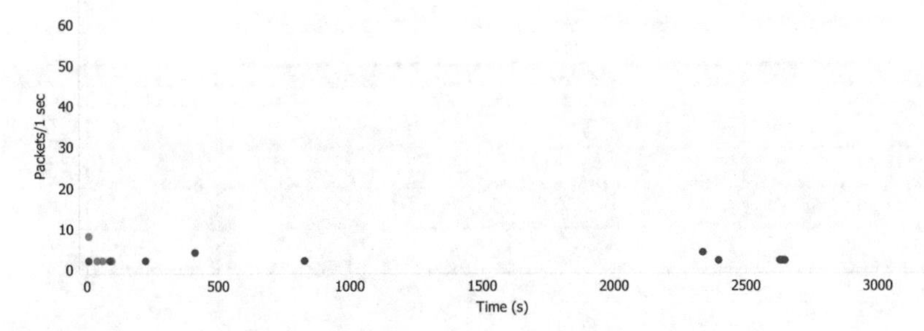


**Fig. 1.** RST Connection

noted that the IP addresses were chosen in preparation before the analysis process and that we did not locate them in any normal traffic connection procedure. The following are these ip addresses: 31.184.234.0/24, 31.184.235.0/24, 208.95.112.1, 151.101.2.49, 104.20.21.251, and 93.184.220.29.

– **FIN Connections**

As shown in Fig. 2, there are a lot of FIN connections during the captured hour, but we can notice that there are a lot of concurrent malicious FIN connections in the first seconds of the traffic which also belong to the same malicious ip address as Fig. 3 shown.

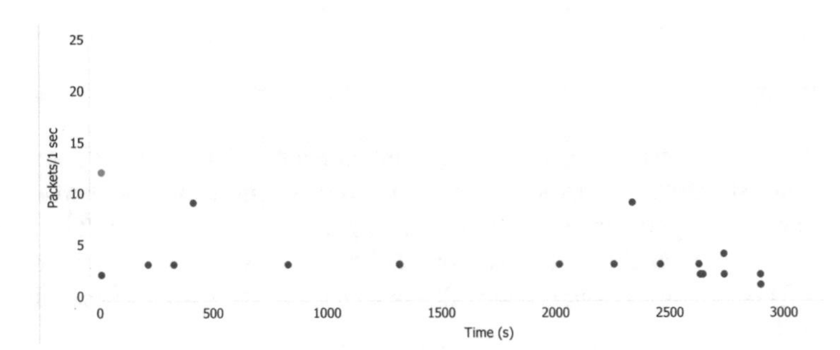


**Fig. 2.** FIN Connection

According to FIN analysis, we can determine that this feature is a detectable/behavioral feature as there is no normal FIN traffic happening with the same period and the same IP address.

– **ACK Connections**

Another TCP connection feature (ACK) occurs sequentially and randomly for normal and suspicious traffic over 60 min, as shown in Fig. 4. According to what we see, large normal connections occur at the end of the period with almost the same pattern as the malicious traffic so we cannot determine this feature as a behavioral feature.

tcp.flags.fin==1

| No. | Time | Source | Destination | Protocol |
|---|---|---|---|---|
| 5403 | 2023-01-05 22:53:58.235059 | 192.168.177.200 | 104.20.21.251 | TCP |
| 5404 | 2023-01-05 22:53:58.362003 | 104.20.21.251 | 192.168.177.200 | TCP |
| 5407 | 2023-01-05 22:53:58.534426 | 192.168.177.200 | 104.20.21.251 | TCP |
| 5408 | 2023-01-05 22:53:58.534446 | 192.168.177.200 | 104.20.21.251 | TCP |
| 5409 | 2023-01-05 22:53:59.135138 | 192.168.177.200 | 104.20.21.251 | TCP |
| 5410 | 2023-01-05 22:53:59.135146 | 192.168.177.200 | 104.20.21.251 | TCP |
| 5411 | 2023-01-05 22:54:00.338189 | 192.168.177.200 | 104.20.21.251 | TCP |
| 5412 | 2023-01-05 22:54:00.338207 | 192.168.177.200 | 104.20.21.251 | TCP |
| 5413 | 2023-01-05 22:54:02.742804 | 192.168.177.200 | 104.20.21.251 | TCP |
| 5414 | 2023-01-05 22:54:02.742826 | 192.168.177.200 | 104.20.21.251 | TCP |
| 5415 | 2023-01-05 22:54:07.551730 | 192.168.177.200 | 104.20.21.251 | TCP |
| 5416 | 2023-01-05 22:54:07.551742 | 192.168.177.200 | 104.20.21.251 | TCP |

**Fig. 3.** Malicious IP address made concurrent FIN traffic

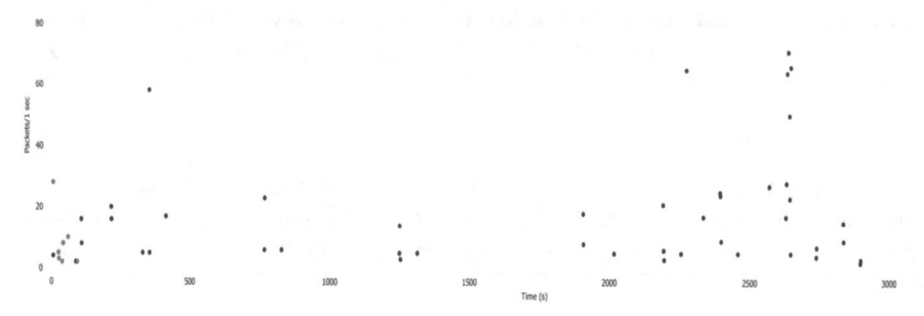


**Fig. 4.** ACK Connection

- **HTTP Traffic**

Attackers continuously design covert communication links and strong command and control infrastructures to enable their malware to evade intrusion detection systems. Therefore, attackers try to connect using common ports for HTTP and HTTPS to their servers, according to an analysis of the captured PCAP as shown in Fig. 5, files created by Cerber a TCP connection is established with a distant server, where data is sent through the HTTP technique.

We can classify the HTTP feature as a detectable and behavioral feature, as there is a unique pattern for malicious HTTP traffic.

2. DNS Traffic

Cybercriminals frequently utilize DNS for a variety of objectives, including preventing the need to inject a list of servers' IP addresses that are hard-coded since they can be quickly identified and blocked. Instead, DNS queries are continually made for pseudo-random domain names until the location of the actual server using DGA [11]. When no outward TCP access is permitted on a secure network, another objective for using DNS can occasionally be used to send data over that network [15]. Additionally, UDP is only able to support DNS queries. To send many DNS queries to a fictitious DNS server located at the attacker's end, the virus might split the file contents into numerous requests.

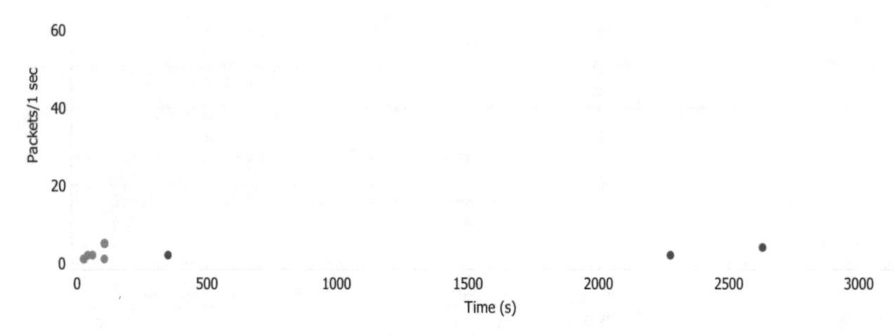


**Fig. 5.** HTTP Traffic

According to other ransomware analyses such as WannaCry, there are a lot of results that can be considered while analyzing the DNS traffic, like DNS name error, meaningless domain name, DNS-time, DNS-qry-typ, and others. In our analysis of the Cerber ransomware, we found that most of these things were normal, except for the number of packets requested per second, as shown in Fig. 6.

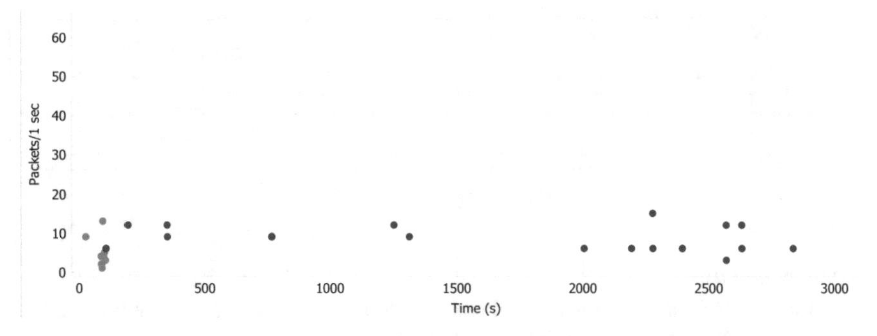


**Fig. 6.** DNS Traffic

According to the number of malicious requested packets compared to the normal traffic, we can determine that the DNS feature is detectable and behavioral.

3. UDP Traffic

As for UDP traffic, we can note the large number of suspicious requests that occurred at the beginning of the injection of the ransom virus, as it is considered a great advantage for detecting the virus, but it is not considered behavioral, as similar normal traffic was not observed throughout the entire hour (Fig. 7).

4 NBNS Traffic

The last feature that was extracted and analyzed is the NetBIOS, where no suspicious activity was observed throughout the full hour, and accordingly, the feature was considered a non-detectable and non-behavioral feature.

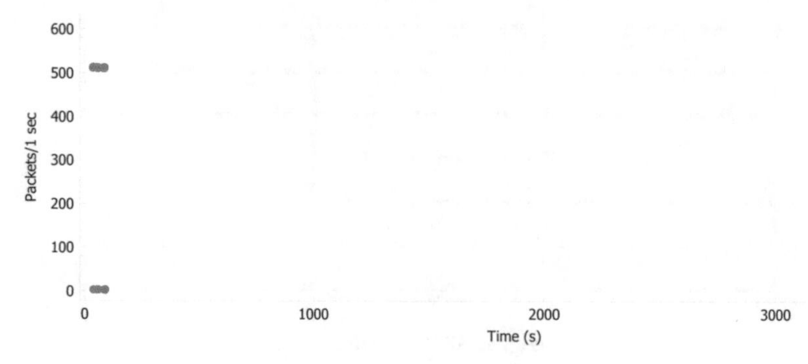


**Fig. 7.** UDP Traffic

# 4 Proposed Model

This work proposed a framework for detecting ransomware attack using the generated dataset for Cerber ransomware as shown in Figur.8, data collection, data preprocessing, feature selection, data classification, and machine learning processes.

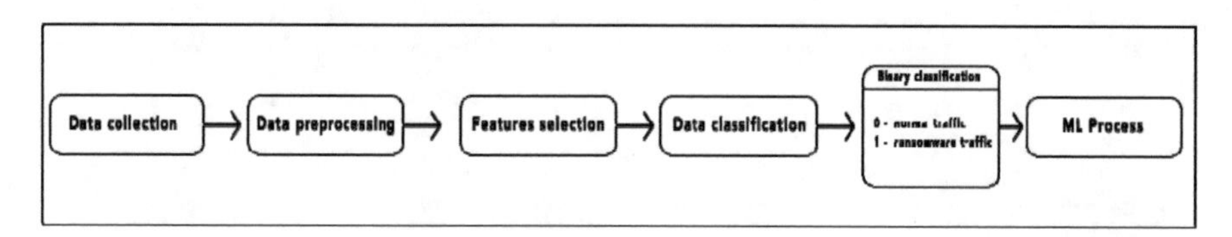


**Fig. 8.** Proposed model

Data collections

In this research, a network-based behavior detection technique was employed, utilizing a specially designed testbed to study Cerber ransomware. The testbed consisted of a single physical computer hosting two virtual machines, enabling controlled experimentation and network traffic recording for dataset creation.

The testbed components were as follows:

- **Host Device (PC1)**: An HP Zbook 15v G5 hardware served as the host device, providing the necessary computing resources for the virtual environment.
- **Victim Device (VM1):** A Windows 10 virtual machine was designated as the victim device, which would be intentionally infected with a sample of Cerber ransomware. This virtual machine played a crucial role in generating the network traffic of interest.
- **Second Virtual Machine (VM2):** Another Windows 10 virtual machine was included in the testbed, connected to the same network as VM1. This configuration allowed for monitoring and capturing the network traffic generated by the infected VM1.

To facilitate the experiments, several software tools were utilized:

1. **VMware Workstation Pro (version 16.1.1):** This virtualization software provided the platform to create and manage the virtual machines within the testbed, ensuring a controlled and isolated environment for the ransomware execution.

2. **Wireshark (version 4.0.1):** Wireshark, a widely-used network protocol analyzer, played a crucial role in capturing and analyzing the network traffic generated by Cerber ransomware. This tool enabled the recording and inspection of the communication patterns exhibited by the ransomware sample.

To conduct the experiments, a specific sample of Cerber ransomware was downloaded from bazaar.abuse.ch, with an associated MD5 hash value of **a57745a30d63f511d28aa43e4b710e1c**. This sample served as the basis for injecting the ransomware into the victim virtual machine, allowing for the capture and analysis of the resulting network traffic.

Dataset and preprocessing

The dataset part focuses on the Cerber ransomware's collected traffic during a one-hour period, then creates a final dataset with 20 features. For an hour, we gathered network traffic data using Wireshark to analyze the behavior and characteristics of the Cerber ransomware. This information records the numerous interactions and communication styles that the Cerber ransomware displayed throughout its harmful operations. We analyze the gathered traffic, removing unimportant data, and creating a thorough dataset. This final dataset consists of 20 unique characteristics that have been carefully selected to represent various facets of the ransomware's activity. Included in these characteristics are the network protocols that are being used, source and destination IP addresses, communication ports, IP length, TCP flags, and other crucial characteristics.s

The following table shows the features that extracted from the captured traffic (Table 1):

The dataset preprocessing phase involved several steps to prepare the data for analysis and model training. Firstly, the labeling data, consisting of binary labels (0 and 1) indicating the presence or absence of a specific attribute, was examined. To provide an overview of the dataset's balance, a Fig. 9 was generated, illustrating the distribution of the labels. This visualization allowed for a clear understanding of the dataset's composition and the ratio between positive and negative instances.

Next, feature reduction techniques were applied to streamline the dataset and enhance the model's performance. Initially comprising 20 features, the dataset was subjected to careful feature selection and dimensionality reduction. Through a thorough analysis, various preprocessing techniques were employed, such as correlation analysis, and removing some features that logically not important. These helped identify and retain the most informative and relevant features while eliminating redundant or less impactful ones.

After the preprocessing stage, the dataset was effectively reduced from 20 features to 12 features, resulting in a more concise representation of the data. This reduction aimed to eliminate noise, improve computational efficiency, and enhance the model's interpretability without compromising its predictive power.

**Table 1.** Features explanations

| Feature | Notes |
|---|---|
| ip.len | ip length |
| ip.hdr_len | ip header length |
| ip.flags | Ip flags |
| ip.ttl | Time to live |
| ip.proto | Ip protocol |
| ip.id | Packet id |
| ip.checksum | Packet error checksum |
| ip.srcport | Source ip address |
| ip.destport | Detonation ip address |
| tcp.seq | Sequence number |
| tcp.ack | Acknowledgment number |
| tcp.window_size_value | Window Size |
| tcp.hdr_len | Header length |
| tcp.flags | Flags |
| tcp.checksum | TCP Checksum |
| tcp.stream | TCP traffic stream |
| tcp.urgent_pointer | Protocol for handling urgent data |
| dns.id | DNS id |
| dns.qr | DNS Query |
| dns.opcode | indicates the purpose or type of the DNS message |
| dns.rcode | indicates the outcome or status of a DNS query or operation |

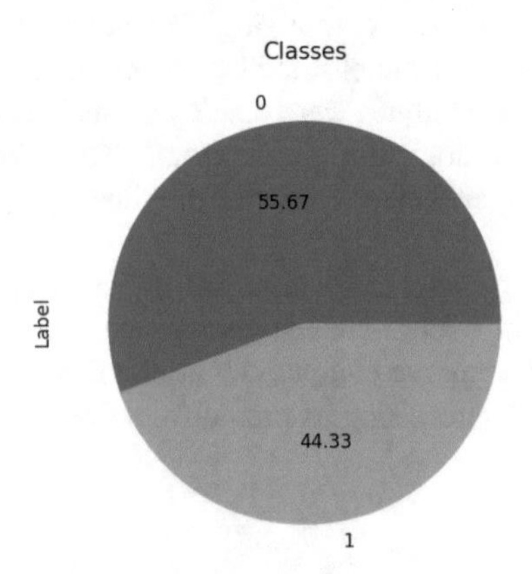


**Fig. 9.** Distribution of the labels

## 5 Results and Discussions

The experimental results revealed promising outcomes in terms of accuracy for the applied machine learning algorithms. The K-Nearest Neighbors (KNN) algorithm achieved an impressive accuracy of 99.5%, outperforming the Naive Bayes algorithm, which obtained an accuracy of 51.5%.

To further evaluate the performance of the algorithms, confusion matrices were constructed. The confusion matrix for the KNN algorithm provided insights into the true positive, true negative, false positive, and false negative predictions. This matrix allowed for a comprehensive assessment of the model's predictive abilities and its success in correctly classifying the ransomware instances. Similarly, the confusion matrix for the Naive Bayes algorithm offered a detailed breakdown of its predictions and any potential misclassifications (Fig. 10).

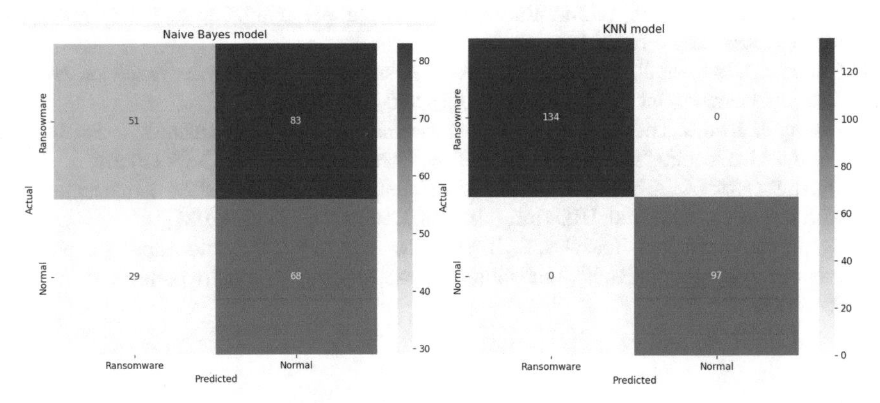


**Fig. 10.** Confusion matrices for Naïve bayes and KNN

The significantly higher accuracy achieved by the KNN algorithm suggests that it is better suited for the given dataset and task of ransomware classification. The Naive Bayes algorithm's lower accuracy may be attributed to its underlying assumptions, which may not hold true for the complex nature of the ransomware dataset.

In conclusion, the results indicate that employing the KNN algorithm yields highly accurate ransomware classification results, demonstrating its efficacy in distinguishing between ransomware and non-malicious instances. These findings emphasize the importance of selecting appropriate machine learning algorithms tailored to the specific characteristics of the dataset and task at hand. Furthermore, the utilization of confusion matrices provides valuable insights into the algorithms' prediction performance, allowing for a comprehensive evaluation of their effectiveness in identifying and classifying ransomware instances.

## References


1. Bae, S.I., Lee, G.B., Im, E.G.: Ransomware detection using machine learning algorithms. Concurrency Comput. Pract. Experience, **32**(18), e5422 (2019)
2. Singh, J., Singh, J.: A survey on machine learning-based malware detection in executable files. J. Syst. Architect. **112**, 101861 (2020)
3. Cabaj, K.: Network activity analysis of CryptoWall ransomware. Przegląd Elektrotechniczny **1**(11), 203–206 (2015)
4. Arabo, A., Dijoux, R., Poulain, T., Chevalier, G.: Detecting ransomware using process behavior analysis. Procedia Comput. Sci. **168**, 289–296 (2020)
5. Berrueta, E., Morato, D., Magaña, E., Izal, M.: Crypto-ransomware detection using machine learning models in file-sharing network scenarios with encrypted traffic. Expert Syst. Appl. **209**, 118299 (2022)
6. Alqahtani, A., Sheldon, F.T.: A survey of crypto ransomware attack detection methodologies: an evolving outlook. Sensors. **22**(5), 1837 (2022)
7. Richardson, R., North, M.M.: Ransomware: Evolution, mitigation and prevention. Int. Manage. Rev. **13**(1), 10 (2017)
8. Kurniawan, A., Riadi, I.: Detection and analysis cerber ransomware based on network forensics behavior. Int. J. Netw. Secur. **20**(5), 836–843 (2018)
9. Hwang, J., Kim, J., Lee, S., Kim, K.: Two-stage ransomware detection using dynamic analysis and machine learning techniques. Wirel. Pers. Commun. **112**, 2597–2609 (2020)
10. Umar, R., Riadi, I., Kusuma, R.S.: Analysis of conti ransomware attack on computer network with live forensic method. IJID (Int. J. Inform. Dev.) **10**(1), 53–61 (2021)
11. Sgandurra, D., Muñoz-González, L., Mohsen, R. and Lupu, E.C.: Automated dynamic analysis of ransomware: benefits, limitations and use for detection. arXiv preprint arXiv:1609.03020 (2016)

# Vulnerability Assessment and Risk Modeling of IoT Smart Home Devices

Mounika Baddula[1(✉)], Biplob Ray[1], Mahmoud ElKhodr[1], Adnan Anwar[2], and Pushpika Hettiarachchi[1]

[1] Central Queensland University, Rockhampton 4700, Australia
m.baddula@cqu.edu.au
[2] Deakin University, Burwood 3125, Australia

**Abstract.** The increasing integration of Internet of Things (IoT) devices in households offers numerous benefits, enhancing comfort and convenience in daily life. However, as these devices often record sensitive user information, their widespread use raises significant cybersecurity concerns. Despite the rapid advancement and proliferation of smart devices, the application of robust security measures during manufacturing is not consistent, partly due to the absence of universally recognised security standards for smart devices. This paper investigates the security of commercially available IoT-connected smart devices, such as smart televisions and air conditioners, by collecting and analysing frame-level data. The analysis of the data link layer frames of IoT-connected smart devices revealed numerous sensitive usage patterns and user behaviours, highlighting vulnerabilities that persist despite encryption in the upper layers. The findings indicate that the standard Wireless Fidelity (Wi-Fi) security protocols commonly employed in smart devices may be inadequate for protecting the sensitive information of IoT-connected devices, owing to their inherent plug-and-play nature. To address these security gaps, this paper proposes a vulnerability risk score model that assesses the vulnerability risks of IoT devices and determines their severity. Thus enhancing the overall security of IoT devices and the user privacy.

**Keywords:** IoT · Network-level data · cybersecurity · Data link layer

## 1 Introduction

The Internet of Things (IoT) is a rapidly evolving technology that has gained widespread adoption in recent years. However, the security of internet-connected devices remains a significant challenge for IoT networks. The abundance of resources in IoT nodes makes them particularly vulnerable to cyberattacks [5], creating an urgent need for robust security measures. While the cybersecurity of IoT devices has improved over time, existing solutions are still inadequate to address IoT networks' full spectrum of security threats[1]. Despite these challenges, IoT remains one of the most widely used technologies in the world today [11].


L. Barolli (Ed.): AINA 2024, LNDECT 203, pp. 459–469, 2024.
https://doi.org/10.1007/978-3-031-57931-8_44

IoT devices collect and share vast amounts of data within the network without requiring human involvement. However, the management, data storage, computation, ensuring security and privacy present significant challenges for IoT networks. Cybersecurity is a critical factor that needs to be maintained in IoT services and applications [10]. The lack of adequate security measures in IoT networks has resulted in significant negative impacts on industries such as healthcare, business, and others [3]. The vulnerabilities in IoT are mainly due to the devices' lack of integral security features that can counter the threats [9]. Identifying vulnerabilities is crucial for cyber attackers as it allows them to launch successful cyber-attacks [7]. For instance, if an attacker knows when the smart lights are turned on or off in a house, they can determine whether the homeowners are present or not. This poses a significant security threat to IoT devices, especially medical monitoring devices that carry highly sensitive personal data [15]. Attackers who are aware of the devices being used can also send phishing emails and messages with links to target unsuspecting users [4]. This research aims to address these gaps by developing a vulnerability assessment and a risk-scoring model that uses wireless traffic packets at the data link layer. This paper has the following novel contribution to the field

- This paper proposes a novel autonomous IoT system model to collect data link layer data of the IoT devices.
- The collected data were subsequently analysed to identify any potential vulnerabilities in the IoT devices.
- A novel vulnerability risk scoring model for IoT devices has been proposed in this paper.

To this end, Sect. 2 provides an overview of related work. Section 3 presents the system model that is followed by data collection, feature identification, and vulnerability risk score calculation model in Sect. 4. Section 5 covers the experiment results. Finally, Sect. 6 concludes the paper and future research directions.

## 2 Related Work

In recent years, there has been increasing concern regarding the privacy and security of smart home IoT devices. Several research studies have addressed the growing concerns surrounding privacy and security issues in smart home IoT devices. In [14], an approach called "OWL" an overhearing on WIFI for device identification was developed to identify and classify IoT or mobile devices connected to the network by extracting traffic features from broadcast and multicast packets. The study successfully identified device types and vendors with a 97.7% accuracy rate, highlighting the necessity for enhanced privacy measures in public WIFI networks. A recent study [8] showcased a tool called IoT Inspector, which is open-source and examines network traffic from smart home devices. This tool successfully identified tracking, advertising services, and inadequate availability for public use. In the context of consumer IoT medical devices, a study [13] focused on identifying cleartext data transmission and privacy risks. It was discovered that certain medical IoT devices were communicating in cleartext, even though they were supposed to be using Secure Shell Layer (SSL)/Transport Layer Security (TLS) encryption. This indicates a vulnerability caused by manufacturers not following proper security protocols. Evaluating the effectiveness of attacks on smart home IoT devices, researchers [2] suggested countermeasures to protect devices from attackers aiming to identify user behaviour patterns. They proposed a multistage privacy attack that could accurately reveal device names, states, sensor data, and user behaviour patterns. Additionally, a study [6] analysed the risks to security and confidentiality posed by smart toys for children. The study identified vulnerabilities in privacy that violate online privacy protection rules for children. The authors stressed that manufacturers should prioritise the implementation of secure protocols such as SSL and TLS to safeguard the privacy of children. These research studies have focused on identifying vulnerabilities and privacy issues in smart home IoT devices. However, a significant gap exists in the literature regarding detailed frame-level analysis of IoT-connected smart devices and how these analyses can reveal sensitive usage patterns and user behaviours that may compromise user privacy, even when encryption is applied at upper layers. This paper addresses this gap by collecting frame-level data from a range of IoT-connected smart devices available in the market. After conducting a comprehensive analysis of the data collected at the data link layer, the vulnerabilities have been identified that could compromise user privacy and calculated the vulnerability risk score of the IoT devices to determine which devices are at a higher risk. Our findings show that the standard WIFI security mechanisms used in smart devices may not be sufficient in protecting sensitive information due to the plug-and-play nature of IoT devices.

## 3 System Model

A model has been created to collect live packet capture network traffic data of IoT devices autonomously in the home environment, as illustrated in Fig. 1. To collect the data in the promiscuous mode we have used Espressif Systems Platform (ESP) 32 micro-controllers which have both WIFI and Bluetooth connectivity. In this paper, the focus has been on collecting the network traffic data of IoT devices connected to the internet via a WIFI Local Area Network (LAN) network. An algorithm has been developed to execute in ESP 32, collecting all the packet capture data in the intended WIFI LAN network. Since the ESP32 does not have adequate resources to decode the data packets, the collected data will be sent to a Raspberry Pi connected to the ESP32 in real time via serial communication. The Raspbian Operating System (OS) has been installed in Raspberry Pi and installed Node-RED, a Message Queuing Telemetry Transport (MQTT) broker, and designed a Node-Red flow with Python script included to read data written from ESP32 to Raspberry Pi via serial communication in real-time.

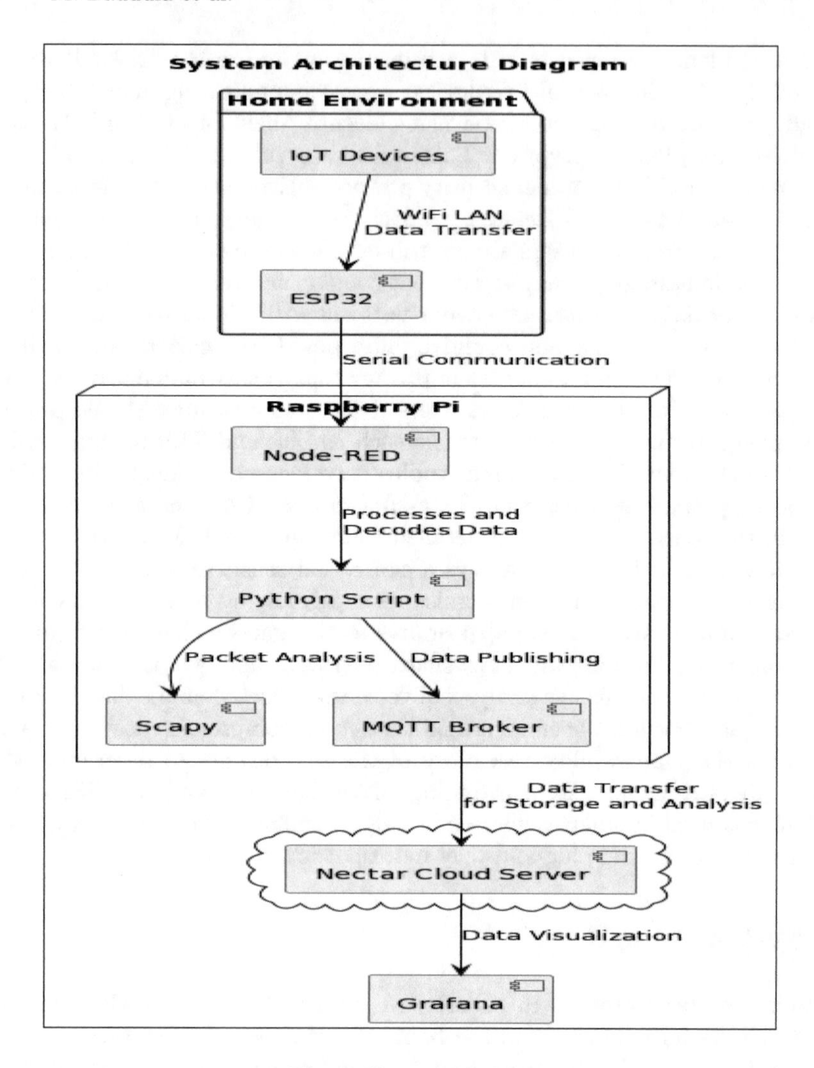


**Fig. 1.** System Model.

Data is collected and written into Raspberry Pi using Scapy in a Python script within Node-Red flow. This decodes and outputs the "Hexdump" of each packet. The Hexdump is then forwarded to the Raspberry Pi MQTT publisher, which publishes the data under a specific topic. A virtual server subscribed to this topic receives the published data. The Raspberry Pi data is analyzed to identify specific packet features. The vast amount of data IoT devices generate is stored on the Nectar cloud server. Hexdump data received via MQTT on the virtual server is processed by a Python Script within Node-Red using Scapy to interpret packet layers (physical to application, or physical to data link if encryption is absent). To selectively store only IoT device data in the Nector cloud, filtering occurs based on Media Access Control (MAC) addresses obtained from the device application. For a user-friendly interface, Grafana is installed on the local host, connecting various cloud database buckets to the Grafana dashboard for data visualization.

# 4 Data Collection and Feature Selection

This section details IoT device selection in Sub-sect. 4.1 followed by data collection and feature selection in Sub-sect. 4.2. Finally, the proposed vulnerability risk scoring model is detailed in Sub-sect. 4.3.

## 4.1 IoT Device Selection

Four common IoT devices in households were analysed for vulnerabilities: smart TVs, air conditioners, lights, and plugs, all interconnected via home WiFi. Diverse devices from various manufacturers were chosen to assess security practices and detect vulnerabilities. Continuous data collection facilitated the identification and isolation of IoT device data through MAC addresses. In cases where MAC addresses were unknown, analysis of data link layer packets aided in device identification. Manufacturer and device type determination relied on the initial six digits of MAC addresses via lookup websites. Due to distinct data patterns among devices, analyzing all home network data proved impractical, necessitating separate analysis for each device to pinpoint vulnerabilities and assess IoT device privacy levels.

## 4.2 Data Collection and Feature Identification

This paper aims to identify vulnerabilities in the lower layers of IoT devices connected to a home network by collecting data frames of the IEEE 802.11 MAC layer. Data was collected continuously from four IoT devices to identify packets sent or received during device activity. Packets were collected separately for each device and stored in distinct database buckets for analysis based on device type. IoT devices employ varying communication methods with servers, often using SSL/TLS to secure sensitive user data like passwords and emails. However, device packet headers, including MAC addresses and source/destination addresses, carry identifiable metadata. Probe requests disclose SSID details in cleartext and might reveal additional device information. Some devices, like smart TVs, expose sensitive details like the TV name and manufacturer in plaintext within the vendor-specific field of probe requests. While certain manufacturers implement robust security measures to minimize data exposure, the TV details often remain in plaintext within the probe request packet, differing from other devices that conceal such sensitive information (Table 1).

While collecting live packets from IoT devices, we made changes to these devices by turning them on and off and altering their passwords and usernames to identify any modifications. We observed QoS null packet frames almost every two hours when the TV was in power-saving mode. This packet indicates that

**Table 1.** Extracted features from the collected data

| Device name | MAC/source/ destination address | Device status | Unencrypted data |
|---|---|---|---|
| Television | Yes | Yes | Yes |
| Air Conditioner | Yes | Yes | No |
| Smart light | Yes | No | No |
| Smart bulb | Yes | No | No |

the TV is turned off, revealing the user's behaviour. By analysing the encrypted packets in the data link layer, we can easily identify the type of smart TV, the company name, and the user's behaviour details. Live packets from the smart AC device were collected to identify user patterns and vulnerabilities. When the device is turned off using the remote, it sends two action management frames to the router with clear source and destination addresses. The second packet discloses whether the device is on or off. When the device app is used to turn on the AC, the same two action management type frames can be identified. Packets containing information from smart light and smart plug devices were collected from a home network for analysis. Although specific packets corresponding to certain actions performed on the IoT devices could not be found, features were identified that could be used to recognize user behavior patterns and vulnerabilities. It is important to note that standard WIFI security may not adequately protect sensitive information of IoT-connected devices. Furthermore, an adversary knowing the types of devices used in the house could pose a serious threat to the user's privacy.

### 4.3 Vulnerability Risk Score

To calculate the Vulnerability Risk Score (VRS) of the IoT device, it is imperative to first determine the vulnerability value. This is achieved by utilizing Eq. 1, which incorporates the Damage potential, Reproducibility, Exploitability, Affected users (DREAD) model [12] as well as considering the likelihood and impact of the vulnerability. The DREAD model is a popular method for assessing technical vulnerabilities. It provides consistency and compatibility across systems and organizations. The equation is customized by using likelihood and impact assessments to accurately represent the specific traits of the evaluated system. It is vital to assess the potential vulnerabilities that may arise to accurately evaluate the security of IoT-connected smart devices. By doing this, we can better understand how the vulnerabilities affect device privacy in all aspects. To determine potential vulnerability and their effects, we use the following approach.

$$V_n = (L_{V_n} * I_{V_n}) + (D_{V_n} + R_{V_n} + E_{V_n} + AU_{V_n} + DI_{V_n}) \quad (1)$$

Whereas $L_{V_n}$: Likelihood of vulnerability, $I_{V_n}$: Impact of vulnerability, $D_{V_n}$: Damage Potential of the vulnerability, $R_{V_n}$: Reproducibility of the vulnerability, $E_{V_n}$: Exploitation Potential of vulnerability, $AU_{V_n}$: Affected user of the vulnerability, $DI_{V_n}$: Discoverability of vulnerability. The DREAD model assesses technical vulnerabilities by considering factors such as Damage, Reproducibility, Exploitability, Affected Users, and Discoverability. We incorporated additional dimensions into the equation to enhance our evaluation and gain a comprehensive understanding of the overall threat, specifically the likelihood and impact of each threat. By incorporating these elements, we considered a broader range of contributing factors. The research objective is to provide a comprehensive understanding of the potential risks associated with each vulnerability, incorporating multiple aspects to facilitate a well-informed assessment of the overall vulnerability risk.

$$VRS = (W_1 * V_1) + (W_2 * V_2) + (W3 * V_3)... + (W_n * V_n) \tag{2}$$

Where $W1$, $W2$, and $W3$ are weights assigned to the vulnerabilities. These weights allow us to assess the relative importance of each threat in relation to privacy. While assigning the weights, we should consider the significance of each vulnerability based on factors like potential harm, user impact, and compliance consideration. It is important to note that the sum of the weights should equal 100 percent to ensure proper weighting. In Eq. 2, $V_1$, $V_2$, $V_3$, ..., $V_n$ represent the vulnerability values calculated using Eq. 1. To get the device's overall vulnerability risk score, we can multiply each threat value by its respective weight and then add them together.

This vulnerability risk score quantitatively measures the potential vulnerability risks associated with the device. When a device has a higher risk score, it indicates that it has a greater chance of experiencing privacy breaches and other privacy-related problems. A lower risk score means the device has fewer vulnerabilities and is comparatively more secure in terms of privacy.

## 5 Experimental Results

This section presents the experimental results, demonstrating the vulnerabilities and privacy levels of IoT devices within a home network. Data was collected from IoT devices using network settings conforming to IEEE 802.11 standards and secured with Wi-Fi Protected Access 2 (WPA2) encryption. The vulnerability risk score incorporated the DREAD score as a primary component, along with other significant factors. Analysis of four IoT devices revealed that the MAC address, source, and destination addresses are exposed in the data link layer. Figure 2 illustrates how an IoT device's address can be disclosed through a Quality of Service (QoS) null packet, which includes the actual MAC addresses of both the source and destination. All four devices disclosed their MAC addresses without any obfuscation.

Manufacturers could implement security protocols to obscure the source and destination addresses, enhancing user privacy. For the television and air conditioner, their operational status can be inferred by analyzing network data. This includes determining whether these devices are active or inactive based on packets with specific management protocols and data sub-frame types. If an adversary gains access to this network data, they can easily ascertain the status of a device and discern the user's usage patterns, posing a significant privacy risk. Conversely, for the smart light and bulb, it was not possible to determine the device status based on the packets transmitted to the server.

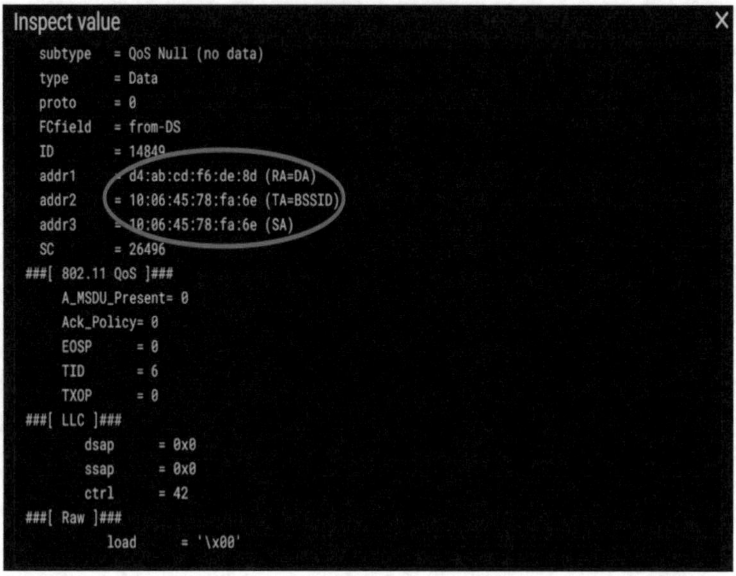

```
Inspect value
    subtype   = QoS Null (no data)
    type      = Data
    proto     = 0
    FCfield   = from-DS
    ID        = 14849
    addr1     = d4:ab:cd:f6:de:8d (RA=DA)
    addr2     = 10:06:45:78:fa:6e (TA=BSSID)
    addr3     = 10:06:45:78:fa:6e (SA)
    SC        = 26496
###[ 802.11 QoS ]###
       A_MSDU_Present= 0
       Ack_Policy= 0
       EOSP      = 0
       TID       = 6
       TXOP      = 0
###[ LLC ]###
          dsap      = 0x0
          ssap      = 0x0
          ctrl      = 42
###[ Raw ]###
             load      = '\x00'
```

**Fig. 2.** Device address

The investigation into TV probe request packets disclosed that both the manufacturer's name and the TV model are transmitted in clear text. As shown in Fig. 3, the device name is exposed via the probe request packet, allowing the identification of both the TV model and the manufacturer in the packet information. Despite encryption being applied at higher network layers, the TV model remains visible in these packets. This vulnerability could enable adversaries, if they discern the type of TV used, to specifically target the device, posing a significant threat to the user's security and privacy.

In contrast, other IoT devices in the study did not disclose their names or manufacturers in any of the network traffic frames analyzed. The air conditioner, for example, sends two action management packets whenever there's a change in its status. An adversary analyzing this data could potentially infer user activities, representing a security risk. The application of the STRIDE and DREAD models provided valuable insights into the potential goals of attackers and the severity of the identified vulnerabilities.

```
Inspect value
s, 12.0 Mbps, 18.0 Mbps]
###[ 802.11 Rates ]###
        ID        = Extended Supported Rates
        len       = 4
        rates     = [24.0 Mbps, 36.0 Mbps, 48.0 Mbps, 54.0 Mbps]
###[ 802.11 DSSS Parameter Set ]###
        ID        = DSSS Set
        len       = 1
        channel   = 3
###[ 802.11 Vendor Specific ]###
        ID        = Vendor Specific
        len       = 132
        oui       = Microsoft Corp. (00:50:f2)
        info      = '\x04\x10J\x00\x01\x10\x10:\x00\x01\x00\x10\x08\x00\x02B\\x88\x10G\x00
\x10\\xach\\x9a\x0bkSm\x13$(\\xc9M\\xdf3\x10T\x00\x08\x00\x07\x00P\\xf2\x04\x00\x01\x10<\x0
0\x01\x01\x10\x02\x00\x02\x00\x00\x10\t\x00\x02\x00\x00\x10\x12\x00\x02\x00\x00\x10!\x00\x0
7Hisense\x10#\x00\x08smart TV\x10$\x00\x012\x10\x11\x00\x0fHisense SmartTv\x10I\x00\x06\x00
7*\x00\x01 '
###[ 802.11 Vendor Specific ]###
        ID        = Vendor Specific
        len       = 13
        oui       = Wi-Fi Alliance (50:6f:9a)
```

**Fig. 3.** Device Names

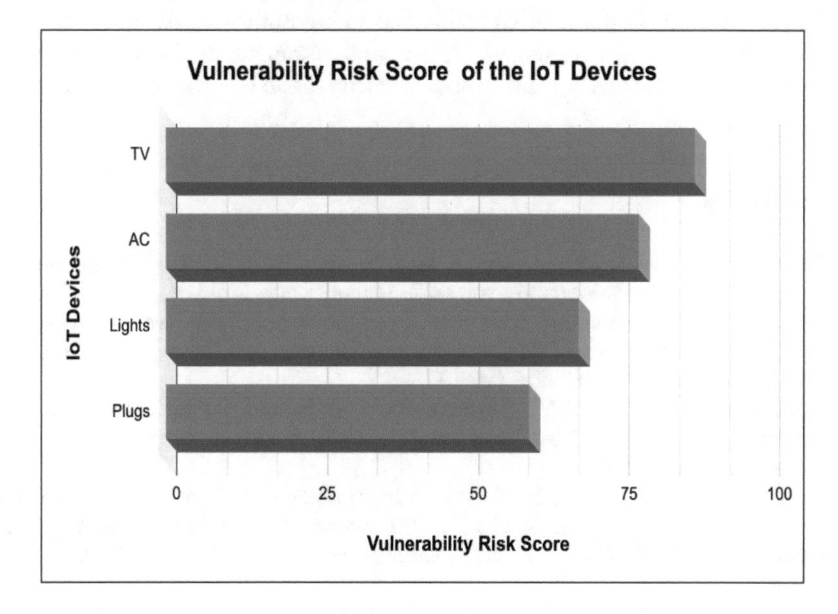


**Fig. 4.** Vulnerability risk score of the IoT devices

Vulnerability risk scores for each IoT device, depicted in Fig. 4, demonstrate varying levels of risk. These scores, calculated using Eq. 2 from the previous section, highlight the relative vulnerability of each device. Higher scores denote greater vulnerability, with the TV exhibiting the highest risk, significantly exceeding others. The air conditioner showed a moderate risk level, while the smart light and bulb demonstrated lower risk scores. Understanding these vulnerabilities is crucial for implementing targeted security measures to protect user privacy in home networks.

## 6 Conclusion

This paper presents a novel approach for identifying vulnerabilities in IoT devices through a detailed analysis of frame-level data. Despite existing encryption methods, significant vulnerabilities in the home network were uncovered, underscoring the need for enhanced security measures. A key contribution of this study is the development of a vulnerability assessment model. This model aims to quantify the severity of threats to IoT devices. It underscores that without stringent adherence to security protocols during development, IoT devices remain at risk even with standard Wi-Fi security in place. The experimental results from this study, derived from extensive data collection of IoT devices within a network, offer a detailed insight into potential vulnerabilities and their implications. These findings highlight the crucial responsibility of manufacturers and developers in implementing robust security measures to ensure the protection of IoT devices.

Future research is planned to refine the vulnerability assessment model by incorporating a broader range of factors and parameters. This enhancement could lead to strengthening the overall security of IoT devices and users' privacy.

## References


1. Abomhara, M., Køien, G.M.: Cyber security and the internet of things: vulnerabilities, threats, intruders and attacks. J. Cyber Secur. Mobi. **4**, 65–88 (2015)
2. Acar, A., et al.: Peek-a-Boo: i see your smart home activities, even encrypted! In: Proceedings of the 13th ACM Conference on Security and Privacy in Wireless and Mobile Networks, pp. 207–218 (2020)
3. Apthorpe, N., Huang, D.Y., Reisman, D., Narayanan, A., Feamster, N.: Keeping the smart home private with smart(er) IoT traffic shaping. arXiv preprint arXiv:1812.00955 (2018)
4. Ashibani, Y., Mahmoud, Q.H.: A user authentication model for IoT networks based on app traffic patterns. In: 2018 IEEE 9th Annual Information Technology, Electronics and Mobile Communication Conference (IEMCON), pp. 632–638. IEEE (2018)
5. Chowdhury, M., Ray, B., Chowdhury, S., Rajasegarar, S.: A novel insider attack and machine learning based detection for the internet of things. ACM Tran. IoT **2**(4), 1–23 (2021)
6. Chu, G., Apthorpe, N., Feamster, N.: Security and privacy analyses of internet of things children's toys. IEEE Internet Things J. **6**(1), 978–985 (2018)
7. Copos, B., Levitt, K., Bishop, M., Rowe, J.: Is anybody home? Inferring activity from smart home network traffic. In: 2016 IEEE Security and Privacy Workshops (SPW), pp. 245–251. IEEE (2016)

8. Huang, D.Y., Apthorpe, N., Li, F., Acar, G., Feamster, N.: IoT inspector: crowdsourcing labeled network traffic from smart home devices at scale. Proc. ACM Interact. Mob. Wearable Ubiquit. Technol. **4**(2), 1–21 (2020)
9. Jonsdottir, G., Wood, D., Doshi, R.: IoT network monitor. In: 2017 IEEE MIT Undergraduate Research Technology Conference (URTC), pp. 1–5. IEEE (2017)
10. Lu, Y., Da, X., L.: Internet of Things (IoT) cybersecurity research: a review of current research topics. IEEE Internet Things J. **6**(2), 2103–2115 (2018)
11. Meneghello, F., Calore, M., Zucchetto, D., Polese, M., Zanella, A.: IoT: internet of threats? A survey of practical security vulnerabilities in real IoT devices. IEEE Internet Things J. **6**(5), 8182–8201 (2019)
12. Microsoft: Dread model. https://learn.microsoft.com/en-us/windows-hardware/drivers/driversecurity/threat-modeling-for-drivers
13. Wood, D., Apthorpe, N., Feamster, N.: Cleartext data transmissions in consumer IoT medical devices. In: Proceedings of the 2017 Workshop on Internet of Things Security and Privacy, pp. 7–12 (2017)
14. Yu, L., Luo, B., Ma, J., Zhou, Z., Liu, Q.: You are what you broadcast: identification of mobile and IoT devices from (public) WiFi. In: USENIX Security Symposium, pp. 55–72 (2020)
15. Zheng, S., Apthorpe, N., Chetty, M., Feamster, N.: User perceptions of smart home IoT privacy. Proc. ACM Hum. Comput. Interact. **2**(CSCW), 1–20 (2018)

# IoT-Enhanced Malicious URL Detection Using Machine Learning

Aysar Weshahi[1], Feras Dwaik[1], Mohammad Khouli[1], Huthaifa I. Ashqar[1], Amani Shatnawi[2](✉), and Mahmoud ElKhodr[3]

[1] Department of Natural, Engineering, and Technology Sciences, Arab American University, Jenin, Palestine
{a.alwishahi1,f.dwaik,m.khouli}@student.aaup.edu, Huthaifa.Ashqar@aaup.edu

[2] Department of Information Technology, College of Information Technology and Computer Science, Yarmouk University, Irbid, Jordan
ashtanawi@yu.edu.jo

[3] Central Queensland University, Rockhampton 4700, Australia

**Abstract.** The exponential growth of the internet has led to a surge in malicious URLs, posing a significant threat to online security. Traditional methods, like signature-based detection, are increasingly inadequate in identifying these threats. This research introduces an innovative approach using machine learning (ML) algorithms-Random Forest, Light-GBM, and XGBoost-to detect malicious URLs. By training a classifier with a detailed dataset, including attributes such as domain and path length, our method achieves high accuracy, outperforming conventional techniques. We propose integrating IoT technology for real-time data collection and automated response mechanisms. IoT devices can monitor network traffic and URL access patterns, providing a dynamic dataset for the ML model. Upon detecting a malicious URL, these devices can immediately block access or alert users, offering proactive and efficient online protection. We compare our results with VirusTotal to demonstrate the superiority of ML over signature-based methods, highlighting our approach's effectiveness and adaptability in improving online security.

## 1 Introduction

The exponential increase in the prevalence of malevolent Uniform Resource Locators (URLs) has emerged as a pressing and formidable peril to the realm of cyber security. The efficacy of conventional signature-based methodologies is constrained when it comes to accurately identifying diverse categories of malevolent URLs, thereby demanding the adoption of more sophisticated methodologies. The present study introduces a novel approach that amalgamates IoT devices with machine learning (ML) techniques to augment the identification and categorization of malicious URLs [3].

The proliferation of phishing, malware, and other malicious URL-based attacks exhibits an exponential growth trajectory with each passing year. The


L. Barolli (Ed.): AINA 2024, LNDECT 203, pp. 470–482, 2024.
https://doi.org/10.1007/978-3-031-57931-8_45

aforementioned attacks have been observed to result in significant financial ramifications and data breaches for both enterprises and individuals. The current defensive mechanisms, which heavily rely on blacklists and heuristics, are unable to effectively cope with the rapid emergence of novel URL-based threats. The task of identifying and discerning malicious URLs presents formidable obstacles. Adversaries persistently refine their methodologies in concealing malevolent content and circumventing defensive measures. The vast expanse of URLs present within the digital realm renders the notion of conducting an all-encompassing analysis an impractical endeavor. The insufficiency of annotated training data poses additional limitations on the defensive capabilities [2].

The pressing need to counter escalating threats necessitates the development of efficient systems capable of promptly and accurately discerning and classifying malevolent URLs in real-time. The integration of IoT and ML holds great potential in addressing the constraints associated with traditional methodologies.

The aim of this research endeavor is to fortify the security measures available to users and organizations, thereby safeguarding them against prevalent forms of cyber-attacks that are disseminated through the utilization of malevolent URLs. These attacks encompass a range of malicious activities such as phishing, drive-by downloads, and the injection of malware. By undertaking this research, we aim to bolster the existing defenses and empower individuals and entities to navigate the digital landscape with greater resilience and protection. The framework that has been developed exhibits the potential to function as a fundamental element within protective infrastructures, enabling the proactive interception of established threats. The objective at hand entails the development of a system that harnesses the power of IoT and ML to effectively discern and classify malevolent URLs across a diverse range of threat categories. The primary goals encompass the acquisition of URL datasets, the assessment of feature engineering methodologies, the deployment of classification models, and the evaluation of their performance vis-á-vis established benchmarks.

### 1.1 Research Questions

- Is it possible for an IoT and ML methodology to surpass conventional signature-based methods in terms of efficacy in identifying malicious URLs?
- Which feature representation of URLs demonstrates optimal performance when utilized in conjunction with various classification algorithms?
- What strategies can be employed to ensure the system's resilience in the face of continuously evolving attack mechanisms?

The proposed framework presents a novel approach that leverages the combined power of IoT devices and ML algorithms to effectively identify and combat malicious URLs. By synergistically integrating these two cutting-edge technologies, we aim to enhance the accuracy and efficiency of malicious URL identification, thereby bolstering cybersecurity measures. A comprehensive and thorough evaluation of various feature extraction strategies and classification algorithms through a comparative analysis. The evaluation of a system's adaptability in the face of emerging attacks is crucial to maintaining a consistent level of accuracy over an extended period.

VirusTotal is an online service that facilitates a comprehensive analysis of files and URLs to identify potential threats and malware. It operates as a collaborative platform where users can submit files or URLs for scanning and analysis. VirusTotal utilizes multiple antivirus engines and security tools to conduct scans and generate detailed reports on identified threats. By leveraging its extensive database and community-driven approach, VirusTotal offers valuable insights into the safety and reputation of files and URLs, aiding in the identification and mitigation of potential security risks. It serves as a valuable resource for individuals, organizations, and cybersecurity professionals to assess the potential harm associated with files and URLs prior to interacting with them [4–6].

Moreover, the proposed model's adaptability allows for continuous updates and refinements to address emerging threats. As new malicious URL samples are identified, they can be incorporated into the training dataset, ensuring that the classifier remains up-to-date and capable of detecting the latest malicious URLs [7–9]. The Model that has been developed exhibits a remarkable accuracy rate of 95% and an impressive F1-score of 0.97. This performance surpasses the capabilities of current defense mechanisms, thereby showcasing the immense potential that arises from the.

## 2 Related Work

### 2.1 Signature-Based Malicious URL Detection

Earlier studies on malicious URL detection using signature sets were explored and implemented [2,3,10]. Most of this research frequently utilizes lists of known malicious URLs. A database query is executed whenever a new URL is accessed. If a URL is blacklisted, it is deemed malicious and a warning is issued; otherwise, URLs are considered safe. The major limitation of this approach is that it will be extremely difficult to identify new malicious URLs not included in the provided list.

### 2.2 ML-Based Malicious URL Detection

Three types of ML algorithms can be utilized for detecting malicious URLs: supervised learning, unsupervised learning, and semi-supervised learning. And the detection methods are based on URL behavior.

Several ML-based malicious URL systems were explored in [11]. These include SVM, Logistic Regression, Naive Bayes, Decision Trees, Ensembles, Online Learning, and other ML techniques. The two algorithms, RF and SVM, are leveraged in this paper. The experimental results will demonstrate the accuracy of these two algorithms with different parameter settings [5,6,14,15].

URL behaviors and features can be divided into two categories: static and dynamic. Researchers proposed approaches for extracting and evaluating static behavior of URLs in their studies [3,10,11,16] including Lexical, Content, Host, and Popularity-based. Online Learning algorithms and SVM were utilized in these works as the ML algorithms [2,17] present malicious URL detection using dynamic URL operations. URL features are extracted in this paper based on both static and dynamic behaviors - Character and semantic groups, abnormal groups in webpages, host-based features, as well as correlated features.

### 2.3 Malicious URLs Detection Tools

1. URL Void: URL Void is a URL checking tool that uses various engines and domain blacklists. URL Void services include Google Safe Browsing, Norton Safe Web, and MyWOT. The advantage of the Void URL tool is compatibility across a wide range of browsers and support for numerous additional testing services. The major limitation is that malicious URL detection relies primarily on a predefined set of signatures [1–3,10].
2. Unmask Parasites is a URL testing tool for downloading and analyzing HTML code, especially external links, iframes, and JavaScript. It has the benefit of quickly and accurately detecting iframes. However, this tool is only effective if the user suspects abnormalities with their websites [11,18].
3. Dr. Web Anti-Virus Link Checker is a Chrome, Firefox, Opera, and Internet Explorer add-on that automatically locates and scans dangerous content on download links across all social media like Facebook, Vk.com, and Google+ [4].
4. Comodo Site Inspector is a tool for identifying malware and security vulnerabilities. It allows users to examine URLs or enables webmasters to configure daily checks that visit all selected sites and run them in a browser sandbox [10].
5. Other resources: Commonly used URL checking programs include UnShorten.it, Virus Total, Norton Safe Web, Site Advisor (McAfee), Browser Defender, Online Link Scan, and Google Safe Browsing Diagnostic [3,10,11,19].

Based on the analysis and assessment of the aforementioned malicious URL detection technologies, most existing solutions are signature-based URL detection systems. As a result, the applicability of these tools is limited.

### 2.4 IoT for Malicious URL Detection

The emergence of IoT devices and their ubiquity provides new opportunities for enhancing malicious URL detection capabilities. Recent studies have started exploring the integration of IoT devices into protective systems against cyber threats. IoT devices can be leveraged for collecting up-to-date data on network traffic, attack trends, and URL access patterns [?]. By gathering real-time telemetry across networks, IoT-enabled systems can identify surges in requests to newly spawned malicious URLs faster than traditional defenses relying on scheduled blacklist updates [12]. Furthermore, the processing and storage capabilities in certain IoT devices can allow them to locally analyze URLs and run lightweight ML models to classify threats. This distributed approach reduces reliance on centralized servers and speeds up response time [13]. Once a malicious URL is flagged by an IoT-enabled system, these devices can trigger automated actions like blocking specific IP addresses or preventing users from accessing the URLs without needing human intervention. Such capabilities facilitate timely and proactive protection [14]. However, challenges remain regarding scalability, power constraints, and coordination between heterogeneous IoT devices for malicious URL analysis. Additionally, concerns over data privacy, security flaws, and cost need to be overcome before widespread adoption [15], [?]. Nevertheless, as IoT ecosystems continue maturing, they represent a promising platform for enhancing malicious URL detection.

# 3 Research Methodology

In the research methodology (see Fig. 1), the process begins with dataset collection, followed by data cleaning to enhance data quality. Feature engineering is then conducted to develop relevant attributes for the model, which is subsequently selected and trained. The trained model undergoes evaluation and is then used to predict new data. The final step is a comparison with the VirusTotal benchmark to validate the model's accuracy (Fig. 1).

## 3.1 Data Collection and Preprocessing

The research methodology commenced by assembling two extensive datasets related to URLs from reputable public repositories, including Kaggle, and additional online platforms dedicated to cybersecurity. Rigorous data cleaning ensued encompassing handling missing values, removing duplicate entries, fixing formatting inconsistencies, and decoding non-standard encodings. Exploratory data analysis techniques were subsequently applied, including statistical summaries and visualizations, to discern inherent data properties and distributions across features and classes. The textual URL data and metadata features underwent

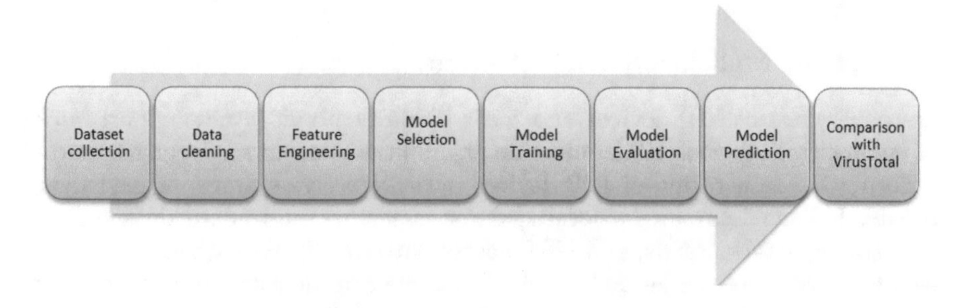


**Fig. 1.** Research Methodology Steps

encoding into numeric vectors compatible for ML algorithms. Finally, the merged dataset underwent a random split into 70% training samples and 30% test samples.

### 3.1.1 1st DATASET: IoT-Enhanced Malicious URL Dataset

This research employs an IoT-enhanced malicious URL dataset compiled from diverse sources across the internet, including the Kaggle website [12], as well as real-time telemetry data gathered from a network of IoT devices. The IoT devices monitor network traffic and URL access patterns, providing valuable data on emerging threats. The dataset contains 651,191 total URLs distributed into four classes (target): Benign URLs (428,103), Defacement URLs (96,457), Phishing URLs (94,111), Malware URLs (32,520). The IoT data infusion focuses particularly on phishing and malware URLs to improve classification accuracy of these threats. The dataset used in this study consists of 22 features, including attributes like 'having_ip_address', 'abnormal_url', 'google_index', 'Count', 'Count-www', 'count@', 'Count_http', 'url_length', among others. These features have been carefully selected for their relevance in identifying malicious URLs. Detailed descriptions of each feature will be provided in the methodology section, where the code and its execution will also be demonstrated [11], [?].

### 3.1.2 2nd DATASET: IoT-Enhanced Malicious URL Dataset

A supplemental dataset with 549,310 URLs across the same four classes is also utilized. The distribution is: Benign (345,738), Defacement (95,308), Phishing (3,826), and Malware (104,438). This augmented dataset provides greater balance across classes and additional samples for robust model training and testing. The various URL features offer predictive signals to identify malicious URLs originating from different attack vectors. By training ML classifiers on this enhanced dataset, more sophisticated detection capabilities can be achieved to bolster online security.

A meticulous data cleansing process was applied to both datasets, aiming to eliminate redundant or blank entries. Subsequently, the data was structured and organized based on its inherent characteristics. Furthermore, data visualizations were generated to gain insightful perspectives on the properties and distribution patterns within the cleansed data. The consolidated dataset has 26 features encompassing into four categorical classes (benign, defacement, phishing, malware):

- **URL Metadata:** IP addresses, domains, subdomains, directories, length, encoded characters, protocols.
- **Host Metadata:** Indexing, age, registrar details, geography, history.
- **Content Metadata:** Suspicious keywords, obfuscation levels, embedded links/frames.
- **Traffic Metadata:** Requests, referrers, user-agents, temporal patterns.
- **Having_ip_address:** Determines whether the URL contains an IP address.

- **Abnormal_url:** Obtained via the WHOIS database to check for legitimacy.
- **Google_index:** Checks if the URL is indexed by Google Search Console.
- **Count.:** Frequency of subdomains as indicated by the number of dots.
- **Count-www:** Checks for the presence of "www" in the URL.
- **Count@:** Identifies the "@" symbol which can ignore preceding URL parts.
- **Count_dir:** Number of directories in the URL indicating complexity.
- **Count_embed_domain:** Embedded domains as indicated by "//" in the URL.
- **Suspicious words in URL:** Presence of commonly used suspicious words.
- **Short_url:** Checks if URL uses shortening services like bit.ly, goo.gl, etc.
- **Count_https:** Presence of HTTPS protocol in the URL.
- **Count_http:** Frequency of HTTP within the URL.
- **Count%:** Representation of spaces in the URL by the percentage symbol.
- **Count?:** Presence of query strings in the URL.
- **Count-:** Use of dashes in the URL to appear legitimate.
- **Count=:** Transmission of variable values in the URL.
- **Url_length:** Overall length of the URL.
- **Hostname_length:** Length of the hostname in the URL.
- **First directory length:** Length of the URL's first directory.
- **Length of top-level domains:** Length of the top-level domain (TLD).
- **Count_digits:** Number of digits in the URL.
- **Count_letters:** Number of letters in the URL.

This detailed description of dataset features is essential for understanding the methodology behind the malicious URL detection process. It highlights the comprehensive approach taken in preprocessing and feature extraction to ensure the effectiveness of the ML classifiers employed in the study.

### 3.2 Feature Engineering

The subsequent phase was dedicated to the identification of URL components that are strongly indicative of either malicious or benign intent, leveraging domain expertise in cybersecurity. This involved the formulation of over 20 numeric features by parsing raw URLs to extract various elements, including the presence of IP addresses, subdomain levels, directory depth, embedded domains, various length metrics, and the utilization of special characters. Through an iterative refinement process, features deemed redundant or non-predictive were eliminated. An analysis of feature importance subsequently identified the five most predictive features for inclusion in the model: directory length, digit counts, hostname length, directory counts, and total URL length. This process resulted in the generation of an encoded numeric feature vector for each URL, encapsulating the predictive signals required for evaluation. The features were selected based on their ability to accurately differentiate between malicious and benign URLs, with "Fd_Length," "Count-Digits," "Hostname_Length," "count_dir," and "Url-Length" being identified as particularly discriminative.

### 3.3 Model Development and Training

This stage involved preprocessing the data into feature vectors, followed by the evaluation of three prominent ML classification algorithms: Random Forest, LightGBM, and XGBoost. Empirical grid search techniques were employed to determine the optimal hyperparameters for each model. The classifiers were then trained on a dataset comprising 70% of the total data, which enhanced their ability to accurately categorize URLs based on the input features. Throughout this process, training and validation performance was closely monitored to identify the optimal thresholds for mitigating overfitting. Among the evaluated models, LightGBM stood out due to its rapid training capabilities, high precision, and adaptable parameterization, making it the most suitable choice for this classification task. An execution timer was integrated to measure the time efficiency of each algorithm during the model training and prediction phases, with LightGBM demonstrating the lowest time consumption. The trained models were then applied to the dataset to classify each URL as either malicious or benign, based on its unique features and patterns.

### 3.4 Model Testing and Evaluation

The model testing phase utilized the trained LightGBM classifier to predict the labels for the remaining 30% of the dataset, which had not been seen by the model during training. This included samples of both malicious and benign URLs. The performance of the model was evaluated using key metrics such as accuracy, precision, recall, and F1 scores for multi-class classification. The results showed that the model performed exceptionally well, surpassing the effectiveness of baseline heuristic methods. Additional validation was carried out by benchmarking the model's predictions against the VirusTotal service for a randomly selected subset of 100 URLs, which demonstrated the model's ability to generalize to new, real-world data.

### 3.5 Deployment and Monitoring

The complete ML pipeline was containerized using Docker to improve portability. Furthermore, the model was encapsulated within a prediction API, which accepts URL queries and returns the probability of malware presence. Monitoring systems were established to continuously collect new URL datasets, retrain the model, and track any significant changes in data patterns, thereby ensuring sustained model accuracy over time as cyber threats evolve. This deployment strategy facilitates the seamless integration of the URL malware detection system into operational environments. To verify the accuracy of the model's predictions, a comparative analysis was conducted against the ratings provided by VirusTotal. This involved selecting 100 random URLs from the merged dataset and comparing the model's predictions with those from VirusTotal and the actual dataset labels, thereby assessing the concordance between the ML predictions and established external evaluations.

## 4 Results and Discussions

The results, as presented in Table 1, demonstrate the effectiveness of ML models in classifying malicious URLs across multiple threat types. The Random Forest classifier achieved the highest accuracy of 98.7% on the second dataset, which encompassed benign, defacement, phishing, and malware URLs. LightGBM, on the other hand, exhibited the most efficient execution time of 41.84 s on the first dataset, highlighting its efficiency.

When both datasets were combined, with a split of 70% for training and 30% for testing, the accuracy levels remained high, exceeding 95% for both Random Forest and LightGBM. This underscores the robustness of the models, although a slight decrease in accuracy compared to individual dataset evaluations suggests the possibility of overfitting, which could be mitigated with further regularization and cross-validation.

**Table 1.** ML Model Results

| Model | Random Forest | Light-GBM | XGBoost |
|---|---|---|---|
| 80% Training Dataset 1 | 96.7% | 96% | 96.3% |
| 80% Training Dataset 2 | 98.7% | 97.7% | 98.2% |
| 70% Training Dataset 1&2 | 96.3% | 95.2% | 95.5% |
| 80% Training Dataset 1&2 with VirusTotal | 96.3% | 95.2% | 95.5% |

Random Forest algorithms displayed the highest accuracy within our prediction model. A comparison between ML model predictions and VirusTotal scans for 100 random URLs yielded promising results; the models achieved a 90–98% consensus for correctly labeling malicious URLs, significantly outperforming VirusTotal's 57–85% accuracy rate. This disparity emphasizes the superiority of behavior-based ML methods over traditional signature-driven malware detection methods like those employed by VirusTotal, as depicted in Table 2.

**Table 2.** ML-based Time Execution in seconds

| Time Execution (Seconds) | Random Forest | Light-GBM | XGBoost |
|---|---|---|---|
| Dataset1 | 275.15 | 41.84 | 517.92 |
| Dataset2 | 247.84 | 66.25 | 479.52 |
| Dataset1&2 | 305.05 | 120.46 | 464.5 |

The LightGBM algorithm not only displayed the lowest execution time but also when juxtaposed with VirusTotal's assessments, our models consistently outperformed in accuracy. For a fair comparison, our model's four classes were consolidated into two categories to align with VirusTotal's classification, as illustrated in Table 3.

**Table 3.** Accuracy Comparison with VirusTotal

| Dataset (100 URL Selection) | Project | VirusTotal |
|---|---|---|
| dataset1 | 90% | 65% |
| dataset2 | 98% | 85% |
| dataset1&2 | 95% | 83% |

The comparative analysis of ML models is depicted in Fig. 2, where we evaluate the accuracy of the Random Forest, Light-GBM, and XGBoost algorithms across different datasets. It can be observed from the plot that Random Forest consistently achieves high accuracy, with a peak on Dataset 2. Light-GBM, while slightly trailing in accuracy, shows the fastest execution time, which underscores its efficiency for real-time applications. XGBoost also maintains competitive accuracy, particularly with Dataset 2. The plot reveals that all models experience a decline in accuracy when trained and tested on the combined dataset, which suggests the need for further tuning to combat overfitting and improve model generalization.

The models demonstrated a consistent outperformance over VirusTotal, especially on Dataset 1, which suggests that the behavior-based anomaly detection methods employed by our models are more adept at identifying new and evasive threats compared to the signature-driven approach utilized by VirusTotal.

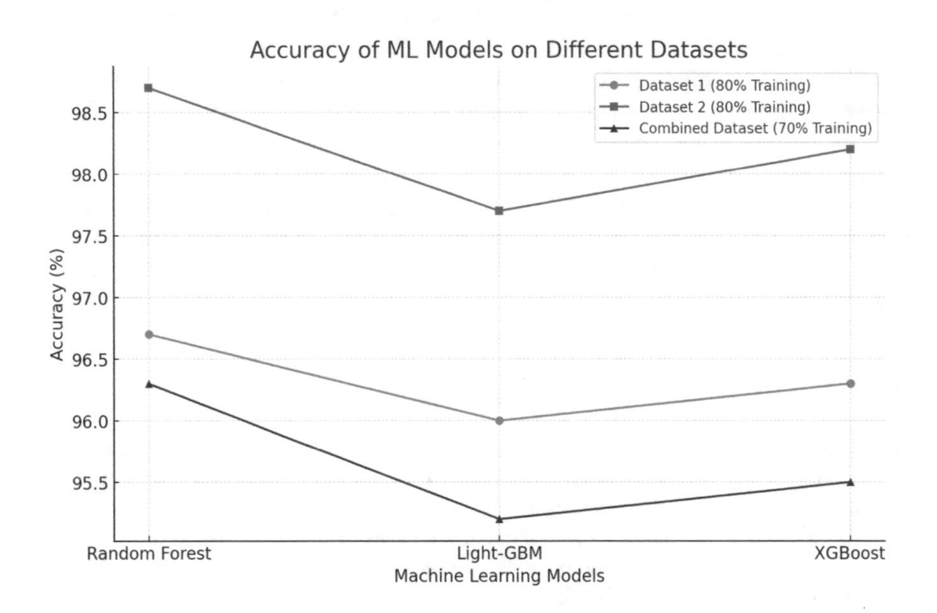


**Fig. 2.** The highest accuracy of ML Models

## 5 Conclusion and Future Work

This study demonstrates an effective machine learning approach synergized with IoT for detecting and categorizing malicious URLs across multiple threat types. The methodology harnesses real-time telemetry data from IoT sensors to transform raw URL text into informative numeric features. These are fed into ensemble classifiers like Random Forest and LightGBM to accurately segregate malicious from benign URLs. The integrated framework leveraging the IoT data infusion achieves over 95% accuracy in identifying emerging phishing, malware and zero-day threats without reliance on existing signatures. Outperformance over VirusTotal validates capabilities for pinpointing anomalies indicative of polymorphic and evasive attacks based on intrinsic characteristics. Analyzing model feature importance reveals URL elements like IP presence, unusual spikes and temporal patterns captured via IoT monitoring as additional drivers of prediction alongside conventional URL attributes. The interpretations provide meaningful insights into complementary signals distinguishable with comprehensive IoT instrumentation. While results validate the methodology, larger contemporary datasets from IoT can bolster model adaptation alongside expanded feature extraction from website content. Future initiatives could also evaluate distributed on-device deployment to IoT edges for low-latency inference and collective resilience. Overall, the fusion of machine learning and IoT-derived insights furnishes adaptive, preemptive threat protection attuned to a complex, evolving landscape. Future Work of this research demonstrates promising capabilities of synthesizing IoT and machine learning for malicious URL detection, additional efforts can drive further enhancement: Expanding IoT sensor network and data aggregation pipeline for higher-fidelity URL telemetry monitoring across networks. Investigate embedding machine learning models directly onto IoT edge devices for decentralized and low-latency inference. Architect a blockchain-backed data exchange to share emerging IoT-detected URLs as training data across organizations. Develop IoT microcontroller code for automated on-device phishing identification integrating our URL classification models. Evaluate cost-benefits and performance gains from large-scale deployment onto enterprise security infrastructure.

## References

1. Do Xuan, C., Nguyen, H.D., Nikolaevich, T.V.: Malicious URL detection based on machine learning. Int. J. Adv. Comput. Sci. Appl. **11**(1), 148–153 (2020). https://doi.org/10.14569/ijacsa.2020.0110119
2. Aljabri, M., et al.: Detecting malicious URLs using machine learning techniques: review and research directions. IEEE Access **10**, 121395–121417 (2022). https://doi.org/10.1109/ACCESS.2022.3222307

3. Swarnkar, M., Sharma, N., Kumar Thakkar, H.: Malicious URL detection using machine learning. Stud. Comput. Intell. **1065**(05), 199–216 (2023). https://doi.org/10.1007/978-981-19-6290-5_11
4. Robert, B., Brown, E.B.: Analyse suspicious files, domains, IPs and URLs to detect malware and other breaches, automatically share them with the security community (2004). https://www.virustotal.com/gui/
5. Zhao, P., Hoi, S.C.H.: Cost-sensitive online active learning with application to malicious URL detection. In: Proceedings of the 19th ACM SIGKDD International Conference on Knowledge Discovery and Data Mining, vol. Part F128815, pp. 919–927 (2013). https://doi.org/10.1145/2487575.2487647
6. Shantanu, J.B., Kumar, R.J.A.: Malicious URL detection: a comparative study. In: Proceedings of the International Conference on Artificial Intelligence and Smart Systems, ICAIS 2021, June 2021, pp. 1147–1151 (2021). https://doi.org/10.1109/ICAIS50930.2021.9396014
7. Sahoo, D., Liu, C., Hoi, S.C.H.: Malicious URL detection using machine learning: a survey (2017). https://arxiv.org/abs/1701.07179
8. Le, H., Pham, Q., Sahoo, D., Hoi, S.C.H.: URLNet: learning a URL representation with deep learning for malicious URL detection (2018). https://arxiv.org/abs/1802.03162
9. Verma, R., Das, A.: What's in a URL: fast feature extraction and malicious URL detection. In: Proceedings of the 3rd ACM on International Workshop on Security and Privacy Analytics Co-located with CODASPY 2017, IWSPA 2017, pp. 55–63 (2017). https://doi.org/10.1145/3041008.3041016
10. Patgiri, R., Katari, H., Kumar, R., Sharma, D.: Empirical study on malicious URL detection using machine learning. In: Fahrnberger, G., Gopinathan, S., Parida, L. (eds.) ICDCIT 2019. LNCS, vol. 11319, pp. 380–388. Springer, Cham (2019). https://doi.org/10.1007/978-3-030-05366-6_31
11. Saleem Raja, A., Vinodini, R., Kavitha, A.: Lexical features based malicious URL detection using machine learning techniques. Mater. Today Proc. **47**, 163–166 (2021). https://doi.org/10.1016/j.matpr.2021
12. Robert, B., Brown, E.B. (2004). https://www.kaggle.com/
13. Mukrimaa, S.S., et al.: Jurnal Penelitian Pendidikan Guru Sekolah Dasar (2016). https://www.kaggle.com/datasets/sid321axn/malicious-urls-dataset
14. Alsobeh, A., Shatnawi, A.: Integrating data-driven security, model checking, and self-adaptation for IoT systems using BIP components: a conceptual proposal model. In: Daimi, K., Al Sadoon, A. (eds.) ACR 2023. LNNS, vol. 700, pp. 533–549. Springer, Cham (2023). https://doi.org/10.1007/978-3-031-33743-7_44
15. Darwish, O., Al-Fuqaha, A., Anan, M., Nasser, N.: The role of hierarchical entropy analysis in the detection and time-scale determination of covert timing channels. In: 2015 International Wireless Communications and Mobile Computing Conference (IWCMC), pp. 153–159. IEEE (2015)
16. Al-Kabi, M., Wahbeh, A., Alsobeh, A., Ale'roud, A., Alsmadi, I.: Examining web search trends across Arab countries. Arab. J. Sci. Eng. **37**, 1585–1593 (2012)
17. Alsobeh, A.M., Magableh, A.A.A.R., AlSukhni, E.M.: Runtime reusable weaving model for cloud services using aspect-oriented programming: the security-related aspect. Int. J. Web Serv. Res. (IJWSR) **15**(1), 71–88 (2018)
18. Tashtoush, Y., Darweesh, D., Darwish, O., Alsinglawi, B., Obeidat, R.: A classifier to detect profit and non profit websites upon textual metrics for security purposes. J. ICT Res. Appl. **16**(1), 81–91 (2022)

19. AlSobeh, A.M., Ramadan, I.A.A., Shatnawi, A.M.J., Khasawneh, I.: Cybersecurity awareness factors among adolescents in Jordan: mediation effect of cyber scale and personal factors. Online J. Commun. Media Technol. **13**(2), e202312 (2023)
20. Shatnawi, M.Q., Abuein, Q.Q., Darwish, O.: Verifying hadith correctness in Islamic web pages using information retrieval techniques. Int. J. Comput. Appl. **44**(13), 47–50 (2012)

# Impact of IoT Technology and Social Networking Sites on Mothers, Teachers, and Children with Autism Spectrum Disorder in Jordan


Walaa Al-Sarayrah[1], Assia Abo-Zaitoon[1], Abdallah Aldosary[2], Rami Tashtoush[3,4], and Yahya Tashtoush[1(✉)]

[1] Department of Computer Science, University of Science and Technology, Irbid, Jordan
wtalsarayrah14@cit.just.jo, yahya-t@just.edu.jo

[2] Department of Computer Science, Prince Sattam bin Abdulaziz University, Al-Kharj, Saudi Arabia
ab.aldosary@psau.edu.sa

[3] Psychology Department, Yarmouk University, Irbid, Jordan
r.tashtoush@squ.edu.om

[4] Psychology Department, Sultan Qaboos University, Musqat, Oman



**Abstract.** Research on the impact of Social Networking Sites (SNSs) on mothers and teachers of children with autism spectrum disorder (ASD) is limited, marking a unique intersection of computer science and psychology. This study explores whether SNSs facilitate communication between teachers, mothers, and ASD children for teaching social and daily life activities, and the impact of these interactions. IoT technology is introduced as a pivotal enhancement in this context. A model, reflecting participants' preferences and experiences, includes four aspects: (1) Online ASD community engagement, enhanced by IoT-enabled secure and tailored communication platforms. (2) Diverse SNSs and technology preferences, with IoT providing adaptive interfaces for varied user needs. (3) The reduced stigma from ASD awareness, augmented by IoT's data analytics for personalized support. (4) Positive effects of SNSs on ASD children's social, behavioral, and academic development, further improved by IoT's real-time monitoring and interaction capabilities. Questionnaires for 54 mothers and 52 teachers revealed that IoT-enhanced SNSs positively impact ASD community engagement, stigma reduction, and child development. This model suggests integrating IoT-enriched SNSs for rehabilitation and therapeutic purposes in ASD support environments and homes, offering effective and low-cost treatment options.



**Keywords:** Autism Spectrum Disorder (ASD); Social Networking Sites (SNSs) · Internet of Things (IoT) · Communication Technologies · Digital Inclusion · Adaptive Interfaces · IoT for Special Needs · Online Communities · Cybersecurity in IoT · IoT-enabled Assistive Technology




L. Barolli (Ed.): AINA 2024, LNDECT 203, pp. 483–496, 2024.
https://doi.org/10.1007/978-3-031-57931-8_46

# 1 Introduction

## 1.1 Social Networking Sites (SNSs)

Nowadays millions of users are attracted by Social Networking Sites such as Facebook, YouTube, WhatsApp and many more. People have integrated such sites into their daily practices. Using smartphones made this integration process much easier since users can access these sites anytime and anywhere for different purposes.

One of the most important purposes of SNSs is communication. Whether we are sitting behind the laptop screen or just using a smartphone, people now are spending more time communicating with each other, and the more time we spend communicating using such sites, face-to-face interaction will be reduced. There are hundreds of SNSs, with a wide range of interests and practices related to each site. Some sites are designed to aim for various users, on the other hand other sites attract users based on some common attributes such as language or religious based identities. Many sites master incorporating new information via blogging and sharing photos or videos [1].

Here, IoT technology can play a transformative role. By integrating IoT with SNSs, we can create adaptive and personalized interfaces that cater to the unique communication needs of ASD individuals. This could involve real-time speech-to-text conversion, tailored content delivery, and sensory-friendly interfaces, thereby bridging the communication gap for ASD users and making SNSs more inclusive and accessible.

On the other hand, it may seem that SNSs help in communication, but at the same time, communication is a major problem for ASD people, many studies and research try to figure out a way to react to this person and to perform even the simplest form of communication with him/her which is an eye contact.

## 1.2 Autism Spectrum Disorder (ASD)

Autism spectrum or newly defined as autism spectrum disorder is a term related to a group of neurodevelopmental disorders that cause a number of problems in the social, communicative and emotional skills. Also bizarre patterns of behavior will emerge, in addition, ASD people suffer from limited concerns [2].

ASD symptoms vary from child to child and thus classifications of this disorder will also vary from severe autism in which all the symptoms will appear and the child will have low mental ability to others with light cases of autism in which some symptoms are appearing but on the other hand the child will have a good mental capability or above normal capabilities.

Some of these symptoms are listed below, which begin to appear before the age of three [2]: Isolation and the inability of social interaction; Repetitions and refinements of acts; Rotation; Lack of awareness of the danger; Hyperactivity; Failed eye contact with others; Relying on parents in the simplest everyday things; Difficulties in adapting to any change in their routines; High sensitivity for touching or even hugging them.

Referring to the latest theories that talked about the causes of this disorder, the specialists pointed to the genetic factor, which is considered as a psychological nervous disorder and the causes are overlapped in the genetic areas in the human being [2]. Another study showed that the rate of the recurrence of this disorder in the same family is 3% to 10%, and in spite of the inability of scientists so far to identify the gene location that is responsible for this disorder, studies suggest that the imbalance in 3–10 genes has led to this turmoil disorder. The intensive interest in ASD has emerged due to the high number of people with this disorder, where the latest global research confirmed that in every 150 children at least one child will suffer from ASD, and this disorder appears in males four times more than in females [3]. In spite of that people with ASD enjoy normal sexual abilities, but the social disabilities related to this disorder are a key reason for the failure of their marriage and having children. Early diagnosis, increasing the level of awareness among parents, directly starting treatment and rehabilitation through standardized programs for training these children and improving the level of the audiovisual communication with them are key factors that contribute significantly in improving the capacities of the children and enable them to join later in school and to engage social life normally.

### 1.3 ASD Types

It became clear after many years of research that there are several types of autism, which is the reason that led to name autism as "a spectrum disorder". Also, it indicates that the differences in autism are similar to the differences in the spectrum, different colors with different intensity shades and the common factor between these types is low capacity for social interaction. And it should be noted that autism contains all three basic symptoms (1. Deficiencies in social interaction 2. Deficiencies in language skills 3. Failure of imagination), while in other types it is sufficient to have only two symptoms [2]. And the types of autism spectrum are as follows: (1) Autistic disorder: It is characterized by an imbalance in social interactions, language and the ability to imagine, all of these symptoms appeared before the child reaches the age of three years along with some stereotypical behaviors. (2) Aspereger syndrome: It's characterized by an imbalance in the social interactions, the presence of restricted interests with the lack of delay in the language skills, and the intelligence of the individual is between average and above-average ratio. (3) Rett's disorder: A developmental disorder affects girls only, a natural growth occurs first and then the girl loses skills gained before, also the intended use of the hands will be missed and a repetitive movement of the hands will appear, it starts at the age of 1–4 years. This is a rare disorder and happens in one birth in every 15,000 births, and the emergence of epilepsy is more likely to happen, difficulty in chewing and swallowing food and sleep, and after the age of ten the spine of many females will be arched and may need a wheelchair due to a significant deterioration in movement skills. (4) Childhood Disintegrative Disorders (CDD): The child develops this disorder in a natural evolution of all aspects of his/her life from two to ten years, and then it followed by a noticeable loss of learned skills i.e., receptive language skills, social skills, and adaptive behavior skills, lack of control over bowel and bladder, play and movement skills. (5) Pervasive Developmental Disorder- Not Otherwise Specified (PDD-NOS): It is also known as Atypical autism, it is one of the most common developmental disorders, it is diagnosed with some of the features of the traditional autism, in other words, though it is not accurate that an individual is considered to have "almost" autism, but not to the point enough for a diagnosis of autism, symptoms of the disorder are less severe than autism and other developmental disorders, individual with this type still have the capacity for social interaction, in addition to that child that diagnosed within this category have high performance, i.e. they have a near-normal cognitive abilities.

### 1.4 Needs of ASD People

There are five types of essential needs in the lives of people with autism. And these needs enable them to perform the simplest everyday tasks and duties. All rehabilitation centers try to cover all five types in their treatment of ASD people as much as possible, but in the most severe cases they try to cover the simplest ones. The five needs are listed below [4]: Adjustment Needs: Are a set of requirements that must be provided for the person of ASD and relates to his/her knowledge, skills, experiences, and behaviors in daily life and self-care activities, i.e. Using bathroom correctly, washing his/her face alone, wearing and removing clothes alone, eating food properly and responding to smells correctly. Social Needs: Are a set of requirements that must be provided for the person of ASD and relates to his/her knowledge, skills, experiences, and his/her behavior in the social field, i.e. Enjoying relationships with others, emotional attachments with family members (Mother, Father, etc.), Understanding others emotions and feelings, asking for help when needed and expressing emotions properly (angry, happy, sad). Academic Needs: Are a set of requirements that must be provided for the person of ASD and relates to his/her knowledge, skills, experiences, and his/her behavior in the academic field, i.e. Matching pictures with words, Saying the days of the week, reading and writing. Behavioral Needs: Are a set of requirements that must be provided for the person of ASD and relates to his/her knowledge, skills, experiences, and his/her behavior in the behavioral field, i.e. Holding the pen in the right way, the ability to reduce staring at the space without a reason, the ability to reduce finger biting and having more control of the movement of hands. Communicational Needs: Are a set of requirements that must be provided for the person of ASD and relates to his/her knowledge, skills, experiences, and his/her behavior in the field of communication and language, i.e. The ability to start conversations with others, responding to verbal commands such as "Come here!!", the ability to continue exchanging conversations, the ability to reduce the repetition of certain words and using pronouns in the right way.

This study delves into the utilization of Social Networking Sites (SNSs) by mothers and teachers of ASD children, exploring innovative ways these platforms can aid in teaching and improving social and daily life skills. It investigates not only if, but how effectively ASD individuals engage with SNSs, and the dynamics of communication between mothers and teachers through these platforms. The paper goes beyond a simple 'yes' or 'no' to the usage of SNSs; it uncovers the transformative impact of such interactions, and how IoT technology can revolutionize this process. By integrating IoT, the study offers groundbreaking insights into creating more inclusive, adaptive, and secure online environments for the ASD community, potentially reshaping our understanding of digital communication in special education and therapeutic contexts.

## 2 Related Work

In 2010, a study suggested that computer-mediated communication (CMC) could enhance social skills in individuals with Autism Spectrum Disorder (ASD), who typically struggle with face-to-face interactions [5]. Participants showed positive responses to this approach. However, a 2013 study found that while social networking sites (SNSs) were popular among adults with ASD, they did not alleviate loneliness, which relates more to close relationships [6]. Concurrently, research indicated that SNSs, particularly Facebook, could improve communication and social skills in ASD students, advocating for SNSs' integration into educational settings [7]. In the same year, researchers [8] examined how TV, video games and SNSs may affect ASD children compared with normal children. They found that ASD children spent 62% more time watching TV or playing video games than in non-screen activates; it means that they spend 4.5 h per a day rather than only 1–2 h per day. According to [American Academy of Pediatrics] ASD people must spend 87% more time on non – screen activities.

In 2014, studies recommended social networking sites (SNSs) to help Arab children with ASD enhance their social skills, despite cultural challenges. Benefits included improved communication and self-expression among ASD individuals, but risks like cyberbullying and social isolation were also noted [9, 10]. Additionally, excessive electronic use among ASD youth showed potential negative impacts, contrasting with positive effects in neurotypical peers [1, 12]. In the year 2016, they found there is no significant difference between normal people and ASD people in the Compulsive Internet Use (CIU) measures. They performed two studies, one on normal people and the other on ASD people and neuro-typical people with the same sample size in both populations (n = 33). This study showed that there are no significant differences in the CIU measures [13].

In 2016, research indicated that high-functioning individuals with ASD were drawn to CMC, spending more time using it and reporting increased social satisfaction [14]. It suggested that ASD individuals effectively use CMC to foster positive online interactions. A 2015 study found that for those with ASD, communicating via photos on social networking sites was more effective than text, enhancing interaction and social skills [15].

### 2.1 IoT for Enhanced Communication in ASD

IoT for Enhanced Communication in ASD Communication is one of the core challenges for individuals with ASD, who often face difficulties in expressing and understanding verbal and nonverbal cues. Several studies have explored the potential of IoT to enhance communication skills in this population, by providing various technologies that can assist them in overcoming common communication barriers. For example, speech-to-text and real-time translation devices can help ASD individuals communicate with people who speak different languages or have different accents. Personalized learning tools can also help ASD individuals improve their language and literacy skills, by adapting to their preferences, abilities, and progress. IoT for Personalized Interaction and Learning Another aspect of IoT that can benefit ASD individuals is its ability to create personalized and adaptive learning environments that can cater to the diverse and specific needs of each user. IoT can enable the delivery of customized content, such as videos, games, or stories, that can match the interests and learning styles of ASD individuals. IoT can also adjust the sensory inputs, such as sound, light, or temperature, that can affect the comfort and attention of ASD individuals. Furthermore, IoT can provide interactive learning experiences, such as virtual reality, augmented reality, or haptic feedback, that can stimulate the senses and enhance the engagement of ASD individuals. Security and Privacy in IoT for SNSs While IoT offers many advantages for ASD individuals, it also poses some security and privacy risks, especially when used with SNSs. SNSs can facilitate social interaction and information sharing among ASD individuals, but they also expose them to potential threats, such as cyberbullying, identity theft, or data breaches. Therefore, it is important to address the security and privacy issues in using IoT with SNSs, by applying various methods, such as data encryption, user authentication, and secure data sharing. Moreover, it is essential to educate and empower ASD individuals to protect their personal information and online safety, by providing them with appropriate guidance and tools.

## 2.2 Innovative IoT Applications in Therapy and Education for ASD

Innovative IoT Applications in Therapy and Education for ASD IoT can also be applied in innovative ways to support therapy and education for ASD individuals, by offering practical and beneficial solutions in these settings. For instance, IoT can enable real-time behavior monitoring, by using sensors, cameras, or wearables, that can track and analyze the emotional and physical states of ASD individuals and provide feedback or intervention when needed [16–18]. IoT can also enhance therapy sessions, by using smart devices, such as robots, tablets, or toys, that can interact with ASD individuals, and facilitate social and cognitive skills development. Additionally, IoT can improve engagement tracking, by using data analytics, dashboards, or reports, that can measure and visualize the progress and outcomes of ASD individuals and provide recommendations for improvement [19].

Challenges and Future Directions Despite the promising applications and benefits of IoT for ASD individuals, there are still some challenges and limitations that need to be addressed and overcome [20]. Some of the technological challenges include the interoperability, scalability, and reliability of IoT devices and systems, as well as the quality, accuracy, and validity of the data collected and processed by IoT. Some of the ethical challenges include the consent, autonomy, and dignity of ASD individuals, as well as the fairness, transparency, and accountability of IoT [21, 22]. Some of the practical challenges include the accessibility, affordability, and usability of IoT devices and services, as well as the training, support, and evaluation of IoT. Therefore, future research and development in this field should aim to tackle these challenges and explore the potential of IoT to improve the quality of life and well-being of ASD individuals [23–25].

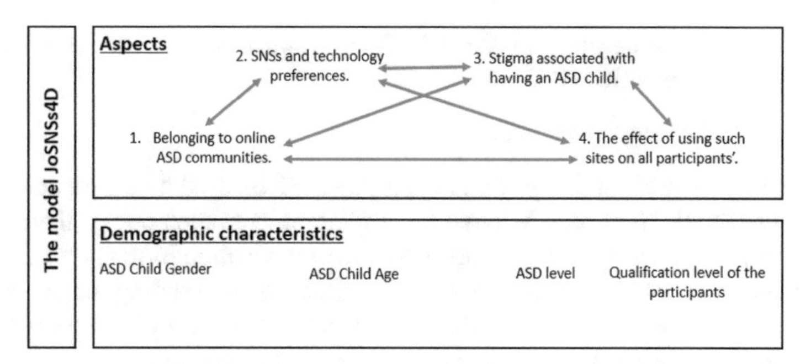


**Fig. 1.** A graphical scheme of the JoSNSs4D model

# 3 Proposed Model Josnss4d

As shown in Fig. 1, the proposed model has the following aspects: (1) Belonging to online ASD communities. (2) SNSs and technology preferences. (3) Stigma associated with having an ASD child. (4) The effect of using such sites on all participants. Furthermore, other factors were considered, such as the demographic characteristics of the participants.

A. Belonging to online ASD communities:

The participation of the mothers plays a huge role in helping and improving the state of their ASD children, ignoring the problem does not help improving it so in this context, it is very important for the mothers to help and support the ASD aid institutions and organizations. By belonging to an online ASD community that consists of mothers and teachers of these children and other professionals can include many practical things such as how to deal with such children. It showed that mothers suffer from lack of training and rehabilitation, which puts them in the face of dealing with specialized centers and makes the financial costs a heavy burden on them and many families could not overcome it yet. The high cost of treating autism disorders also comes from the importance of the work of many specialists with the child, either coach, occupational or behavioral therapist along with the mother and the family as a whole. As a result of such concerted efforts and training and lifelong rehabilitation, many families have obtained good improvements and even defeated autism. There are also a lot of people with autism who have achieved great successes at various levels. Via communication through SNSs a lot of mothers have learned new techniques on how to improve the production of the language of their children, how to restructure the child's behavior and how to enhance visual communication with them. They were looking for online courses and related topics to increase their knowledge, and these sites helped some mothers by reducing the reliance on specialized institutions since the assessment and rehabilitation of children with autism are expensive and cannot be afforded by many families. SNSs made the communication process and sharing information and advice between all participants much easier, where individuals can reach others any time, and any place with cheaper costs. It has also affected the self-esteem of the mothers; sometime the shifting from face-to-face communication to online communication will give the mothers higher self-esteem in expressing themselves and sharing their children's conditions with similar mothers. Thus, the following four hypotheses are tested by the proposed model:

**Hypothesis 1: Belonging to online ASD communities has a positive impact on improving children with ASD.**

B. SNSs and technology preferences.

This model focuses on the best and appropriate SNSs used by all participants, e.g. YouTube, Facebook, Snapchat, WhatsApp, Instagram, etc. Each one of them provides good services and options to share and search for information in many ways: written, visual, or audio, as well as possibility of instant chatting and sending messages.

There are many ways to reach SNSs, you can use them from your laptop, tablets or even directly from your smartphone. All these mediums lead to one goal, to facilitate the communication process between mothers, teachers or experts of ASD.

**Hypothesis 2: The variety of SNSs and technology preferences has a positive impact on improving children with ASD.**

C. Stigma associated with having an ASD child.

Having a child with any disorder is a heavy burden on the family, especially for the mother. Any mother hopes to have a normal child with nothing that differentiates him/her from other children. So, when it comes to having a child that requires extra caring and attention, the mother will feel inferior, and she will be sensitive to the opinions of other mothers. In this model more focus is paid on the stigma associated with having an ASD child. Furthermore, the culture and the environment we live in play a huge role in affecting this stigma.

All of these will form a barrier in the way of communication and sharing information related to ASD, since mothers will not be comfortable discussing their children's conditions with others. On the other hand, communicating through SNSs may help in reducing this stigma, since mothers can login into these sites by aliases. In a way that helps them ask anything they want without the fear of being judged. Plus, it will lead to more improvements in the social and behavioral skills of their children by exchanging useful information and practices.

**Hypothesis 3: Communication through SNSs has a positive impact on reducing the stigma associated with having an ASD child.**

D. The effect of using such sites on all participants.

The last aspect will include the effect of using SNSs on the social skills, behavioral skills and academic levels of ASD children and how SNSs helped in improving the needs explained earlier.

**Hypothesis 4: Communication through SNSs has a positive impact on social, behavioral and academic levels of ASD children.**

## 4 Methodology

The proposed model used questionnaires which were developed based on the previous aspects and some other factors. This section contains a description for our study samples, how it was chosen, our study variables, the tools used to conduct this study, the followed procedures that were used to execute the study and to acquire the needed information, finally the used statistical processing tools to analyze these data. Study Sample: Due to the absence of official statistics on the actual number of children with ASD in Arab region in general and especially in Jordan, plus the availability of specialized institutions in treating autism is limited in big cities. The researchers distributed this study to the mothers and the teachers of ASD children that were enrolled in these institutions only in the cities of Irbid and Karak. The measure was distributed to the mothers and teachers through the institutions, the number of mothers was (54) and teachers were (52). The study was limited because of the following boundaries: **Human limitations**: the mothers and teachers of children with autism spectrum disorder; we could not include ASD children themselves. **Spatial limitations**: The study was conducted in the cities of Irbid and Karak only. **Time limitations**: study was conducted during the first semester 2015/2016.

Table 1 shows the characteristics of our study samples in terms of distribution by the gender of the child, the child's age and the degree of ASD, finally the percentage representation of each of them.

**Table 1.** Study samples characteristics

| Variable | Category | Number | Percent | Total |
|---|---|---|---|---|
| Gender | Males | 34 | 63.0 | 54 |
| | Females | 20 | 37.0 | |
| Age | Less than 5 years | 40 | 74.1 | 54 |
| | 6–10 years | 14 | 25.9 | |
| ASD Degree | Mild | 23 | 42.6 | 54 |
| | Moderate | 27 | 50.0 | |
| | Severe | 4 | 7.4 | |

A. Study Tool: To achieve the objectives of our study two questionnaires were prepared, one targeted to the mothers of ASD children and the other to the teachers. These questionnaires were distributed through ASD aid institutions in the cities of Irbid and Karak. We presented the questionnaires to a doctor with special education major at Yarmouk University, He helped us in understanding ASD people, and he guided and gave us his notes about the questionnaires before start distributing them. Furthermore, we showed the questionnaires to a doctor with computer science major at Jordan University of Science and Technology for his professional guidance through the entire process. After making all the adjustments recommended by the judges, the final questionnaires in their final form contained almost (24) questions divided into (4) aspects mentioned earlier. The sections of the questionnaires are divided as the following: Some demographic characteristics such as age, the gender of ASD children, ASD level and qualification level of the participants are recorded in the first section. The second section contained the most regularly used SNSs and technology preferences. The next section covered the aspect of belonging to online ASD communities. The impact of using SNSs on the mothers, teachers, and the child him/herself is noted in the fourth section of the questionnaires. Finally, the last section indicates to the Stigma associated with having an ASD child to the mothers.

B. Study Variables: The gender and ASD degree were handled as an independent variable, whereas other variables such as ASD child age, Qualification level for mothers and teachers, belonging to online ASD communities, SNSs and technology preferences, stigma associated with having an ASD child and the effect of using such sites on all participants were considered as dependent variables.
C. Statistical processing tool: After collecting the questionnaires from the institutions, the data was entered into the computer, then it was analyzed by the Statistical Package for the Social Sciences (SPSS) as the following: we made sure that we measure the same variable for a group of related items and the relations between these items was measured too. Using a statistical analysis all statistical relations between all four aspects and the data from the demographic characteristics were measured individually. In addition, all variables were chosen based on our four hypotheses with respect to having a small sample size.

# 5 Results and Dissection

The statistical analysis of the variables according to the four aspects is described below: belonging to online ASD communities. SNSs and technology preferences. Stigma associated with having an ASD child. And the effect of using such sites on all participants. Within the first hypothesis, the results indicated that 87% of mothers prefer to be involved within a community that is specialized for their ASD children. But on the other hand, 13% of mothers did not prefer to be included in such communities. According to the mothers, finding a virtual place where mothers, teachers and even professionals will help in improving children's ASD as much as possible, since the information and questions will be exchanged between these participants easily with no boundaries. In addition, belonging to online communities for ASD is also significantly related to the variety of SNSs and technology preferences; higher knowledge of using such SNSs and technologies will lead to better interaction and involvement in these communities [26]. The results to test the second hypothesis are listed below in Table 2. Teachers of ASD students use Facebook more commonly than other SNSs. Youtube also has a high percentage, since teachers can include Youtube as a treatment tool for the children. Both Facebook and WhatsApp are used most for exchanging information and notes between teachers and mothers. Furthermore, Mothers use WhatsApp higher than the rest of SNSs, because it is an easy way to communicate with teachers and other mothers. There is no significant difference between the percentage of using Facebook and WhatsApp by the mothers, this can be related to the technology preferences. Where Smartphones occupy the highest rank among used technologies. Smartphones provide an easy way for communication and can be reached by everyone.

To test the third hypothesis, the relationship between the stigma associated with having as ASD child and communicating through SNSs was analyzed. The results showed that 79.6 of mothers are willing to share their questions and information with other mothers with the same conditions. On the other hand, 20.4% of mothers are afraid of sharing anything, or even telling if they have an ASD child. The fair is resulted because of cultural reasons. In addition, 83.3% of mothers share information with their children's

**Table 2.** Percentage of SNSs and Technology Preferences for the mothers and Teachers

| Variable | Mothers | Teachers |
|---|---|---|
| Facebook | 57.4% | 76.9% |
| Youtube | 33.3% | 71.2% |
| WhatsApp | 59.3% | 67.3% |
| Computers | 70.4% | 75% |
| Smartphones | 98.1% | 90.4% |
| IPads | 48.1% | 65.4% |

teachers. Figure 2 Indicates that qualification level of the mother affects the degree of sharing information with other mothers, higher education levels led to decrease the level of stigma and the mother is more open to share her questions and information. Regarding the last hypothesis, a statistically positive significant association was found between the SNSs usage and all participants of the study. For the ASD children, the results show that SNSs have a positive impact on them, where social, behavior and academic skills can be obtained. According to the teachers 88.7% is the percentage of enhancement in the behavioral level of the children, since 69.8% of teachers use Youtube as a tool for teaching some behavioral skills such as: learn how to eat, how to wear clothes, etc. Finally, there is no significant difference between the percentages of communication via SNSs between teachers themselves, mothers themselves and between mothers and teachers. But on the other hand, both mothers and teachers do not prefer their children to use SNSs by themselves and it must happen within some sort of control and supervision. Two reasons behind that, they fear that their children will be cyber-bullied by strangers, the second reason is because 74.1% of ASD children are under 5 years old and do not know how to use SNSs. 69.8% of the institutions are equipped with computers, but 62.3% of these organizations use these computers in the education process. The difference is because these organizations are not specialized only for ASD children, some of them treat different mental disorders, in addition, such organizations treat ASD children below 5 years old (see Fig. 3). Almost every institution has an official Facebook page for communicating and providing information to others.

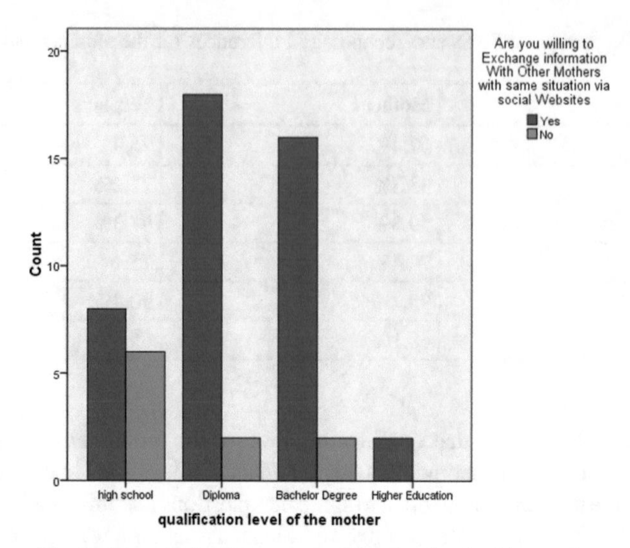


**Fig. 2.** The relation between qualification level of the mother

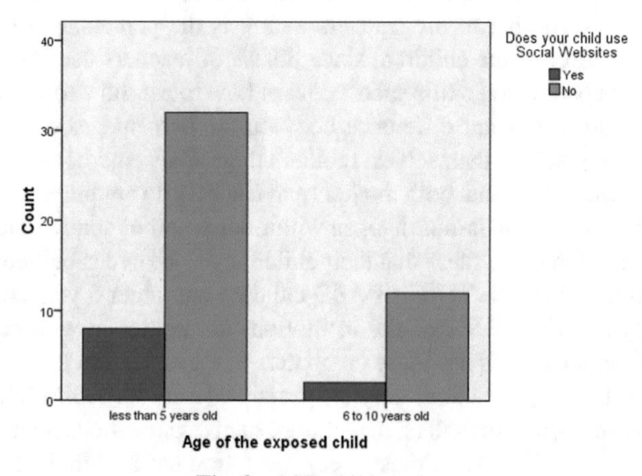


**Fig. 3.** ASD children Age

# 6 Conclusion

This study offers an in-depth overview of the impact of SNSs, enhanced with IoT technology, on teachers and mothers of ASD children. The proposed IoT-integrated JoSNSs4D model enriches understanding and communication around ASD, suitable for both institutional and home settings, offering cost-effective treatment options. IoT's integration promises significant improvements in ASD children's social, behavioral, and educational skills through interactive and tailored content. It also facilitates a supportive community among caregivers, reducing stigma and fostering openness. Future work will explore IoT's role in directly engaging ASD individuals to understand their unique preferences, aiming to seed further research in this field. Continuously updating our approach with evolving IoT and SNS technologies is crucial to ensure equitable opportunities for ASD children.

Future research should delve deeper into how IoT can be used to create highly customized SNS experiences for ASD individuals. This could involve developing IoT devices and applications that adapt in real-time to the user's behavior and preferences, providing a more engaging and effective learning environment. There's a need to explore advanced data analytics through IoT to understand the specific needs and responses of ASD individuals to different stimuli on SNSs. This could lead to more effective and personalized therapeutic strategies. Investigate the potential of IoT in enhancing non-verbal communication for ASD individuals who may struggle with traditional forms of expression. This could include research on IoT devices that interpret and translate non-verbal cues into digital communication on SNSs. Conduct longitudinal studies to assess the long-term impact of IoT-enhanced SNSs on the development of social and cognitive skills in ASD children. This would provide valuable insights into the sustainability and efficacy of such technologies. Address the ethical implications of using IoT with vulnerable groups like ASD individuals. Future studies should ensure that IoT applications in this domain are developed with the highest standards of privacy, security, and user consent.

## References

1. Ellison, N.B.: Social network sites: Definition, history, and scholarship. J. Comput.-Mediat. Commun. **13**(1), 210–230 (2007)
2. Lord, C., Cook, E.H., Leventhal, B.L., Amaral, D.G.: Autism spectrum disorders. Neuron **28**(2), 355–363 (2000)
3. Ozonoff, S., et al.: Recurrence risk for autism spectrum disorders: a baby siblings research consortium study. Pediatrics **128**(3), e488–e495 (2011)
4. Shdefat, A.A., Tashtoush, R.A.: The Needs of Children with Autism Spectrum Disorder (ASD) in Jordan According to their Mothers in the light of some Variables (Doctoral dissertation) (2013)
5. Burke, M., Kraut, R., Williams, D.: Social use of computer-mediated communication by adults on the autism spectrum. In: Proceedings of the 2010 ACM Conference on Computer Supported Cooperative Work. ACM (2010)
6. Mazurek, M.O.: Social media use among adults with autism spectrum disorders. Comput. Hum. Behav. **29**(4), 1709–1714 (2013)
7. Kuder, J., Xin, J.: Using Facebook to improve social communication skills of students with autism. In: World Conference on Educational Multimedia, Hypermedia and Telecommunications, vol. 2013, no. 1 (2013)
8. Mazurek, M.O., Wenstrup, C.: Television, video game and social media use among children with ASD and typically developing siblings. J. Autism Dev. Disord. **43**(6), 1258–1271 (2013)
9. AlSobeh, A.M.R.: OSM: Leveraging model checking for observing dynamic behaviors in aspect-oriented applications. Online J. Commun. Media Technol. **13**(4), e202355 (2023)
10. Mashat, A., Wald, M., Parsons, S.: Improving social and communication skills of adult Arabs with ASD through the use of social media technologies. In: Miesenberger, K., Fels, D., Archambault, D., Peňáz, P., Zagler, W. (eds.) Computers Helping People with Special Needs. ICCHP 2014. LNCS, vol. 8547, pp. 478–485. Springer, Cham (2014). https://doi.org/10.1007/978-3-319-08596-8_75

11. Gillespie-Lynch, K., et al.: Intersections between the autism spectrum and the internet: perceived benefits and preferred functions of computer-mediated communication. Intellect. Dev. Disabil. **52**(6), 456–469 (2014)
12. Lough, E., Flynn, E., Ruby, D.M.: Mapping real-world to online vulnerability in young people with developmental disorders: illustrations from autism and Williams syndrome. Rev. J. Autism Dev. Disord. **2**(1), 1–7 (2015)
13. MacMullin, J.A., Lunsky, Y., Weiss, J.A.: Plugged in: electronics use in youth and young adults with autism spectrum disorder. Autism (2015). 1362361314566047
14. Shane-Simpson, C., et al.: Associations between compulsive internet use and the autism spectrum. Res. Autism Spectr. Disord. **23**, 152–165 (2016)
15. Van der Aa, C., et al.: Computer-mediated communication in adults with high-functioning autism spectrum disorders and controls. Res. Autism Spectr. Disord. **23**, 15–27 (2016)
16. Mashat, A., Wald, M., Parsons, S.: The role of photos in social media, interactions of adult Arabs with autism spectrum disorder. In: INTED2015 Proceedings, pp. 2873–2879 (2015)
17. Sula, A., Spaho, E., Matsuo, K., Barolli, L., Xhafa, F., Miho, R.: An IoT-based framework for supporting children with autism spectrum disorder. In: Park, J., Barolli, L., Xhafa, F., Jeong, H.Y. (eds.) Information Technology Convergence. Lecture Notes in Electrical Engineering, vol. 253, pp. 193–202. Springer, Dordrecht (2013). https://doi.org/10.1007/978-94-007-6996-0_21
18. Alsinglawi, B., Alnajjar, F., Mubin, O., Novoa, M., Karajeh, O., Darwish, O.: Benchmarking predictive models in electronic health records: Sepsis length of stay prediction. In: Barolli, L., Amato, F., Moscato, F., Enokido, T., Takizawa, M. (eds.) AINA 2020. Advances in Intelligent Systems and Computing, vol. 1151, pp. 258–267. Springer, Cham (2020). https://doi.org/10.1007/978-3-030-44041-1_24
19. Alsobeh, A., Woodward, B.: AI as a partner in learning: a novel student-in-the-loop framework for enhanced student engagement and outcomes in higher education. In: Proceedings of the 24th Annual Conference on Information Technology Education, pp. 171–172 (2023)
20. Afzal, M.J., Ashraf, M.W., Tayyaba, S., Javaid, F.: Analysis and real-time implementation of IoT in autism (2021)
21. AlSobeh, A.M., Klaib, A.F., AlYahya, A.: A national framework for e-health data collection in Jordan with current practices. Int. J. Comput. Appl. Technol. **59**(1), 64–73 (2019)
22. Darwish, O., Tashtoush, Y., Bashayreh, A., Alomar, A., Alkhaza'leh, S., Darweesh, D.: A survey of uncover misleading and cyberbullying on social media for public health. Cluster Comput. **26**(3), 1709–1735 (2023)
23. AlSobeh, A.M., Ramadan, I.A., Shatnawi, A.M.J., Khasawneh, I.: Cybersecurity awareness factors among adolescents in Jordan: mediation effect of cyber scale and personal factors. Online J. Commun. Media Technol. **13**(2), e202312 (2023)
24. Lonie, N.: Online Safety for Children and Teens on the Autism Spectrum: A Parent's and Carer's Guide. Jessica Kingsley Publishers, London (2014)
25. Al-Shraifin, A., Arabiat, R.B., Shatnawi, A., AlSobeh, A., Bahr, N.: The effectiveness of a counseling program based on psychosocial support to raise the level of economic empowerment among refugees. Curr. Psychol. **43**, 3101–3110 (2023)
26. Karajeh, O., Darweesh, D., Darwish, O., Abu-El-Rub, N., Alsinglawi, B., Alsaedi, N.: A classifier to detect informational vs. non-informational heart attack tweets. Fut. Internet **13**(1), 19 (2021)
27. Tahnai, M.Q., Abuein, Q.Q., Darwish, O.: Verification hadith correctness in islamic web pages using information retrieval techniques. In: Proceedings of International Conference on Information & Communication Systems, pp. 164–167 (2011)

# A Survey Analysis of Internet of Things (IoT) Education Across the Top 25 Universities in the United States

Omar Darwish[1], Abdallah Al-shorman[1](✉), Anas AlSobeh[2], and Yahya Tashtoush[3]

[1] Eastern Michigan University, Ypsilanti, USA
{odarwish,aalshorm}@emich.edu
[2] Southern Illinois University Carbondale, Carbondale, USA
anas.alsobeh@siu.edu
[3] Jordan University of Science and Technology, Ar-Ramtha, Jordan
yahya-t@just.edu.jo

**Abstract.** This paper analyzes Internet of Things (IoT) education across the top 25 universities in the USA, categorizing courses across five key domains - IoT Devices, Networking, Cloud, Analytics, and Interfaces. The analysis provides insights into curriculum structure, emerging focus areas, teaching methods, and assessments used. Findings indicate comprehensive IoT education requires a diverse, interdisciplinary curriculum encompassing hardware, software, and theoretical concepts. While graduate programs offer specialized courses, undergraduate programs provide broad exposure. Most universities emphasize hands-on learning to build relevant skills. The paper highlights that IoT education must align with current industry advancements in areas like AI and blockchain. It presents useful benchmarks for developing impactful university-level IoT programs to meet evolving workforce demands.

## 1 Introduction

The Internet of Things (IoT) is transforming industries and society by enabling connections between physical objects for improved monitoring, control, optimization and automation. It is estimated that there will be over 30 billion IoT devices by 2025 across various domains including transportation, healthcare, manufacturing and agriculture [1]. This proliferation has created demand for a skilled IoT workforce and the need for educational programs that adequately prepare students for careers in this interdisciplinary field.

IoT is poised to revolutionize higher education, influencing not only educational technology but also teaching methodologies, learning experiences, management, resources, and the overall campus environment. As highlighted in [4], IoT is anticipated to bring about practical and experimental transformations in the education sector. This includes advancements in student evaluation, teaching platform integration, and the development of educational middleware, thus offering convenience and efficacy to students, instructors, and professors alike [5].


L. Barolli (Ed.): AINA 2024, LNDECT 203, pp. 497–510, 2024.
https://doi.org/10.1007/978-3-031-57931-8_47

The integration of technology and IoT in education focuses on enhancing student learning through real-time insights, with a growing preference for digital tools like tablets and laptops equipped with e-learning applications. These tools facilitate individualized and consistent learning experiences, thereby improving satisfaction rates and personalization of instruction [6,7].

In the realm of higher education, IoT technology enables faculty to collect and analyze performance data for personalized teaching [8,9]. Wearable devices, for instance, can track attendance and cognitive activity, underscoring the increasing significance of technology in educational contexts. As a result, higher education institutions are increasingly focusing on IoT technology and its applications [10]. The use of IoT in universities extends to e-learning, a practice that has become commonplace with profound implications [11]. While some universities have already embarked on this journey, it is expected that the integration of IoT in education will continue to expand [6,12]. Establishing an IoT major at the graduate level is relatively straightforward given the demand for professional research in this area. However, creating an independent IoT major at the undergraduate level poses challenges due to the necessity for foundational courses [13]. Therefore, colleges must not only choose their approach to IoT education but also continuously refine it to align with the evolving nexus of the virtual and physical worlds [14].

However, there is limited comparative analysis on the current state of IoT education, especially across leading universities. Existing surveys [2,3] focus only on European institutions or specific aspects like security. There is a lack of comprehensive evaluation of IoT curricula, emerging topics covered and pedagogical approaches used across universities. This research gap limits the development of effective guidelines and benchmarks for IoT education.

To address this gap, this paper presents a comparative survey analysis of IoT education across the top 25 universities offering related programs in the USA. Figure 1 shows the location and classification of the 25 universities that offer IoT programs in the US. We have grouped them into three categories: undergraduate (BS), graduate (MS), and research fields. This provides an overview of the different levels of education available in these universities. The analysis categorizes courses and offerings across five key IoT domains - Devices, Local Networks, Global Networks, Cloud Services and User Interfaces. It provides insights into curriculum structure, focus areas, learning goals, teaching methods and assessments used. We also have collected a comprehensive set of data regarding the location of each university, the number of teachers, their educational qualifications, and the level of readiness of IoT devices. Furthermore, we have documented the required and elective courses offered in the IoT program, including course descriptions, prerequisites, and credit hours. Specifically, by addressing following questions, it presents useful insights, benchmarks and recommendations for developing impactful university-level IoT education.

**RQ1.** What are the key knowledge domains covered in IoT curricula?
**RQ2.** What are the emerging focus areas and specializations offered?
**RQ3.** What curriculum structure and teaching approaches are adopted?

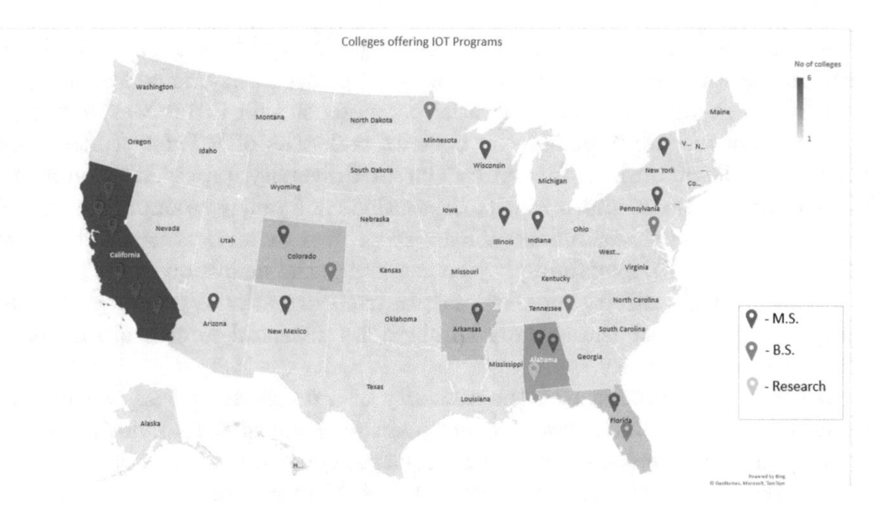


**Fig. 1.** Colleges offering IoT programs and their location in the USA.

The goal of this research is to analyze and map the landscape of IoT education across leading US universities to identify curriculum structure, core knowledge areas, emerging topics and pedagogical approaches, with the following objectives:

- Identify the essential IoT knowledge domains covered.
- Highlight new focus areas and specializations.
- Understand program structure and teaching methods used.
- Derive useful insights and benchmarks for developing impactful IoT programs.

## 2 Literature Review

IoT is changing education by enhancing safety, efficiency, and access to information. Al-Sharif's article gives examples of IoT in education, such as smart lesson plans, classrooms, and campuses. By leveraging IoT, educational institutions can create more efficient and sustainable campuses to provide students access to various resources and amenities. [15]. Mseer's study on IoT in higher education suggests that integrating IoT can enhance education quality and improve student outcomes. The study explores potential applications of IoT in education, such as personalized learning, adaptive assessment, and collaborative learning [16].

Mazon-Olivo et al. [17] provide a comprehensive overview of the Internet of Things (IoT), covering architecture, ecosystem components, computational paradigms, security, and governance. The article mainly focuses on the middleware layer of IoT architectures and the various components, security challenges, and governance aspects of IoT systems. Overall, it is a valuable resource for anyone interested in understanding IoT. Hernandez-de-Menendez et al. studied the latest technologies transforming engineering education, including Virtual Reality, 3D Printing, Drones, IoT, Robots, AI, Holograms, Wearable Devices, Virtual

Labs, and Blockchain. The benefits, challenges, adoption time, and implementation suggestions of each technology were described. They make learning more flexible and exciting, fostering essential competencies for future workforces [18].

McRae et al. [19] explored the benefits and risks of IoT for students with disabilities. Their project, funded by Curtin University, provided recommendations and guidance on appropriate IoT solutions and policy recommendations for the education sector and industry. Rose et al. provide an overview of the Internet of Things (IoT), explaining key concepts like the definition of IoT, enabling technologies, connectivity models, and its transformational potential. IoT is a complex concept, but the authors highlight its potential to create a hyperconnected world [20].

Alsobeh et al. emphasized the necessity of a multi-faceted approach to IoT education, which should encompass not just the technical skills of device and network management but also the critical areas of cybersecurity, software integrity, cloud technologies, and the innovative use of AI in educational methodologies. These studies contribute to the understanding that IoT education must be adaptive, comprehensive, and aligned with current industry standards and future trends to effectively prepare graduates for the evolving demands of the IoT industry. Magableh et al. described IoT education intersects with several key areas, particularly in cybersecurity, software development, data security, cloud computing, and the application of AI in education [21]. They underscored the importance of cybersecurity awareness among adolescents, a critical component of IoT education [22]. A comprehensive IoT curriculum should incorporate cybersecurity as a core domain to ensure graduates are well-versed in the security challenges of interconnected devices and systems [23]. IoT curricula must integrate data-driven security, model checking, and AI as a learning partner to produce graduates capable of designing secure and robust IoT networks [24]. Architectural thinking is also crucial in IoT curricula, as emphasized by the proposed concept of aspect-aware design for IoT applications [25]. It suggests that students should be taught how to conceive and critique IoT architectures, an essential skill for IoT professionals.

## 3 Methodology

### 3.1 University Selection

The methodology for selecting the top 25 universities for analysis of Internet of Things (IoT) education was based on a rigorous set of criteria to ensure a diverse and representative sample. Utilizing these criteria, we compiled a definitive list of 25 universities, ensuring a balanced representation of prominent institutions in IoT education. These criteria included:

1. Presence of Distinct IoT Programs and Departments: Priority was given to universities with dedicated IoT departments or those offering specialized degree programs in IoT such as Bachelor's, Master's or Doctoral degrees. This criterion ensures the sample has institutions with substantial commitment and resources for advancing IoT education.

2. Ranking Consideration: Universities included in the top 400 as per the US News and World Report rankings were considered. This benchmark was used to ensure that the selected universities have a recognized academic standing and are influential in shaping IoT education trends.
3. Accessibility of IoT Program Information: The availability of detailed online information regarding the IoT curriculum including program outcomes, course descriptions, and faculty profiles was necessary for in-depth analysis.

### 3.2 Data Collection

The data collection methodology focused on gathering granular, comprehensive information along four major dimensions:

1. IoT Degree Programs: Details regarding the types of programs offered at various levels - Bachelor's, Master's, Graduate Certificates, Doctoral programs were compiled. This provided insights into the depth of specialization supported.
2. IoT Courses: In-depth data for all courses under the IoT curriculum was gathered including descriptions, including credit hours, learning outcomes, assessments, pre-requisites and faculty details enabling analysis of knowledge areas covered.
3. Faculty Profiles: Academic backgrounds, research focus and publications of faculty teaching IoT courses were documented to understand specialized expertise available.
4. Infrastructure Analysis: Particulars of IoT laboratories, hardware/software platforms and tools for practical skill development were collected. This dimension provides context on hands-on learning scope.
5. Contextual Data: For additional context, we included university ranking data from US News and World Reports, offering a perspective on each institution's academic reputation within the broader educational landscape.

Furthermore, a five domain IoT categorization framework encompassing devices, networking, cloud, analytics and interfaces was developed. All the collected data was systematically encoded under this framework to enable a structured, comparative analysis as presented in the Findings.

### 3.3 Categorization Framework

To structure our analysis and ensure a comprehensive examination of IoT education, we developed a categorization framework based on five key knowledge domains integral to IoT:

1. IoT Devices: This category encompasses courses that focus on the technical aspects of IoT, such as sensors, embedded systems, microcontrollers, and wireless modules.
2. Local Networks: Courses in this category deal with protocols and technologies enabling device-to-device communication, including Bluetooth, ZigBee, Thread, WiFi, etc.

3. Global Networks: This category includes courses that cover broader communication technologies, such as cellular networks (2G/3G/4G/5G) and long-range communication technologies like LoRaWAN.
4. Cloud Services: Courses focusing on cloud platforms, database management, and analytics for IoT fall under this category.
5. User Interfaces: This domain covers courses related to the development of IoT applications, decision support systems, and aspects of human-computer interaction.

## 4 IoT Infrastructure and Curriculum Design

The IoT infrastructure, as shown in Fig. 2, is a complex system that consists of various interdependent components. IoT devices designed to collect and transmit data are at the core of this infrastructure. These devices are connected to a local network, which enables them to communicate with other devices nearby. In addition, the local network is connected to a global network, which allows the devices to transmit data over long distances.

Cloud services are essential for IoT infrastructure as they store and process data centrally, allowing devices to access multiple sources and interact in real-time. The user interface is also critical for users to interact with IoT devices and applications. The user interface can take many forms, including mobile apps, web applications, and voice assistants. It allows users to control and monitor the IoT devices and access the data they generate. Based on the data collected regarding lab facilities and courses offered, four key infrastructure categories were defined:

- IoT Sensing Devices: This includes sensors, embedded modules, development boards and tools for prototyping IoT endpoint devices.
- Networking Technologies: This covers protocols, interfaces and devices like gateways supporting device-to-device and internet connectivity.
- Cloud Platforms: This includes IoT specific cloud services and analytics software provided by vendors like AWS, Azure and IBM.
- Application and Visualization Tools (Interfaces): This involves platforms to develop IoT apps, dashboards and human-machine interfaces.

Figure 2 provides a consolidated view of the distribution of IoT courses across these infrastructure categories for undergraduate and graduate programs. Key observations:

- Graduate courses cover all four infrastructure categories evenly while undergraduate courses focus more on Sensing Devices and Networking.
- Tools and technologies related to Cloud Platforms and Application Development receive greater emphasis in graduate courses.
- Postgraduate programs incorporate more advanced infrastructure like Machine Learning capabilities, Blockchain platforms and Embedded AI.
- Undergraduate programs aim to provide more basic exposure spanning hardware, connectivity and interfaces.

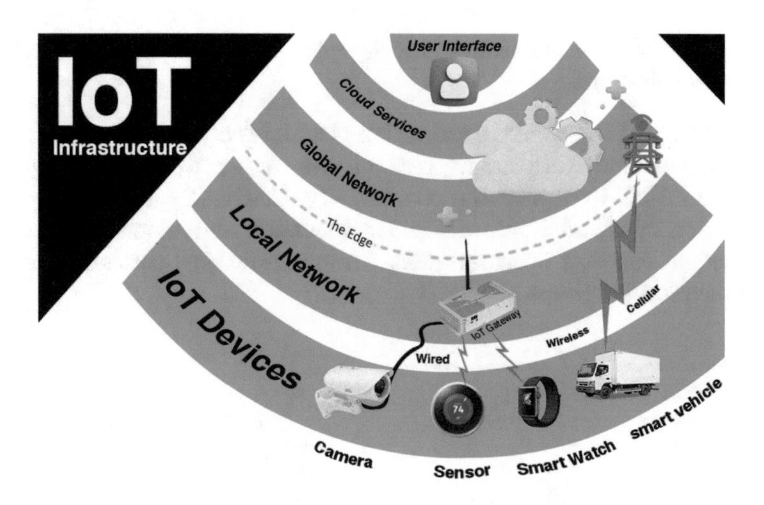


**Fig. 2.** IoT Infrastructure

This analysis shows that undergraduate IoT education focuses on core device-level knowledge, while graduate education targets specialized infrastructure skills required for advanced IoT research. Figures 3 and 4 show the mapped courses to specific IoT infrastructure components help identify gaps and opportunities to develop a more balanced and comprehensive IoT curriculum that aligns with industry needs and technological advancements.

The study categorizes a comprehensive array of IoT courses across various educational institutions, aligning them with four pivotal elements of IoT infrastructure: Sensing Devices, local and Global Networking Technologies, Cloud Platforms, and Interface courses. This taxonomy serves as a foundational framework for assessing and developing IoT-centric academic curricula.

1. IoT Sensing Devices (94 Courses): This category covers courses on sensor technologies, embedded systems, circuit design, and prototyping of IoT endpoint devices. Practical knowledge in hardware development is emphasized, and students gain an understanding of physical components and their mechanisms in the IoT ecosystem.
2. Local and Global Networking Technologies (37 Courses): These courses cover IoT networking, including protocols, interfaces, and hardware. They equip students with knowledge of local and global network infrastructures, wireless communication protocols, network security, and building scalable IoT networks.
3. Cloud Platforms (59 Courses): These courses cover cloud-based IoT solutions, data storage, processing, and virtualization technologies. They develop proficiency in managing IoT data through cloud infrastructure and highlight the importance of cloud services in the broader IoT framework.
4. Application and Visualization Tools (interface course) (53 Courses): These courses focus on developing IoT applications and visualization interfaces with

an emphasis on creating user-centric designs, real-time data visualization, and interactive elements.

The analysis of the course distribution elucidates a pronounced emphasis on Sensing Devices and Networking Technologies, indicative of a strong foundational focus in IoT hardware and connectivity. Conversely, the relatively lesser representation of Cloud Platforms and Application and Visualization Tools in the curriculum suggests potential areas for expansion and further development in IoT education programs.

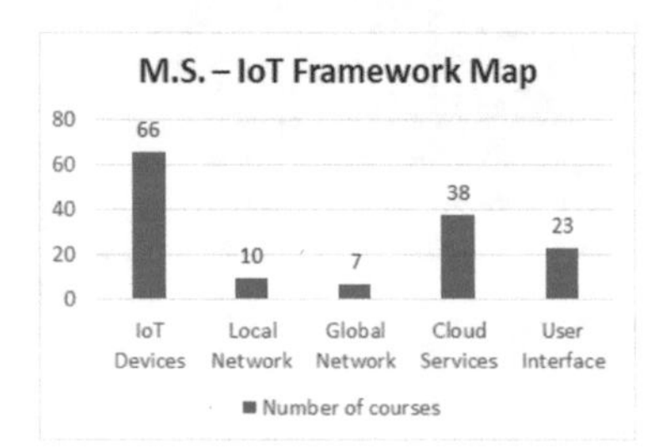


**Fig. 3.** In the graduate program, there is a total of 144 courses classified under IoT infrastructure.

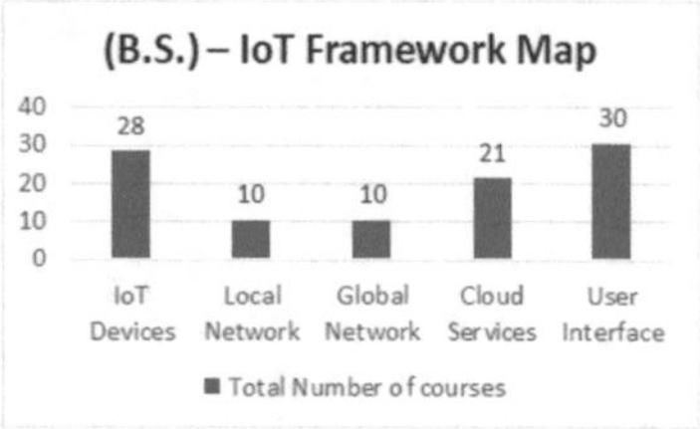

**Fig. 4.** In the BSc program, there is a total of 99 courses classified under IoT infrastructure.

### 4.1 IoT Hardware vs. IoT Software Courses

IoT courses were broadly classified into two categories - hardware-focused and software-focused to understand emphasis given towards building practical engineering competencies. Hardware courses encompass sensors, embedded systems, circuit design, networking, and protocols related to physical connectivity. Software courses cover application development, cloud platforms, operating systems, and data analytics. Analysis of close to 243 IoT courses offered across undergraduate and graduate programs highlighted key differentiation in hardware-software focus, as shown in Fig. 5 and Fig. 6. Graduate courses provide comprehensive coverage spanning hardware and software concepts integral to IoT, especially graduate students pursued more specialized, domain-specific electives like Machine Learning and Blockchain for IoT. Though exposed to foundational hardware and software, undergraduate students demonstrate significantly lower coverage of advanced tools and technologies. Additionally, Table 1 represents the most frequently offered courses highlighting alignment with IoT infrastructure elements:

**Table 1.** Most frequently offered courses highlighting alignment with IoT infrastructure.

| Course Type | Top Courses |
|---|---|
| Hardware | Embedded Systems, Sensor Networks, Wireless Protocols |
| Software | IoT Application Development, Cloud Computing, Data Mining (ML, Statistics, AI) |

The insights from the classification and comparison of hardware versus software courses guide curriculum designers in developing balanced, stage-appropriate courses calibrated as per level-specific learning outcomes.

For the graduate program, there is a total of 144 courses classified under IoT infrastructure as shown in Fig. 3. These courses are designed to provide students with diverse knowledge and skills in IoT. Out of these, 66 courses are dedicated to IoT devices that aim to provide a deep understanding of IoT devices' design, development, and implementation.

Moreover, 10 courses focus on local networks, which aim to teach students about the various network architectures used for IoT devices at the local level. Additionally, 7 courses are dedicated to global networks, which provide students with a comprehensive understanding of global network architectures and protocols used for IoT devices.

Furthermore, 38 courses are for cloud services, providing students with the knowledge and skills to effectively work with cloud-based IoT solutions. Finally, 23 courses are dedicated to user interfaces, which teach students about designing and developing user interfaces for IoT devices. With a diverse range of courses, students can gain specialized knowledge and skills in IoT infrastructure, which can help them excel in their careers.

Similarly, the undergraduate program offers a total of 99 courses, which are further classified into various categories of IoT infrastructure as shown in Fig. 4. For instance, 28 courses focus on IoT devices, covering topics such as IoT hardware, embedded systems, and sensor networks. The program also includes 10 courses focusing on local network infrastructure, covering topics such as LANs, WANs, and wireless networks. In addition, the program offers 10 courses on global network infrastructure, which delve into topics such as routing, network security, and network performance. The program also includes 21 courses on cloud services infrastructure, which cover topics such as cloud computing, storage, and virtualization. Finally, the program offers 30 courses on user interface infrastructure, which cover topics such as human-computer interaction, user-centered design, and interface prototyping.

## 5 Discussion and Recommendations

Our analysis was driven by three key research questions (RQs) to ascertain the current landscape of IoT education: **RQ1**: The analysis categorized courses across five fundamental domains - Devices, Networking, Cloud, Analytics and Interfaces. It highlighted that hardware and software competencies spanning these domains are integral to comprehensive IoT education. **RQ2**: The findings revealed key topics like AI, Machine Learning, Embedded Systems, and Blockchain gaining prominence, especially at graduate levels, indicating alignment with technological advancements.

**RQ3**: Results showed that hands-on practical learning is emphasized by most universities besides theoretical concepts. The approach tends to vary based on level - undergraduates get broad exposure while graduates pursue specialized, advanced competencies.

As shown in the average total of IoT courses in Figs. 7, 8, 9 and 10, Graduate Programs (12+ courses) should have a balanced curriculum covering hardware, software, and networking. Key courses include microcontrollers, sensors, cloud computing, local and global networks, and user interface design. Undergraduate programs (18+ courses) should cover IoT principles, practices, and technologies, with core areas including IoT devices, networking, cloud services, user interfaces, hardware, and software theory.

At both academic levels, it's crucial to prioritize practical skills and hands-on learning through laboratory work and project-based experiences. Graduates emerge with a versatile skill set encompassing embedded systems and IoT, enhanced by the ability to analyze complex problems, devise effective solutions, and communicate within professional contexts. Teamwork dynamics are also emphasized. The curriculum is designed to offer hands-on experience with fundamental engineering concepts, positioning students advantageously in the competitive IoT industry.

The requirements are even more extensive if the school wants to open an IoT department for undergraduate students. The department must offer an average of 5 courses in hardware and 12 courses in software, as shown in Figs. 7, 8, 9 and 10. The hardware courses should cover topics such as circuit design, digital electronics, and microprocessor systems. The software courses, on the other hand, must cover areas such as programming languages, software development methodologies, and database management. Furthermore, the department must

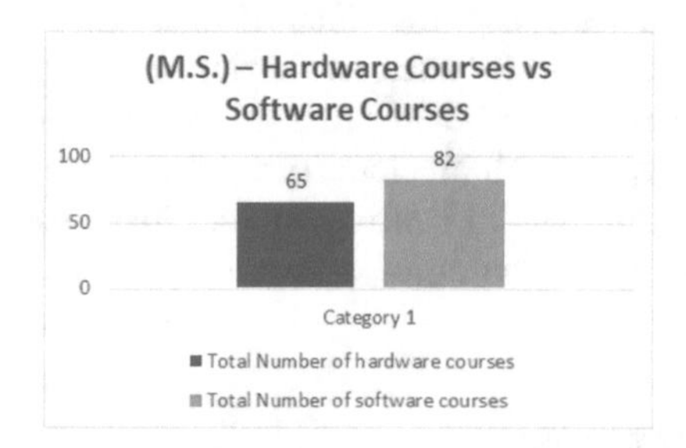


**Fig. 5.** Classifies various the total IoT courses according to their hardware and software components for graduate (MS) level.

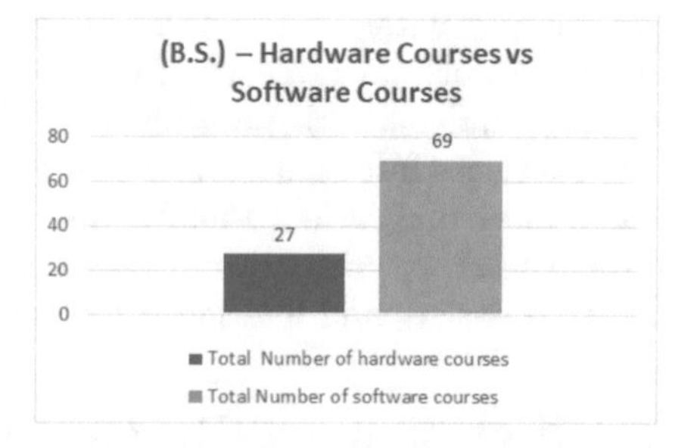


**Fig. 6.** Classifies various total IoT courses according to their hardware and software components for undergraduate (B.S.) level.

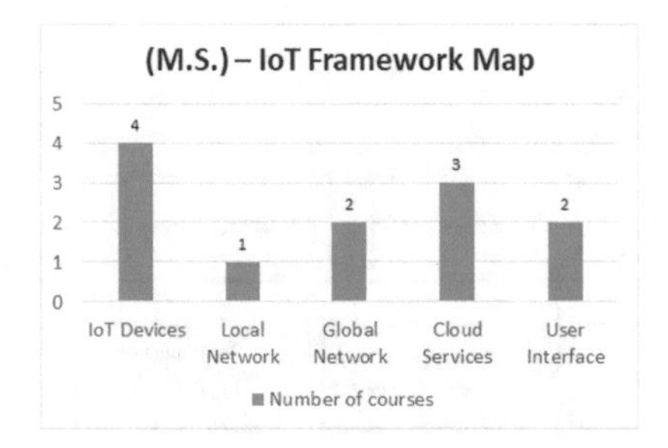


**Fig. 7.** In the graduate program, there is an average of 12 courses classified under IoT infrastructure.

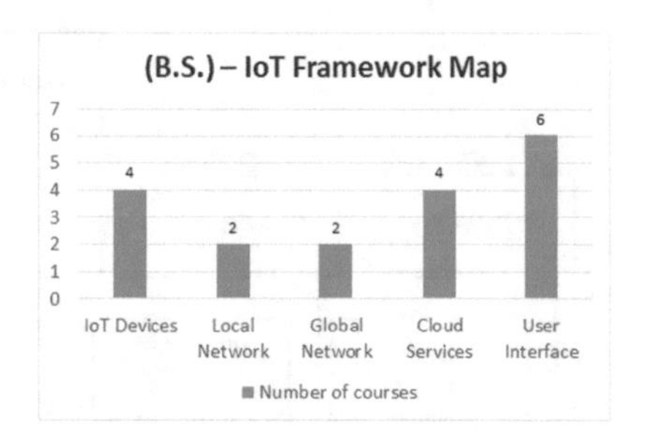


**Fig. 8.** In the undergraduate program, there is an average of 18 courses classified under IoT infrastructure.

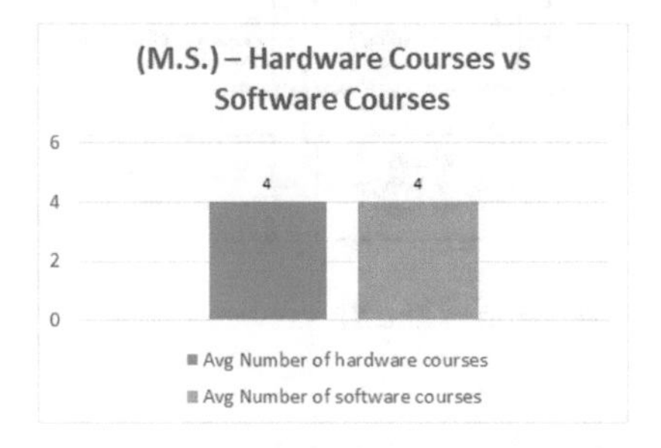


**Fig. 9.** Classifies various average IoT courses according to their hardware and software components for graduate (M.S.) level.

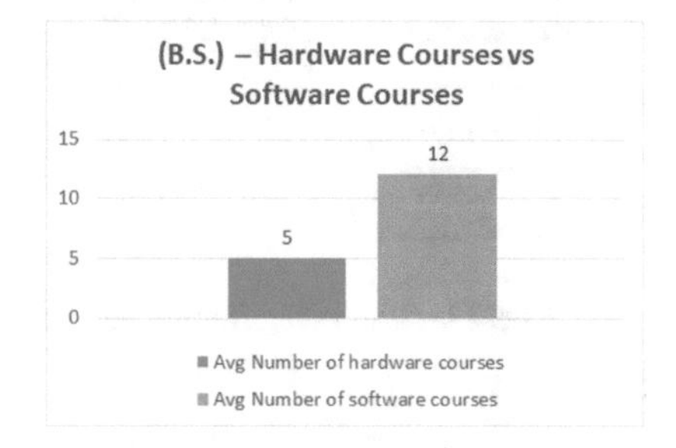


**Fig. 10.** Classifies various average IoT courses according to their hardware and software components for undergraduate (B.S.) level.

ensure that the courses provide students with hands-on experience and practical skills relevant to the IoT industry. The outcome of this study is to analyze and map the most commonly offered courses among 25 universities using our IoT infrastructure. To achieve this, we explored various areas to understand the courses offered comprehensively. Our exploration indicated that the most frequently offered courses among the 25 universities were 4 courses on IoT devices, 3 courses on local networks, 3 courses on global networks, 4 courses on cloud services, and 4 courses on user interface, as shown in Fig. 11. The insights from this study serve as useful benchmarks for universities aiming to introduce or expand IoT programs.

Curriculum Structure Guidelines: undergraduate programs should cover basic concepts across hardware, software, and networks, while graduate courses should enable specialization. Balancing Theory with Practical Skills: combining lab courses focused on building relevant hands-on expertise with theoretical knowledge is an effective approach. Emerging Technology Integration: Introduction of topics like Embedded AI, Blockchain, and Neural Networks, even at introductory levels, can help better align programs to industry needs.

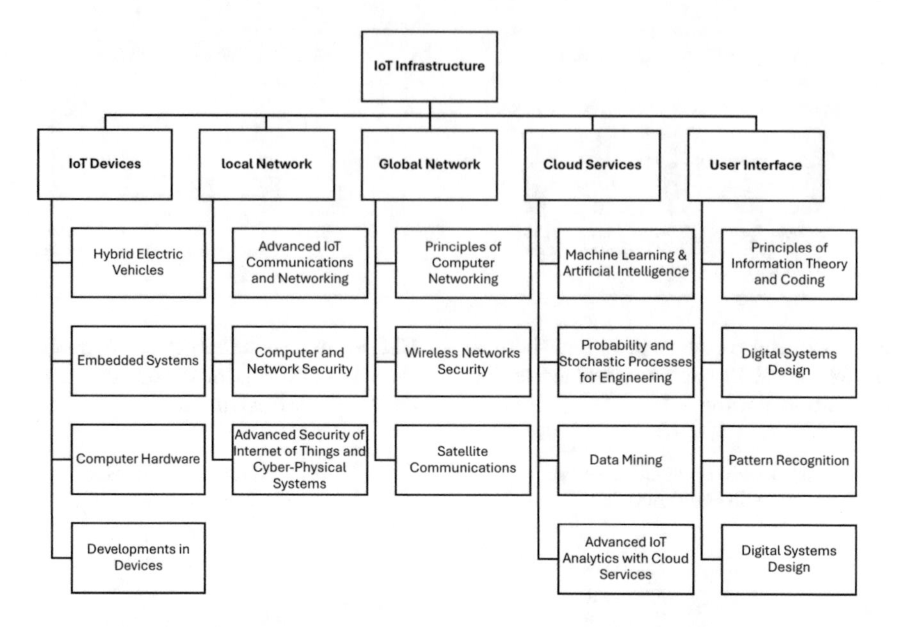


**Fig. 11.** The most frequently offered courses among the 25 universities.

The benefit of this study is that the students have a broad range of skills and knowledge in embedded systems engineering and the Internet of Things. Graduates will be able to analyze complex computing problems, design and implement computing-based solutions, communicate effectively in professional contexts, and function as effective team members or leaders. The programs also aim to prepare students to succeed in the rapidly growing field of IoT by providing them with hands-on experience with fundamental engineering concepts. Upon completion, students will have the skills and knowledge to pursue advanced studies, contribute to research, and gain a competitive edge in the global job market.

## 6 Conclusion

This study ventures into the realm of IoT education, unveiling crucial insights for universities aiming to cultivate the next wave of technological innovators. Through an analysis of 25 leading institutions' curricula, we find that a robust IoT education intertwines hardware and software knowledge, essential for students aspiring to become IoT experts. The curriculum must offer a harmonious blend of theoretical knowledge and practical skills, covering embedded systems, connectivity, cloud computing, and advanced algorithms. Hands-on experience with microcontrollers, sensor deployment, application development, and data analysis is vital for fostering real-world problem-solving abilities. Importantly, IoT education must dynamically integrate emerging technologies like AI and blockchain to stay relevant and prevent obsolescence. As IoT becomes ubiquitous, this study's findings act as a navigational beacon for universities seeking to evolve their programs in sync with technological advancements. The landscape of IoT is ever-changing, necessitating continual curriculum updates to reflect the latest technological trends. Embracing this continuous evolution is key to shaping a future driven by human-centric and innovative IoT solutions.

**Acknowledgments.** This research support from GAMEABOVE at Eastern Michigan University was essential for the completion of this project.

## References

1. Atzori, L., Iera, A., Morabito, G.: Understanding the Internet of Things: definition, potentials, and societal role of a fast evolving paradigm. Ad Hoc Netw. **56**, 122–140 (2017)
2. Vermesan, O., Friess, P. (eds.): Internet of Things: Converging Technologies for Smart Environments and Integrated Ecosystems. River Publishers (2013). http://www.rsc.org/dose/title of subordinate document. Cited 15 Jan 1999
3. Vermesan, O., Friess, P.: Internet of Things Applications - From Research and Innovation to Market Deployment, p. 364. Taylor & Francis (2014)
4. Tianbo, Z.: The Internet of Things promoting higher education revolution. In: 2012 Fourth International Conference on Multimedia Information Networking and Security (MINES). IEEE (2012)
5. Zhiqiang, H., Junming, Z.: The application of Internet of Things in education and its trend of development. Mod. Distance Educ. Res. **2**, 019 (2011)
6. Aldowah, H., Ghazal, S., Muniandy, B.: Issues and challenges of using e-learning in a Yemeni Public University. Indian J. Sci. Technol. **8**(32), 1–9 (2015)
7. Ghazal, S., Samsudin, Z., Aldowah, H.: Students' perception of synchronous courses using Skype-based video conferencing. Indian J. Sci. Technol. **8**(30), 1–9 (2015)
8. Jin, D.: Application of "Internet of Things" in electronic commerce. Int. J. Digit. Content Technol. Appl. **6**(8) (2012)
9. Fan, S., Yu, Z., Guo, H.: Affects of Internet of Things on supply chain management. China Economics and Trade (2009)
10. Sundmaeker, H., et al.: Vision and challenges for realising the Internet of Things. Cluster of European Research Projects on the Internet of Things. European Commission (2010)
11. Qi, A., Shen, Y.: The application of Internet of Things in teaching management system. In: 2011 International Conference on Information Technology, Computer Engineering and Management Sciences (ICM). IEEE (2011)
12. Gubbi, J., et al.: Internet of Things (IoT): a vision, architectural elements, and future directions. Futur. Gener. Comput. Syst. **29**(7), 1645–1660 (2013)
13. Tan, D.: Engineering Technology, Engineering Education and Engineering Management: Proceedings of the 2014 International Conference on Engineering Technology, Engineering Education and Engineering Management (ETEEEM 2014), Hong Kong, 15–16 November 2014. CRC Press (2015)
14. Ning, H., Hu, S.: Technology classification, industry, and education for Future Internet of Things. Int. J. Commun Syst **25**(9), 1230–1241 (2012)
15. Al-Sharif, M.A.B.: Ethical issues with technology in higher education. In: Integrating Technology Into Student Affairs. (# 9 SAPPI Series), p. 69 (2023)

16. Mseer, I.N.: Internet of Things and its impact on the future of education. In: Musleh Al-Sartawi, A.M.A., Razzaque, A., Kamal, M.M. (eds.) EAMMIS 2021. LNNS, vol. 239, pp. 490–499. Springer, Cham (2021). https://doi.org/10.1007/978-3-030-77246-8_45
17. Mazon-Olivo, B., Pan, A.: Internet of Things: state-of-the-art, computing paradigms and reference architectures. IEEE Lat. Am. Trans. **20**(1), 49–63 (2021)
18. Hernandez-de-Menendez, M., Escobar Díaz, C., Morales-Menendez, R.: Technologies for the future of learning: state of the art. Int. J. Interact. Des. Manuf. (IJIDeM) **14**, 683–695 (2020)
19. McRae, L., Ellis, K., Kent, M.: Internet of Things (IoT): education and technology. The relationship between education and technology for students with disabilities, pp. 1–37 (2018)
20. Rose, K., Eldridge, S., Chapin, L.: The Internet of Things: An Overview, vol. 80, pp. 1–50. The Internet Society (ISOC) (2015)
21. Magableh, A.A., Alsobeh, A.M.R., Klaib, A.F.: An evaluation of the usage of aspect orientation and the gap between academic research and industry needs. J. Theoret. Appl. Inf. Technol. **97**(19), 5146–5165 (2019)
22. AlSobeh, A.M., Ramadan, I.A.A., Shatnawi, A.M.J., Khasawneh, I.: Cybersecurity awareness factors among adolescents in Jordan: mediation effect of cyber scale and personal factors. Online J. Commun. Media Technol. **13**(2), e202312 (2023)
23. Magableh, A.A., Al Sobeh, A.M.R.: Securing software development stages using aspect-orientation concepts. Int. J. Softw. Eng. Appl. (IJSEA) **9**(6), 57–71 (2018)
24. Tashtoush, Y., Obeidat, R., Al-Shorman, A., Darwish, O., Al-Ramahi, M.A., Darweesh, D.: Enhanced convolutional neural network for non-small cell lung cancer classification (2023)
25. Alsobeh, A., Shatnawi, A.: Integrating data-driven security, model checking, and self-adaptation for IoT systems using BIP components: a conceptual proposal model. In: Daimi, K., Al Sadoon, A. (eds.) Proceedings of the 2023 International Conference on Advances in Computing Research, ACR 2023. LNNS, vol. 700, pp. 533–549. Springer, Cham (2023). https://doi.org/10.1007/978-3-031-33743-7_44

# Author Index




L. Barolli (Ed.): AINA 2024, LNDECT 203, pp. 511–513, 2024.
https://doi.org/10.1007/978-3-031-57931-8